D0161218

INTRODUCTION TO CIRCUITS WITH ELECTRONICS

AN INTEGRATED APPROACH

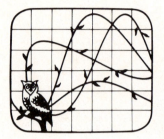

**HRW
Series in
Electrical and
Computer Engineering**

M. E. Van Valkenburg, Series Editor

P. R. Bélanger

E. L. Adler

N. C. Rumin
McGILL UNIVERSITY, MONTREAL

INTRODUCTION TO CIRCUITS WITH ELECTRONICS

AN INTEGRATED APPROACH

HOLT, RINEHART AND WINSTON
New York   Chicago   San Francisco   Philadelphia
Montreal   Toronto   London   Sydney   Tokyo
Mexico City   Rio de Janeiro   Madrid

Acquisitions Editor	Deborah L. Moore
Production Manager	Paul Nardi
Project Editor	Lynn Contrucci, Cobb/Dunlop
Interior Design	Rita Naughton
Cover Design	Scott Chelius
Cover Photo	Marc David Cohen
Illustrations	Scientific Illustrators

Copyright © 1985 CBS College Publishing
All rights reserved.
Address correspondence to:
383 Madison Avenue, New York NY 10017

Library of Congress Cataloging in Publication Data

Bélanger, P. R.
  Introduction to circuits with electronics.

  (HRW series in electrical and computer engineering)
  Includes index.
  1. Electronic circuits.   I. Adler, E. L.
II. Rumin, N. C.   III. Title.   IV. Series.
TK7867.B37   1985   621.3815'3   85-760

ISBN 0-03-064008-3

Printed in the United States of America

Printed simultaneously in Canada

5 6 7 8   016   9 8 7 6 5 4 3 2 1

CBS COLLEGE PUBLISHING
Holt, Rinehart and Winston
The Dryden Press
Saunders College Publishing

Contents

Preface

The traditional approach to circuits and electronics is to begin with a course on circuit theory and to follow with a course on electronics. The circuit theory course is usually taught to students who are simultaneously taking a first course in differential equations, so that the dynamic response concepts must be built upon a fresh and still precarious mathematical foundation. In our experience many students find it difficult to absorb simultaneously two bodies of knowledge as concept heavy as differential equations and circuit theory. This is especially so in view of the fact that, without electronics, it is not easy to motivate the study of circuit theory. Many engineering students must get a sense of the application in order to learn the theory; to such students circuit theory appears dry and abstract.

This book takes a new approach to the subjects of circuits and electronics by treating both in an integrated fashion. The first eight chapters of the book constitute enough material for a one-semester course. These chapters require only elementary calculus, plus some knowledge of linear algebra; the linear algebra required is also elementary and includes the expression of linear equations in vector-matrix form and the inversion of small (2×2) matrices. No modern physics is required, since electronic devices are described by their terminal characteristics rather than by their physics. This part of the book covers resistive networks, nodal and mesh analysis, power, resistive two-ports, nonlinear nondynamic two-ports, diode networks, the bipolar and field-effect transistors in some elementary configurations, and logic circuits.

The last seven chapters form the basis for a second one-semester course. A first course in differential equations is assumed, plus some familiarity with complex numbers. Networks with a single time constant are treated using classical differential equations, but Laplace transforms are introduced to study the more general case. This half of the book covers transient response, the sinusoidal steady state, ac power, op amp circuits, transistor analog circuits, and some power circuits.

At McGill, we teach this material in 2 one-semester courses to electrical engineering sophmores. The second course has as prerequisites the first course plus a course on differential equations. The book contains very little physical electronics. At McGill, we have a course in devices at the junior-year level, when the student has acquired enough physics to study this topic in more appropriate depth.

To describe the contents in greater detail, Chapter 1 presents basic definitions, branch laws, and Kirchhoff's laws. Because it is important for students to start solving networks as quickly as possible, an inspection method is presented. This is followed by a section on energy and power.

The formal development of network equations is covered in Chapter 2. Some topology is presented to justify the choice of the KCL and KVL equations to be retained. Node and mesh variables are introduced and inspection procedures are derived for nodal and mesh analysis. A special section deals with networks with controlled sources; such networks, which many students find difficult, occur throughout the text. Tellegen's theorem is discussed next, and the chapter closes with the principle of superposition.

The subject of Chapter 3 is equivalent networks. This includes reduction techniques for resistive networks and the Thevenin and Norton equivalent circuits. Source splitting, plus the Thevenin and Norton equivalents, are used to transform current sources to voltage sources and vice versa. The substitution theorem is presented and used to analyze symmetric circuits in terms of common-mode and differential-mode half-circuits.

Chapter 4 deals with linear two-ports. The various characterizations are taught as outcomes of superposition with a source at each port. Bridges, lattices, and tee and pi networks are seen as special cases, as are ideal amplifiers, ideal transformers, mutual inductance elements, and gyrators. The ideal operational amplifier is presented as a limiting case of the differential amplifier. The chapter concludes with a discussion of terminated two-ports.

Chapter 5 is an introduction to nonlinear circuits. Resistive circuits with a single nonlinear, no-memory one-port are solved graphically using load lines, and the idea of small-signal modeling is introduced. Diode circuits are introduced, and piecewise-linear analysis is discussed. A parallel development is carried out for nonlinear two-ports, including load lines and small-signal models.

The metal-oxide semiconductor (MOS) field-effect transistor (FET) is introduced in Chapter 6. Following a very brief discussion of the device's physics, the operating modes of the MOSFET are described. The simple inverter with a resistor load is then used to illustrate dc, small-signal and large-signal analysis. This is followed by an analysis of the inverter with an active load. The chapter concludes with a similar but less extensive treatment of the JFET. Both analytical and graphical methods of analysis are illustrated.

Chapter 7 parallels the development of Chapter 6, for the bipolar junction transistor (BJT). Basic versions of the common-emitter and common-collector amplifier are discussed. The piecewise-linear model of the BJT is used as a basis for discussing the BJT inverter.

Chapter 8 is concerned with logic families. It begins with some elementary Boolean arithmetic and logic functions. Simple combinatorial and sequential circuits are described in order to provide motivation for the rest of the chapter and to emphasize the gate as the basic logic element. A simple BJT inverter is used to define signal representation, logic gain, noise margin, and propagation delay time. This is followed by an exposition of several logic families, including transistor-transistor logic (TTL), emitter-coupled logic (ECL), integrated-injection logic (I^2L), NMOS logic, and complementary MOS (CMOS) logic.

Chapter 9 contains some review material on classical differential equations, including homogeneous and particular solutions and natural frequencies. The mathematics are motivated with circuits, however. Special signals such as the step, the ramp, and the impulse are introduced, and complex exponentials are used to generate a broad class of signals. The system function (transfer function) is presented as the ratio of the particular solution to the excitations for exponential excitations. Inspection techniques are developed to analyze networks with a single time constant, and applied to networks with switches and diodes.

Chapter 10 begins by extending the classical approach of Chapter 9 to second-order networks, with particular stress on the underdamped case. The Laplace transform is introduced with its principal theorems and used to solve differential equations. This leads to the concepts of impedance, transform networks, and network stability. The chapter closes with a treatment of networks not initially in the zero state, where the initial conditions are taken into account by the introduction of sources in the models of the inductances and capacitances.

Chapter 11 is concerned with the sinusoidal steady state. It begins with the phasor concept and continues with the study of equivalent networks at a single frequency. The frequency response is related to the network function poles and zeros and is derived and analyzed in some depth for a few simple networks. Resonance is discussed as a special case. There are some applications to electronic circuits.

Power in the sinusoidal steady state is the subject of Chapter 12. The concepts of root-mean-square (rms) values and average power, including complex power, are introduced. This is followed by some applications drawn from power engineering, including balanced three-phase systems. Superposition of average power is demonstrated for sinusoids and dc and is used to analyze the power flow in a BJT amplifier. This is followed by a discussion of impedance matching and two-port scattering parameters.

Chapter 13 is dedicated to the study of analog circuits using op amps. Several standard linear circuits are presented including the switched-capacitor integrator and the negative-impedance converter. Nonlinear applications such as rectification, envelope detection and limiting are given. The op amp comparator is also discussed, including the case with hysteresis.

Chapter 14 is an introduction to transistor amplifier design. Biasing of the discrete BJT single-stage amplifier is discussed. This is followed by descriptions of the three basic amplifier configurations, including their approximate high-frequency behavior. Biasing in integrated circuits is presented next via the current source. Finally, multiple-transistor circuits, including differential amplifiers and the Darlington and cascode connections, are analyzed, and examples of practical implementations are shown.

Power electronic circuits are the subject of Chapter 15, beginning with unregulated and regulated power supplies. Power amplifiers follow, and the chapter closes with a discussion of silicon-controlled rectifier circuits.

There is an appendix on complex numbers, intended as a review vehicle.

It is very easy to use this text for a traditional circuits course by covering Chapters 1 to 4 and 9 to 12. These chapters are not devoid of electronics; on the contrary, they contain many applications to electronic circuits. However, the circuit models are given and no knowledge of the other chapters is required. For students who have had an introductory circuits course, this text can be used for an electronics course by covering most of Chapters 4 to 8 and 13 to 15.

The reader will find that the topics in this book do not necessarily fall into neat, mutually exclusive packages. Some op amp circuits, for example, are treated in Chapter 4 as well as in Chapter 13. Transistor amplifiers are discussed in Chapters 6 and 7 as well as in Chapter 14. The frequency response of such amplifiers is given as an example in Chapter 11 and taken up again in Chapter 14. Such repetitions are only apparent, because the teaching objectives differ. In Chapter 4, the op amp is presented as an example of a linear two-port, while in Chapter 13 the objective is to teach op amp circuits. The treatment of transistors in Chapters 6 and 7 concentrates on developing in the student skills in quiescent, small-signal, large-signal, and first-order analyses, while dynamic effects and practical circuits are treated in Chapter 14. We have found pedagogical advantages in touching the same material twice, because the initial encounter is at a simpler level than the second; this allows the student to advance in steps.

Most sections conclude with one or several drill exercises of a rather straightforward nature, with answers provided. There is a study guide at the end of each chapter, intended to warn students against certain pitfalls which, in our experience, are particularly insidious.

In closing, we wish to acknowledge the help of our colleagues Barry Howarth, who worked out some of the solutions. We are indebted to Nevine Nassif, a doctoral graduate student in our department, for typing the manuscript; we discovered the value of an expert typist who also understands the text and can correct the odd technical mistake. The professional indexing services of Christine Jacobs are also gratefully acknowledged. Finally, we wish to thank our reviewers.

Sydney R. Parker	Naval Postgraduate School
Dr. Glen C. Gerhard	University of New Hampshire
M. D. Calhoun	Mississippi State University
Andrew P. Sage	George Mason University
Charles L. Alley	University of Utah
Douglas J. Hamilton	University of Arizona
Ronald E. Guentzler	Ohio Northern University
Patricia D. Daniels	University of Washington
Ray W. Palmer	Milwaukee School of Engineering
Darrel Vines	Texas Tech University
Omar Wing	Columbia University
Martha Sloan	Michigan Technological University

They will find in the text several of the modifications which they suggested.

P. R. Bélanger
E. L. Adler
N. C. Rumin

CHAPTER 1

Basic Elements

1.1 ANALYSIS AND SYNTHESIS

By the time the study of circuit theory is started, the reader has accumulated several years' worth of background in mathematics and the physical sciences. During those years of study, the student has become familiar with the process of *analysis*; engineers, on the other hand, are mostly concerned with *synthesis*.

Analysis seeks to answer the question: What is the behavior of a given physical system? Synthesis, on the other hand, addresses the following problem: given a desired behavior, describe a physical system that achieves it.

There are basically two analytical methods: the experimental and the theoretical. The experimental approach requires construction of the object (or a scaled replica) and the use of observations to describe the behavior. Sometimes this is the only available approach; for example, wind tunnel experiments are a necessity in the prediction of aerodynamic effects on complex structures such as aircraft. The experimental approach looks upon a physical system as a whole: the theoretical,

1

on the other hand, seeks to break it down into smaller pieces. The theoretical approach consists of the decomposition of a system into components, each describable by known physical laws. The simultaneous expression of these laws by mathematical relations defines a problem in mathematics whose solution gives a prediction of the system behavior. This set of mathematical relations is usually called a *system model,* and formulating models is one of the skills which an engineer must develop.

It is possible to do synthesis by experiment alone: craftsmen do this all the time, and have done so since the beginning of time. This is basically a trial-and-error procedure, and it works quite well if the desired behavior is close to that of a known system. To design a new system, however, it is usually better to rely on theory. Building a system is expensive: if the trial-and-error procedure is to be used, it is better to solve sets of equations several times than to build, say, a number of aircraft until one works. Furthermore, the theory often leads directly to synthesis procedures; given the desired answer, the mathematics can lead to a set of components. In effect, the synthesis problem is just the inverse of the analysis problem.

Since the goal of analysis is to predict the behavior of a physical system, the result should be unique. On the other hand, there are many answers to the synthesis problem, with different performances and costs.

It is clear that a good understanding of the process of analysis is necessary to do synthesis: in order to design for a given objective, one must be able to predict the behavior of a given arrangement of components.

In this book, analysis is the main concern, since it is the necessary first step. The student will come to grips with simple design problems in the exercises, but more formal and extensive encounters with synthesis will be left to more advanced courses. The first step in analyzing a system is to write mathematical descriptions of the basic phenomena: this leads directly to the issue of models.

1.2 MODELS

Most likely, the student has an unsuspected familiarity with models and modeling. Consider the following problem: given the initial direction and velocity of a projectile, predict its trajectory. The simplest model of the situation would neglect air resistance and use Newton's laws; the resulting prediction will be useful in some instances. A more accurate model can be constructed by assuming air resistance to be proportional to the square of the velocity. Still greater accuracy can be achieved by considering wind direction and velocity, a more realistic formula for air resistance, etc.

Clearly, a whole hierarchy of models can be generated for a single physical situation. None of the models is a "true" model: reality may be circumscribed as closely as one likes (at the expense of added complexity), but the physical phenomenon in its entirety is forever beyond reach. One often speaks of model "validity," but this is a misnomer. Models are neither valid nor invalid: they simply do, or do not, give useful predictions.

When modeling a physical system, care must be taken to ensure that the essential elements are included. In some cases, the set of equations resulting from a model may have no solution: the model is too simple and fails to capture an essential mechanism of the physical situation. This means that, unfortunately, one cannot ignore such mathematical niceties as existence of solutions. The oft-heard argument "but this is a physical system and there must be a solution" may not be used; remember, models are not reality, but some abstract, ideal version of it.

In the text, the words *resistance*, *capacitance*, and *inductance* are used to denote models: the terms *resistor*, *capacitor*, and *inductor* refer to the physical devices.

1.3   VARIABLES AND ELEMENTS

1.3.1 Voltage and Current

Models will be represented by boxes with two or more terminals. Associated with each terminal is a *current*, flowing into or out of the box. *Voltages* are defined to exist between terminals.

From physics, the student will associate current with the rate of flow of charge: if i is the current and q is the charge, then $i = dq/dt$. Voltage is basically a difference in electric potential energy. Charge is expressed in coulombs (C), voltage is expressed in volts (V), current in amperes (A).

Voltage and current are directional quantities. Current may flow into a terminal, or out of it; a given terminal may be at a higher or lower voltage than another. A reference direction is assigned to each voltage and current: an arrow for current, plus (+) and minus (−) signs for voltage. This only means that current is called positive if it goes in the direction of the arrow, negative if not; or, that a voltage is termed positive if the terminal labeled + is in fact positive with respect to the terminal labeled −, negative if not. The same thing is done in mechanics when reference directions are assigned to, say, forces and velocities: representation of a force by an upward arrow does not mean that the force is in the upward direction, only that it is in that direction if it is positive.

For example, if $i_1 = +1$ in Fig. 1.1, then current flows into terminal 1; if $i_1 = -1$, then current flows out of terminal 1. Similarly, $v_1 = 1$ indicates that terminal 1 is positive with respect to terminal 4, while $v_1 = -1$ means that terminal 1 is negative with respect to terminal 4.

1.3.2 Lumped Elements

A *lumped element* has the property that the algebraic sum of currents entering the terminals is zero at all instants of time. The term *algebraic sum* means that the currents are treated as signed variables. In Fig. 1.1, this condition implies

$$i_1 + i_2 + i_3 - i_4 = 0$$

A minus sign is needed with i_4 because its reference direction is away from the element: if i_4 is positive, current leaves, resulting in a negative contribution to the total current entering.

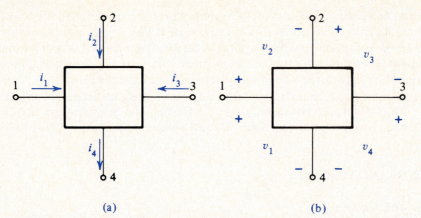

Figure 1.1 A lumped element with reference directions for (a) currents and (b) voltages.

This condition on the current implies conservation of charge: at any time, the charge going into the element is equal to the charge leaving it.

A lumped element is specified by its *terminal characteristics,* i.e., by a set of equations relating the terminal voltages and currents. The terminal characteristics can also be described graphically. In many cases, the terminals may be paired in such a way that the current entering one terminal is always equal to the current leaving the other. These special pairs of terminals are called *ports,* and the element is then called a one-, two-, three-, . . . , port network. Figure 1.2 illustrates a two-port element.

1.3.3 Branches

A *branch* is a lumped element with two terminals, as shown in Fig. 1.3. Since a branch is a lumped element, the current leaving 1' must equal the current entering 1, otherwise the algebraic sum of currents entering the element would not be zero. Clearly, a branch is also a one-port network.

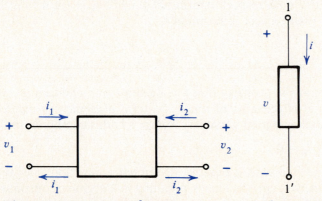

Figure 1.2 A two-port element.   **Figure 1.3** A branch.

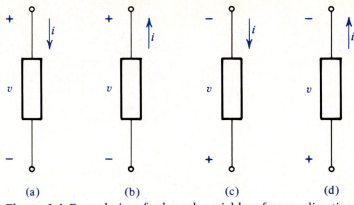

(a)   (b)   (c)   (d)

Figure 1.4 Four choices for branch variable reference directions.

Obviously, for a branch element it is not necessary to specify two currents: it is enough to specify the current at one of the two terminals.

Since reference directions may be freely selected, any of the four definitions of Fig. 1.4 could be used. However, it is conventional to choose reference directions where the current enters the + terminal and leaves the − terminal, i.e., as in (a) and (d) of Fig. 1.4. Using this convention, only v or i need be specified and the polarity of the other follows.

A branch is described by a single relationship between its voltage and current, the v-i, or branch relationship. This can be presented as an equation or as a curve in the v-i plane. It can be algebraic or it may involve derivatives and integrals. It may be linear (i.e., v and i appear only to the first power) or nonlinear.

Several special branch elements are described in Sec. 1.4. These branches will be used throughout the text as building blocks in the models of networks.

1.4 BASIC LINEAR BRANCH ELEMENTS

1.4.1 Resistance

A *resistance* is a branch element described by the linear algebraic v-i relationship

$$v = Ri \tag{1.1}$$

Figure 1.5 shows the symbol for a resistance and the reference polarities for v and i in Eq. (1.1). If v is in volts and i is in amperes, then R is expressed in ohms (Ω). Equation (1.1) is known as Ohm's law. Provided $R \neq 0$, Eq. (1.1) can be rewritten as

$$i = \frac{1}{R}v \tag{1.2}$$

For convenience, define

$$G = \frac{1}{R} \tag{1.3}$$

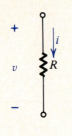

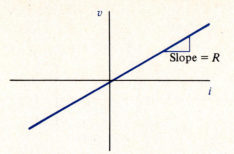

Figure 1.5 Symbol and polarity definition for a resistance.

Figure 1.6 Voltage-current relationship
for a resistance.

The constant G is the *conductance*, which, today, is usually expressed in siemens (S), although the older *mhos* (℧) are still encountered.

Equation (1.1) holds even if i and v vary with time. For example, if $i(t) = \sin 3t$, then $v(t) = R \sin 3t$. Figure 1.6 shows a graph of v vs i (or i vs v). Clearly, the graph is a straight line through the origin: R and G are linear branch elements.

1.4.2 Capacitance

A *capacitance* is a branch element described by the linear differential v-i relationship

$$i = C \frac{dv}{dt} \tag{1.4}$$

Figure 1.7 shows the symbol for a capacitance and the reference polarities for v and i in Eq. (1.4). If v is in volts and i is in amperes, then C is in farads (F). Equation (1.4) may be inverted by integrating both sides:

$$\int_{t_0}^{t} i(\tau)\, d\tau = C \int_{t_0}^{t} \frac{dv}{d\tau}\, d\tau$$

$$= C[v(t) - v(t_0)]$$

or, assuming $C \neq 0$,

$$v(t) = v(t_0) + \frac{1}{C} \int_{t_0}^{t} i(\tau)\, d\tau \tag{1.5}$$

If $v(t_0) = 0$, then Eq. (1.5) becomes

$$v(t) = \frac{1}{C} \int_{t_0}^{t} i(\tau)\, d\tau$$

But, since $i = dq/dt$, the integral of the current is the charge delivered to the capacitance; therefore,

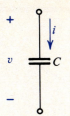

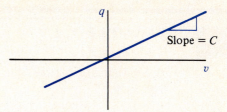

Figure 1.7 Symbol and polarity definition for a capacitance.

Figure 1.8 Charge-voltage relationship for a capacitance.

$$v(t) = \frac{q(t)}{C}$$

or

$$q(t) = Cv(t) \tag{1.6}$$

Equation (1.6) has the q-v graph shown in Fig. 1.8.

It is not possible to draw a graph of v vs i, since i is not related to v but to its rate of change.

EXAMPLE 1.1

The voltage across a 2-F capacitance is as shown in Fig. E1.1(a). Find the current in the capacitance, for $t \geq 0$.

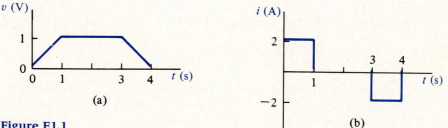

(a)

Figure E1.1

(b)

Solution

Since dv/dt is the slope of the curve v vs t, then by inspection

$$\frac{dv}{dt} = \begin{cases} 1 & 0 \leq t < 1 \\ 0 & 1 \leq t < 3 \\ -1 & 3 \leq t < 4 \\ 0 & t > 4 \end{cases}$$

since $i(t) = 2 \, dv/dt$, the solution curve of Fig. E1.1(b) follows.

EXAMPLE 1.2

Given that $i(t) = 2 \cos 3t$ is the current in a 2-F capacitance, find $v(t)$ for $t > 0$, if $v(0) = 1$ V.

Solution

$$v(t) = v(t_0) + \frac{1}{C} \int_{t_0}^{t} i(\tau)\, d\tau$$

$$= 1 + \frac{1}{2} \int_{0}^{t} 2 \cos 3\tau\, d\tau$$

$$= 1 + \tfrac{1}{2}(\tfrac{2}{3}) \sin 3t \big|_{0}^{t}$$

$$= 1 + (\tfrac{1}{3}) \sin 3t$$

1.4.3 Inductance

An *inductance* is a branch element described by the linear differential *v-i* relationship

$$v = L \frac{di}{dt} \tag{1.7}$$

Figure 1.9 shows the symbol for an inductance and the reference polarities of v and i in Eq. (1.7). If v is in volts and i is in amperes, then L is expressed in henrys (H). Equation (1.6) may be inverted in a manner exactly analogous to Eq. (1.4), to yield

$$i(t) = i(t_0) + \frac{1}{L} \int_{t_0}^{t} v(\tau)\, d\tau \tag{1.8}$$

Equations (1.4) and (1.7) may be obtained from each other simply by interchanging i and v. Because of this, L and C are called *dual* elements: statements made about the one can be made about the other if i and v are interchanged. Thus, the voltage across an inductance depends on the rate of change of current

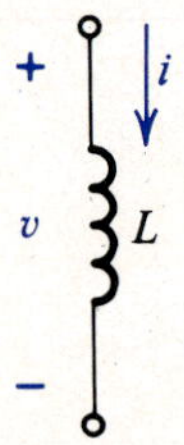

Figure 1.9 Symbol and polarity definition for an inductance.

(compare with the corresponding statement about capacitance). If $i(t_0) = 0$, then Eq. (1.8) becomes

$$i(t) = \frac{1}{L} \int_{t_0}^{t} v(\tau)\, d\tau$$

The integral of the voltage across an inductance is called the *flux* and is given the symbol Λ. It is expressed in webers (W). Thus

$$\Lambda(t) = Li(t) \tag{1.9}$$

The flux in an inductance is the dual of the charge in a capacitance, and Eq. (1.9) is represented by a straight line in the plane of Λ vs i.

EXAMPLE 1.3

The voltage across a 3-H inductance is shown in Fig. E1.3. Find the current $i(t)$, $t > 0$, given that $i(0) = 0$.

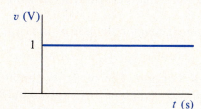

Figure E1.3

Solution

$$i(t) = i(0) + \frac{1}{3} \int_0^t v(\tau)\, d\tau$$

$$= 0 + \frac{1}{3} \int_0^t 1\, d\tau$$

$$= \frac{t}{3}$$

1.4.4 Independent Sources

An *independent voltage source* is a branch element defined by the *v-i* relationship

$$v = v_S \tag{1.10}$$

where v_S is a constant or function of time. It is important that v_S be known a priori.

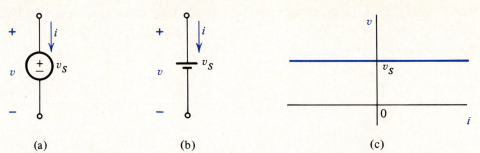

Figure 1.10 (a) Symbol for an independent voltage source. (b) Special symbol used for constant v_S. (c) Voltage-current relationship for the independent voltage source.

The symbol of Fig. 1.10(a) is used in general, although that of Fig. 1.10(b) is often used for constant v_S. In the latter case, the long dash is assumed to indicate the positive terminal.

The v vs i graph of an independent voltage source is shown in Fig. 1.10(a). Equation (1.10) places no constraint at all on the current.

If $v_S = 0$, the independent voltage source is called a *short circuit*: it forces the voltage between its terminals to be zero, whatever the current, and can be replaced by a straight piece of "wire."

An *independent current source* is a branch element defined by the v-i relationship

$$i = i_S \tag{1.11}$$

where i_S is a known constant or function of time. The symbol for an independent current source is shown in Fig. 1.11(a), and a graph of v vs i is given in Fig. 1.11(b). No constraint is placed on the voltage.

If $i_S = 0$, the independent current source becomes an open circuit, since no current can flow between the terminals for any value of applied voltage. The source symbol can be removed entirely.

1.4.5 Controlled Sources

Controlled sources are voltage or current sources where the source voltage or current is not a quantity known a priori, but depends on some controlling voltage or current elsewhere in the network. It will be seen that controlled sources are a necessary ingredient in the modeling of transistors and other devices.

Figure 1.12 shows the four types of linear controlled sources. In each case, v_c (or i_c) is a voltage (or current) in some other branch, also shown in Fig. 1.12. The four types are

1. The *voltage-controlled voltage source* (VCVS), where k_v is a dimensionless constant

2. The *current-controlled voltage source* (CCVS), where the constant k_r is in ohms (the term volts/ampere is often used to emphasize the nature of this element)

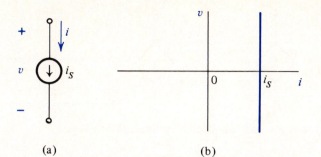

(a)   (b)

Figure 1.11 (a) Independent current source symbol. (b) Voltage-current relationship for the independent current source.

3. The *voltage-controlled current source* (VCCS), where the constant k_g is in siemens (the term ampere/volt is also used)

4. The *current-controlled current source* (CCCS), where k_i is a dimensionless constant

It may help the student to think of a controlled source as being adjusted by someone reading the value of the controlling variable on a voltmeter or ammeter, in such a way that the controlled source value is proportional to the reading.

It is fair to say that a controlled source is a generalized resistance. A resistance imposes a constraint between the voltage and current in the same branch; a controlled source enforces a relationship between any pair of network variables. It is possible to represent a resistance by a controlled source, a fact that will be used in the analysis of certain electronic circuits. For example, the three branches of Fig. 1.13 all have the identical terminal characteristics $v = Ri$, where $R = 1/G$.

By way of illustration, Fig. 1.14 is a network which, under certain assumptions, can be used to model a transistor amplifier. The controlled source is of CCCS type, and the current i is the controlling variable.

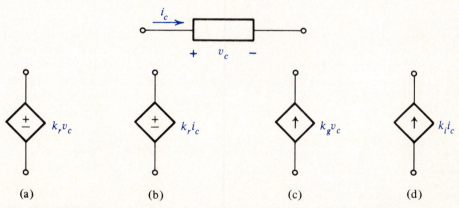

(a)   (b)   (c)   (d)

Figure 1.12 The four types of controlled sources: (a) VCVS, (b) CCVS, (c) VCCS, (d) CCCS.

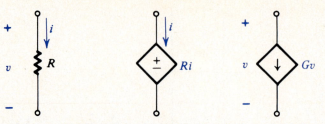

Figure 1.13 Three branches with identical terminal characteristics.

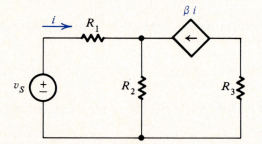

Figure 1.14 A network with a controlled source.

DRILL EXERCISES

1.1 Solve for i in Fig. D1.1.

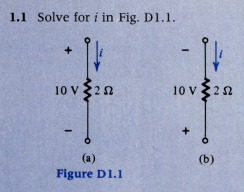

(a)   (b)

Figure D1.1

Ans. (a) 5A, (b) −5A.

1.2 Calculate the current in (a) a 2-F capacitance, (b) a 1-H inductance if the applied voltage is as shown in Fig. D1.2. The inductance current at $t = 0$ is zero.

Ans. (a) $i(t) = \begin{cases} 2 \text{ A} & 0 < t \leq 1 \\ -2 \text{ A} & 1 < t \leq 2 \\ 0 & t > 2 \end{cases}$

(b) $i(t) = \begin{cases} \dfrac{t^2}{2} & 0 < t \leq 1 \\ -1 + 2t - \dfrac{t^2}{2} & 1 < t \leq 2 \\ 1 \text{ A} & t > 2 \end{cases}$

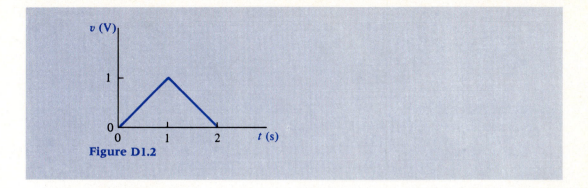

Figure D1.2

1.5 SCALED UNITS

It is usually convenient to work in scaled units, to avoid the manipulation of numbers that are awkwardly small or large. In electronic circuits, resistance values are usually in the thousands to millions of ohms and capacitances are in millionths or billionths of a farad.

The standard prefixes and abbreviations for the more common multipliers are tabulated below.

Prefix	Abbreviation	Multiplier
femto	f	10^{-15}
pico	p	10^{-12}
nano	n	10^{-9}
micro	μ	10^{-6}
milli	m	10^{-3}
kilo	k	10^{3}
mega	M	10^{6}
giga	G	10^{9}

When using scaling, it is convenient to scale the variables in such a way that the branch relationships defined in Sec. 1.4 hold with the scaled variables. For example, consider Ohm's law

$$v(V) = R(\Omega)i(A)$$

and suppose that current is to be expressed in milliamperes and voltage in volts. It is possible to write an equation between the units, just as in physics. Thus,

$$[V] = [\Omega][A]$$

$$[V] = [\Omega][10^3 \text{ mA}]$$

$$[V] = [10^3 \text{ } \Omega][\text{mA}]$$

or

$$[V] = [k\Omega][\text{mA}] \tag{1.12}$$

The conclusion is that Ohm's law may be used directly if resistances are expressed in kiloohms, voltages in volts, and currents in milliamperes.

Time may also be scaled, and usually is in electronics. For example, consider the branch law

$$i = C \frac{dv}{dt}$$

The corresponding equation between the units is

$$[A] = [F][V][s]^{-1} \tag{1.13}$$

Suppose current is to be expressed in milliamperes, voltage in volts and time in milliseconds. Equation (1.13) becomes

$$[10^3 \text{ mA}] = [F][V][10^3 \text{ ms}]^{-1}$$

$$[\text{mA}] = [10^{-6} \text{ F}][V][\text{ms}]^{-1}$$

or

$$[\text{mA}] = [\mu\text{F}][V][\text{ms}]^{-1} \tag{1.14}$$

It is concluded that the capacitance branch equation need not be modified if one uses milliamperes, volts, milliseconds, and microfarads.

DRILL EXERCISE

1.3 Suppose v is in volts, i in milliamperes, and t in microseconds. What units of R, L, and C should be used in order that the branch laws hold as defined in this chapter?
Ans. kΩ, mH, nF.

1.6 KIRCHHOFF'S LAWS

Figure 1.15 represents a *network*, i.e., an interconnection of branches. Each branch imposes a relationship between its voltage and its current. There are also constraints dependent on the interconnection, known as *Kirchhoff's laws*.

A *node* is a connecting point of two or more branches. The network of Fig. 1.15 has four nodes, labeled A, B, C, D. All points connected by short circuits are deemed to constitute a single node.

The *Kirchhoff current law* (KCL) states that the algebraic sum of currents entering (or leaving) a node is zero.

The following KCL equations apply to the network of Fig. 1.15, taking currents entering as positive:

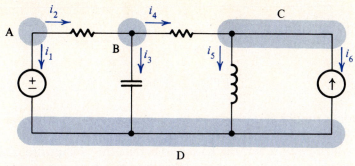

Figure 1.15 Nodes of a network.

Node A: $-i_1 - i_2 = 0$
Node B: $i_2 - i_3 - i_4 = 0$
Node C: $i_4 - i_5 - i_6 = 0$
Node D: $i_1 + i_3 + i_5 + i_6 = 0$

An equivalent way of writing KCL is to equate the sum of the currents entering the node to the sum of the currents leaving it. In the above equations, this amounts to transferring to the right-hand side the currents preceded by negative signs.

Figure 1.16 represents the same network as Fig. 1.15; the voltage and current labels are consistent between the two figures, when the conventional choice is made.

A *loop* is a closed path in which no node is traversed more than once. A *mesh* is a loop that contains no other loop. The network of Fig. 1.16 has three meshes labeled A, B, and C. There are three other loops, identifiable by giving the voltage subscripts of the branches traversed: 1245, 1246, 346.

The *Kirchhoff voltage law* (KVL) states that the algebraic sum of voltages around any loop is zero. Voltages may be taken as positive either if traversed from + to − (voltage drop), or from − to + (voltage rise): that, and the direction of traversal (clockwise or counterclockwise), is a matter of taste. It is important, however, to be consistent.

The following KVL equations apply to the network of Fig. 1.16, taking voltage drops as positive and going clockwise:

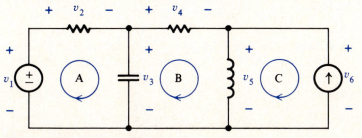

Figure 1.16 Meshes of a network.

Loop 123 (mesh A): $-v_1 + v_2 + v_3 = 0$
Loop 345 (mesh B): $-v_3 + v_4 + v_5 = 0$
Loop 56 (mesh C): $-v_5 + v_6 = 0$
Loop 1245: $-v_1 + v_2 + v_4 + v_5 = 0$
Loop 1246: $-v_1 + v_2 + v_4 + v_6 = 0$
Loop 346: $-v_3 + v_4 + v_6 = 0$

The student should verify that counting voltage rises as positive or traversing counterclockwise merely multiplies each KVL equation above by -1.

An equivalent way of writing KVL is to equate the voltage drops around the loop to the voltage rises. In the above equations, this amounts to transferring to the right-hand side the voltages preceded by negative signs.

EXAMPLE 1.4

In Fig. E1.4, rectangles represent branches, which may in fact contain any type of element. Use KVL to solve for v.

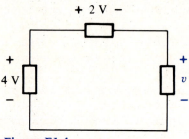

Figure E1.4

Solution
By KVL,

$$-4 + 2 + v = 0$$

$$v = 2 \text{ V}$$

EXAMPLE 1.5

Solve for i_1, i_2, v_1, and v_3 in Fig. E1.5.

Solution
By KCL at node A,

$$3 - 1 - i_1 = 0$$

$$i_1 = 2 \text{ A}$$

By KCL at node B,

$$-3 + i_2 = 0$$

$$i_2 = 3 \text{ A}$$

By KVL around mesh A,

$$-2 + v_3 + 1 = 0$$

$$v_3 = 1 \text{ V}$$

By KVL around mesh B,

$$+v_1 - v_3 = 0$$

$$v_1 = v_3 = 1 \text{ V}$$

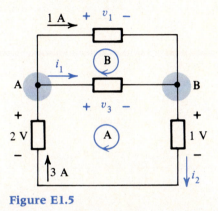

Figure E1.5

DRILL EXERCISE

1.4 Solve for the voltages v_1 and v_2 and the current i_1 in Fig. D1.4(a) and (b).

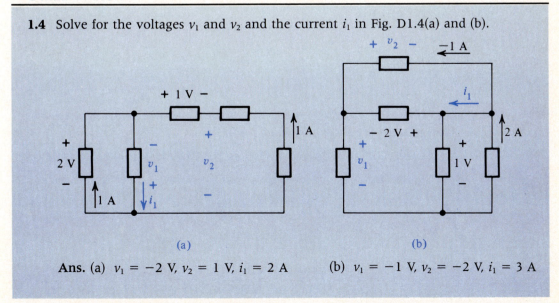

(a)

(b)

Ans. (a) $v_1 = -2$ V, $v_2 = 1$ V, $i_1 = 2$ A   (b) $v_1 = -1$ V, $v_2 = -2$ V, $i_1 = 3$ A

1.7 SOLUTION OF SIMPLE NETWORKS BY INSPECTION

A systematic approach to the solution of network problems will be developed in Chap. 2. However, it is desirable to develop quick and intuitive means to solve simple networks; most engineers work this way, reserving the more ponderous systematic methods for more difficult problems.

Intuition cannot be taught, only developed by repeated exercise. However, the examples presented here do have a common thread. In each case, (1) an unknown network variable (voltage or current) is chosen and (2) one attempts, by repeated use of KCL, KVL, and branch relationships, to express a known variable (i.e., a source voltage or current) in terms of the unknown. This done, it is not difficult to solve for the unknown.

Several examples are given, using networks with resistances and sources.

EXAMPLE 1.6

Solve for i in the network of Fig. E1.6(a).

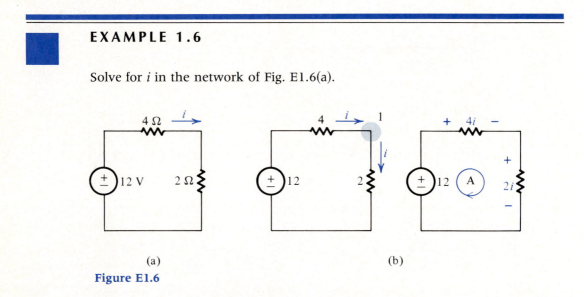

(a)   (b)

Figure E1.6

Solution
The solution steps are shown in Fig. E1.6(b).

Step 1: Use KCL at node 1 to write i as the current in the 4-Ω resistance.

Step 2: Use Ohm's law to obtain the voltages across the two resistances.

Step 3: Use KVL around mesh A to write

$$-12 + 4i + 2i = 0$$

or

$$i = 2 \text{ A}$$

EXAMPLE 1.7

Solve for v in Fig. E1.7(a) (v_S is variable, but given).

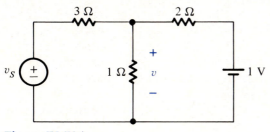

Figure E1.7(a)

Solution

Step 1: Use KVL around mesh A to arrive at $(v - 1)$ for the voltage across the 2-Ω resistance.

Step 2: Use Ohm's law to write the two currents shown.

Step 3: By KCL at node 1, the current in the 3-Ω resistance is $v + (v - 1)/2 = (3v - 1)/2$.

Step 4: Use Ohm's law to write the voltage across the 3-Ω resistance.

Step 5: Using KVL around mesh B, write

$$-v_S + \frac{3(3v - 1)}{2} + v = 0$$

$$v = \frac{2v_S + 3}{11}$$

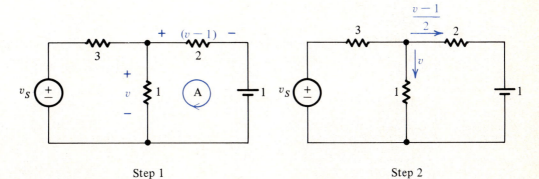

Step 1   Step 2

Figure E1.7(b)

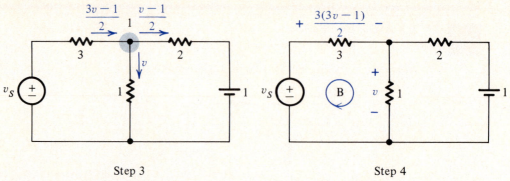

Step 3   Step 4

Figure E1.7(b) continued

EXAMPLE 1.8

Solve for v in Fig. E1.8(a).

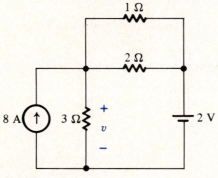

Figure E1.8(a)

Solution

Step 1: Use KVL around mesh A to deduce that $(v - 2)$ is the voltage across the 2-Ω resistance, and KVL around mesh B to write the same voltage across the 1-Ω resistance.

Step 2: Use Ohm's law to write the currents shown.

Step 3: By KCL at node 1,

$$8 - \frac{v}{3} - \frac{v-2}{2} - (v - 2) = 0$$

$$v = 6 \text{ V}$$

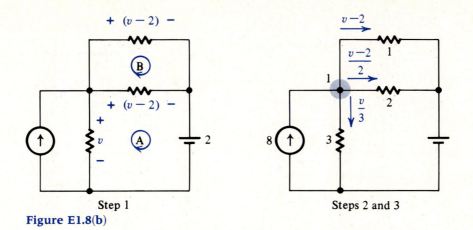

Figure E1.8(b)

EXAMPLE 1.9

Solve for v in Fig. 1.9(a) (v_S is in volts). (Under certain assumptions, this circuit is a model of a bipolar transistor amplifier.)

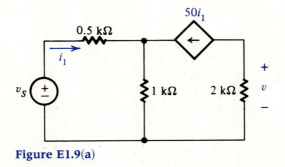

Figure E1.9(a)

Solution

Since v_S is in volts and resistances are in kiloohms (kΩ), work with currents in milliamperes (mA) [see Eq. (1.12)].

In problems with controlled sources, it is often a good idea to use the controlling variable (i_1 in this case) as the unknown. The reason is that one can immediately also write the controlled voltage or current. The solution steps are shown in Fig. E1.9(b).

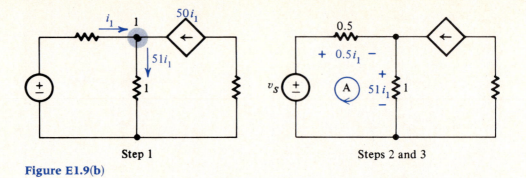

Step 1   Steps 2 and 3

Figure E1.9(b)

Step 1: Use KCL at node 1 to deduce that the current in the 1-kΩ resistance is $51i_1$.

Step 2: Use Ohm's law to write the voltages, as shown.

Step 3: By KVL around mesh A,

$$-v_S + 0.5i_1 + 51i_1 = 0$$

$$i_1 = \frac{v_S}{51.5} \text{ mA}$$

Step 4: To obtain v, use Ohm's law:

$$v = -50i_1(2)$$

$$v = -100i_1 = -1.94v_S$$

DRILL EXERCISES

1.5 Solve for v in Fig. D1.5.

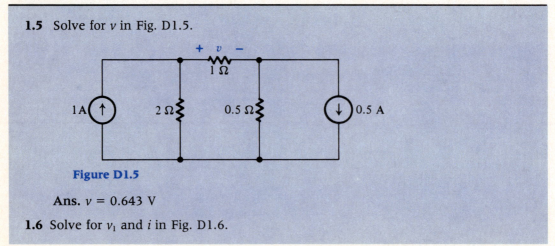

Figure D1.5

Ans. $v = 0.643$ V

1.6 Solve for v_1 and i in Fig. D1.6.

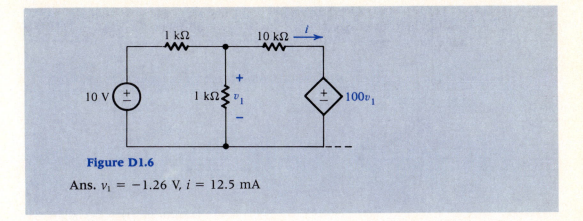

Figure D1.6

Ans. $v_1 = -1.26$ V, $i = 12.5$ mA

1.8 ENERGY AND POWER

The power supplied to a branch is defined as

$$p(t) = v(t)i(t) \tag{1.15}$$

with v and i defined in Fig. 1.17. It can be shown from electromagnetics that $p(t)$ does in fact correspond to the physical notion of power. If v is in volts and i is in amperes, p is in *watts* (W).

If $p(t)$ is negative, power is taken from the branch, i.e., the branch supplies power.

The energy supplied to a branch between times t_0 and t_1 is

$$E(t_0, t_1) = \int_{t_0}^{t_1} p(t) \, dt \tag{1.16}$$

For a resistance,

$$p(t) = Ri(t)i(t) = Ri^2(t) \tag{1.17}$$

Since $i^2(t) \geq 0$, the power supplied to a resistance is always nonnegative.

The energy supplied to a resistance is

$$E(t_0, t_1) = \int_{t_0}^{t_1} Ri^2(t) \, dt \tag{1.18}$$

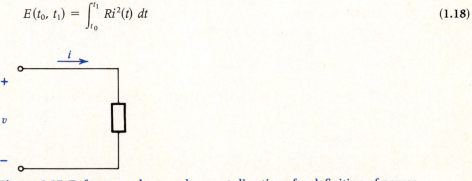

Figure 1.17 Reference voltage and current directions for definition of power.

Since $E(t_0, t_1) \geq 0$, the resistance is said to dissipate energy because it cannot give back any of the energy it receives.

The power dissipated in a resistance may also be written as

$$p(t) = v(t)i(t)$$

$$= \frac{v(t)v(t)}{R}$$

$$= \frac{v^2(t)}{R}$$

or

$$p(t) = Gv^2(t) \tag{1.19}$$

For a capacitance,

$$p(t) = v(t)C\frac{dv}{dt}$$

$$= Cv(t)\frac{dv}{dt}$$

$$p(t) = \frac{C}{2}\frac{d(v^2(t))}{dt} \tag{1.20}$$

The power supplied to a capacitance may be either positive or negative, depending on whether the magnitude of v is increasing or decreasing. The energy supplied to a capacitance between times t_0 and t_1 is

$$E(t_0, t_1) = \int_{t_0}^{t_1} \frac{C}{2}\frac{d(v^2(t))}{dt}\,dt$$

$$= \tfrac{1}{2}CV^2(t_1) - \tfrac{1}{2}CV^2(t_0) \tag{1.21}$$

The energy supplied to a capacitance can be recovered. To see this, suppose $v(t_0) = v(t_2) = 0$, $v(t_1) \neq 0$, $t_0 < t_1 < t_2$. Then, using Eq. (1.21),

$$E(t_0, t_1) = \tfrac{1}{2}Cv^2(t_1)$$

$$E(t_1, t_2) = -\tfrac{1}{2}Cv^2(t_1)$$

The energy supplied to the capacitance during the interval t_0 to t_1 is taken back during the interval t_1 to t_2. Because it can be recovered, the energy is said to be *stored*. Taking $v(t_0) = 0$ as a reference, the energy stored in a capacitance is therefore given by

$$E_C(t) = \tfrac{1}{2}Cv^2(t) \tag{1.22}$$

For an inductance,

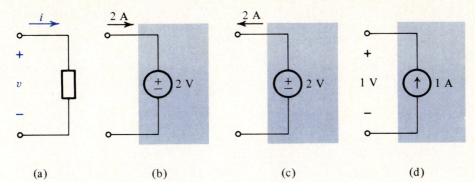

| (a) | (b) | (c) | (d) |

Figure 1.18 Illustration of sources supplying and absorbing power.

$$p(t) = Li(t)\frac{di}{dt}$$

$$= \frac{L}{2}\frac{d(i^2(t))}{dt} \tag{1.23}$$

and

$$E(t_1, t_0) = \tfrac{1}{2}Li^2(t_1) - \tfrac{1}{2}Li^2(t_0) \tag{1.24}$$

Because of the similarity in the equations, it is seen that the energy in an inductance is also stored. Taking $i(t_0) = 0$ as a reference, the stored energy is written as

$$E_L(t) = \tfrac{1}{2}Li^2(t) \tag{1.25}$$

Sources may either supply or absorb power. Figure 1.18(a) reproduces Fig. 1.17 for convenience, to recall the reference directions of voltage and current. In Fig. 1.18(b), power is supplied to the source, since both v and i are positive: $p = 2\text{ V} \times 2\text{ A} = 4\text{ W}$, and the source absorbs power.

In Fig. 1.18(c) and (d), power is taken from the sources: in each case v is positive, but i is negative since the current is opposite the reference direction. To say that power is taken from a source is to say that it supplies, or generates, power.

EXAMPLE 1.10

For the network of Fig. E1.10, find the power dissipated in each resistance, and the power supplied or absorbed by each source.

Solution
This network is the network of Example 1.7. The solution for v was

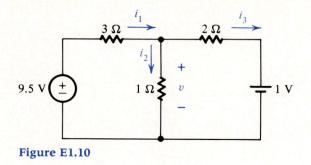

Figure E1.10

$$v = \frac{2v_S + 3}{11}$$

hence, with $v_S = 9.5$ V,

$$v = \frac{2(9.5) + 3}{11} = 2 \text{ V}$$

From step 3 in Fig. E1.7(b), there results

$$i_1 = \frac{3v - 1}{2} = 2.5 \text{ A}$$

$$i_2 = v = 2 \text{ A}$$

$$i_3 = \frac{v - 1}{2} = 0.5 \text{ A}$$

The resistances dissipate power, as follows:

1-Ω resistance: $p = 2^2 \times 1 = 4$ W
2-Ω resistance: $p = 0.5^2 \times 2 = 0.5$ W
3-Ω resistance: $p = 2.5^2 \times 3 = 18.75$ W

The power absorbed by the 1-V source is

$$p = 0.5 \times 1 = 0.5 \text{ W}$$

The power absorbed by the 9.5-V source is

$$p = 9.5(-i_1)$$

because the reference direction of i_1 is opposite that of Fig. 1.17. Thus,

$$p = -9.5 \times 2.5 = -23.75 \text{ W}$$

This source generates 23.75 W of power. The student can verify that this is equal to the sum of the dissipated and absorbed power.

EXAMPLE 1.11

In Example 1.1, the voltage and current across a 2-F capacitance were as shown in Fig. E1.11(a). Calculate, as a function of time, the power delivered to the capacitance, and find the energy delivered between $t = 0$ and $t = 4$.

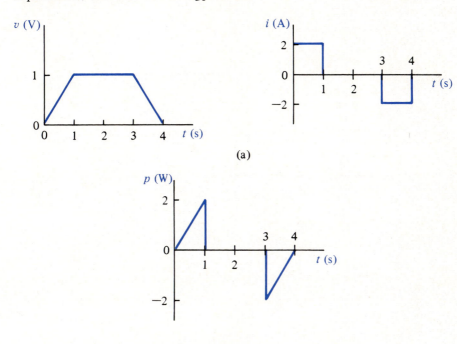

(a)

(b)

Figure E1.11

Solution

Since $p(t) = v(t)i(t)$, one need only multiply the graphs of Fig. 1.11(a). The result is shown in Fig. 1.11(b).

The energy delivered is the integral of the power, i.e., the area under the curve between $t = 0$ and $t = 4$. Clearly, this is zero, and there is no stored energy at $t = 4$. Since $v(4) = 0$, this agrees with the expression $E_c = Cv^2(t)/2$.

DRILL EXERCISES

1.7 For the network of Fig. D1.7, calculate the power dissipated in the resistance and the power supplied by each source.

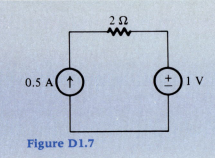

Figure D1.7

Ans. 0.5 W dissipated in R, -0.5 W supplied by the voltage source, 1 W by the current source.

1.8 For the network of Fig. D1.8, calculate the power dissipated in each resistance and the power supplied by each source.

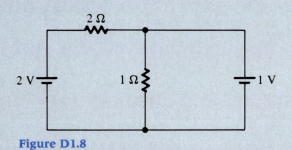

Figure D1.8

Ans. Power dissipated: 0.5 W in the 2-Ω resistance, 1 W in the 1-Ω resistance. Power supplied: 1 W by the 2-V source, 0.5 W by the 1-V source.

1.9 Calculate, as a function of time, the power applied to a 2-F capacitance for the applied voltage waveform of Exercise 1.2. What is the energy stored in the capacitance at $t = 1$ s?

Ans. $p(t) = \begin{cases} 2t & 0 < t \le 1 \\ 2(t - 2) & 1 < t \le 2 \\ 0 & t > 2 \end{cases}$

The energy stored at $t = 1$ is 1 J.

1.9 SUMMARY AND STUDY GUIDE

The main purpose of this chapter has been the presentation of some basic concepts in network analysis. Following a few sections of general introduction, the concept of reference polarities for voltages and currents was introduced. This was followed by a description of the basic branch elements. The interconnection laws, KCL and

KVL, were discussed next and were used to solve simple networks. The chapter concluded with an introduction to energy and power.

The attention of the student is drawn to the following points.

Reference polarities only indicate when a variable is taken to be positive or negative. Assignment of, say, an arrow to a current does not imply that this is the direction of current flow.

The current through a voltage source (voltage across a current source) can have any value and is determined by the rest of the network.

Kirchhoff's laws depend only on the interconnection pattern of the branches and are independent of the nature of the branches in the network.

One way of solving for a branch variable in a simple network is to assume it to be known and to use KCL, KVL, and the branch laws to obtain an equation for the desired variable.

Power can be absorbed or supplied; the student must be capable of distinguishing between those two situations.

PROBLEMS

1.1 In Fig. P1.1, solve for v or i, as the case may be.

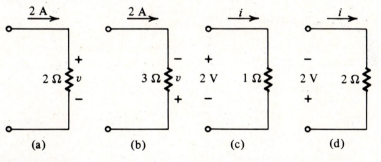

Figure P1.1

1.2 (a) For the branches of Fig. P1.2, write down the v-i relationship. (b) Sketch v vs i.

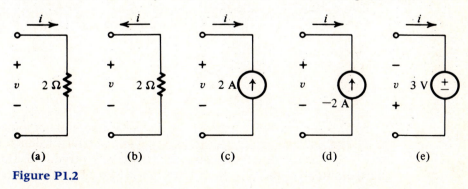

Figure P1.2

1.3 For the inductance of Fig. P1.3(a), find $v(t)$ for the currents given in Fig. P1.3(b) and (c).

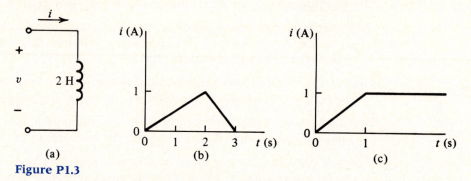

(a) (b) (c)

Figure P1.3

1.4 The current in the inductance of Fig. P1.4(a) is zero at $t = 0$. Find $i(t)$ for the waveforms $v(t)$ given in Fig. P1.4(b) and (c).

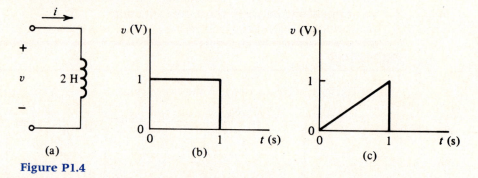

(a) (b) (c)

Figure P1.4

1.5 (a) Find the current $i(t)$ in a 0.5-F capacitance if the applied voltage $v(t) = 2 \cos 4t$.
(b) Find $v(t)$ if $i(t) = 4 \cos 2t$, given that $v(0) = 2$ V.

1.6 The magnetic field produced by an inductance is proportional to the current in the inductance. In a television picture tube, a magnetic field is used to deflect an electron beam horizontally. The beam deflection is proportional to the field, and the field, in turn, is proportional to the current in a coil, which varies as in Fig. P1.6. The beam is swept linearly in time during the first 60 μs, then swept back during the next 5 μs. If the coil is modeled as an inductance with the value 0.01 H, what voltage waveform must be applied to the terminals of the coil to yield the current of Fig. P1.6?

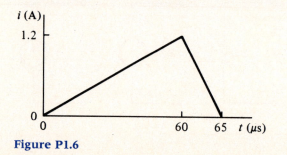

Figure P1.6

1.7 Calculate and sketch $v(t)$ across the capacitance in Fig. P1.7(a), given that $v(0) = 0$ and that $i_s(t)$ is the waveform shown in Fig. P1.7(b).

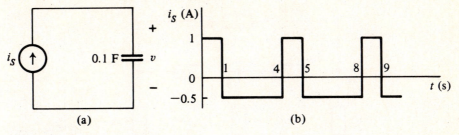

(a)   (b)

Figure P1.7

1.8 Given the following units of voltage, current, and time, find the units of R, C, and L such that the branch relationships of the elements apply in the scaled units given:

(**a**) V, mA, ns   (**c**) kV, kA, ms
(**b**) V, μA, ms   (**d**) mV, μA, μs

1.9 Show that, if time is in milliseconds, the unit of frequency is the kilohertz. Calculate the current in milliamperes flowing in a 3-μF capacitance for an applied voltage $v(t) = 2 \cos 2\pi f_0 t$ volts, where f_0 is 2 kHz.

1.10 Solve for the labeled voltages and currents in the networks of Fig. P1.10. (*Note*: v_1 may be thought of as the voltage across an open circuit, e.g., a current source of value zero.)

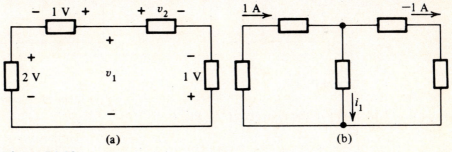

(a)   (b)

Figure P1.10

1.11 Repeat Problem 1.10 for the network of Fig. P1.11.

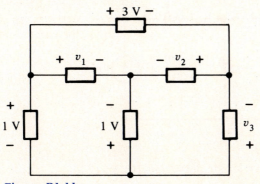

Figure P1.11

1.12 Repeat Problem 1.10 for the network of Fig. P1.12.

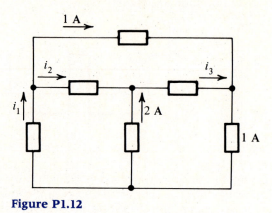

Figure P1.12

1.13 Find *i* in the networks of Fig. P1.13.

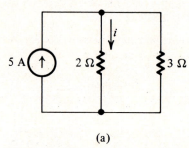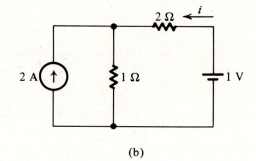

(a)   (b)

Figure P1.13

1.14 Find *v* in the network of Fig. P1.14.

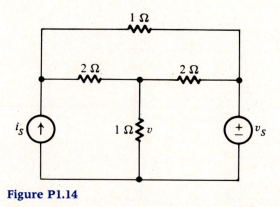

Figure P1.14

1.15 Repeat Problem 1.14 for the network of Fig. P1.15.

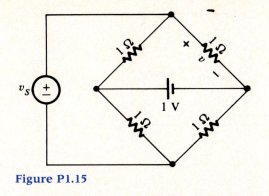

Figure P1.15

1.16 In Fig. P1.16, calculate first the controlling variable i_1, then calculate v (v_s is in volts).

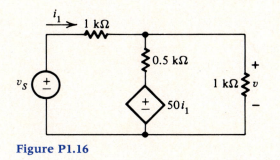

Figure P1.16

1.17 In Fig. P1.17, calculate first the controlling variable v_1, then calculate v (v_s is in volts).

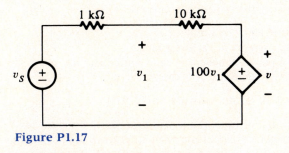

Figure P1.17

1.18 Under certain conditions, the network of Fig. P1.18 is a model of a field-effect transistor amplifier. Calculate the controlling variable v_1 and voltage v_0 (v_s is in volts and the controlled current is in milliamperes).

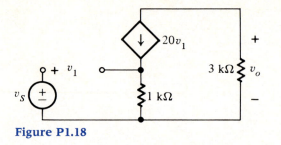

Figure P1.18

1.19 Under certain conditions, the network of Fig. P1.19 is a model of a bipolar junction transistor amplifier. Calculate the controlling variable i_1 and the voltage v_o. (v_s is in volts.)

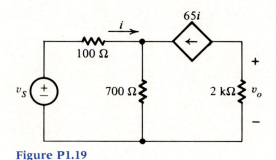

Figure P1.19

1.20 Find the power dissipated if 10 V are applied across a resistance of (a) 1 Ω, (b) 10 Ω, (c) 100 Ω, (d) 1 kΩ, (e) 10 kΩ.

1.21 For the network of Fig. P1.13(a),
   (a) Calculate the voltage across the current source.
   (b) Calculate the power supplied by the source and the power dissipated by each resistance. Verify that power is conserved.

1.22 For the network of Fig. P1.13(b),
   (a) Calculate the currents in each resistance.
   (b) Calculate the power supplied by each source and the power dissipated by each resistance. Verify that power is conserved.

1.23 Repeat Problem 1.22 for the network of Fig. P1.19.

1.24 Sketch, as a function of time, the power supplied to the inductance in Problem 1.4 when the voltage of Fig. P1.4(b) is applied. Integrate to find the energy supplied to the inductance between $t = 0$ and $t = 1$ and use Eq. (1.25) to verify your answer.

1.25 The voltage across a 2-F capacitance is $2 \cos \pi t$. Calculate, as a function of time, the power supplied to the capacitance between $t = 0$ and $t = 2$ s. Calculate the energy supplied to the capacitance during that time interval.

1.26 The voltage across a 2-μF capacitance is as shown in Fig. P1.26. Sketch, as functions of time, the current through the capacitance, the charge, the power supplied to the capacitance and the stored energy.

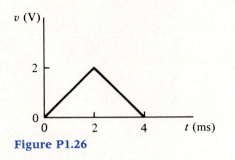

Figure P1.26

CHAPTER

Formulation of Network Equations

2.1 INTRODUCTION

Many useful networks are too complex to be solved by "intuitive" methods and require a more systematic approach. This is also a necessity if a computer is to be used for network analysis: computers can only follow steps in a well-defined procedure.

The first problem addressed in this chapter arises from the fact that writing KCL, KVL, and the branch laws usually results in more equations than variables. Since, for b branches, the number of variables is $2b$ (b voltages and b currents), it is necessary to find exactly $2b$ independent equations to solve uniquely for the $2b$ unknowns.

The second problem addressed is the reduction of the number of equations; with five branches (a modest number), $2b = 10$, a rather large number of equations. Special variables, called node voltages and mesh currents, are introduced so that, in effect, many of the equations are easily solved by inspection. Mesh currents are also used to derive Tellegen's theorem, which leads directly to the principle of conservation of power.

The chapter closes with an exposition of the principle of superposition, a key concept in the analysis of networks and all other linear systems.

2.2 NETWORK GRAPHS

2.2.1 Definition and Properties

A network *graph* is used to emphasize the interconnections between the branches, often referred to as the *topology* of the network. A graph is constructed by reducing each node to a point and replacing each branch by a line. A graph becomes an *oriented graph* by putting arrows on the branches. Figure 2.1 shows a network and its oriented graph. The arrow constitutes a reference direction for the current; by the convention of Chap. 1, the current direction in also the "+" to "−" direction, i.e., voltage drops in the arrow direction. One says that a branch enters (or leaves) when the arrow points toward (or away from) the node. For example, in Fig. 2.1(b), branch 5 leaves node A and enters node D.

A graph is called *planar* if it can be drawn on a plane in such a way that the branches intersect at nodes only.

In some cases, a graph must be redrawn in order to appear planar: this is the case for the graph of Fig. 2.2(a), redrawn in Fig. 2.2(b). On the other hand, a nonplanar graph cannot be so redrawn [the student is invited to try for the graph of Fig. 2.2(c)]. Only planar graphs are considered in this text.

In a planar graph, the following relationship exists:

$$l = b - n + 1 \tag{2.1}$$

where l = number of meshes
$\quad n$ = number of nodes
$\quad b$ = number of branches

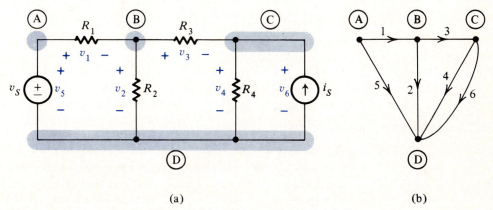

(a)   (b)

Figure 2.1 Example of a network (a) and its oriented graph (b).

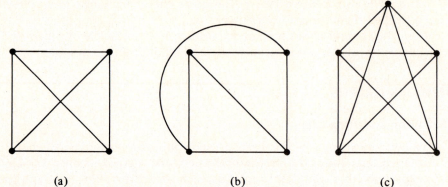

Figure 2.2 Planar [(a) and (b)] and nonplanar (c) graphs.

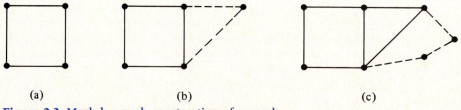

Figure 2.3 Mesh-by-mesh construction of a graph.

To prove Eq. (2.1), consider the construction of a graph, one mesh at a time, as in Fig. 2.3. The first mesh [Fig. 2.3(a)] includes as many nodes as branches. In creating the second mesh [Fig. 2.3(b)], a new node is included with each new branch, except for the last one, which joins up with a node already present. With two meshes in place, the number of nodes is equal to the number of branches, less one. The third mesh also brings in one fewer new nodes than new branches: the node count after three meshes is thus *two* less than the number of branches. After *l* meshes, the node count is $l - 1$ less than the number of branches. Since b is the number of branches in the network, then

$$n = b - (l - 1)$$

or

$$l = b - n + 1$$

2.2.2 Use of the Graph to Write the Kirchhoff Laws

Since KCL and KVL depend only on the network topology, they can be written using only the oriented graph. For the graph of Fig. 2.1(b),

KCL:

Node A: $-i_1 - i_5 = 0$
Node B: $i_1 - i_2 - i_3 = 0$

Node C: $i_3 - i_4 - i_6 = 0$
Node D: $i_2 + i_4 + i_5 + i_6 = 0$

KVL:

Mesh 125: $v_1 + v_2 - v_5 = 0$
Mesh 234: $-v_2 + v_3 + v_4 = 0$
Mesh 46: $-v_4 + v_6 = 0$
Loop 1345: $v_1 + v_3 + v_4 - v_5 = 0$
Loop 1365: $v_1 + v_3 + v_6 - v_5 = 0$
Loop 236: $-v_2 + v_3 + v_6 = 0$

The remaining equations are the branch laws. In order to write them, one must go to the network and use the element $v = i$ relationships:

$$v_1 - R_1 i_1 = 0$$

$$v_2 - R_2 i_2 = 0$$

$$v_3 - R_3 i_3 = 0$$

$$v_4 - R_4 i_4 = 0$$

$$v_5 = v_S$$

$$i_6 = -i_S$$

There are 12 variables in this network, i.e., one voltage and one current for each of six branches. There are 16 equations; the first problem to be tackled is to choose the "right" set of 12 equations for the 12 unknowns. In order to do this efficiently, it is useful to recall briefly some basic ideas on systems of linear equations.

DRILL EXERCISE

2.1 For the graphs of Fig. D2.1 give the number of branches, nodes, and meshes. Verify the relation $b = n - 1 + l$.

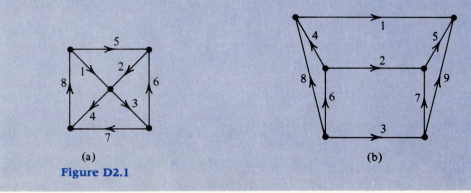

(a)   (b)

Figure D2.1

Ans. (a) $b = 8, n = 5, l = 4$
(b) $b = 9, n = 6, l = 4$

2.3 REVIEW OF SOME CONCEPTS ON LINEAR EQUATIONS

Consider the equations

$$2x - y + z = 0$$

$$-x + 3y - 2z = 0$$

$$3x + y \quad\ = 0$$

The reader can verify that the third equation is twice the first plus the second. These equations are termed *dependent*, a term used to indicate that one (or more) left-hand sides can be generated as a linear combination of the others.

In this case, the third equation is implied by the first two: if the first two hold, then the third is found to hold also and is therefore redundant.

If the right-hand sides are nonzero, a set of dependent equations may or may not have a solution. Consider, for example,

$$2x - y + z = 1$$

$$-x + 3y - 2z = -2$$

$$3x + y \quad\ = 0$$

Here, the third equation can again be implied from the first two; this set of equations is called *consistent* and has an infinite number of solutions, since there are really only two equations and three unknowns. If the right-hand side of the third equation is anything but zero, the set of equations is called *inconsistent* and no solution exists.

Finally, recall that a set of n independent equations in n unknowns has one, and only one, solution. The network problem has $2b$ unknowns, one voltage and one current for each branch. Since there are more than $2b$ equations, the objective is to find a set of $2b$ independent equations, and also to show that the remaining equations are consistent with the first $2b$.

2.4 FINDING $2b$ INDEPENDENT EQUATIONS

2.4.1 The Kirchhoff Current Laws

The graph of Fig. 2.1(b) is repeated in Fig. 2.4 for convenience. Recall the KCL equations:

Node A: $-i_1 - i_5 = 0$

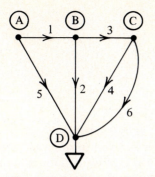

Figure 2.4 A network-oriented graph showing nodes.

Node B: $i_1 - i_2 - i_3 = 0$

Node C: $i_3 - i_4 - i_6 = 0$

Node D: $i_2 + i_4 + i_5 + i_6 = 0$ (2.2)

In matrix form, these are written

$$
\begin{array}{c}
\text{Branch} \\
\text{Node}
\end{array}
\begin{array}{cccccc}
1 & 2 & 3 & 4 & 5 & 6
\end{array}
$$

$$
\begin{array}{c}
A \\ B \\ C \\ D
\end{array}
\begin{bmatrix}
-1 & 0 & 0 & 0 & -1 & 0 \\
1 & -1 & -1 & 0 & 0 & 0 \\
0 & 0 & 1 & -1 & 0 & -1 \\
0 & 1 & 0 & 1 & 1 & 1
\end{bmatrix}
\begin{bmatrix}
i_1 \\ i_2 \\ i_3 \\ i_4 \\ i_5 \\ i_6
\end{bmatrix}
=
\begin{bmatrix}
0 \\ 0 \\ 0 \\ 0
\end{bmatrix}
\qquad (2.3)
$$

The matrix in Eq. (2.3) is called the *incidence* matrix of the network. Each column corresponds to a branch, each row to a node. A column contains only two nonzero entries: a +1 in the row corresponding to the node entered by a branch, a −1 in the row corresponding to the node left by the branch. For example, branch 3 leaves node B and enters node C; therefore, the third column has a −1 in row B and a +1 in row C.

This column property is characteristic of the incidence matrix of any oriented graph. Conversely, a matrix with this column property can always be interpreted as the incidence matrix of some oriented graph; the number of rows is the number of nodes in the graph, and the columns specify where each branch starts and ends.

Since every column of the incidence matrix has a +1 and a −1, the sum of its rows is zero. This implies that the sum of the KCL equations is zero, i.e., that any KCL equation is the negative sum of all the others. Thus, the KCL equations are not independent.

This being the case, remove one equation from the set, say the one pertaining to node D; node D becomes the so-called *datum node*, and is identified by the triangle symbol in Fig. 2.4.

The three remaining equations are

$$
\begin{aligned}
-i_1 \qquad\qquad\quad - i_5 \qquad &= 0 \\
i_1 - i_2 - i_3 \qquad\qquad\quad &= 0 \\
i_3 - i_4 \quad - i_6 &= 0
\end{aligned}
$$

(2.4)

Since the first equation is the only one to have the term i_5, it surely cannot be expressed by combining the other two. The second and third equations also have isolated terms, i.e., terms that are absent from the other equations: i_2 in the case of the second equation, i_4 and i_6 for the third. Clearly, these three equations are independent.

The isolated variables are precisely the currents of branches connected to the datum node. In terms of the incidence matrix, removal of the last row leaves columns 2, 4, 5, and 6 with only one nonzero entry. Each of those columns corresponds to a branch that is connected to the datum node, and the nonzero entry appears in the row corresponding to the node connected to the other end of that branch. For example, branch 2 connects B to the datum node, so the -1 is in the second row.

In this case, nodes A, B and C are all connected to D so that removal of the KCL equations for D places isolated terms in each of the other three equations. In general, not all nodes are connected to the datum node; nevertheless, it can still be shown that the set of $n - 1$ equations that remains when one node KCL equation is removed is a set of $n - 1$ independent equations.

The set of n KCL equations is consistent, because the datum node KCL equation is the negative sum of the $n - 1$ others.

It is not difficult to see that any node could have been picked as the datum node; in general, given n KCL equations,

1. Any $n - 1$ KCL equations are independent.

2. The remaining KCL equation is consistent with the other $n - 1$.

2.4.2 The Kirchhoff Voltage Laws

Figure 2.5 represents the same graph as Fig. 2.4, but shows the mesh definition. The arrow directions indicate the direction of traversal of each mesh. It is assumed that the mesh directions are all the same.

The mesh KVL equations for the graph of Fig. 2.4 are

Mesh A: $v_1 + v_2 - v_5 = 0$

Mesh B: $-v_2 + v_3 + v_4 = 0$ (2.5)

Mesh C: $-v_4 + v_6 = 0$

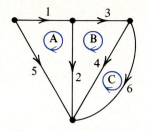

Figure 2.5 A network-oriented graph showing mesh traversal directions.

In matrix form, these become

$$
\begin{array}{c}
\underline{\text{Branch}} \\
\text{Mesh} \\
\begin{array}{c} A \\ B \\ C \end{array}
\end{array}
\begin{array}{cccccc}
1 & 2 & 3 & 4 & 5 & 6
\end{array}
\left[\begin{array}{cccccc}
1 & 1 & 0 & 0 & -1 & 0 \\
0 & -1 & 1 & 1 & 0 & 0 \\
0 & 0 & 0 & -1 & 0 & 1
\end{array}\right]
\left[\begin{array}{c}
v_1 \\ v_2 \\ v_3 \\ v_4 \\ v_5 \\ v_6
\end{array}\right]
=
\left[\begin{array}{c}
0 \\ 0 \\ 0
\end{array}\right]
\qquad (2.6)
$$

The matrix in Eq. (2.6) is called the *mesh matrix*. Each column corresponds to a branch, each row to a mesh. A column contains a single nonzero entry if the corresponding branch belongs to only one mesh, and that entry appears in the row corresponding to the mesh. The entry is $+1$ if the branch reference direction is the same as the mesh direction, -1 otherwise. If a branch belongs to two meshes, it is traversed positively in one mesh and a $+1$ appears in the row corresponding to that mesh, and negatively in the other mesh which generates a -1 in the corresponding row.

For example, in Fig. 2.5, branch 1 is an exterior branch traversed only by mesh A in the positive direction: column 1 has only one nonzero entry, a $+1$ in row A. Branch 2, an interior branch, belongs to meshes A and B. It is traversed positively in mesh A, negatively in mesh B; column 2 has a $+1$ in row A, a -1 in row B.

A row may be added to the mesh matrix in such a way that each column contains a $+1$ and a -1:

$$
\left[\begin{array}{cccccc}
1 & 1 & 0 & 0 & -1 & 0 \\
0 & -1 & 1 & 1 & 0 & 0 \\
0 & 0 & 0 & -1 & 0 & 1 \\
\hdashline
-1 & 0 & -1 & 0 & 1 & -1
\end{array}\right]
$$

This matrix has the column property of an incidence matrix; it is the incidence matrix of some other graph, called the *dual graph*. The rows of the matrix represent the KCL equations for the dual graph. In particular, the first three KCL equations for the dual graph are the same as the mesh KVL Eqs. (2.5), except that the branch voltages of the graph of Fig. 2.5 are replaced by the branch currents of the dual graph. Since any three KCL equations for the dual graph are independent, it follows that the three mesh KVL Eqs. (2.5) are independent. This argument carries over to the general case, and the l KVL equations are independent.

There are other KVL equations, for the loops that are not meshes. Might there not be some among them that are independent of the mesh equations? The answer is, no. For example, consider loop 1345 in Fig. 2.4. This loop contains two meshes with the KVL equations

$$v_1 + v_2 - v_5 = 0$$

and

$$-v_2 + v_3 + v_4 = 0$$

Adding these equations yields

$$v_1 + v_3 + v_4 - v_5 = 0$$

which is precisely the KVL equation for loop 1345. Because branch 2 is interior to this loop, v_2 enters positively in one mesh equation and negatively in the other: it is canceled out in the sum, leaving only the branches belonging to the loop. Thus, the loop KVL equation is implied by the mesh KVL equations. The same reasoning holds for all other loops. To summarize:

1. The l mesh KVL equations are independent.

2. The other loop KVL equations are consistent with the mesh KVL equations.

2.4.3 The Network Equations

Since the KCL and KVL equations involve different variables (currents in one case, voltages in the other), they are independent. Therefore, the $n - 1$ node KCL equations and l mesh KVL equations constitute a set of $n - 1 + l = b$ independent equations.

Each of the b branches has its own branch law equation. The b branch equations are independent because they involve different voltages and currents.

EXAMPLE 2.1

Write the equations describing the network of Fig. E2.1.

Solution
KCL:

Node a: $\quad -i_1 - i_2 = 0$ $\hfill (1)$

Node b: $\quad i_2 - i_3 - i_4 = 0$ $\hfill (2)$

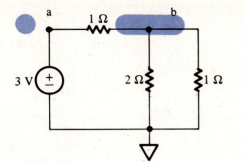

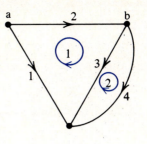

Figure E2.1

KVL:

   Mesh 1: $-v_1 + v_2 + v_3 = 0$ (3)

   Mesh 2: $-v_3 + v_4 = 0$ (4)

Branch equations:

   $v_1 \qquad = 3$ (5)

   $v_2 - \quad i_2 = 0$ (6)

   $v_3 - 2i_3 = 0$ (7)

   $v_4 - \quad i_4 = 0$ (8)

To solve this set of equations, use (5) and (3), then (4) to write

   $v_2 = 3 - v_3$

   $v_4 = v_3$

which, using (6), (7), and (8) become

   $i_2 = 3 - 2i_3$

   $i_4 = 2i_3$

Insert these two equations into (2). There results

   $3 - 2i_3 - i_3 - 2i_3 = 0$

   $i_3 = \frac{3}{5}$ A

The complete solution unravels at this point, and the result is

   $i_1 = -\frac{9}{5}$ A $\qquad v_1 = 3$ V

   $i_2 = \frac{9}{5}$ A $\qquad v_2 = \frac{9}{5}$ V

   $i_3 = \frac{3}{5}$ A $\qquad v_3 = \frac{6}{5}$ V

   $i_4 = \frac{6}{5}$ A $\qquad v_4 = \frac{6}{5}$ V

EXAMPLE 2.2

Write the equations for the network of Fig. E2.2.

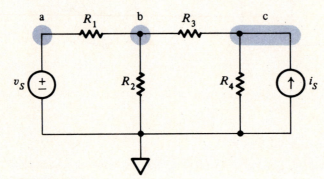 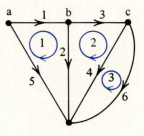

Figure E2.2

Solution

KCL:

Node a: $-i_1 - i_5 = 0$

Node b: $i_1 - i_2 - i_3 = 0$

Node c: $i_3 - i_4 - i_6 = 0$

KVL:

Mesh 1: $v_1 + v_2 - v_5 = 0$

Mesh 2: $-v_2 + v_3 + v_4 = 0$

Mesh 3: $-v_4 + v_6 = 0$

Branch equations:

$v_1 - R_1 i_1 = 0$

$v_2 - R_2 i_2 = 0$

$v_3 - R_3 i_3 = 0$

$v_4 - R_4 i_4 = 0$

$v_5 \qquad = v_S$

$i_6 \qquad = -i_S$

There are 12 equations and 12 unknowns. The solution will not be attempted: the number of equations is rather forbidding, at least for a hand calculation. Systematic procedures for the elimination of variables are obviously needed, and are developed in the next section.

DRILL EXERCISES

2.2 For the oriented graph of Fig. D2.2, find the incidence matrix.

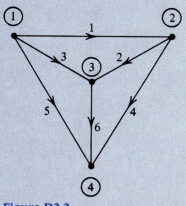

Figure D2.2

Ans.

$$\begin{bmatrix} -1 & 0 & -1 & 0 & -1 & 0 \\ 1 & -1 & 0 & -1 & 0 & 0 \\ 0 & 1 & 1 & 0 & 0 & -1 \\ 0 & 0 & 0 & 1 & 1 & 1 \end{bmatrix}$$

2.3 Give the oriented graph for which the matrix

$$\begin{bmatrix} 1 & 0 & -1 & -1 \\ 0 & 1 & 0 & 1 \\ -1 & -1 & 1 & 0 \end{bmatrix}$$

is the incidence matrix.
Ans. See Fig. D2.3.

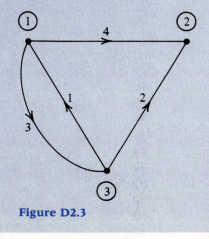

Figure D2.3

2.4 Draw the oriented graph for the network of Fig. D2.4. Write a full set of KCL, KVL, and branch equations.

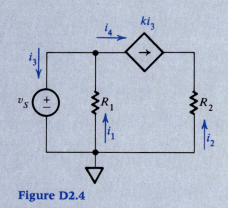

Figure D2.4

Ans. The graph is the same as in Exercise 2.3.

KCL:
$$i_1 - i_3 - i_4 = 0$$
$$i_2 + i_4 = 0$$

KVL:
$$-v_1 - v_3 = 0$$
$$v_1 - v_2 + v_4 = 0$$

Branch:
$$v_1 - R_1 i_1 = 0$$
$$v_2 - R_2 i_2 = 0$$
$$v_3 = v_S$$
$$i_4 - K i_3 = 0$$

2.5 NODE AND MESH EQUATIONS

2.5.1 Node Voltages

The rather large number of equations generated by a network of relatively modest size calls for a systematic way of eliminating variables. To introduce the problem, consider the mesh KVL equations of Example 2.2:

$$v_1 + v_2 - v_5 = 0$$
$$-v_2 + v_3 + v_4 = 0 \tag{2.7}$$
$$-v_4 + v_6 = 0$$

With three independent equations and six unknowns, it should be possible to express three variables in terms of the remaining three. To express v_1, v_2, and v_3 in terms of v_4, v_5, and v_6 is to solve the equations

$$v_1 + v_2 = v_5$$
$$-v_2 + v_3 = -v_4 \qquad\qquad (2.8)$$
$$0 = v_4 - v_6$$

It is not possible to solve Eq. (2.8) uniquely for v_1, v_2, and v_3, because the three unknowns appear in only two equations. On the other hand, try to solve for v_4, v_5, and v_6 in terms of v_1, v_2, v_3:

$$-v_5 = -v_1 - v_2$$
$$v_4 = v_2 - v_3 \qquad\qquad (2.9)$$
$$-v_4 + v_6 = 0$$

Equations (2.9) have the unique solution

$$v_4 = v_2 - v_3 \qquad v_5 = v_1 + v_2 \qquad v_6 = v_2 - v_3$$

This example shows that the choice of dependent and independent variables is not arbitrary. It must, in fact, be such that the dependent variables are uniquely determined by the independent variables.

In general, there are l independent mesh equations for the b branch voltages: it must be possible to solve for l branch voltages in terms of the $b - l = n - 1$ others. The problem is to choose a "good" set of $n - 1$ independent variables.

The graph of Fig. 2.6(a) is reproduced in (b), but with the (dotted) branch 5 added. Branch 5 is an open-circuit branch (i.e., a current source of value 0) so that the network is unchanged. It is introduced so that, for the time being, all nodes are directly connected to the datum.

In Fig. 2.6(b), $b = 5$ and $l = 2$: it should be possible to express two voltages

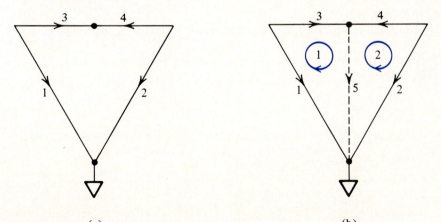

(a)   (b)

Figure 2.6 (a) Graph (b) augmented with open-circuit branch (dashed).

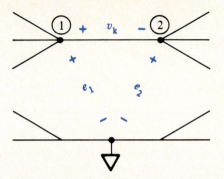

Figure 2.7 A branch and its node voltages.

in terms of the other three. A choice of independent variables is v_1, v_2, v_5, i.e., the voltages of the branches connecting the datum node D to the other nodes.

The mesh KVL equations are

Mesh 1: $-v_1 + v_3 + v_5 = 0$ or $v_3 = v_1 - v_5$

Mesh 2: $-v_5 - v_4 + v_2 = 0$ or $v_4 = v_2 - v_5$

Since v_3 and v_4 are uniquely specified, the choice of v_1, v_2, and v_5 is valid.

The *node voltages* are the voltages of the nodes with respect to the datum. They may or may not be branch voltages. Since there are n nodes, there are $n - 1$ node voltages. Any branch voltage is uniquely expressed as the difference between the voltages of adjacent nodes; for example, in Fig. 2.7,

$$v_k = e_1 - e_2 \qquad (2.10)$$

To illustrate, if node 1 is 10 V above the datum node and node 2 is 8 V above the datum, then node 1 is 2 V above node 2, and the branch voltage v_k is 2 V.

EXAMPLE 2.3

In Fig. E2.3, express the branch voltages in terms of the node voltages.

Solution
By inspection

$v_1 = e_3$

$v_2 = e_4$

$v_3 = e_5$

$v_4 = e_3 - e_4$

$v_5 = e_5 - e_4$

$v_6 = e_3 - e_1$

$v_7 = e_5 - e_2$

$v_8 = e_2 - e_1$

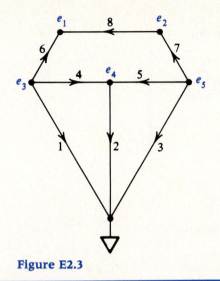

Figure E2.3

2.5.2 Nodal Analysis for Networks with Current Sources

The use of node voltages is nothing more than the expression, through KVL, of some voltages in terms of others. In effect, the KVL equations have been "used up." The basic idea in *nodal analysis* is to apply the remaining equations, KCL and the branch laws, simultaneously. Branch voltages are written by inspection in terms of node voltages: for resistive branches, the branch current follows by multiplying by the conductance. For networks containing only resistances and current sources, the node KCL equations are written by inspection, since all branch currents are either given (for sources) or easily written in terms of the node voltages.

It is convenient to write KCL in the form

$$\sum \text{ currents leaving node} = 0$$

From Fig. 2.8(a), the current i_1 leaving node 1 is

$$i_1 = Gv_1 = G(e_1 - e_2) \tag{2.11}$$

while the current i_2 leaving node 2 in Fig. 2.8(b) is

$$i_2 = Gv_2 = G(e_2 - e_1) \tag{2.12}$$

(a)   (b)

Figure 2.8 Two ways of writing a branch current.

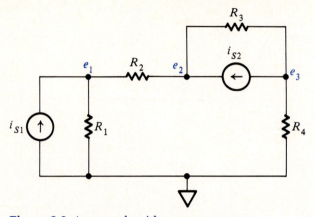

Figure 2.9 A network with current sources.

Of course, $i_2 = -i_1$; Eq. (2.11) is used when writing KCL at node 1 and Eq. (2.12) when writing KCL at node 2. In either case, the branch relationship has been easily incorporated.

The KCL equations for the network of Fig. 2.9 are

Node 1: $\quad -i_{S1} + G_1 e_1 + G_2(e_1 - e_2) = 0$

Node 2: $\quad G_2(e_2 - e_1) - i_{S2} + G_3(e_2 - e_3) = 0$

Node 3: $\quad i_{S2} + G_3(e_3 - e_2) + G_4 e_3 = 0$ $\qquad$ (2.13)

In matrix form, Eq. (2.13) becomes

$$\begin{bmatrix} G_1 + G_2 & -G_2 & 0 \\ -G_2 & G_2 + G_3 & -G_3 \\ 0 & -G_3 & G_3 + G_4 \end{bmatrix} \begin{bmatrix} e_1 \\ e_2 \\ e_3 \end{bmatrix} = \begin{bmatrix} i_{S1} \\ i_{S2} \\ -i_{S2} \end{bmatrix} \qquad (2.14)$$

Equations (2.13) or (2.14) are called *node equations*. The matrix in (2.14) is the *node conductance matrix*.

EXAMPLE 2.4

Write the node equations for the network of Fig. E2.4.

Solution

Node 1: $\quad -i_{S1} + G_1 e_1 - i_{S2} + G_3(e_1 - e_2) + G_2(e_1 - e_3) = 0$

Node 2: $\quad i_{S2} + G_3(e_2 - e_1) + G_4(e_2 - e_3) = 0$ $\qquad$ (2.15)

Node 3: $\quad G_4(e_3 - e_2) + G_2(e_3 - e_1) + G_5 e_3 = 0$

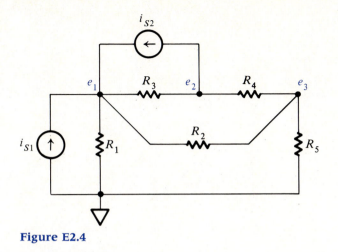

Figure E2.4

In matrix form,

$$
\begin{bmatrix}
G_1 + G_2 + G_3 & -G_3 & -G_2 \\
-G_3 & G_3 + G_4 & -G_4 \\
-G_2 & -G_4 & G_2 + G_4 + G_5
\end{bmatrix}
\begin{bmatrix}
e_1 \\
e_2 \\
e_3
\end{bmatrix}
=
\begin{bmatrix}
i_{S1} + i_{S2} \\
-i_{S2} \\
0
\end{bmatrix}
\tag{2.16}
$$

The analyses of these two networks suggest an inspection method to generate the node voltage equations in matrix form.

1. The ith diagonal entry of the node conductance matrix is the sum of the conductances connected to the ith node; for example, the 11 element in Example 2.4 is the sum of the three conductances, G_1, G_2, and G_3, connected to node 1.

2. The ijth off-diagonal entry is the negative sum of the conductances connected between nodes i and j; for example, the 13 element in Example 2.4 is $-G_2$ because that is the conductance connected between nodes 1 and 3.

3. The ith element of the right-hand-side vector is the sum of source currents entering the ith node; for example, the second element in Example 2.4 is $-i_{S2}$ because the source current i_{S2} leaves node 2.

This systematic procedure is independent of branch reference polarities. Voltage sources are excluded from nodal analysis at this point. As shown in the next chapter, they are best handled by transformation to current sources.

EXAMPLE 2.5

Solve for the node voltages in the network of Fig. E2.5.

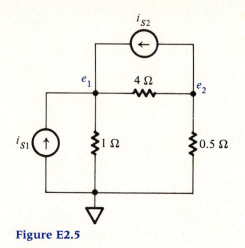

Figure E2.5

Solution

By inspection,

$$\begin{bmatrix} 1 + \frac{1}{4} & -\frac{1}{4} \\ -\frac{1}{4} & \frac{1}{4} + 1/0.5 \end{bmatrix}\begin{bmatrix} e_1 \\ e_2 \end{bmatrix} = \begin{bmatrix} i_{S1} + i_{S2} \\ -i_{S2} \end{bmatrix}$$

or

$$\begin{bmatrix} 1.25 & -0.25 \\ -0.25 & 2.25 \end{bmatrix}\begin{bmatrix} e_1 \\ e_2 \end{bmatrix} = \begin{bmatrix} i_{S1} + i_{S2} \\ -i_{S2} \end{bmatrix}$$

The solution is

$$e_1 = 0.818 i_{S1} + 0.727 i_{S2}$$

$$i_2 = 0.091 i_{S1} - 0.364 i_{S2}$$

2.5.3 Mesh Currents

The n-1 node KCL equations are used to solve for $n - 1$ branch currents in terms of the other $b - (n - 1) = l$ branch currents. Node voltages were used as independent variables in nodal analysis despite the fact that they are not necessarily branch voltages. It is also expedient in choosing independent current variables to define auxiliary current variables known as *mesh currents*.

As shown in Fig. 2.10, the mesh currents j_1, j_2, and j_3 are imagined to circulate in each mesh. If a branch belongs to two meshes, the branch current is taken to be the algebraic sum of the two mesh currents. The branch currents in Fig. 2.10 are as follows:

$$i_1 = j_1$$

$$i_2 = -j_1 + j_3$$

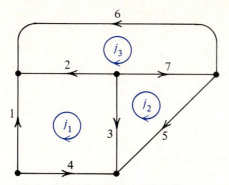

Figure 2.10 Definition of mesh currents.

$$i_3 = j_1 - j_2$$
$$i_4 = -j_1$$
$$i_5 = j_2 \tag{2.17}$$
$$i_6 = -j_3$$
$$i_7 = j_2 - j_3$$

Clearly, the branch currents are uniquely expressed in terms of the mesh currents.

2.5.4 Mesh Analysis for Networks with Voltage Sources

The expression of branch currents in terms of mesh currents is, in effect, an application of the KCL equation. The basic idea in mesh analysis is to apply the remaining equations, KVL and the branch laws, simultaneously. Branch currents are written by inspection in terms of mesh current: for resistive branches, the voltage follows by multiplying the current by the resistance. For networks containing only resistances and voltage sources, the mesh KVL equations are written by inspection, since all branch voltages are either given (for sources) or easily written in terms of the mesh currents.

It is convenient to write KVL by traversing each mesh in the direction of the mesh current, counting voltage drops as positive. From Fig. 2.11(a), the voltage drop across R when traversing mesh 1 is

$$v_1 = Ri_1 = R(j_1 - j_2) \tag{2.18}$$

while the voltage drop across R in the traversing direction of mesh 2 is

$$v_2 = Ri_2 = R(j_2 - j_1) \tag{2.19}$$

Of course, $v_2 = -v_1$; Eq. (2.18) is used when writing KVL around mesh 1 and Eq. (2.19) when writing KVL around mesh 2.

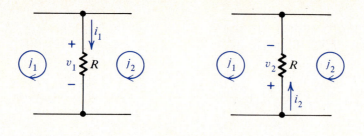

(a)   (b)

Figure 2.11 Two ways of writing a branch voltage.

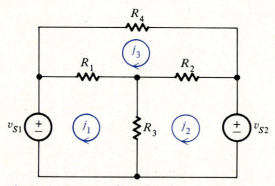

Figure 2.12 Network with voltage sources.

The mesh equations for the network of Fig. 2.12 are

Mesh 1: $-v_{S1} + R_1(j_1 - j_3) + R_3(j_1 - j_2) = 0$

Mesh 2: $R_3(j_2 - j_1) + R_2(j_2 - j_3) + v_{S2} = 0$ (2.20)

Mesh 3: $R_1(j_3 - j_1) + R_4 j_3 + R_2(j_3 - j_2) = 0$

In matrix form,

$$\begin{bmatrix} R_1 + R_3 & -R_3 & -R_1 \\ -R_3 & R_2 + R_3 & -R_2 \\ -R_1 & -R_2 & R_1 + R_2 + R_4 \end{bmatrix} \begin{bmatrix} j_1 \\ j_2 \\ j_3 \end{bmatrix} = \begin{bmatrix} v_{S1} \\ -v_{S2} \\ 0 \end{bmatrix}$$ (2.21)

Equations (2.20) and (2.21) are the *mesh equations,* and the matrix is the *mesh resistance matrix.*

EXAMPLE 2.6

Write the mesh equations for the network of Fig. E2.6.

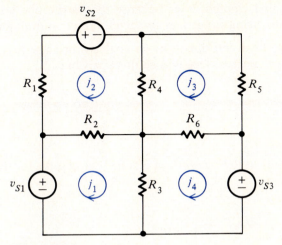

Figure E2.6

Solution

Mesh 1: $-v_{S1} + R_2(j_1 - j_2) + R_3(j_1 - j_4) = 0$

Mesh 2: $R_1 j_2 + v_{S2} + R_4(j_2 - j_3) + R_2(j_2 - j_1) = 0$

Mesh 3: $R_4(j_3 - j_2) + R_5 j_3 + R_6(j_3 - j_4) = 0$

Mesh 4: $R_3(j_4 - j_1) + R_6(j_4 - j_3) + v_{S3} = 0$

In matrix form,

$$\begin{bmatrix} R_2 + R_3 & -R_2 & 0 & -R_3 \\ -R_2 & R_1 + R_2 + R_4 & -R_4 & 0 \\ 0 & -R_4 & R_4 + R_5 + R_6 & -R_6 \\ -R_3 & 0 & -R_6 & R_3 + R_6 \end{bmatrix} \begin{bmatrix} j_1 \\ j_2 \\ j_3 \\ j_4 \end{bmatrix} = \begin{bmatrix} v_{S1} \\ -v_{S2} \\ 0 \\ -v_{S3} \end{bmatrix}$$

Here again, a general pattern emerges from the two analyses.

1. The ith diagonal entry of the mesh resistance matrix is the sum of the resistances in the ith mesh; for example, the 11 element in Example 2.6 is $R_2 + R_3$ because mesh 1 contains those two resistances.

2. The ijth off-diagonal entry is the negative sum of the resistances common to meshes i and j, when all mesh current reference directions are identical. (If the reference directions of meshes i and j are not the same, the entry ij is positive.) For example, since R_2 is common to meshes 1 and 2 in Example 2.6, the 12 element of the matrix is $-R_2$.

3. The ith element of the right-hand-side vector is the sum of voltage *rises* encountered when traversing mesh i in the direction of the mesh current;

for example, the first element is v_{S1} because that is the one voltage rise encountered in traversing mesh 1 in the given direction in Example 2.6.

Current sources have been left out of mesh analysis, because it is easier to transform them to voltage sources. This will be done in Chap. 3.

EXAMPLE 2.7

Solve for the mesh currents in the network of Fig. E2.7, and solve for the branch current i.

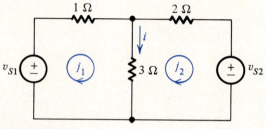

Figure E2.7

Solution

By inspection,

$$\begin{bmatrix} 1 + 3 & -3 \\ -3 & 2 + 3 \end{bmatrix} \begin{bmatrix} j_1 \\ j_2 \end{bmatrix} = \begin{bmatrix} v_{S1} \\ -v_{S2} \end{bmatrix}$$

or

$$\begin{bmatrix} 4 & -3 \\ -3 & 5 \end{bmatrix} \begin{bmatrix} j_1 \\ j_2 \end{bmatrix} = \begin{bmatrix} v_{S1} \\ -v_{S2} \end{bmatrix}$$

The solution is

$$j_1 = \tfrac{5}{11} v_{S1} - \tfrac{3}{11} v_{S2}$$

$$j_2 = \tfrac{3}{11} v_{S1} - \tfrac{4}{11} v_{S2}$$

The branch current i is

$$i = j_1 - j_2$$

$$= \tfrac{2}{11} v_{S1} + \tfrac{1}{11} v_{S2}$$

DRILL EXERCISES

2.5 Use nodal analysis to solve for e_1 and e_2 in Fig. D2.5.

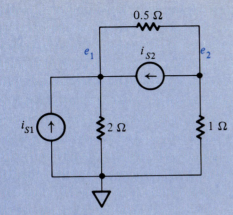

Figure D2.5

Ans. $e_1 = 0.857i_{S1} + 0.286i_{S2}$

$e_2 = 0.571i_{S1} - 0.143i_{S2}$

2.6 Use mesh analysis to solve for j_1 and j_2 in Fig. D2.6.

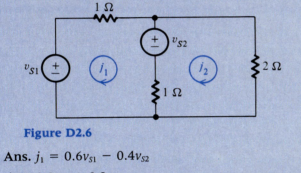

Figure D2.6

Ans. $j_1 = 0.6v_{S1} - 0.4v_{S2}$

$j_2 = 0.2v_{S1} + 0.2v_{S2}$

2.6 NETWORKS WITH CONTROLLED SOURCES

The network of Fig. 2.13 has a current-controlled current source and is to be solved for v_0. At this stage in the development, the presence of the two current sources imposes the use of the nodal method.

In the case of networks with controlled sources, there is a preliminary step to be carried out before the node or mesh method is applied.

Preliminary Step: Express the controlling variables in terms of node voltages (or mesh currents) and independent source voltages and currents.

In this case,

$$i_1 = G_1 e_1 \tag{2.22}$$

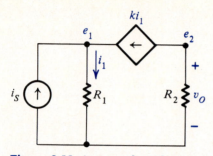

Figure 2.13 A network with a controlled source.

The network is now redrawn, with the controlled source expressed in terms of node voltages (or mesh currents) and, possibly, independent source values. This leads to Fig. 2.14.

Nodal analysis is now applied by inspection:

$$\begin{bmatrix} G_1 & 0 \\ 0 & G_2 \end{bmatrix} \begin{bmatrix} e_1 \\ e_2 \end{bmatrix} = \begin{bmatrix} i_s + KG_1e_1 \\ -KG_1e_1 \end{bmatrix} \qquad (2.23)$$

The right-hand side contains the unknown e_1, which needs to be brought to the left. This yields

$$\begin{bmatrix} G_1 - KG_1 & 0 \\ KG_1 & G_2 \end{bmatrix} \begin{bmatrix} e_1 \\ e_2 \end{bmatrix} = \begin{bmatrix} i_s \\ 0 \end{bmatrix} \qquad (2.24)$$

The solution is

$$e_1 = \frac{i_s}{G_1(1 - K)}$$

$$e_2 = \frac{-Ki_s}{G_2(1 - K)}$$

If the branch voltage v_0 is to be solved for, then

$$v_0 = \frac{-Ki_s}{G_2(1 - K)} \qquad (2.25)$$

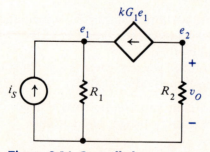

Figure 2.14 Controlled source expressed in terms of node voltage.

This network is also easily solved by the inspection method of Chap. 1. Applying KCL at node 1 in Fig. 2.13,

$$i_1 - Ki_1 = i_S$$

or

$$i_1 = \frac{i_S}{1 - k}$$

Solving for v_0,

$$v_0 = -KR_2 i_1$$

$$= \frac{-KR_2 i_S}{1 - K} \tag{2.26}$$

which is identical to the previous result.

EXAMPLE 2.8

Solve for v_0 in the network of Fig. E2.8(a).

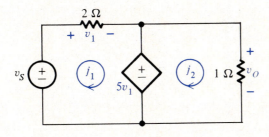

(a)

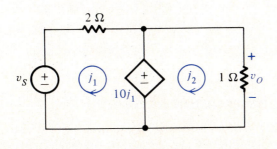

(b)

Figure E2.8

Solution

Since $v_1 = 2j_1$, the controlled source is expressed in terms of j_1 in Fig. E2.8(b). By inspection,

$$\begin{bmatrix} 2 & 0 \\ 0 & 1 \end{bmatrix}\begin{bmatrix} j_1 \\ j_2 \end{bmatrix} = \begin{bmatrix} v_S - 10j_1 \\ 10j_1 \end{bmatrix}$$

Transferring the terms in j_1 to the left,

$$\begin{bmatrix} 12 & 0 \\ -10 & 1 \end{bmatrix}\begin{bmatrix} j_1 \\ j_2 \end{bmatrix} = \begin{bmatrix} v_S \\ 0 \end{bmatrix}$$

which yields

$$j_1 = \frac{v_S}{12}$$

$$j_2 = \tfrac{5}{6}v_S$$

The branch voltage v_0 is

$$v_0 = j_2 = \tfrac{5}{6}v_S$$

DRILL EXERCISE

2.7 Solve for the mesh currents in Fig. D2.7.

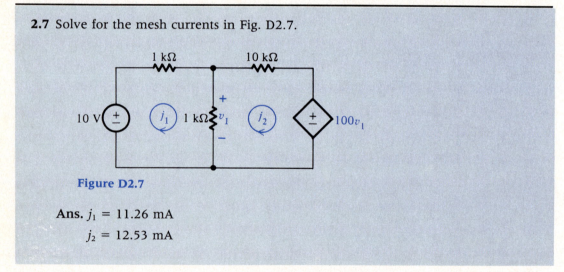

Figure D2.7

Ans. $j_1 = 11.26$ mA

$j_2 = 12.53$ mA

2.7 TELLEGEN'S THEOREM AND CONSERVATION OF POWER

Tellegen's theorem states that given a network whose graph has b branches, and given arbitrary sets of branch voltages $v_1, v_2, \ldots, v_b$ and branch currents $i_1, i_2, \ldots, i_b$ that satisfy KVL and KCL, respectively, then

$$\sum_{k=1}^{b} v_k i_k = 0 \tag{2.27}$$

The theorem is now proved for the graph of Fig. 2.15. First, write the mesh KVL equations:

$$v_1 + v_2 + v_3 = 0$$
$$-v_3 + v_4 + v_5 = 0$$
$$-v_2 - v_4 + v_6 = 0 \tag{2.28}$$

Next, multiply each mesh KVL equation by its mesh current:

$$j_1 v_1 + j_1 v_2 + j_1 v_3 = 0$$
$$-j_2 v_3 + j_2 v_4 + j_2 v_5 = 0 \tag{2.29}$$
$$-j_3 v_2 - i_3 v_4 + j_3 v_6 = 0$$

Now, add those three equations:

$$j_1 v_1 + (j_1 - j_3)v_2 + (j_1 - j_2)v_3 + (j_2 - j_3)v_4 + j_2 v_5 + j_3 v_6 = 0 \tag{2.30}$$

By KCL, the branch currents are

$$i_1 = j_1$$
$$i_2 = j_1 - j_3$$
$$i_3 = j_1 - j_2$$
$$i_4 = j_2 - j_3$$
$$i_5 = j_2$$
$$i_6 = j_3$$

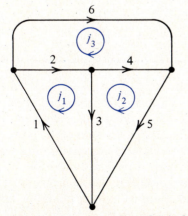

Figure 2.15 A graph with mesh currents.

Substituting in Eq. (2.30) yields

$$i_1 v_1 + i_2 v_2 + i_3 v_3 + i_4 v_4 + i_5 v_5 + i_6 v_6 = 0$$

which is Tellegen's theorem applied to this graph.

This method of proof is extendible to the general case. Consider an outer branch of the graph, for example branch 1. It belongs only to mesh 1, so that $j_1 = i_1$. The voltage v_1 appears only in one mesh KVL equation, and is multiplied only by i_1. Now consider an inner branch, for example branch 2, which belongs to meshes 1 and 3. The voltage v_2 is traversed positively in mesh 1, negatively in mesh 3 and therefore appears with a "$+$" in the first KVL equation and with a "$-$" in the third. Multiplication by mesh currents results in a term $(j_1 - j_3)v_2$ in the sum; because of the reference directions, $(j_1 - j_3)$ is immediately identified as i_2.

EXAMPLE 2.9

To drive home the point that Tellegen's theorem depends only on Kirchhoff's laws and not at all on the branch laws, choose arbitrary sets of branch voltages and currents satisfying Kirchhoff's laws for Fig. 2.15, and show that they satisfy Tellegen's theorem.

Solution
To choose voltages, the KVL equations (2.28) are used. A solution (by no means unique) is $v_1 = 1$, $v_2 = 2$, $v_3 = -3$, $v_4 = -1$, $v_5 = -2$, $v_6 = 1$.

The node KCL equations are used to pick currents: $i_1 = 1$, $i_2 = 2$, $i_3 = 1$, $i_4 = 1$, $i_5 = 0$, $i_6 = -1$.

To apply Tellegen's theorem, write

$$v_1 i_1 + v_2 i_2 + v_3 i_3 + v_4 i_4 + v_5 i_5 + v_6 i_6$$

$$= (1)(1) + (2)(2) + (-3)(1) + (-1)(1) + (-2)(0) + (1)(-1)$$

$$= 1 + 4 - 3 - 1 - 1 = 0$$

When the voltages and currents are solutions to the network, they also satisfy the branch relationships, and the product $v_k i_k$ is the power delivered to branch k. In that case, Tellegen's theorem is a statement that the sum of the power delivered to all branches of a network is zero, a statement of conservation of power.

It is quite remarkable that conservation of power should follow only from Kirchhoff's laws, without the need to add anything to the problem formulation. Indeed, Tellegen's theorem is a general theorem which holds for any lumped network, regardless of the nature of the branches.

DRILL EXERCISE

2.8 For the network in Fig. D2.8, show that the following values of branch voltages and currents satisfy Kirchhoff's laws and Tellegen's theorem: $v_1 = 2$, $v_2 = 1$, $v_3 = -3$, $v_4 = 2$, $v_5 = 3$, $i_1 = 1$, $i_2 = -1$, $i_3 = 3$, $i_4 = -2$, $i_5 = 4$.

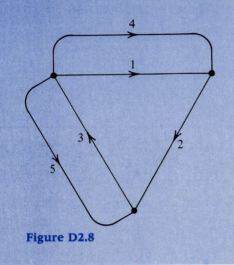

Figure D2.8

2.8   THE PRINCIPLE OF SUPERPOSITION

The *principle of superposition* is a key idea in the study of all linear systems. It has two distinct parts, corresponding to two important properties.

1. The *scaling property*. Given branch voltages and currents resulting from the application of a single independent source, if the independent source voltage or current is multiplied by a constant, all branch currents and voltages are multiplied by the same constant.

2. The *additivity property*. In a network with several independent sources, any branch voltage (or current) is the sum of the voltages (or currents) obtained when each independent source acts alone, with the others set to zero.

The scaling property implies, for example, that doubling the independent source voltage or current will double all branch voltages and currents. If the branch voltage or current resulting from the application of one independent source is called the contribution of that source, the additivity property states that, in order to calculate the branch voltage or current resulting from several independent sources, one need only add the contributions of the individual sources.

The principle of superposition is demonstrated for the specific example of Fig. 2.16.

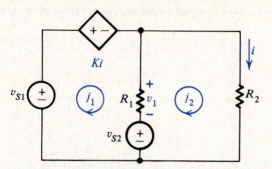

Figure 2.16 Network example to illustrate superposition.

The network of Fig. 2.16 is described by the following mesh equations:

$$\begin{bmatrix} R_1 & -R_1 \\ -R_1 & R_1 + R_2 \end{bmatrix} \begin{bmatrix} j_1 \\ j_2 \end{bmatrix} = \begin{bmatrix} v_{S1} - Ki - v_{S2} \\ v_{S2} \end{bmatrix} \tag{2.31}$$

Since $i = j_2$ and $Ki = Kj_2$, Eq. (2.31) is, therefore, rewritten as

$$\begin{bmatrix} R_1 & -R_1 + K \\ -R_1 & R_1 + R_2 \end{bmatrix} \begin{bmatrix} j_1 \\ j_2 \end{bmatrix} = \begin{bmatrix} v_{S1} - v_{S2} \\ v_{S2} \end{bmatrix} \tag{2.32}$$

By Cramer's rule,

$$
j_1 = \frac{\det \begin{bmatrix} v_{S1} - v_{S2} & -R_1 + K \\ v_{S2} & R_1 + R_2 \end{bmatrix}}{\det \begin{bmatrix} R_1 & -R_1 + K \\ -R_1 & R_1 + R_2 \end{bmatrix}}
$$

$$
= \frac{(R_1 + R_2)v_{S1} - (R_2 + K)v_{S2}}{(R_2 + K)R_1} \tag{2.33}
$$

Similarly,

$$j_2 = \frac{R_1 v_{S1}}{(R_2 + K)R_1} \tag{2.34}$$

Now, recall from Sec. 2.5.3 that any branch current can be written as a linear combination of j_1 and j_2. Furthermore, any resistance voltage is also a linear combination of j_1 and j_2 because it is calculated by multiplying the branch current by a constant. Since j_1 and j_2 are both of the form $A_1 v_{S1} + A_2 v_{S2}$, where A_1, A_2 are constants, it follows that *any* branch voltage or current is also of that form.

For example, using Eqs. (2.33) and (2.34), the voltage v_1 in Fig. 2.16 is given by

$$v_1 = R_1(j_1 - j_2)$$

$$= \frac{R_2}{R_2 + K} v_{S1} - v_{S2} \tag{2.35}$$

which is of the form $A_1 v_{S1} + A_2 v_{S2}$.

In the calculation of a mesh current by Cramer's rule, it is seen that the numerator determinant has one single column consisting of sums and differences of independent source voltages. If one calculates the determinant by expansion about that column, there results a sum of terms of the column (source voltages) times their respective (constant) cofactors: this is linear in the independent source voltages.

The same reasoning applies, via nodal analysis, to a network with current sources. In fact, the result is valid for any network composed of linear elements: *all* branch currents and voltages are linear combinations of the independent source voltages and currents.

Using the network of Fig. 2.16 as an example, consider Eq. (2.35), rewritten for brevity as

$$v_1 = A_1 v_{S1} + A_2 v_{S2} \tag{2.36}$$

If $v_{S2} = 0$, then

$$v_1 = A_1 v_{S1} \tag{2.37}$$

The scaling property is clear from Eq. (2.37): if v_{S1} is multiplied by a constant, so is v_1.

The term $A_1 v_{S1}$ is the solution v_1 when v_{S1} acts alone, i.e., when $v_{S2} = 0$: it is the contribution of v_{S1} to v_1. Similarly, $A_2 v_{S2}$ is the contribution of v_{S2}. It is obvious that the additivity property is satisfied, since by Eq. (2.36), v_1 is the sum of the individual contributions of v_{S1} and v_{S2}.

Since all branch variables of a network composed of linear elements are linear combinations of the independent source voltages and currents, it is clear that such networks satisfy both superposition properties.

EXAMPLE 2.10

Use superposition to solve for i in the network of Fig. E2.10(a).

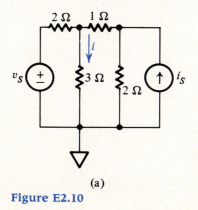

(a)

Figure E2.10

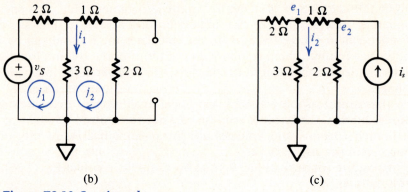

(b)   (c)

Figure E2.10 Continued

Solution

Making $i_s = 0$ means replacing the current source by an open circuit, as in Fig. E2.10(b). The mesh current equation for Fig. E2.10(b) is

$$\begin{bmatrix} 2+3 & -3 \\ -3 & 2+3+1 \end{bmatrix}\begin{bmatrix} j_1 \\ j_2 \end{bmatrix} = \begin{bmatrix} v_s \\ 0 \end{bmatrix}$$

leading to

$$j_1 = \frac{6v_s}{21} = \frac{2v_s}{7}$$

$$j_2 = \frac{3v_s}{21} = \frac{v_s}{7}$$

The contribution of v_s to i is

$$i_1 = j_1 - j_2 = \frac{v_s}{7}$$

Making $v_s = 0$ means replacing the voltage source by a short circuit, as in Fig. E 2.10(c). Solving Fig. E2.10(b) by the node method,

$$\begin{bmatrix} \frac{1}{3} + \frac{1}{2} + 1 & -1 \\ -1 & 1 + \frac{1}{2} \end{bmatrix}\begin{bmatrix} e_1 \\ e_2 \end{bmatrix} = \begin{bmatrix} 0 \\ i_s \end{bmatrix}$$

leading to

$$e_1 = \frac{i_s}{\frac{7}{4}} = \frac{4i_s}{7}$$

$$e_2 = \frac{\frac{11}{6}i_s}{\frac{7}{4}} = \frac{22i_s}{21}$$

The contribution of i_s to i is

$$i_2 = \frac{e_1}{3} = \frac{4i_s}{21}$$

Adding the two contributions,

$$i = i_1 + i_2 = \frac{2v_S}{7} + \frac{4i_S}{21}$$

By superposition, any branch variable is equal to a linear combination of the *independent* source voltages and currents. The contribution of each source can be calculated separately, by turning each source "on" one at a time. A naive application of superposition would lead one to include controlled sources in this procedure. This does not work, for the following reason. If all independent sources are set to zero, the network only has resistances and controlled sources; all branch currents and voltages are equal to zero, since the network is not driven [see Eq. (2.32) with $v_{S1} = v_{S2} = 0$]. Therefore, when using superpositon, it is only the independent sources that are turned on one at a time; the controlled sources are never turned "off" and enter every calculation.

DRILL EXERCISE

2.9 Solve for j_1 and j_2 in the network of Exercise 2.6 when (a) v_{S1} acts alone and (b) v_{S2} acts alone. Using the answer of Exercise 2.6 show that superposition holds.

2.9   SUMMARY AND STUDY GUIDE

The purpose of this chapter was to introduce systematic methods for the formulation of network equations. From the network graph, there followed a relation between the number of branches (b), nodes (n), and meshes (l). It was shown that writing all the KCL and KVL equations, plus the branch laws, results in more equations than unknowns. After a brief recall of some ideas from linear algebra, a set of $2b$ independent equations was extracted from KCL, KVL, and the branch laws. Nodal and mesh analysis were introduced, essentially to enable network voltages and currents to be expressed in terms of relatively few independent variables by inspection. A simple modification was introduced, to handle networks with controlled sources. Tellegen's theorem was shown to follow Kirchhoff's laws. Finally, the principle of superposition was demonstrated to be a consequence of linearity.

There are certain key points that should be clear in the student's mind.

The KCL and KVL equations depend only on the network graph, not on the branch laws.

Any $n - 1$ KCL equations are independent and so are the l mesh KVL equations; the addition of the b branch law equations completes the set of required network equations.

Node voltages are variables used to solve the KVL equations for the branch voltages in terms of $n - 1$ independent voltage variables. The node voltages are defined to make this possible by inspection. The same remark applies with respect to mesh currents.

The node voltage equations are written by inspection for networks with *current* sources and resistances, the mesh current equations are also written by inspection for networks with *voltage* sources and resistances.

Controlled sources can be handled in nodal and mesh analysis, provided two extra steps are taken: (1) the controlling variable (hence the controlled voltage or current) must be expressed in terms of node voltages or mesh currents and independent source voltages and currents; (2) the systematic equation formulation will introduce node voltage or mesh current terms on the right-hand side that must be shifted to the left before proceeding with the solution.

Tellegen's theorem is independent of the branch laws.

Superposition is a direct consequence of linearity and includes both the scaling and the additivity properties.

When solving a network by superposition, it is necessary to turn the sources "on" one at a time, setting the others to zero.

When applying superposition to a network with controlled sources, it is only the independent sources that are turned "on" one at a time; the controlled sources are never turned "off."

PROBLEMS

2.1 For the network of Fig. P2.1
(a) Define reference directions for each branch and draw the network graph.
(b) Write down all KCL and KVL equations.
(c) Show that all KVL equations for loops other than meshes are consistent with the mesh KVL equations.

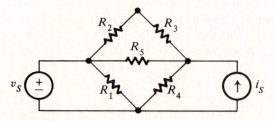

Figure P2.1

2.2 Give the graph of a network whose incidence matrix is

$$\begin{bmatrix} 0 & 1 & -1 & 0 \\ -1 & -1 & 0 & 0 \\ 1 & 0 & 0 & -1 \\ 0 & 0 & 1 & 1 \end{bmatrix}$$

2.3 Repeat Problem 2.2 for the incidence matrix

$$\begin{bmatrix} 1 & -1 & 1 & -1 \\ 0 & 1 & -1 & 0 \\ -1 & 0 & 0 & 1 \end{bmatrix}$$

2.4 For the network graph of Fig. P2.4, write the KCL equations at all nodes but the datum node and the mesh KVL equations.

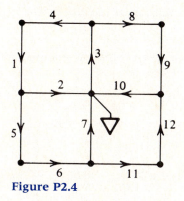

Figure P2.4

2.5 Repeat Problem 2.4 for the network graph of Fig. P2.5.

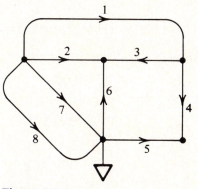

Figure P2.5

2.6 For the network of Fig. P2.1, write a complete set of independent equations for the branch voltages and currents.

2.7 Repeat Problem 2.6 for the network of Fig. P2.7.

2.8 Figure P2.8 is meant to model a physical situation where two batteries are used in parallel to power a resistive load R.

(a) In the network of Fig. P2.8(a), the batteries are modeled by ideal voltage sources. Write a full set of KVL, KCL, and branch equations. Show that the equations have either no solution or an infinite number of solutions.

(b) In Fig. P2.8(b), the network is modified to include resistances R_1 and R_2 in the models of the two batteries. Write the KVL, KCL, and branch laws as in part (a),

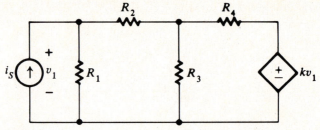

Figure P2.7

and show that a unique solution always exists if R_1 and/or $R_2 \neq 0$. (This problem shows that overly simplified models can lead to a situation without a unique mathematical solution.)

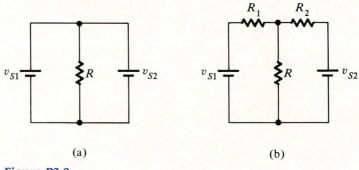

(a)   (b)

Figure P2.8

2.9 Write a set of node voltage equations for the network of Fig. P2.9.

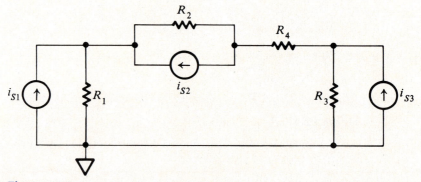

Figure P2.9

2.10 Repeat Problem 2.9 for the network of Fig. P2.10.

2.11 For the network of Fig. P2.11:
   (a) Write a set of node voltage equations.
   (b) Solve for the node voltages.
   (c) Find the current i.
   (d) Calculate the power supplied by each current source and the power dissipated in each resistance. Verify that the power supplied equals the power dissipated.

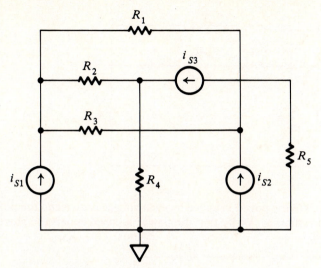

Figure P2.10

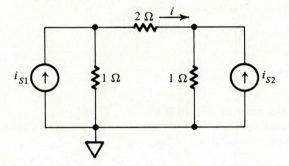

Figure P2.11

2.12 Write a set of mesh current equations for the network of Fig. P2.12.

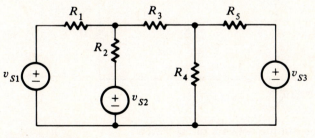

Figure P2.12

2.13 Repeat Problem 2.12 for the network of Fig. P2.13.

2.14 For the network of Fig. P2.14:
   (a) Write a set of mesh current equations.
   (b) Solve for the mesh currents.
   (c) Find v.

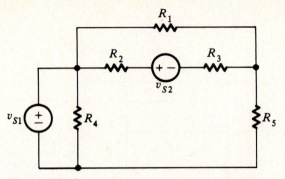

Figure P2.13

(d) Calculate the power supplied by each voltage source and the power dissipated in each resistance. Verify that the power supplied equals the power dissipated.

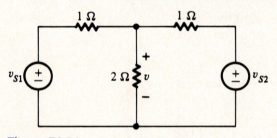

Figure P2.14

2.15 Solve for v_0 in the network of Fig. P2.15:
(a) Using the mesh method. (Do not forget to express v_1 in terms of mesh currents.)
(b) Using the inspection method of Chap. 1.

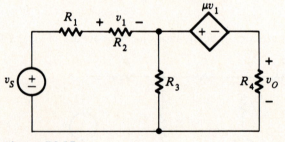

Figure P2.15

2.16 Solve for v in the network of Fig. P2.16:
(a) Using the node method.
(b) Using the inspection method of Chap. 1.

2.17 Using the mesh method, solve for v_0 in the network of Fig. P2.17. (*Hint*: This network has only *one* mesh. Express v_1 in terms of v_s and the mesh current.)

2.18 Using the node method, solve for v_0 in the network of Fig. P2.18.

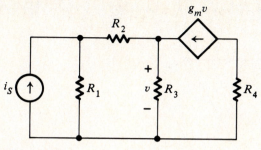

Figure P2.16

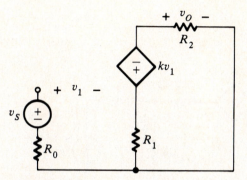

Figure P2.17

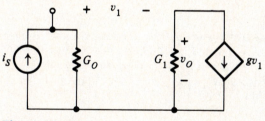

Figure P2.18

2.19 For the graph of Fig. P2.19, find a set of branch voltages and currents that satisfy Kirchhoff's laws (there are many such sets). Verify that Tellegen's theorem holds.

2.20 For the network of Fig. P2.11, calculate i due to i_{s1} acting alone and i_{s2} acting alone. Verify that superposition holds.

2.21 For the network of Fig. P2.14, calculate v due to v_{s1} acting alone and v_{s2} acting alone. Verify that superposition holds.

2.22 For the network of Fig. P2.22:
(a) Calculate i and i_1.
(b) Calculate i and i_1 for each *independent* source acting by itself.
(c) Verify that superposition holds.

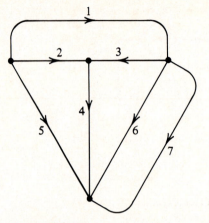

Figure P2.19

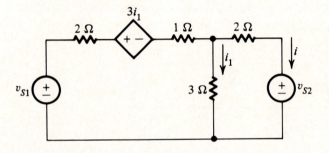

Figure P2.22

2.23 Network N in Fig. P2.23 contains only resistances and controlled sources. It is observed that (i) If $v_S = 2$ V and $i_S = 0$, then $i = 4$ A and $v = -1$ V. (ii) If $v_S = 0$ and $i_S = 1$ A, then $i = -1$ A and $v = 2$ V. What are v and i if

    (a) $v_S = 2$ V and $i_S = 1$ A?
    (b) $v_S = -2$ V and $i_S = 1$ A?
    (c) $v_S = 4$ V and $i_S = 2$ A?
    (d) $v_S = 4$ V and $i_S = -1$ A?

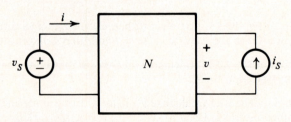

Figure P2.23

2.24 For the network of Fig. P2.23, it is observed that (i) If $v_S = 1$ V and $i_S = -1$ A, then $i = 3$ A and $v = 2$ V. (ii) If $v_S = -2$ V and $i_S = -2$ A, then $i = -4$ A and $v = -1$ V. Find v and i for those values of v_S and i_S in parts (a) to (d) of Problem P2.23.

2.25 A network is driven by two independent sources. The current through a given resistance R may be expressed as $i_1 + i_2$, where i_1 and i_2 are the contributions of sources 1 and 2, respectively. Show that power does not obey superposition in general, i.e., show that the power dissipated in R is not the sum of the power dissipated with source 1 acting alone and of the power dissipated with source 2 acting alone.

2.26 The element N in Fig. P2.26 is a nonlinear element, described by the relationship $v_1 = i_1^3$. Write, if they apply, the following equations:
   (a) The three node KCL, the mesh KVL, and the branch laws
   (b) The three node KCL equations, expressed in terms of node voltages
   (c) The node voltage equations written in matrix form by inspection
   (d) The branch voltage v_1 as a sum of individual contributions of the two current sources

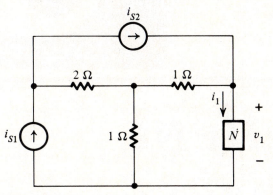

Figure P2.26

CHAPTER

Equivalent Networks

3.1 EQUIVALENCE

There are many instances in network analysis where it is advantageous to replace a network by a more simple equivalent circuit. The concept of equivalence and its applications are the subjects of this chapter.

For example, consider the problem of designing an amplifier that is to be connected to a given power supply, as in Fig. 3.1(a). A power supply is a fairly complex piece of circuitry: the combined power supply–amplifier system will be more complex still. Is it necessary to consider the detailed power supply circuit in the analysis and design of the amplifier?

Mercifully, the answer is no. Seen from its terminals, the power supply is just a network element, completely specified by its terminal characteristics. It is possible to replace the complex power supply network by a much simpler network having the same, i.e., *equivalent*, terminal *v-i* charactersitics. This simple network is used by the amplifier designer.

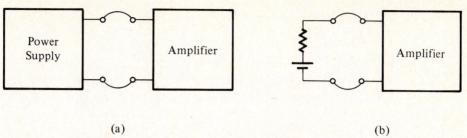

(a)   (b)

Figure 3.1 (a) Power supply connected to an amplifier. (b) The same interconnection, with the power supply replaced by an equivalent network.

For a power supply, the equivalent network is generally a voltage source with a very small series resistance as shown in Fig. 3.1(b). The fact that such a simple equivalent circuit exists does not mean that designing a power supply is that simple; ideal voltage sources are theoretical elements, and are approximated in practice only with relatively complex networks.

Two networks are said to be *equivalent* with respect to a set of terminals if, for the same terminal voltage or current excitations, the terminal currents and voltages of the two are identical. For the special case of a two-terminal network, this reduces to the requirement that the *v-i* characteristics be identical.

3.2 EQUIVALENT RESISTANCE

3.2.1 Networks with Resistances and Controlled Sources

The object of this section is the derivation of an equivalent circuit for a network composed of resistances and controlled sources. Only linear elements are allowed because the derivation rests on the principle of superposition.

The network N of Fig. 3.2 is composed only of resistances and linear controlled sources. Its *v-i* relationship is obtained by either of the two "thought experiments" of Fig. 3.2(b) or (c). Since the only independent sources in both (b) and (c) are those driving the terminals, and since any network variable is a linear combination of the independent source values,

$$i = K_1 v \tag{3.1}$$

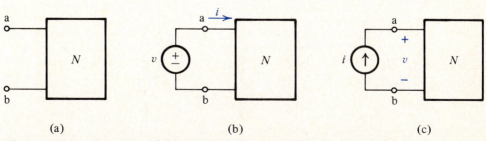

(a)   (b)   (c)

Figure 3.2 Measurement of the terminal characteristics of network N.

for Fig. 3.2(b) and

$$v = K_2 i \tag{3.2}$$

for Fig. 3.2(c).

It is not difficult to think of a simple network with the *v-i* relationship of Eqs. (3.1) or (3.2); a simple resistance R_{eq} will do, with $R_{eq} = K_2$ or $G_{eq} = K_1 = 1/K_2$. Thus,

$$v = R_{eq} i \tag{3.3}$$

and

$$i = G_{eq} v \tag{3.4}$$

The quantities R_{eq} and G_{eq} are the *driving point* resistance and conductance, respectively, of network N. The term *driving point* refers to the fact that the voltage and current in Eqs. (3.3) and (3.4) refer to the same terminals.

Equations (3.3) and (3.4) are statements of a rather surprising result: a one-port network composed of resistances and controlled sources is equivalent to a single resistance. This is true regardless of the complexity of the network.

The next few sections are devoted to the study of special resistive networks, one goal being the development of procedures to reduce a network of resistances to a single equivalent resistive element.

3.2.2 Resistances in Series

The two resistances in the network of Fig. 3.3 are connected in *series*. A voltage source *v* is applied between *a* and *b*, and the current *i* is to be calculated. The network mesh equation is

$$-v + (R_1 + R_2)i = 0 \tag{3.5}$$

or

$$v = (R_1 + R_2)i \tag{3.6}$$

which is written as

$$v = R_{eq} i \tag{3.7}$$

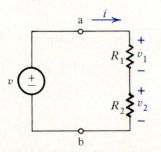

Figure 3.3 Resistances in series.

where

$$R_{eq} = R_1 + R_2 \tag{3.8}$$

This is the *v-i* relationship of the network seen at terminals *ab*. In words, two resistances in series can be replaced by a single resistance R_{eq}, with value equal to the sum of the two. Equation (3.8) is also written as

$$\frac{1}{G_{eq}} = \frac{1}{G_1} + \frac{1}{G_2} \tag{3.9}$$

The results are easily extended to the more general case of *n* resistances in series:

$$R_{eq} = R_1 + R_2 + \cdots + R_n \tag{3.10}$$

$$\frac{1}{G_{eq}} = \frac{1}{G_1} + \frac{1}{G_2} + \cdots + \frac{1}{G_n} \tag{3.11}$$

In words, the equivalent resistance of resistances in series is the sum of the resistances; the reciprocal of the equivalent conductance is the sum of the reciprocal conductances.

Assuming $R_1, R_2, \ldots, R_n > 0$, R_{eq} is obviously always greater than the largest series resistance.

Another useful expression follows from Fig. 3.3. Since $v_1 = R_1 i$ and $v_2 = R_2 i$, it is clear that

$$\frac{v_1}{v} = \frac{R_1 i}{(R_1 + R_2)i} = \frac{R_1}{R_1 + R_2}$$

and

$$\frac{v_2}{v} = \frac{R_2 i}{(R_1 + R_2)i} = \frac{R_2}{R_1 + R_2} \tag{3.12}$$

These two equations are known as the *voltage divider* relationships.

The key fact leading to the series relationships is that the same current flows through all the elements. This fact can be used to determine whether or not certain elements in a complex network are in series.

3.2.3 Resistances in Parallel

The resistances in the network of Fig. 3.4 are connected in parallel. A current source *i* is applied between *a* and *b*, and the voltage *v* is calculated. The network node equation is

$$-i + (G_1 + G_2)v = 0 \tag{3.13}$$

or

$$i = G_{eq}v \tag{3.14}$$

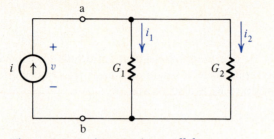

Figure 3.4 Resistances in parallel.

where

$$G_{eq} = G_1 + G_2 \tag{3.15}$$

This is the v-i relationship of the network at terminals ab. In words, the two conductances in parallel can be replaced by a single conductance, G_{eq}, with value equal to the sum of the two. This is also written as

$$\frac{1}{R_{eq}} = \frac{1}{R_1} + \frac{1}{R_2} \tag{3.16}$$

Here again, the generalization comes easily; for n conductances in parallel

$$G_{eq} = G_1 + G_2 + \cdots + G_n \tag{3.17}$$

or

$$\frac{1}{R_{eq}} = \frac{1}{R_1} + \frac{1}{R_2} + \cdots + \frac{1}{R_n} \tag{3.18}$$

In words, the equivalent conductance of conductances in parallel is the sum of the conductances; the reciprocal of the equivalent resistances is the sum of the reciprocal resistances. For only two resistances, another useful expression follows:

$$\begin{aligned} R_{eq} &= \frac{1}{G_{eq}} = \frac{1}{1/R_1 + 1/R_2} \\ &= \frac{1}{(R_1 + R_2)/R_1 R_2} \\ &= \frac{R_1 R_2}{R_1 + R_2} \end{aligned} \tag{3.19}$$

For the case $G_1, G_2, \ldots, G_n > 0$, G_{eq} is greater than the largest conductance. This implies that R_{eq} is less than the smallest of the resistances.

For Fig. 3.4, the branch laws are $i_1 = G_1 v$ and $i_2 = G_2 v$, hence

$$\frac{i_1}{i} = \frac{G_1 v}{(G_1 + G_2)v} = \frac{G_1}{G_1 + G_2} \tag{3.20}$$

and

$$\frac{i_2}{i} = \frac{G_2 v}{(G_1 + G_2)v} = \frac{G_2}{G_1 + G_2} \tag{3.21}$$

These in turn are written in terms of the resistances by multiplying Eqs. (3.20) and (3.21) by $R_1 R_2$:

$$\frac{i_1}{i} = \frac{R_2}{R_1 + R_2} \tag{3.22}$$

and

$$\frac{i_2}{i} = \frac{R_1}{R_1 + R_2} \tag{3.23}$$

These relationships, in either the conductance or resistance forms, are known as the *current divider* relationships.

The parallel relationships are all predicated on one key fact: the voltage across all parallel elements is the same. This fact can be used to determine whether or not certain elements in a complex network are in parallel.

EXAMPLE 3.1

Find the equivalent resistance at the terminals *ab* for the network of Fig.E3.1(a).

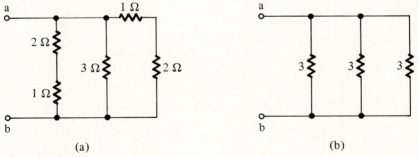

(a)   (b)

Figure E3.1

Solution

1. First, use the series rule to reduce the two series combinations as shown in Fig. E3.1(b).

2. Then, use the parallel rule,

$$G_{eq} = \tfrac{1}{3} + \tfrac{1}{3} + \tfrac{1}{3} = 1 \text{ S}$$

$$R_{eq} = \frac{1}{G_{eq}} = 1\ \Omega$$

EXAMPLE 3.2

Find the equivalent resistance at terminals *ab* for the network of Fig. E3.2(a).

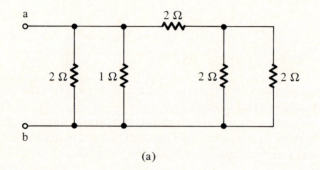

(a)

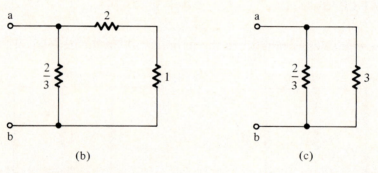

(b)   (c)

Figure E3.2

Solution

To reduce the network of Fig. E3.2(a) proceed as follows:

1. Use the parallel rule to obtain Fig. 3.2(b).

2. Next, use the series rule to obtain Fig. E3.2(c).

3. Finally, use the parallel rule again:

$$R_{eq} = \frac{\frac{2}{3} \times 3}{\frac{2}{3} + 3} = \frac{6}{11}\,\Omega.$$

EXAMPLE 3.3

Find the voltage v in Fig. E3.3(a).

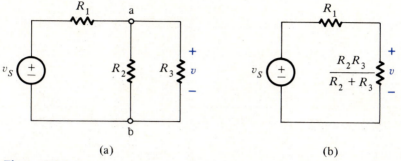

(a)   (b)

Figure E3.3

Solution

Although this can be done by the methods of Chap. 2, it is more efficient to proceed using the relationships of the present section.

1. Use the parallel rule to get the equivalent resistance between a and b. There results Fig. E3.3(b).

2. Since the resistances R_1 and $R_2 R_3/(R_2 + R_3)$ are in series, the voltage divider relation is used:

$$\frac{v}{v_S} = \frac{R_2 R_3/(R_2 + R_3)}{R_1 + R_2 R_3/(R_2 + R_3)}$$

$$= \frac{R_2 R_3}{R_1 R_2 + R_1 R_3 + R_2 R_3}$$

EXAMPLE 3.4

Find the voltage v in Fig. E3.4(a).

Solution

1. Use the voltage divider on the series combination of the 1- and 2-Ω resistances:

$$\frac{v}{v_1} = \frac{2}{2 + 1} = \frac{2}{3}$$

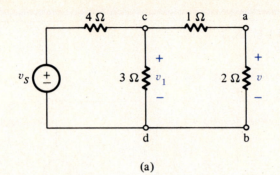

(a)

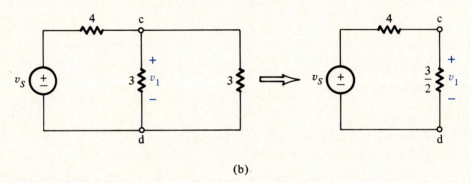

(b)

Figure E3.4

or

$$v = \tfrac{2}{3}v_1$$

2. Use the series and parallel rules to get the equivalent resistance between c and d, as shown in Fig. E3.4(b).

3. Use the voltage divider relation:

$$\frac{v_1}{v_S} = \frac{\tfrac{3}{2}}{4 + \tfrac{3}{2}} = \frac{3}{11}$$

or

$$v_1 = \tfrac{3}{11}v_S$$

Finally,

$$v = \tfrac{2}{3}v_1 = \tfrac{2}{11}v_S$$

It is worth pointing out a common error in the use of the voltage divider relations. It is tempting to write, from Fig. E3.4(a),

$$v_1 = \frac{3v_S}{3 + 4} = \tfrac{3}{7}v_S$$

(3.24)

$$v = \frac{2v_1}{2 + 3} = \tfrac{2}{5}v_1$$

and

$$v = \tfrac{6}{35}v_S$$

This result is wrong because Eq. (3.24) is incorrect. The 3- and 4-Ω resistances are not in series, as assumed in writing Eq. (3.24). The current flowing through the 4-Ω resistance is not the same as the current in the 3-Ω resistance.

EXAMPLE 3.5

Find the branch current i in the network of Fig. E3.5(a).

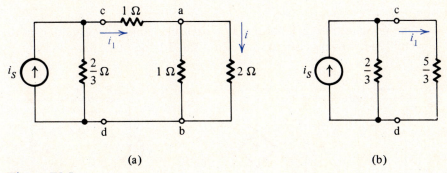

(a)

(b)

Figure E3.5

Solution

1. Use the current divider relation with the 1- and 2-Ω resistances between a and b:

$$\frac{i}{i_1} = \frac{1}{1 + 2} = \frac{1}{3}$$

or

$$i = \tfrac{1}{3}i_1$$

2. Reduce the network to the right of cd to an equivalent resistance, as shown in Fig. E3.5(b).

3. Use the current divider relation:

$$\frac{i_1}{i_S} = \frac{\frac{2}{3}}{\frac{2}{3} + \frac{5}{3}} = \frac{2}{7}$$

Finally,

$$i = \frac{i_1}{3} = \tfrac{2}{21} i_S$$

For networks containing controlled sources as well as resistances, it is not possible to apply these reduction techniques. It is then necessary to drive the network with a voltage (current) source and to calculate the resulting driving point current (voltage). Since the driving point v-i relationship is of the form $i = G_{eq}v$ ($v = R_{eq}i$), a source of unit magnitude yields a current (voltage) numerically equal to $G_{eq}(R_{eq})$. This fact simplifies calculations.

EXAMPLE 3.6

The network of Fig. E3.6(a) may represent a transistor amplifier, in the so-called emitter-follower configuration. Find the driving point resistance at terminals ab.

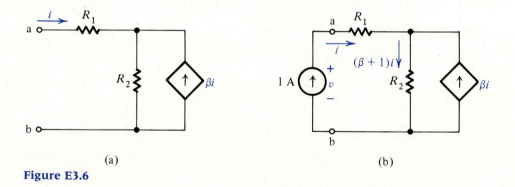

(a)   (b)

Figure E3.6

Solution
In Fig. E3.6(b), the network is driven with a 1-A source, so that $i = 1$. By KCL, the current through R_2 is $(\beta + 1)i$. Applying KVL,

$$v = R_1 + (\beta + 1)R_2$$

Since a unit current source was used,

$$R_{eq} = R_1 + (\beta + 1)R_2$$

DRILL EXERCISES

3.1 Calculate the driving point resistance at *ab* in the networks of Fig. D3.1.

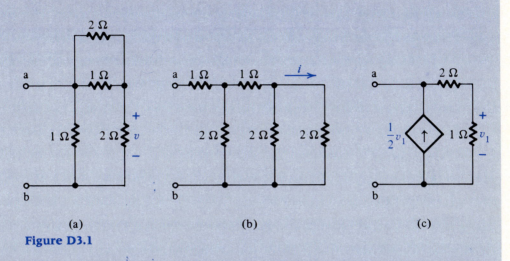

Figure D3.1

Ans. (a) $8/11\,\Omega$ (b) $2\,\Omega$ (c) $6\,\Omega$

3.2 The networks of Fig. D3.1 are driven with 2-V voltage sources (*a* positive with respect to *b*). Find the voltage *v* in Fig. D3.1(a) and the current *i* in Fig. D3.1(b).
Ans. $v = \frac{3}{2}$ V, $i = \frac{1}{4}$ A

3.3 Repeat Exercise 3.2, but with the networks driven by 4-A current sources (positive direction into terminal *a*).
Ans. $v = \frac{24}{11}$ V, $i = 1$ A

3.3   THE THEVENIN AND NORTON THEOREMS

3.3.1 Statement of the Theorems and Examples

Section 3.2.1 dealt with equivalent circuits for networks composed of resistances and controlled sources. In this section, the networks contain independent sources as well.

The network N in Fig. 3.5 contains resistances, controlled sources, independent voltage sources and independent current sources. The terminal characteristics are to be calculated, using either of the two "experiments" of Fig. 3.5.

Assume, for the sake of simplicity, that the network contains two independent voltage sources, v_{S1} and v_{S2}, and one independent current source i_{S1}. By super-

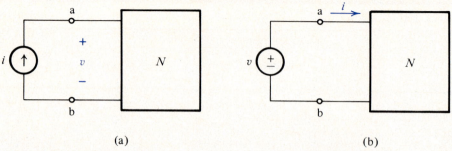

(a)   (b)

Figure 3.5 (a) Current driven network and (b) voltage driven network

position, any network variable in Fig. 3.5(a), including v, may be written as a linear combination (weighted sum) of all the *independent* sources, including the external source i. Thus,

$$v = K_1 i + A_1 v_{S1} + A_2 v_{S2} + A_3 i_{S1} \tag{3.25}$$

where K_1, A_1, A_2, and A_3 are constants. Grouping the sum of the internal independent source contributions into a single term, K_2, Eq. (3.25) simplifies to

$$v = K_1 i + K_2 \tag{3.26}$$

where $K_2 = A_1 v_{S1} + A_2 v_{S2} + A_3 i_{S1}$.

When all internal independent sources are set to zero, $K_2 = 0$ and the equivalent circuit with respect to terminals ab is known to be just a resistance, R_{eq}, i.e.,

$$v = K_1 i = R_{eq} i \tag{3.27}$$

Therefore, K_1 is the equivalent resistance R_{eq} seen at ab with all internal independent sources set to zero.

When $i = 0$, i.e., the terminals ab are left open, then $v = K_2$. For this reason, K_2 is called the *open-circuit voltage,* denoted by v_T (T for Thevenin).

With these interpretations, Eq. (3.26) is written as

$$v = R_{eq} i + v_T \tag{3.28}$$

It is not difficult to find a simple network with the v-i characteristics of Eq. (3.28). The student can easily verify that the network of Fig. 3.6 does indeed have the required characteristics; this network is known as the *Thevenin equivalent* of the network N, with respect to terminals ab.

If the same network is driven by a voltage source v as in Fig. 3.5(b), superposition yields

$$i = K_3 v + K_4 \tag{3.29}$$

With all internal independent sources turned off, $K_4 = 0$ and the v-i relationship becomes

$$i = K_3 v = G_{eq} v \tag{3.30}$$

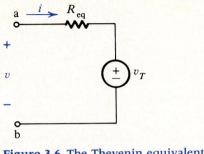

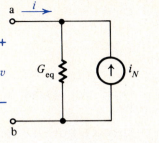

Figure 3.6 The Thevenin equivalent network.

Figure 3.7 The Norton equivalent network.

Thus, K_3 is the equivalent conductance $G_{eq} = 1/R_{eq}$ seen at *ab* with all internal independent sources set to zero.

If $v = 0$, i.e., the terminals *ab* are short-circuited, then $i = K_4$. This is the short-circuit current going *into* terminal *a* in Fig. 3.5(b); it is customary to define i_N as the short-circuit current coming *out* of that terminal, so $K_4 = -i_N$ and (3.29) becomes

$$i = G_{eq}v - i_N \qquad (3.31)$$

Here again, a simple circuit can be synthesized with the *v-i* characteristics of Eq. (3.31). Figure 3.7 is such a circuit, known as the *Norton equivalent* of the network N, with respect to terminals *ab*.

The terminal relationships of Eqs. (3.28) and (3.31) are represented by the graph shown in Fig. 3.8. Note that *any* network composed of resistances and sources has a *v-i* graph of this form. It follows that the driving point relationship of any linear one-port network containing sources and resistances is completely specified by just two parameters.

The Thevenin and Norton equivalents for the same network must themselves be equivalent. Provided both R_{eq} and G_{eq} exist, it is easy to verify by comparing Eqs. (3.28) and (3.31) that

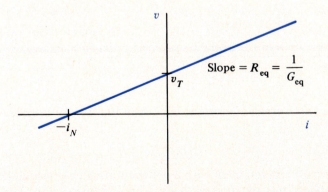

Figure 3.8 Voltage-current graph for the Thevenin and Norton equivalent circuits.

$$G_{eq} = \frac{1}{R_{eq}} \tag{3.32}$$

$$v_T = R_{eq} i_N \tag{3.33}$$

$$i_N = G_{eq} v_T \tag{3.34}$$

Three methods may be used to calculate the Norton and Thevenin equivalent circuits.

Method 1: Drive the network externally with a current (or voltage) source and calculate the driving point voltage (or current). The resulting expression is of the form $v = K_1 i + K_2$ (or $i = K_3 v + K_4$) and the result $R_{eq} = K_1$, $v_T = K_2$ (or $G_{eq} = K_3$, $i_N = -K_4$) follows immediately.

Method 2: Calculate the voltage across the terminals when these are left open-circuited (v_T) or the current flowing through the terminals when they are short-circuited (i_N). Then, set the internal independent sources (*not* the controlled sources) to zero, and calculate the driving point resistance R_{eq} (or conductance G_{eq}).

Method 3: Calculate both v_T and i_N, as in method 2. Use Eq. (3.33) or (3.34) to calculate

$$R_{eq} = \frac{v_T}{i_N} \text{ or } G_{eq} = \frac{i_N}{v_T}$$

The choice of method is a matter of experience; the student must be familiar with all three. If the network is sufficiently complex to warrant the use of nodal or mesh analysis, method 1 is suggested since the other two methods each require the solution of two network problems instead of just one. Methods 2 and 3 are useful to the extent that those two problems are simple enough to be solved by inspection. In particular, method 2 is a poor choice if the network has controlled sources because the driving point resistance cannot usually be obtained by inspection.

EXAMPLE 3.7

Find the Thevenin and Norton equivalent circuits of the network of Fig. E3.7(a), from terminals *ab*.

Solution
Method 1. The network is driven at *ab* with a voltage source *v*. Although the mesh method can be applied, this network is easily solved by inspection. The branch voltages are indicated in Fig. E3.7(b), and KCL at node *c* yields

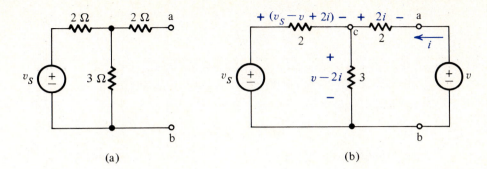

(a)

(b)

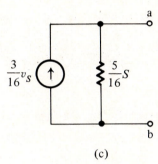

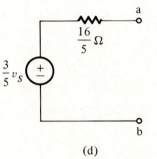

(c)

(d)

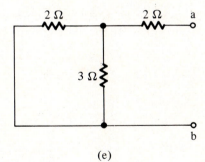

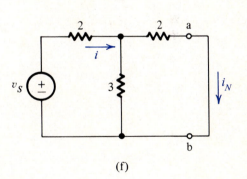

(e)

(f)

Figure E3.7

$$\frac{v_S - v + 2i}{2} + i = \frac{v - 2i}{3}$$

Solving for i,

$$i = \tfrac{5}{16}v - \tfrac{3}{16}v_S \qquad\qquad\qquad (3.35)$$

As expected, the result is a linear combination of two sources. Identifying terms with Eq. (3.31) yields

$$G_{eq} = \tfrac{5}{16} S$$

$$i_N = \tfrac{3}{16} v_S$$

and the Norton equivalent is shown below in Fig. E3.7(c).

To obtain the Thevenin circuit, solve Eq. (3.35) for v,

$$v = \tfrac{16}{5} i + \tfrac{3}{5} v_S$$

Matching terms with Eq. (3.28),

$$R_{eq} = \tfrac{16}{5} \, \Omega$$

$$v_T = \tfrac{3}{5} v_S$$

and the Thevenin equivalent is shown in Fig. E3.7(d).

Method 2. The open-circuit voltage across *ab* is also the voltage across the 3-Ω resistance, since there is no voltage drop across the 2-Ω resistance connected to *a* (current = zero). By the voltage divider rule,

$$v_T = \frac{3v_S}{3 + 2} = \tfrac{3}{5} v_S$$

The equivalent resistance, calculated from Fig. E3.7(e), is

$$R_{eq} = 2 + \frac{2 \times 3}{2 + 3} = \tfrac{16}{5} \Omega$$

This completes the calculation of the two parameters of the Thevenin equivalent circuit.

If, instead, the Norton equivalent is to be obtained (assuming the Thevenin circuit is not available), it is necessary to solve for the short-circuit current as in Fig. E3.7(f).

By the current divider rule,

$$i_N = \frac{3i}{2 + 3} = \tfrac{3}{5} i$$

and

$$i = \frac{v_S}{2 + 2 \times 3/(2 + 3)} = \tfrac{5}{16} v_S$$

so that

$$i_N = \tfrac{3}{5} i = \tfrac{3}{16} v_S$$

The calculation of the equivalent resistance is the same as with the Thevenin circuit.

Method 3. In method 3, both v_T and i_N are calculated, as above. Then

$$R_{eq} = \frac{v_T}{i_N} = \frac{\tfrac{3}{5} v_S}{\tfrac{3}{16} v_S} = \tfrac{16}{5} \, \Omega$$

EXAMPLE 3.8

The network of Fig. E3.8(a) may represent a bipolar junction transistor amplifier, over a small range of signal amplitude. Find the Thevenin and Norton equivalents at terminals ab.

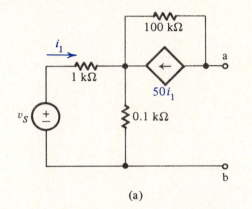

(a)

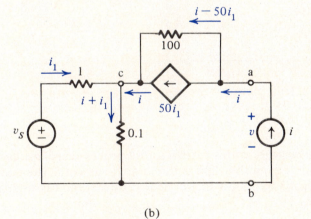

(b)

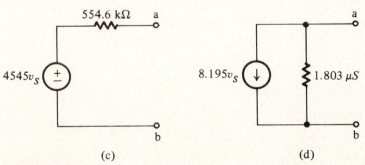

(c)   (d)

Figure E3.8

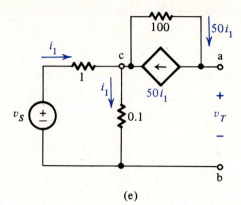

(e)

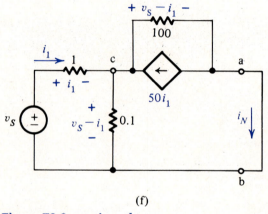

(f)

Figure E3.8 continued

Solution

Method 1. Driving the network with a current source proves expedient, because the current entering node c from the right is the known current i. The currents are entered directly in Fig. E3.8(b) using KCL at node c. By KVL, and assuming current to be in milliamperes,

$$v_S = i_1 + 0.1(i_1 + i)$$

and, therefore,

$$i_1 = \frac{v_S}{1.1} - \frac{0.1i}{1.1}$$

$$= 0.909v_S - 0.0909i$$

The driving point voltage is

$$v = 100(i - 50i_1) + 0.1(i_1 + i)$$

$$= 100.1i - 4999.9i_1$$

Eliminating i_1,

$$v = 554.6i - 4545v_S$$

and the Thevenin equivalent circuit is shown in Fig. E3.8(c).

The Norton conductance is simply

$$G_{eq} = \frac{1}{(554.6 \text{ k}\Omega)} = 1.803 \ \mu S$$

As for the Norton current i_N, it is found from

$$i_N = \frac{v_T}{R_{eq}} = -8.195v_S$$

so that the Norton equivalent circuit is as shown in Fig. E3.8(d).

Method 2 is not recommended in this case because the presence of the controlled source precludes the use of reduction methods to find R_{eq} by inspection.

Method 3. In Fig. E3.8(e), the current entering node c from the right is zero, so the 1- and 0.1-Ω resistances are effectively in series. Thus,

$$v_S = 1.1i_1$$

or

$$i_1 = 0.909v_S$$

and

$$v_T = 0.1i_1 - 100 \times 50i_l$$

$$= -4999.9i_1 = -4545v_S$$

After using KVL to label the branch voltages, write KCL at node c in Fig. E3.8(f),

$$i_1 + 50i_1 = \frac{v_S - i_1}{0.1} + \frac{v_S - i_1}{100}$$

which simplifies to

$$i_1 = 0.164v_S$$

and

$$i_N = -50i_1 + \frac{v_S - i_1}{100}$$

$$= -8.195v_S$$

The equivalent resistance is

$$R_{eq} = \frac{v_T}{i_N} = 554.6 \text{ k}\Omega$$

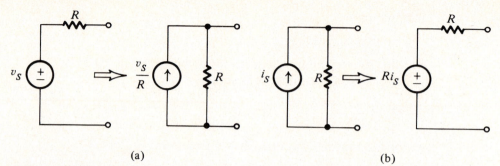

(a)   (b)

Figure 3.9 Transformation of (a) voltage source to current source and (b) current source to voltage source.

3.3.2 Source Transformations

The Thevenin and Norton theorems are often used to transform a current source into a voltage source, and vice versa. A voltage source must be accompanied by a series resistance and a current source by a parallel resistance, as illustrated in Fig. 3.9.

The two transformations in Fig. 3.9 follow from the facts that (1) R_{eq} in the Thevenin circuit is equal to $1/G_{eq}$ in the Norton equivalent and (2) the Thevenin voltage and Norton current are related through $v_T = R_{eq}i_N$.

EXAMPLE 3.9

Apply the source transformations to the network of Fig. E3.9(a) to obtain (1) a network with two voltage sources and (2) a network with two current sources.

Solution
For part 1, apply the current-to-voltage source transformation to the combination of i_S and R_4 to obtain Fig. E3.9(b).

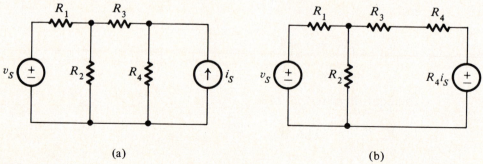

(a)   (b)

Figure E3.9

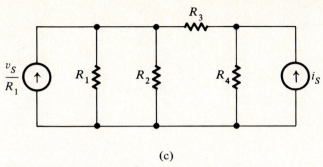

(c)

Figure E3.9 continued

For part 2, apply the voltage-to-current source transformation to the combination of v_S and R_1; this leads to Fig. E3.9(c).

EXAMPLE 3.10

Use source transformations to solve for v in the ladder network of Fig. E3.10(a).

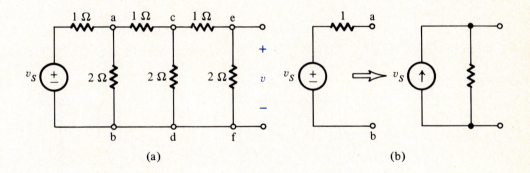

(a)   (b)

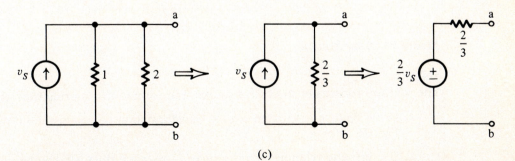

(c)

Figure E3.10

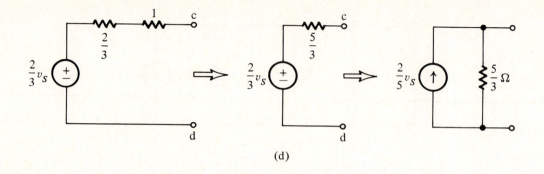

(d)

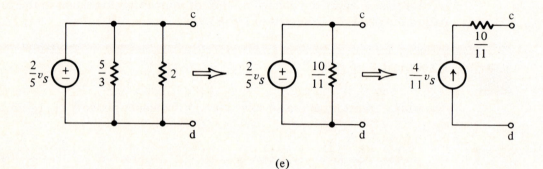

(e)

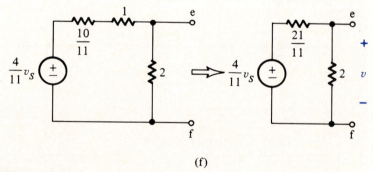

(f)

Figure E3.10 continued

Solution

The solution consists of successive source transformations, starting from the left.

1. Use a voltage-to-current source transformation to obtain the network of Fig. E3.10(b).

2. Use first the relation for resistances in parallel, then a current-to-voltage source transformation. This leads to Fig. E3.10(c).

3. Use first the series resistance relation, then a voltage-to-current source transformation, as shown in Fig. E3.10(d).

4. A step similar to 2 yields Fig. E3.10(e).

5. A final step leads to the diagram shown in Fig. E3.10(f).

By the voltage divider relation,

$$v = \frac{2}{2 + \frac{21}{11}} \frac{4}{11} v_S$$

$$= \frac{8}{43} v_S$$

EXAMPLE 3.11

The network of Fig. E3.11(a) is a so-called R/2R digital-to-analog (D/A) converter. A four-bit number is written as $b_3 b_2 b_1 b_0$, where the bits b_i are either 0 or 1. The number is to be interpreted as a number in base 2, so that the quantity represented is $b_0 + b_1 2^1 + b_2 2^2 + b_3 2^4$. In the network, the voltage sources v_0, v_1, v_2, v_3 are 4 V if the corresponding bit is 1, 0 V if it is a 0, i.e., $v_i = 4b_i$, $i = 1, 2, 3, 4$. Calculate the Norton equivalent seen at ab.

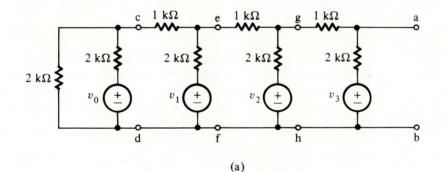

(a)

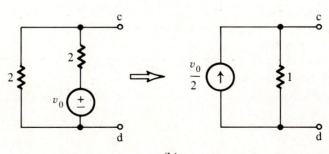

(b)

Figure E3.11

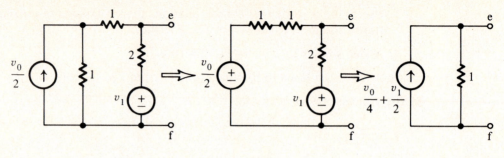

(c)

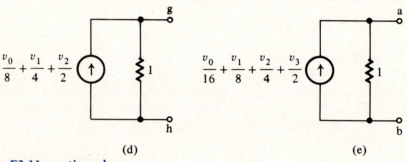

(d)   (e)

Figure E3.11 continued

Solution
The network to the left of *cd* is transformed as in Fig. E3.11(b).

The network to the left of *ef* is now transformed as in Fig. E3.11(c).

Note the similarity between the last network in Fig. E3.11(c) and the last network in Fig. E3.11(b). This similarity is exploited to yield the Norton equivalents in Fig. E3.11(d) and (e).

Since $v_i = 4b_i$ volts, the Norton current in Fig. E3.11(e) is

$$i_N = \frac{b_0 + 2b_1 + 4b_2 + 3b_3}{4}$$

which is proportional to the quantity represented by the binary number. The voltage across a resistance load connected between *a* and *b* is also proportional to that same quantity.

Because of its special structure, this particular D/A converter is extendible to any desired number of bits.

Source transformations are used to prepare a network for the application of nodal or mesh analysis. Nodal analysis requires that all sources be current sources, while voltage sources are needed for mesh analysis.

EXAMPLE 3.12

Transform the voltage source in Fig. E3.12(a) to a current source, and prepare the network for nodal analysis by expressing the controlled source in terms of node and independent voltages.

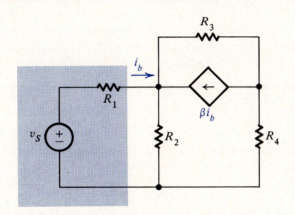

(a)

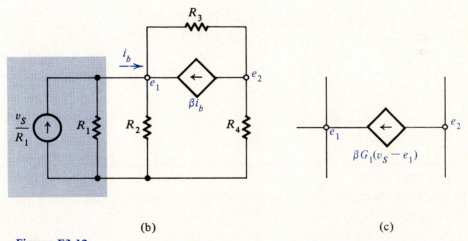

(b)   (c)

Figure E3.12

Solution

The voltage-to-current transformation is used to convert the voltage source. The dotted lines in Fig. 3.12(a) and (b) should make it clear that i_b is *not* the current through R_1 in Fig. E3.12(b).

The next step is to express the controlling variable i_b in terms of e_1, e_2, and v_S, using KVL and branch laws

$$i_b = \frac{v_S}{R_1} - \frac{e_1}{R_1} = G_1(v_S - e_1)$$

The result is shown in Fig. E3.12(c).

The source tranformation can also be applied to a part of a network containing a controlled source, provided that the controlling variable is in another part of the network. This is demonstrated in the following example.

EXAMPLE 3.13

Prepare the network of Fig. E3.12(a) for the application of the mesh method by a current-to-voltage source transformation.

Solution

The transformation yields Fig. E3.13 and the mesh method can be applied directly since i_b is a mesh current.

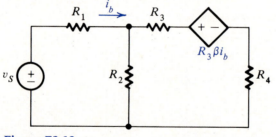

Figure E3.13

DRILL EXERCISES

3.4 For the networks of Fig. D3.4, find the Thevenin and Norton equivalents at terminals *ab* (try all three methods).
Ans. (a) $v_T = i_S/3$, $R_{eq} = \frac{4}{3}\,\Omega$, $i_N = i_S/4$, $G_{eq} = \frac{3}{4}\,S$ (b) $v_T = 9v_S/11$, $R_{eq} = \frac{10}{11}\,\Omega$, $i_N = 9v_S/10$, $G_{eq} = \frac{11}{10}\,S$

3.5 Calculate the Thevenin and Norton equivalent circuits at terminals *ab* for the network of Fig. D3.5.
Ans. $v_T = 0.0826v_S$, $R_{eq} = 82.6\,\Omega$, $i_N = v_S$, $G_{eq} = 0.0121\,S$

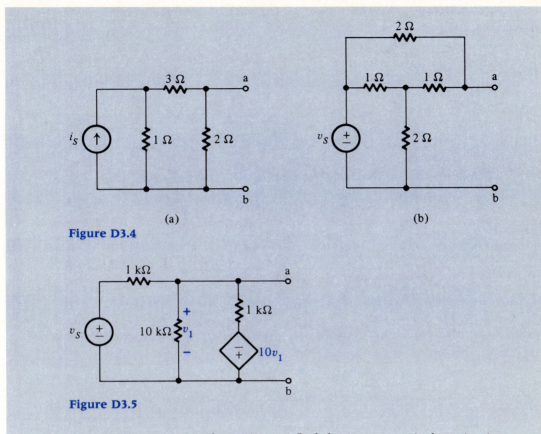

Figure D3.4

(a)   (b)

Figure D3.5

3.6 Use successive source transformations to find the Norton equivalent circuit at *ab* in Fig. D3.4(a).

3.4   SOURCE SPLITTING

The Thevenin and Norton theorems were used in Sec. 3.3.2 to transform voltage sources into current sources, and vice versa. The transformation is possible if the voltage source appears in series with a resistance, or the current source in parallel with a resistance. Figure 3.10 shows two examples where these conditions do not hold. Such cases are handled by a procedure known as *source splitting*.

The network of Fig. 3.11(a) is described by terminal equations

$$v_1 = R_1 i_1 + v_S \tag{3.36}$$

$$v_2 = R_2 i_2 + v_S \tag{3.37}$$

It may be verified that the network of Fig. 3.11(b) is also described by the same two equations, so that the two networks are equivalent. Each voltage source in the

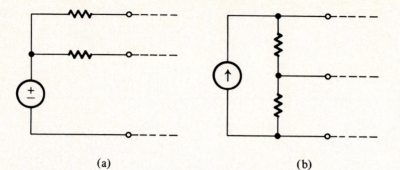

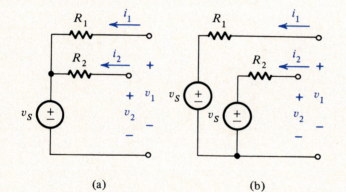

(a)   (b)

Figure 3.10 Cases of (a) a voltage source and (b) a current source that require a source splitting procedure prior to transformation.

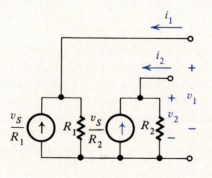

(c)

Figure 3.11 (a) Voltage source with two resistances, (b) the split-source, (c) Norton representation.

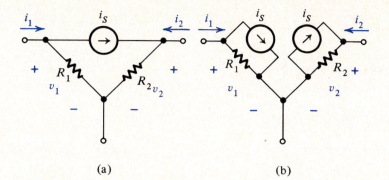

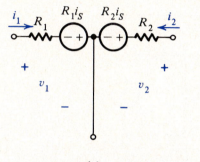

Figure 3.12 (a) Current source with two resistances, (b) the split-source, (c) Thevenin representation.

network of Fig. 3.11(b) is in series with a resistance, and Norton's theorem may be used to derive appropriate current sources, as shown in Fig. 3.11(c).

The network of Fig. 3.12(a) has the terminal relationships

$$i_1 = \frac{v_1}{R_1} + i_S \tag{3.38}$$

$$i_2 = \frac{v_2}{R_2} - i_S \tag{3.39}$$

Since these also hold for the network of Fig. 3.12(b), the two networks are equivalent. Thevenin's theorem may be brought to bear to transform the current sources in Fig. 3.12(b) to voltage sources as in Fig. 3.12(c).

EXAMPLE 3.14

Transform the network of Fig. E3.14(a) to a network with only current sources.

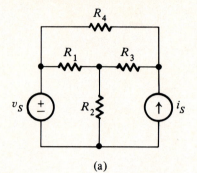

(a)

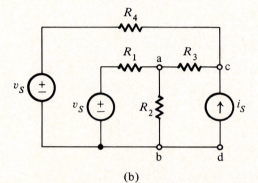

(b)

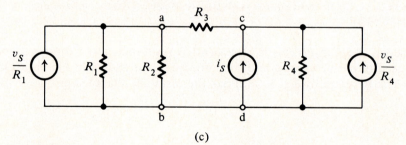

(c)

Figure E3.14

Solution

1. Split the voltage source as in Fig. E3.14(b).

2. Transform the voltage sources to obtain Fig. E3.14(c).

EXAMPLE 3.15

Split the controlled source and transform all sources to voltage sources in the network of Fig. E3.15(a).

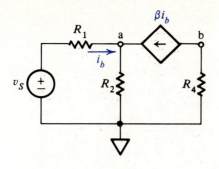

(a)

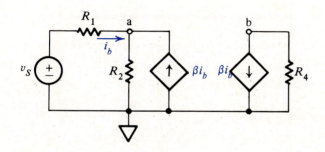

(b)

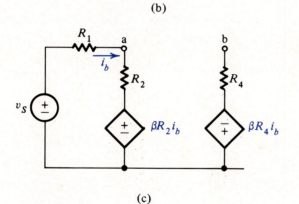

(c)

Figure E3.15

Solution

1. Split the current source, as in Fig. E3.15(b).

2. Transform the current sources; this yields Fig. E3.15(c).

It is useful at this point to summarize the steps to be taken when solving a network by the mesh or nodal analysis.

Step 1 Transform the sources to obtain all voltage sources (mesh analysis) or all current sources (nodal analysis). This may or may not require source splitting.

Step 2 If there are controlled sources, express the controlling variables in terms of either mesh currents (mesh analysis) or node voltages (nodal analysis), and independent source voltages and currents.

Step 3 Write the mesh or node equations by inspection.

Step 4 Transfer to the left-hand side any unknown variables appearing on the right. Solve the equations.

DRILL EXERCISES

3.7 Use source splitting and source tranformations to reduce the network of Fig. D3.4(b) to its Norton equivalent at terminals *ab*.

3.8 Use source splitting and a source transformation to find the Thevenin equivalent circuit at *ab* in Fig. D3.8.

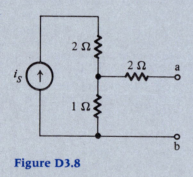

Figure D3.8

Ans. $v_T = i_S$, $R_{eq} = 3\ \Omega$

3.5 THE SUBSTITUTION THEOREM

The *substitution theorem* is based on the simple observation that replacing a variable in a set of linear equations by its solution value does not affect the rest of the solution. In a network, this means replacing a branch having a known voltage by a voltage source, or a branch having a known current by a current source.

EXAMPLE 3.16

Show that, by the substitution theorem, the 2-Ω resistance in the network of Fig. E3.16(a) may be replaced either by a short circuit or by an open circuit.

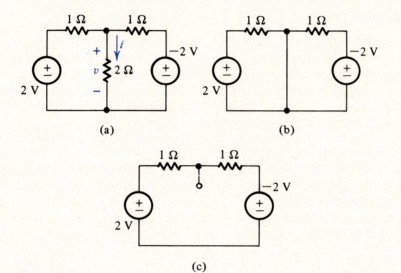

(a)

(b)

(c)

Figure E3.16

Solution

By superposition and use of the voltage divider rule,

$$v = \frac{\frac{2}{3}}{1 + \frac{2}{3}}(2) + \frac{\frac{2}{3}}{1 + \frac{2}{3}}(-2)$$

$$= 0$$

Since $v = 0$, the current in the 2-Ω resistance is also zero. Invoking the substitution theorem leads to the networks of Fig. E3.16(b) (since $v = 0$) and (c) (since $i = 0$).

In both networks, a current of 2 A flows through the 1-Ω resistances, as it does in the network of Fig. E3.16(a).

DRILL EXERCISE

3.9 In Fig. E3.16(a), change the value of the right-hand source in the diagram from -2 to 2 V. Calculate the branch voltage v and the current i for the 2-Ω resistance, and show that the voltages and currents in the other branches remain the same if the 2-Ω resistance is replaced by a voltage source of value v or a current source of value i.

3.6 SYMMETRIC NETWORKS

Figure 3.13(a) shows an example of a symmetric network, with two mirror-image halves separated by the dotted line. Figure 3.13(b) represents the general case, although only three connections are shown for simplicity of presentation. Networks N_1 and N_2 are mirror images of each other and contain only resistances and controlled sources. The independent voltage sources are shown for purpose of illustration, and could just as well be current sources.

This type of network is used in several common electronic designs, notably in the design of differential amplifiers.

The symmetry is exploited to simplify the circuit analysis, which can be reduced to the analysis of two simple networks called the *common-mode* and *differential-mode half-circuits*.

In order to introduce symmetry in the sources, the *common-mode* excitation voltage and the *differential-mode* excitation voltage are introduced.

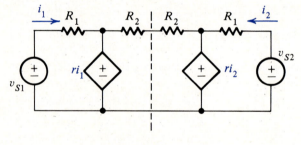

(a)

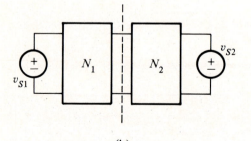

(b)

Figure 3.13 (a) Example of a symmetric network, (b) the general case.

The common-mode excitation voltage is the average of v_{S1} and v_{S2}. It is defined as

$$v_C = \frac{v_{S1} + v_{S2}}{2} \tag{3.40}$$

The differential-mode excitation voltage is just the difference between the two source voltages,

$$v_D = v_{S1} - v_{S2} \tag{3.41}$$

The two source voltages v_{S1} and v_{S2} are expressed in terms of v_C and v_D, as follows:

$$v_{S1} = v_C + \frac{v_D}{2} \tag{3.42}$$

$$v_{S2} = v_C - \frac{v_D}{2} \tag{3.43}$$

thus the network of Fig. 3.13(b) is replaced by that of Fig. 3.14.

Superposition is used to replace the solution of the network of Fig. 3.14 by the sum of solutions of the two networks of Fig. 3.15.

The network with common-mode excitation is further decomposed, by superposition, into the two networks of Fig. 3.16.

Because of symmetry, it is clear that $i_1'' = i_1'$, $i_2'' = i_2'$, $i_3'' = i_3'$. Applying superposition to calculate the currents in Fig. 3.15(a).

$$i_1 = i_1' - i_1'' = 0$$

$$i_2 = i_2' - i_2'' = 0$$

$$i_3 = i_3' - i_3'' = 0$$

The substitution theorem is now applied: the branches joining N_1 and N_2 in Fig. 3.15(a) are replaced by current sources of zero value, i.e., by open circuits. There results two identical half-circuits known as *common-mode* half-circuits, shown in Fig. 3.17.

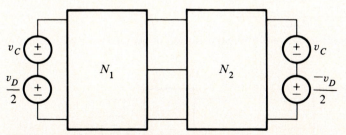

Figure 3.14 Symmetric network with sources expressed in terms of common-mode and differential-mode components.

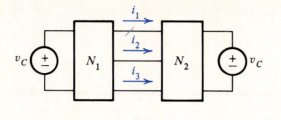

(a)

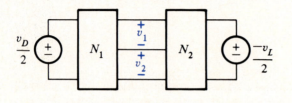

(b)

Figure 3.15 Network with (a) common-mode excitation and (b) differential-mode excitation.

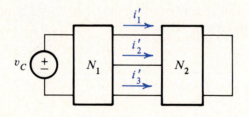

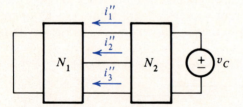

Figure 3.16 Decomposition, by superposition, of the network with common-mode excitation.

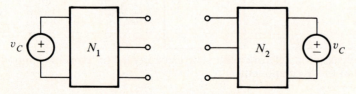

Figure 3.17 The two common-mode half-circuits.

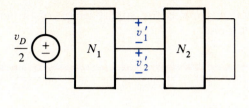

(a)

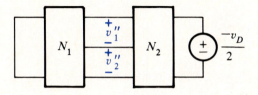

(b)

Figure 3.18 Decomposition, by superposition, of the network with differential-mode excitation.

The two common-mode half-circuits are identical and are identically driven. Every branch voltage or current in N_2 is equal to the voltage or current for the corresponding branch in N_1.

To analyze the network with differential-mode excitation, use superposition of the two networks in Fig. 3.18.

Because of symmetry and the opposite polarities of the two sources, $v_1'' = -v_1'$, $v_2'' = v_2'$. The voltages in Fig. 3.15(b) are calculated by superposition,

$$v_1 = v_1' + v_1'' = 0$$

$$v_2 = v_2' + v_2'' = 0$$

The substitution theorem implies that the branches between N_1 and N_2 can all be joined by short circuits. Here again, the outcome consists of two half-circuits, identical except for the source polarities, known as *differential-mode* half-circuits. These are shown in Fig. 3.19.

Because the sources driving the two differential mode half-circuits are of equal but opposite polarity, every branch voltage or current in N_2 is equal to the *negative* of the branch voltage or current for the corresponding branch in N_1.

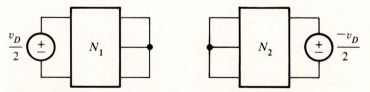

Figure 3.19 The two differential-mode half-circuits.

The solution of the original network is the superposition of the common-mode and differential-mode solutions.

EXAMPLE 3.17

In the network of Fig. E3.17(a), solve for v_1 and v_2.

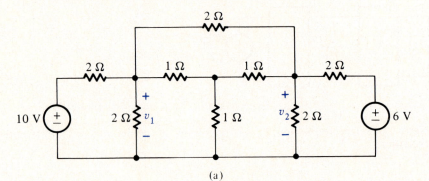

(a)

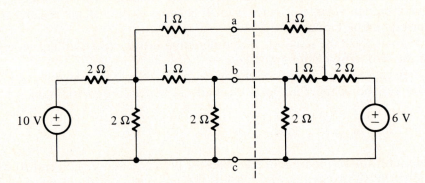

(b)

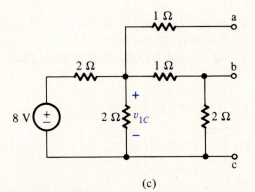

(c)

Figure E3.17

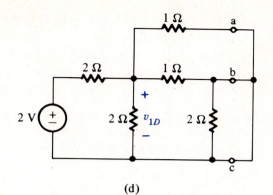

(d)

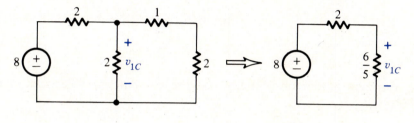

(e)

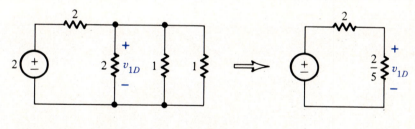

(f)

Figure E3.17 continued

Solution

First, redraw the circuit to exhibit the symmetry, as in Fig. E3.17(b). Note that series and parallel equivalents have been used to split two of the resistances.

Next, calculate the common-mode and differential-mode voltages,

$$v_C = \frac{10 + 6}{2} = 8 \text{ V}$$

$$v_D = 10 - 6 = 4 \text{ V}$$

The two half-circuits are shown in Fig. E3.17(c) and (d). In the common-mode half-circuit, the terminals a, b, and c, are left open-circuited; those same three terminals are joined by a short circuit to obtain the differential-mode half-circuit.

Now, analyze the common-mode half-circuit. Since no current flows out of a, the 1-Ω resistance connected to a is ignored. The remainder of the circuit is shown in Fig. E3.17(e).

By the voltage divider rule,

$$v_{1C} = \frac{\frac{6}{5}}{2 + \frac{6}{5}}(8) = 3 \text{ V}$$

The 2-Ω resistance connected between b and c in the differential-mode half-circuit is in parallel with a short circuit and is therefore removed. The two 1-Ω resistances are effectively connected to c, and are so drawn in Fig. E3.17(f).

By the voltage divider rule,

$$v_{1D} = \frac{\frac{2}{5}}{2 + \frac{2}{5}}2 = \tfrac{1}{3} \text{ V}$$

The branch voltage v_1 is the sum of the common-mode and differential-mode contributions,

$$v_1 = v_{1C} + v_{1D} = 3\tfrac{1}{3} \text{ V}$$

The common-mode contributions to v_2 is v_{1C}, but the differential-mode contribution is $-v_{1D}$ so that

$$v_2 = v_{1C} - v_{1D} = 2\tfrac{2}{3} \text{ V}$$

DRILL EXERCISES

3.10 In the network of Fig. E3.16(a), change the two source values to v_{S1} (left of diagram) and v_{S2} (right of diagram). Find the common-mode and differential-mode components of v.
Ans. Common-mode: $v = \tfrac{2}{5}(v_{S1} + v_{S2})$
Differential-mode: $v = 0$

3.11 Solve for the common-mode and differential-mode components of v_1 in Fig. D3.11, and solve for v_1 and v_2.

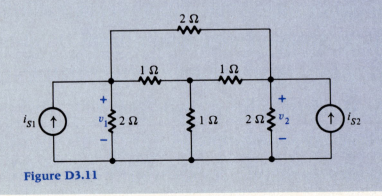

Figure D3.11

Ans. Common-mode: $v_1 = \frac{3}{5}(i_{S1} + i_{S2})$
Differential-mode: $v_1 = \frac{1}{5}(i_{S1} - i_{S2})$
$v_1 = \frac{4}{5}i_{S1} + \frac{2}{5}i_{S2}$
$v_2 = \frac{2}{5}i_{S1} + \frac{4}{5}i_{S2}$

3.7 SUMMARY AND STUDY GUIDE

This chapter was dedicated to the study of equivalent networks. Linear one-port networks were reduced either to a single resistance or to their Thevenin or Norton equivalents. Source transformations were used to convert between the two types of sources. Finally, the symmetry of certain networks was exploited to simplify their solution.

The student should pay special attention to the following points.

Equivalence must be defined with respect to a given pair of terminals.

Equivalent networks have identical terminal v-i characteristics and therefore have the same effect on any network to which they are connected.

Networks containing only resistances can often be reduced to a single equivalent resistance by repeated application of the series and parallel rules. (Another useful procedure for the reduction of resistances, the tee-pi transformation, is covered in the next chapter.)

It is necessary to use an external excitation source to calculate the equivalent resistance of a network containing resistances and controlled sources.

Three methods can be used to compute the Thevenin and Norton equivalent circuits. Methods 2 and 3 each require the solution of two network problems, and are usually more efficient for simple networks than method 1.

When a network is replaced by an equivalent, its internal branch voltages and currents no longer appear in the calculations; they cannot be found unless the equivalent network is removed and replaced by the original network.

Many network problems can be solved efficiently by successive source transformations.

Source transformations using the Thevenin or Norton theorems are used to convert all sources to a single type (current or voltage) in preparation for the application of nodal or mesh analysis. This may require source splitting as a prior step.

The solution of symmetric network problems is greatly simplified if reduced to the solution of the two half-circuit problems. All branch variables are expressed as sums (or differences) of the common-mode and differential-mode components.

PROBLEMS

3.1 Find the equivalent driving point resistance at *ab* for the following networks of Fig. P3.1.

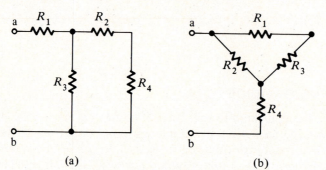

(a)   (b)

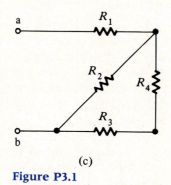

(c)

Figure P3.1

3.2 Repeat Prob. 3.1 for the networks of Fig. P3.2.

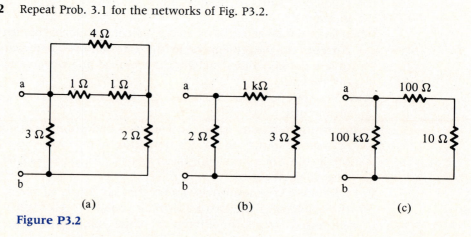

(a)   (b)   (c)

Figure P3.2

3.3 Solve for *v* in be the network of Fig. P3.3.

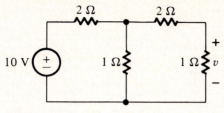

Figure P3.3

3.4   Solve for i in the network of Fig. P3.4.

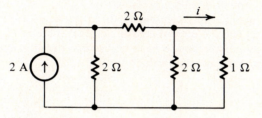

Figure P3.4

3.5   Solve for v in the network of Fig. P3.5.

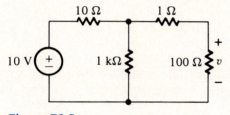

Figure P3.5

3.6   Solve for i in the network of Fig. P3.6.

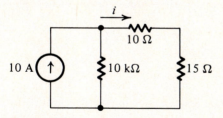

Figure P3.6

3.7   For the network of Fig. P3.7,
   (a) Calculate v/v_S.
   (b) If $R_L \gg R_2$, the ratio $R_2/(R_1 + R_2)$ is a good approximation to v/v_S. The percent error incurred by using the approximation is defined as

$$100 \frac{|v/v_S - R_2/(R_1 + R_2)|}{|v/v_S|}$$

Calculate the percent error for $R_L = 100R_2,\ 10R_2,\ R_2$.

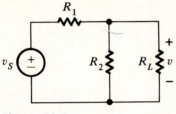

Figure P3.7

3.8 Suppose R_1 in Fig. P3.8 can vary between $(1 - \sigma)R_1$ and $(1 + \sigma)R_1$, and R_2 can vary between $(1 - \sigma)R_2$ and $(1 + \sigma)R_2$. (a) Show that $R_2/(R_1 + R_2)$ decreases with increasing R_1 and increases with increasing R_2. (b) What range of values can be taken by the ratio v/v_S? (c) For the case where the nominal value of $R_2/(R_1 + R_2)$ is 0.5, what must the fractional tolerance σ be to ensure that the ratio does not change by more than $\pm 2\%$ from the nominal value?

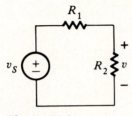

Figure P3.8

3.9 Figure P3.9(a) represents a meter movement that reads full scale for a current $i = 100$ mA. The internal resistance R_i of the meter is 0.5 Ω. (The circle with the arrow is meant to represent the meter, and behaves as a short circuit.)
(a) Find R_s in Fig. P3.9(b) so that the meter reads full scale for $v = 1$ V.
(b) Repeat for $v = 100$ V.

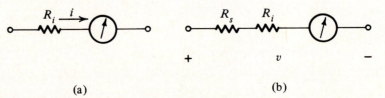

         (a)   (b)

Figure P3.9

3.10 This time, the meter of Prob. 3.9(a) is used to measure current. Find R_n in Fig. P3.10 so that the meter reads full scale for $i = 1$ A. How does this value of R_n change if $R_i = 1\Omega$?

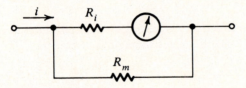

Figure P3.10

3.11 Calculate the driving point resistance at *ab* for the network of Fig. P3.11.

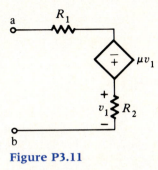

Figure P3.11

3.12 Repeat Prob. 3.11 for the network of Fig. P3.12. (Under certain assumptions, this network models a common-collector bipolar transistor amplifier.)

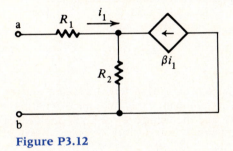

Figure P3.12

3.13 Repeat Prob. 3.11 for the network of Fig. P3.13. (This is also the common-collector amplifier, but seen from a different pair of terminals.)

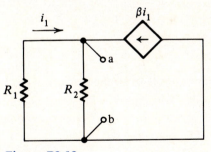

Figure P3.13

3.14 Repeat Prob. 3.11 for the network of Fig. P3.14, which, under appropriate conditions, models a common-drain, field-effect transistor amplifier.

3.15 Find the Thevenin and Norton equivalent circuits of the device of Fig. P3.15(a) given its *v-i* characteristics in Fig. P3.15(b).

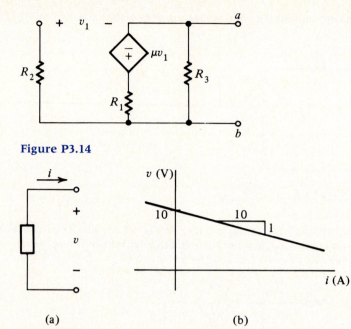

Figure P3.14

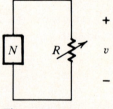

(a)

(b)

Figure P3.15

3.16 The branch N in Fig. P3.16 is composed of resistances and sources. A variable resistance, R, is connected to its terminals. For $R = \infty$ (open-circuit), $v = 5$ V. The value of R is decreased gradually, and the voltage $v = 2.5$ V is attained for $R = 3\ \Omega$. Find the Thevenin equivalent of N.

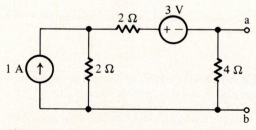

Figure P3.16

3.17 Find the Thevenin and Norton equivalents for the network of Fig. P3.17, looking into ab.

Figure P3.17

3.18 Repeat Problem 3.17 for the network of Fig. P3.18.

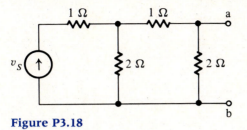

Figure P3.18

3.19 (a) Find the Thevenin equivalent of the network of Fig. P3.19, looking into *ab*.
   (b) A resistance R_s is connected between *a* and *b*. What conditions must be satisfied to ensure that the current through R_s is zero given that $v_s \neq 0$?
   (c) This circuit is known as a bridge and is used to measure resistance. A current detector is connected between *a* and *b*. Resistances R_1 and R_3 are fixed, R_2 is variable, and R_4 is the unkown resistance. R_2 is varied until the current detector shows a null. How can R_4 be calculated?

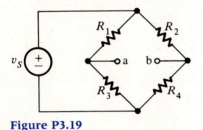

Figure P3.19

3.20 Find, if they exist, the Thevenin and Norton equivalent circuits at *ab* for the networks of Fig. P3.20.

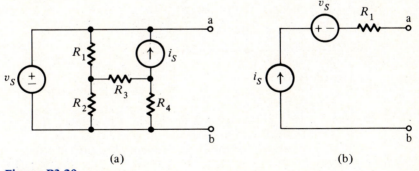

(a)   (b)

Figure P3.20

3.21 Find the Thevenin and Norton equivalents seen at *ab* for the network of Fig. P3.21. (This network models a form of the field-effect transistor common-source amplifier.)

3.22 Repeat Prob. 3.21 for the network of Fig. P3.22. (This network is a model of a form of the bipolar transistor common-emitter amplifier.)

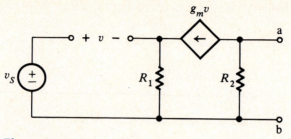

Figure P3.21

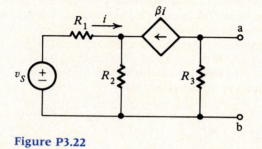

Figure P3.22

3.23 Repeat Prob. 3.21 for the network of Fig. P3.23. (This network models a bipolar transistor common-collector amplifier.)

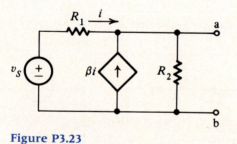

Figure P3.23

3.24 (a) Solve for v in the network of Fig. P3.24 by splitting the voltage source, transforming to current sources, and using nodal analysis.
(b) Repeat part (a), but transform the controlled current source and use mesh analysis.

3.25 Solve for v in the network of Fig. P3.25 by splitting the current source, transforming to voltage sources, and using mesh analysis.

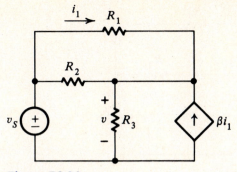

Figure P3.24

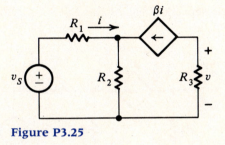

Figure P3.25

3.26 Find the Thevenin and Norton equivalents looking into *ab* in the network of Fig. P3.26.

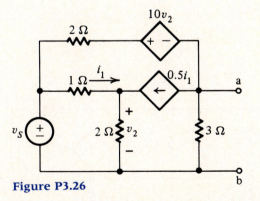

Figure P3.26

3.27 Solve for v_1 and v_2 in the symmetric network of Fig. P3.27. (*Hint:* Consider the right-hand-side resistance R_1 as being in parallel with a current source of zero value.)

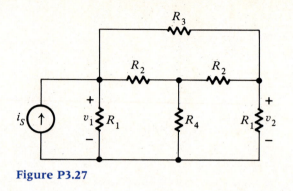

Figure P3.27

3.28 Solve for i in the symmetric network of Fig. P3.28.

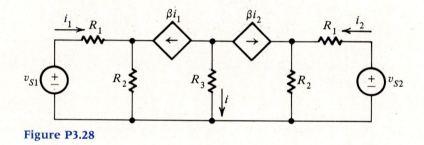

Figure P3.28

3.29 The network of Fig. P3.29 is a model for a bipolar junction transistor differential amplifier. Calculate v_1 and v_2.

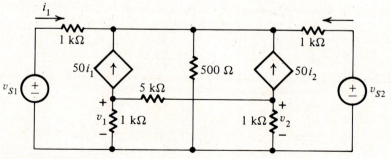

Figure P3.29

3.30 The network of Fig. P3.30 is a model for a field-effect transistor differential amplifier. Calculate v_{o1} and v_{o2}, if $g_m = 4$ mS.

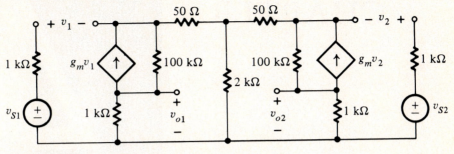

Figure P3.30

CHAPTER

4

Linear Two-Port Networks

4.1 INTRODUCTION

The notion of a port as a pair of terminals (or nodes) of a network was introduced in Chap. 1, and in Chap. 3 it was shown that a linear one-port has a Thevenin or a Norton representation. It was also briefly mentioned in Chap. 1 that a four-terminal network is called a two-port when its terminals can be paired in such a way that the current entering one terminal of each pair always equals the current leaving the other. In this chapter the description and the characterization of linear two-port networks are discussed. These network formulations form the basis for many of the linear circuit models for transistors and other devices of technological importance.

4.2 TERMINAL CHARACTERISTICS OF TWO-PORT NETWORKS

In Fig. 4.1(a) the two-port network with no independent sources is indicated by a box and the electrical terminal or port variables are labeled using the con-

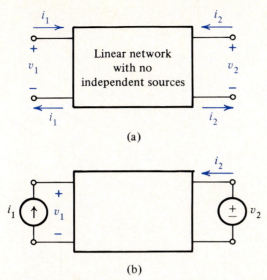

(a)

(b)

Figure 4.1 (a) Two-port showing electrical variables. (b) Two-port driven by two sources.

ventional polarity designations. There are two voltages and two currents as shown, and it is usual to refer to port 1 as the input port and port 2 as the output port. Of these four variables only two can be independent, meaning that if any two are specified then the other two are determined. For example, if in Fig. 4.1(a) v_2 and i_1 are both specified, this is equivalent to connecting an independent voltage source to the port 2 terminals and an independent current source to the port 1 terminals as shown in Fig. 4.1(b). Since the box is a linear network with no internal independent sources, *all* the currents and voltages, in particular i_2 and v_1, can be determined using the method of superposition discussed in Chap. 2. For the arbitrary choice made in the above example, this is equivalent to saying that v_1 and i_2 are dependent on v_2 and i_1. Hence, because the network is linear and there are no internal independent sources, superposition applies and v_1 and i_2 are given by the linear combinations

$$v_1 = h_{11}i_1 + h_{12}v_2 \tag{4.1a}$$

$$i_2 = h_{21}i_1 + h_{22}v_2 \tag{4.1b}$$

or, in matrix form,

$$\begin{bmatrix} v_1 \\ i_2 \end{bmatrix} = \begin{bmatrix} h_{11} & h_{12} \\ h_{21} & h_{22} \end{bmatrix} \begin{bmatrix} i_1 \\ v_2 \end{bmatrix}$$

4.2.1 The Hybrid or *h* Parameters

The *h* parameters or coefficients in Eqs. (4.1) characterize the linear relations between the dependent variables v_1, i_2 and the independent variables i_1, v_2. These equations have a graphical and a network interpretation. Recall that in a linear

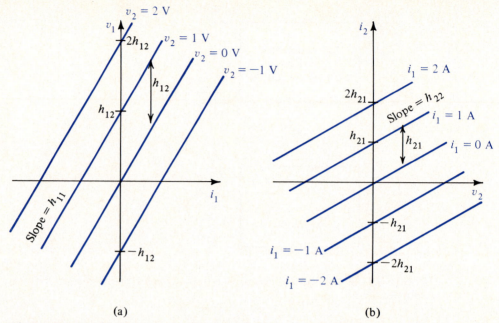

(a)   (b)

Figure 4.2 Graphical representation for Eq. (4.1).

one-port (or branch) the one-port relation is a straight line in the voltage-current plane and the one-port has a Thevenin or Norton network representation. In a two-port, the relations (4.1) define a family of parallel straight lines in the v_1, i_1 plane with v_2 as a parameter, and a family of parallel straight lines in the i_2, v_2 plane with i_1 as a parameter. The graphical interpretations, shown in Fig. 4.2(a) and (b), called the input and the output characteristics, respectively, are sets of parallel lines with equidistant spacings for equal increments of the running parameter variable. Equations (4.1) also specify a network representation for each of the ports: Eq. (4.1a) defines a Thevenin representation for port 1, consisting of an $R_{eq} = h_{11}$ and a Thevenin voltage source controlled by v_2, and Eq. (4.1b) defines a Norton representation for port 2, consisting of $G_{eq} = h_{22}$ and a Norton current source controlled by i_1. This representation, shown in Fig. 4.3, emphasizes the fact that h_{11} and h_{22} have units of resistance and conductance, respectively, and that

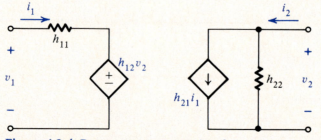

Figure 4.3 h-Parameter representation for a two-port.

h_{12}, h_{21} are dimensionless; these parameters are mixed or hybrid, hence the designation h.

An examination of either Eq. (4.1) or Fig. 4.3 provides an interpretation for each of the four parameters and defines procedures for their measurement. They are measured by one open-circuit test and one short-circuit test. Setting $v_2 = 0$

$$h_{11} = \frac{v_1}{i_1}$$

is the driving point resistance at port 1 when port 2 is short-circuited and

$$h_{21} = \frac{i_2}{i_1}$$

is the forward current ratio with port 2 shorted. Setting $i_1 = 0$,

$$h_{22} = \frac{i_2}{v_2}$$

is the driving point conductance at port 2 with port 1 open-circuited and

$$h_{12} = \frac{v_1}{v_2}$$

is the reverse voltage ratio with port 1 open.

EXAMPLE 4.1

Calculate the h parameters for the circuit of Fig. E4.1 and hence draw the model corresponding to Fig. 4.3.

Solution
The two pertinent circuits, one with $i_1 = 0$ and one with $v_2 = 0$, are shown in Fig.

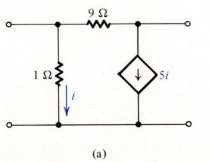

(a)

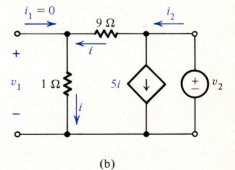

(b)

Figure E4.1

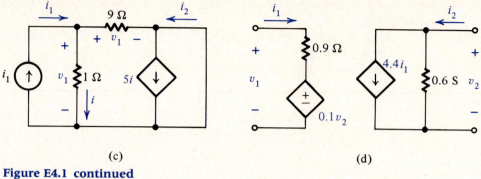

(c)   (d)

Figure E4.1 continued

E4.1(b) and (c). From (a), expressing all variables in terms of the controlling variable i (a procedure recommended in Chap. 3),

$$i_2 = 6i \qquad \text{by KCL}$$

$$v_2 = (9 + 1)i = 10i$$

$$v_2 = i$$

Therefore, by definition,

$$h_{22} = \frac{i_2}{v_2} = 0.6 \text{ S}$$

$$h_{12} = \frac{v_1}{v_2} = 0.1 \text{ V/V}$$

From (b), using the same strategy,

$$v_1 = i$$

$$i_1 = i + \frac{v_1}{9} = \frac{10i}{9}$$

$$i_2 = 5i - \frac{i}{9} = \frac{44i}{9}$$

Therefore, by definition,

$$h_{11} = \frac{v_1}{i_1} = 0.9 \text{ } \Omega \qquad \text{(can be seen by inspection)}$$

$$h_{21} = \frac{i_2}{i_1} = 4.4 \text{ A/A}$$

The corresponding model is shown in Fig. E4.1(d).

Table 4.1   LINEAR TWO-PORT PARAMETERS

| Designation | Port Variables | | Equations |
	Dependent	Independent	
Hybrid (h)	v_1, i_2	i_1, v_2	$v_1 = h_{11}i_1 + h_{12}v_2$ (4.1a)
			$i_2 = h_{21}i_1 + h_{22}v_2$ (4.1b)
Inverse hybrid (g)	i_1, v_2	v_1, i_2	$i_1 = g_{11}v_1 + g_{12}i_2$ (4.2a)
			$v_2 = g_{21}v_1 + g_{22}i_2$ (4.2b)
Impedance (z)	v_1, v_2	i_1, i_2	$v_1 = z_{11}i_1 + z_{12}i_2$ (4.3a)
			$v_2 = z_{21}i_1 + z_{22}i_2$ (4.3b)
Admittance (y)	i_1, i_2	v_1, v_2	$i_1 = y_{11}v_1 + y_{12}v_2$ (4.4a)
			$i_2 = y_{21}v_1 + y_{22}v_2$ (4.4b)
Transmission (A, B, C, D)	v_1, i_1	v_2, i_2	$v_1 = Av_2 - Bi_2$ (4.5a)
			$i_1 = Cv_2 - Di_2$ (4.5b)
Reverse transmission ($\hat{A}, \hat{B}, \hat{C}, \hat{D}$)	v_2, i_2	v_1, i_1	$v_2 = \hat{A}v_1 - \hat{B}i_1$ (4.6a)
			$i_2 = \hat{C}v_1 - \hat{D}_1i_1$ (4.6b)

The h-parameter description of a linear two-port network is only one of the six possible ways of choosing two of the four port variables as independent. These six ways are summarized in Table 4.1 which also gives the two linear equations defining each of the parameter choices. The last two ways, referred to as transmission parameters, are included for completeness and will not be discussed in this book.

In the next two subsections two of these two-port relations are discussed in detail, the remainder are left as exercises for the reader. The designations *impedance* and *admittance* are generalizations of resistance and conductance for networks which contain inductance and capacitance elements. Such networks are discussed in later chapters.

4.2.2 The Admittance or *y* Parameters

The general linear relations corresponding to the circuit of Fig. 4.1, when using Eq. (4.4) of Table 4.1, are

$$i_1 = y_{11}v_1 + y_{12}v_2 \tag{4.4a}$$

$$i_2 = y_{21}v_1 + y_{22}v_2 \tag{4.4b}$$

or, in matrix form,

$$\begin{bmatrix} i_1 \\ i_2 \end{bmatrix} = \begin{bmatrix} y_{11} & y_{12} \\ y_{21} & y_{22} \end{bmatrix} \begin{bmatrix} v_1 \\ v_2 \end{bmatrix}$$

The graphical representation is similar to that in Fig. 4.2 with all the families consisting of lines parallel to each other. The four coefficients are called the *y parameters* or *short-circuit admittance parameters* of the two-port. These parameters are measured by means of two short-circuit tests as follows:

$$y_{11} = \frac{i_1}{v_1} \quad \text{with } v_2 = 0$$

is the driving point conductance at port 1 when port 2 is short-circuited; it is also the slope of the input family of characteristics. The second parameter

$$y_{21} = \frac{i_2}{v_1} \quad \text{with } v_2 = 0$$

is the forward transconductance with port 2 shorted. A similar procedure yields

$$y_{22} = \frac{i_2}{v_2} \quad \text{with } v_1 = 0$$

for the driving point conductance at port 2 when port 1 is short-circuited, which is the slope of the output family of characteristics. The final parameter

$$y_{12} = \frac{i_1}{v_2} \quad \text{with } v_1 = 0$$

is the reverse transconductance with port 1 shorted. The equivalent circuit corresponding to this *y*-parameter representation, which contains two voltage-controlled current sources, is shown in Fig. 4.4.

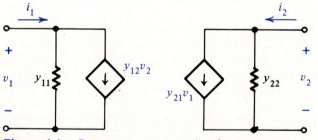

Figure 4.4 *y*-Parameter representation for a two-port.

EXAMPLE 4.2

Calculate the *y* parameters for the circuit shown in Fig. E4.2(a).

Solution

The two pertinent short-circuit conditions, one with $v_1 = 0$ and one with $v_2 = 0$, are shown in Fig. E4.2(b) and (c). From (b) it is evident that $i = 0$.

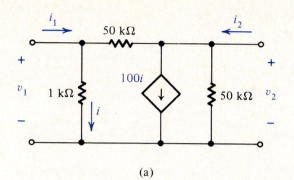

(a)

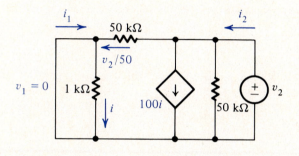

(b)

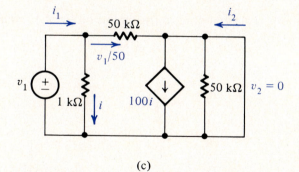

(c)

Figure E4.2

Hence the controlled source is zero and

$$i_1 = \frac{-v_2}{50} \text{ mA}$$

$$i_2 = \frac{v_2}{50} + \frac{v_2}{50} = \frac{v_2}{25} \text{ mA}$$

Therefore, by definition,

$$y_{12} = \frac{i_1}{v_2} = -20 \times 10^{-6} \text{ S}$$

$$y_{22} = \frac{i_2}{v_2} = 40 \times 10^{-6} \text{ S}$$

From (c),

$$i = v_1 \text{ mA}$$

$$i_1 = v_1 + \frac{v_1}{50} = \frac{51v_1}{50} \text{ mA}$$

$$i_2 = 100i - \frac{v_1}{50} = 99.98v_i \text{ mA}$$

Therefore, by definition,

$$y_{11} = \frac{i_1}{v_1} = \frac{51}{50} \text{ mA/V} = 1.02 \times 10^{-3} \text{ S}$$

$$y_{21} = \frac{i_2}{v_1} = 0.09998 \text{ S}$$

4.2.3 The Impedance or z Parameters

The general linear relations corresponding to the circuit of Fig. 4.1, when using Eq. (4.3) of Table 4.1, are

$$v_1 = z_{11}i_1 + z_{12}i_2 \tag{4.3a}$$

$$v_2 = z_{12}i_1 + z_{22}i_2 \tag{4.3b}$$

or, in matrix form,

$$\begin{bmatrix} v_1 \\ v_2 \end{bmatrix} = \begin{bmatrix} z_{11} & z_{12} \\ z_{21} & z_{22} \end{bmatrix} \begin{bmatrix} i_1 \\ i_2 \end{bmatrix}$$

where the four coefficients are called the z parameters or open-circuit impedance parameters of the two-port. These parameters are measured by means of two open-circuit tests as follows:

$$z_{11} = \frac{v_1}{i_1} \qquad \text{with } i_2 = 0$$

is the driving point resistance at port 1 with port 2 open. The second parameter

$$z_{21} = \frac{v_2}{i_1} \qquad \text{with } i_2 = 0$$

is the forward transresistance with port 2 open. A similar procedure yields the

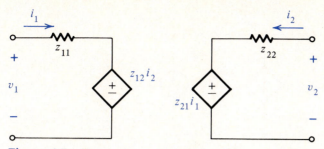

Figure 4.5 z-Parameter representation for a two-port.

driving point resistance at port 2 with port 1 open

$$z_{22} = \frac{v_2}{i_2} \qquad \text{with } i_1 = 0$$

and the reverse transresistance

$$z_{12} = \frac{v_1}{i_2} \qquad \text{with } i_1 = 0$$

The corresponding circuit representation is shown in Fig. 4.5 and contains two current-controlled voltage sources.

It should be pointed out that the six formulations given in Table 4.1 are interrelated; the only difference is in the arbitrary choice of which variables are called dependent, which ones independent. The z and y formulations are complementary in the sense that the dependent-independent variables are interchanged, with the consequence that the **Z** matrix is the inverse of the **Y** matrix and vice versa provided the matrices are nonsingular.

EXAMPLE 4.3

Calculate the z parameters for the circuit of Example 4.2.

Solution
The two pertinent open-circuit conditions, one with $i_1 = 0$ and one with $i_2 = 0$ are shown in Fig. 4.3(a), and (b). From (a), expressing i in milliamperes, and noting the KCL constraints labeled on the circuit,

$$v_1 = i \text{ V}$$

$$v_2 = 51i \text{ V}$$

$$i_2 = 101i_1 + \frac{v_2}{50} = \left(101 + \frac{51}{50}\right)i \text{ mA}$$

$$\simeq 102i \text{ mA}$$

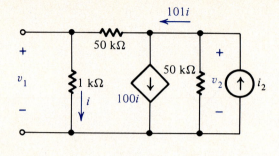

(a)

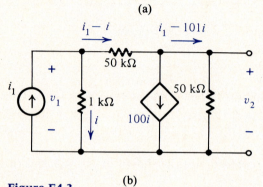

(b)

Figure E4.3

Therefore, by definition,

$$z_{12} = \frac{v_1}{i_2} = 0.0098 \text{ k}\Omega = 9.8 \ \Omega$$

$$z_{22} = \frac{v_2}{i_2} = \frac{51}{102} = 0.5 \text{ k}\Omega = 500 \ \Omega$$

From (b), and noting the KCL constraints labeled on the circuit,

$$v_1 = i \text{ V}$$

$$v_1 = 50(i_1 - i) + 50(i_1 - 101i)$$

$$= 100i_1 - 50 \times 102v_1$$

which reduces to

$$v_1 = 0.0196i$$

$$v_2 = v_1 - 50(i_1 - i)$$

$$= v_1 - 50i_1 + 50v_1$$

$$= 0.0196i_1 - 50i_1 + 50 \times 0.0196i_1$$

$$= -49i_1$$

Therefore, by definition,

$$z_{11} = \frac{v_1}{i_1} = 0.0196 \text{ k}\Omega = 19.6 \text{ }\Omega$$

$$z_{12} = \frac{v_2}{i_1} = -49 \text{ k}\Omega$$

Since **Y** and **Z** matrices are inverses of each other, the above calculations can be checked by showing that their matrix product gives a unit matrix.

4.2.4 The Pi, Tee, and Symmetric Lattice Networks

These three networks are very frequently encountered in many different situations and the pi-tee (T) or (delta-star) transformation was discussed in Chap. 3. The first two are three terminal networks and are shown in Fig. 4.6; note that there are no controlled sources. The two mesh equations for the T network of Fig. 4.6(a) are, by inspection,

$$v_1 = (R_1 + R_3)i_1 + R_3 i_2 \tag{4.7a}$$

$$v_2 = R_3 i_1 + (R_2 + R_3)i_2 \tag{4.7b}$$

which means that the z parameters for a T network are

$$z_{11} = R_1 + R_3 \qquad \text{(sum of mesh 1 elements)}$$

$$z_{12} = z_{21} = R_3 \qquad \text{(common element)}$$

$$z_{22} = R_2 + R_3 \qquad \text{(the sum of mesh 2 elements)}$$

and can thus be written by inspection. Conversely, given the z parameters of a three-terminal two-port with the special property $z_{12} = z_{21}$, the two-port has the T representation of Fig. 4.6(a).

The two node equations for the pi network of Fig. 4.6(b) are, by inspection,

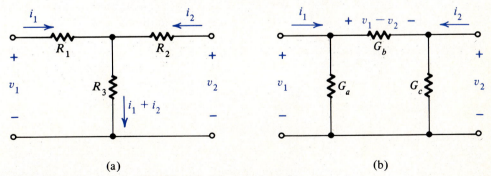

(a)   (b)

Figure 4.6 The (a) tee and (b) pi two-port networks.

$$i_1 = G_a v_1 + G_b(v_1 - v_2) \tag{4.8a}$$

$$i_2 = G_c v_2 + G_b(v_2 - v_1) \tag{4.8b}$$

or

$$i_1 = (G_a + G_b)v_1 - G_b v_2 \tag{4.9a}$$

$$i_2 = -G_b v_1 + (G_c + G_b)v_2 \tag{4.9b}$$

Thus the y parameters for a pi network are

$$y_{11} = G_a + G_b \qquad \text{(sum of conductances)}$$

connected to node 1,

$$y_{22} = G_c + G_b \qquad \text{(sum of conductances)}$$

connected to node 2, and

$$y_{12} = y_{21} = -G_b$$

the negative of the conductance connecting node 1 to node 2. These can, therefore, be written by inspection for a pi network. Conversely, given the y parameters of a three terminal two-port with the special property $y_{12} = y_{21}$, the two-port has the pi representation of Fig. 4.6(b).

In some applications it is sometimes necessary to replace a T network by a pi equivalent (or the converse). This equivalent transformation is called the T-pi (or star-delta) transformation and simply requires finding the y parameters given the z parameters, namely calculating the inverse of **Z**. The two transformations are summarized below.

TEE-PI TRANSFORMATION
Calculating

$$\mathbf{Y} = \mathbf{Z}^{-1} = \begin{bmatrix} z_{11} & z_{12} \\ z_{21} & z_{22} \end{bmatrix}^{-1}$$

the network element values for Fig. 4.6(b) are obtained in terms of the network element values of Fig. 4.6(a) expressed as conductances:

$$G_a = \frac{G_1 G_3}{G_1 + G_2 + G_3} \tag{4.10a}$$

$$G_b = \frac{G_1 G_2}{G_1 + G_2 + G_3} \tag{4.10b}$$

$$G_c = \frac{G_3 G_2}{G_1 + G_2 + G_3} \tag{4.10c}$$

a form convenient for calculations, and reminiscent of the product over sum formula for conductances in series.

PI-TEE TRANSFORMATION
Calculating

$$\mathbf{Z} = \mathbf{Y}^{-1} = \begin{bmatrix} y_{11} & y_{12} \\ y_{21} & y_{22} \end{bmatrix}^{-1}$$

the network element values for Fig. 4.6(a) are obtained in terms of the network element values of Fig. 4.6(b), expressed for convenience as resistances:

$$R_1 = \frac{R_a R_b}{R_a + R_b + R_c} \tag{4.11a}$$

$$R_2 = \frac{R_b R_c}{R_a + R_b + R_c} \tag{4.11b}$$

$$R_3 = \frac{R_a R_c}{R_a + R_b + R_c} \tag{4.11c}$$

EXAMPLE 4.4

Find the equivalent resistance seen at the terminals for the network of Fig. E4.4(a).

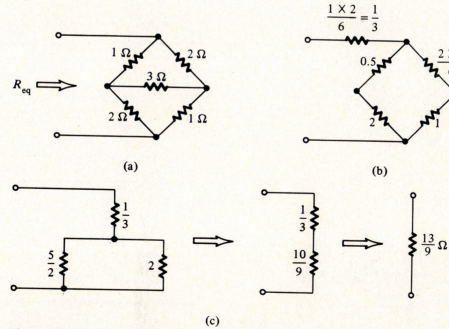

(a)

(b)

(c)

Figure E4.4

Solution

First use the Pi-Tee transformation, Eq. (4.11), to obtain Fig. E4.4(b). Then use the series and parallel rules to obtain the sequence of reductions shown in Fig. E4.4(c).

In Fig. 4.7(b) a symmetric lattice network is shown for which the z parameters are to be found. In Fig. 4.7(a) the same network is drawn as a bridge. The z parameters are obtained using the appropriate open-circuit definitions given in Sec. 4.2.3. Open circuiting port 2 ($i_2 = 0$), the circuit reduces to R_1 in series with R_2, shunted by R_1 in series with R_2. Hence

$$z_{11} = \frac{R_1 + R_2}{2} = z_{22}$$

by symmetry. By definition,

$$z_{21} = \frac{v_2}{i_1} = \frac{v_{21'} - v_{2'1'}}{i_1}$$

where the double subscript indicates the nodes between which the voltages are defined as shown in Fig. 4.7(a). The current i_1 devides equally into the two branches; hence

$$v_{21'} = \frac{R_2 i_1}{2}$$

$$v_{2'1'} = \frac{R_1 i_1}{2}$$

and

$$z_{21} = \frac{R_2 - R_1}{2} = z_{12}$$

by symmetry.

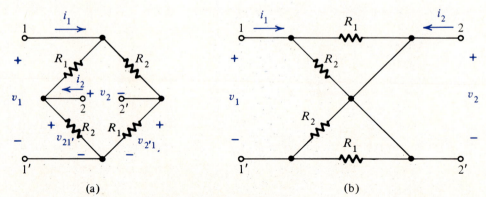

(a)   (b)

Figure 4.7 Bridge (a) or lattice (b) two-port.

EXAMPLE 4.5

Calculate the voltage v_{ab} in the network of Fig. E4.5 and hence find what combination of R_1, R_2, R_3, R_4 make it zero.

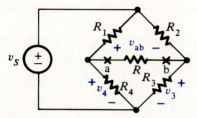

Figure E4.5 A bridge network.

Solution

Use Thevenin's theorem to find $v_{ab} = v_4 - v_3$ when R is removed (the open-circuit voltage), and R_{eq} (looking into terminals ab) when $v_s = 0$. Then calculate v_{ab} using the voltage divider rule.

By the voltage divider rule with R removed

$$v_{ab}(oc) = v_4 - v_3 = \left(\frac{R_4}{R_1 + R_4} - \frac{R_3}{R_2 + R_3} \right) v_s$$

Setting $v_s = 0$,

$$R_{eq} = \frac{R_1 R_4}{R_1 + R_4} + \frac{R_2 R_3}{R_2 + R_3}$$

Hence

$$v_{ab} = \frac{R v_{ab}(oc)}{R + R_{eq}}$$

$$= \frac{R(R_2 R_4 - R_1 R_3)v_s}{(R + R_{eq})(R_1 + R_4)(R_2 + R_3)}$$

This will be zero irrespective of R if

$$R_2 R_4 = R_1 R_3 \qquad \text{(balance equation)}$$

When this holds the bridge is said to be balanced. This type of network is basic in the precision measurement of circuit components; for example, R_3 can be measured accurately by balancing the bridge using high-precision variable resistances for R_1, R_2, and R_4 and using the balance equation to calculate R_3.

To conclude the development in Sec. 4.2 the student should work out the details for the two-port description when using the other three representations summarized in Table 4.1. In particular the student should note that calculating or measuring the transmission and reverse transmission parameters, Eqs. (4.5) and (4.6), requires shorting and opening the same port, and applying independent sources at the other.

DRILL EXERCISES

4.1 Draw the input and output characteristics for the y and z parameters.

4.2 A two-port has $z_{11} = 2\ \Omega$, $z_{22} = 3\ \Omega$, $z_{12} = z_{21} = 1\ \Omega$. Find the pi-equivalent network.
Ans. $2.5\ \Omega$, $5\ \Omega$, $5\ \Omega$

4.3 In the circuit of Fig. E4.5, if $R_1 = 1\ \text{k}\Omega$, $R_2 = 2\ \text{k}\Omega$, and $R_3 = 5\ \text{k}\Omega$, what value of R_4 will balance the bridge? What is the resulting value of R_{eq}?
Ans. $2.5\ \text{k}\Omega$, $2.14\ \text{k}\Omega$

4.2.5 Reciprocal Networks

Many networks have the property that the response produced at one point of the network due to an excitation at another point is invariant if the excitation and response are interchanged. For example in Fig. 4.8(a) the excitation is v_1 and the response is i_2 whereas in Fig. 4.8(b) the excitation is $\hat{v}_2$ and the response is $\hat{i}_1$. A reciprocal network will have the property that

$$\hat{i}_1 = i_2 \qquad \text{when } \hat{v}_2 = v_1$$

or, since only linear networks are involved, this may be stated as

$$\frac{i_2}{v_1} = \frac{\hat{i}_1}{\hat{v}_2} \tag{4.12}$$

For two-port networks the left-hand side of Eq. (4.12) is, by definition, the short-circuit admittance parameter y_{21} and the right-hand side is, by definition, y_{12}. Thus an equivalent statement of reciprocity for two-ports is

$$y_{12} = y_{21} \tag{4.13}$$

and, since the open-circuit impedance matrix is just the inverse of the short-circuit admittance matrix, it also follows that

$$z_{12} = z_{21} \tag{4.14}$$

for reciprocal networks. The generalization of these equalities for multiport networks is simply that the nodal admittance and the mesh impedance matrices are

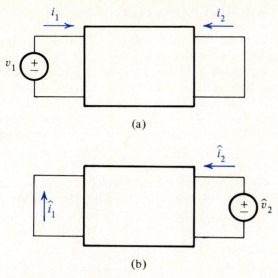

(a)

(b)

Figure 4.8 Reciprocity condition.

symmetric. Clearly from the discussion of T and pi networks it follows that three-terminal reciprocal networks have T and pi representations.

EXAMPLE 4.6

Express the h parameters of a network in terms of the z parameters and hence show that for a reciprocal two-port $h_{12} = -h_{21}$.

Solution
The z description is

$$v_1 = z_{11}i_1 + z_{12}i_2 \tag{a}$$

$$v_2 = z_{21}i_1 + z_{22}i_2 \tag{b}$$

where, by definition, $h_{21} = i_2/i_1$ when $v_2 = 0$. Hence, from Eq. (b),

$$h_{12} = \frac{-z_{21}}{z_{22}}$$

Again, by definition, $h_{12} = v_1/v_2$ when $i_1 = 0$. Hence, from Eqs. (a) and (b),

$$h_{12} = \frac{z_{12}i_2}{z_{22}i_2} = \frac{z_{12}}{z_{22}}$$

and, since for reciprocal networks $z_{12} = z_{21}$, it follows that $h_{12} = -h_{21}$.

4.3 INTERRELATIONS BETWEEN TWO-PORT PARAMETERS

As mentioned previously, the six formulations given in Table 4.1 are interrelated. In fact, given any one of the six formulations, the other five are obtained by algebraic manipulations. Tables of formulas, expressing each parameter set in terms of each of the other five, are found in many network theory textbooks. There is a practical need to be able to calculate one set from another because in practice, due to measuring equipment technology and experimental considerations, it is often easier and more accurate to measure one set of parameters, whereas a different set may be more convenient for design or analysis purposes. In this section some of the ways of converting from one parameter set to another are described. One relatively straightforward way is to rearrange the variables so as to express the dependent variables for the desired parameter set in terms of the given set by carrying out the required matrix algebra. For example, to find the h set given the y set as

$$i_1 = y_{11}v_1 + y_{12}v_2 \tag{4.15a}$$

$$i_2 = y_{21}v_1 + y_{22}v_2 \tag{4.15b}$$

the first step is to reorder the terms so that v_1, i_2 are on the left side, i.e.,

$$-y_{11}v_1 = -i_1 + y_{12}v_2 \tag{4.16a}$$

$$-y_{21}v_1 + i_2 = y_{22}v_2 \tag{4.16b}$$

Now rewrite (4.16) in the matrix form

$$\begin{bmatrix} -y_{11} & 0 \\ -y_{21} & 1 \end{bmatrix} \begin{bmatrix} v_1 \\ i_2 \end{bmatrix} = \begin{bmatrix} -1 & y_{12} \\ 0 & y_{22} \end{bmatrix} \begin{bmatrix} i_1 \\ v_2 \end{bmatrix} \tag{4.17}$$

which corresponds to the h-parameter choice of dependent/independent variables. Since 2×2 matrices are involved, the required matrix inverse is

$$\begin{bmatrix} -y_{11} & 0 \\ -y_{21} & 1 \end{bmatrix}^{-1} = \begin{bmatrix} \dfrac{-1}{y_{11}} & 0 \\ \dfrac{-y_{21}}{y_{11}} & 1 \end{bmatrix} \tag{4.18}$$

(this inverse would not exist if $y_{11} = 0$, in which case there would be no h-parameter description). Thus (4.17) becomes

$$\begin{bmatrix} v_1 \\ i_2 \end{bmatrix} = \begin{bmatrix} \dfrac{-1}{y_{11}} & 0 \\ \dfrac{-y_{21}}{y_{11}} & 1 \end{bmatrix} \begin{bmatrix} -1 & y_{12} \\ 0 & y_{22} \end{bmatrix} \begin{bmatrix} i_1 \\ v_2 \end{bmatrix}$$

from which one obtains the desired h-parameter matrix

$$[h] = \begin{bmatrix} \dfrac{1}{y_{11}} & \dfrac{-y_{12}}{y_{11}} \\ \dfrac{y_{21}}{y_{11}} & y_{22} - \dfrac{y_{12}y_{21}}{y_{11}} \end{bmatrix}$$

(4.19)

Another way is to use the measurement definitions and Eqs. (4.15a) and (4.15b) directly.

EXAMPLE 4.7

A two-port network has the following y parameters, all in siemens; $y_{11} = 4$, $y_{21} = 1$, $y_{12} = 0.5$, $y_{22} = 0.25$. Draw the controlled-source representation for the network and also obtain its h-parameter representation.

Solution
The network is obtained by inspection and is shown in Fig. E4.7.

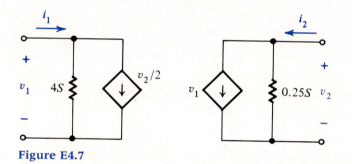

Figure E4.7

From the h-parameter definition, $v_2 = 0$ and $i_1 = 0$ are the appropriate conditions for measuring the parameters. Directly from the diagram then, with port 2 shorted, $v_2 = 0$ and

$$i_1 = 4v_1$$

$$i_2 = v_1 = \frac{i_1}{4}$$

therefore,

$$h_{11} = \frac{v_1}{i_1} = 0.25 \ \Omega$$

$$h_{21} = \frac{i_2}{i_1} = 0.25$$

when $i_1 = 0$,

$$v_1 = \frac{-v_2}{8} = -0.125v_2$$

$$i_2 = v_1 + 0.25v_2 = 0.125v_2$$

therefore,

$$h_{12} = \frac{v_1}{v_2} = -0.125$$

$$h_{22} = 0.125 \text{ S}$$

DRILL EXERCISES

4.4 Find the z parameters for the network in Example 4.7 using the conceptual measuring method.
Ans. $z_{11} = 0.5\ \Omega$, $z_{12} = -1\ \Omega$, $z_{21} = -2\ \Omega$, $z_{22} = 8\ \Omega$

4.5 A two-port network has all its four y parameters equal to 1 S. Obtain the z and h parameters for the two-port.
Ans. z description does not exist; $h_{11} = 1\ \Omega$, $h_{12} = -1$, $h_{21} = 1$, $h_{22} = 0$.

4.4   SPECIAL TWO-PORT NETWORK BUILDING BLOCKS

4.4.1 Ideal Amplifiers

In the preceding sections it was convenient to introduce four types of controlled sources, namely,

CCCS: Current controlled current source

VCVS: Voltage controlled voltage source

CCVS: Current controlled voltage source

VCCS: Voltage controlled current source

These four controlled sources form the basis of two-port ideal amplifier networks with the circuit models shown in Fig. 4.9.

These four ideal amplifier circuits are considered basic two-ports and are important in the development of small-signal models for active circuits containing transistors and other semiconductor devices. They are considered ideal in two ways: first, none of these four circuits require any power input to control port 1 since either v_1 or i_1 are zero; second, their controlled output variable value is independent of the branch element connected to port 2. Their output graphical characteristics are shown in Fig. 4.10.

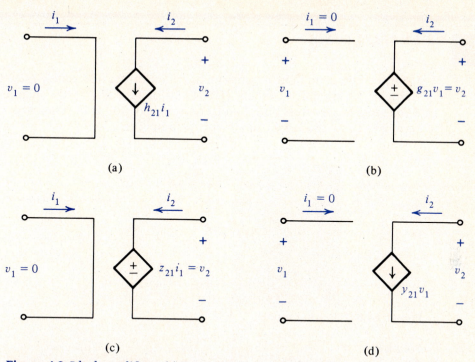

(a)

(b)

(c)

(d)

Figure 4.9 Ideal amplifiers. (a) Transcurrent amplifier. (b) Transvoltage amplifier. (c) Transimpedance amplifier. (d) Transadmittance amplifier.

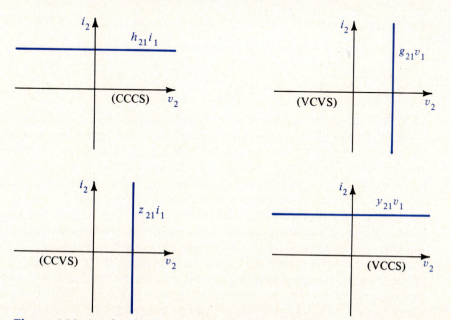

Figure 4.10 Graphical characteristics of ideal amplifiers.

EXAMPLE 4.8

Under certain conditions a transistor amplifier can be adequately modeled by the transcurrent amplifier element [Fig. 4.8(a)] having $h_{21} = 100$, augmented by a 25-kΩ resistance across port 2 and a resistance of 500 Ω in series with one port 1 terminal. Draw the circuit model for the amplifier.

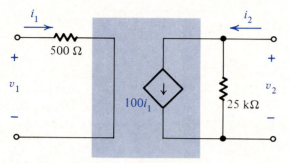

Figure E4.8 Transistor model.

Solution

The desired circuit is shown in Fig. E4.8.

4.4.2 The Ideal Transformer

This is a two-port network having the following defining equations:

$$v_1 = nv_2 \tag{4.20a}$$

$$i_2 = -ni_1 \tag{4.20b}$$

where n is the *transformation ratio*. Note that the two-port is defined uniquely by a single parameter and has the controlled-source representation shown in Fig. 4.11(a). In Fig. 4.11(b) the symbol which is normally used to represent this two-port is shown; the dots indicate the polarity of v_2 with respect to v_1. When *both* dots are at the plus (minus) terminals then v_2 and v_1 have the same polarity. Otherwise they are of opposite polarity and $v_1 = -nv_2$ and $i_2 = ni_1$ are written instead of Eqs. (4.20a) and (4.20b). If no dots are shown it is understood that v_1 and v_2 have the same polarity and Eqs. (4.20a) and (4.20b) are used. It is unfortunate that the symbol for the ideal transformer contains two inductancelike symbols since this sometimes gives rise to some uncertainty in the early stages of studying electric circuits. By always writing "ideal" next to the symbol it is hoped that no confusion will arise. The ideal transformer is an element which does not absorb power since $v_1i_1 + v_2i_2 = 0$. All the power entering port 1 leaves at port 2. It clearly has a hybrid description as is evident from the definition with $h_{12} = n$ and $h_{21} = -n$ as the only nonzero h parameters.

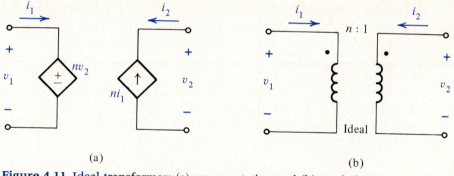

(a)   (b)

Figure 4.11 Ideal transformer: (a) representation and (b) symbol.

An equally valid controlled-source representation for Eq. (4.20) consists of a CCCS for port 1 as $i_1 = -i_2/n$, and a VCVS for port 2 as $v_2 = v_1/n$; this form is used in Fig. E4.9 of Example 4.9.

EXAMPLE 4.9

An ideal transformer defined by Eq. (4.20) has a capacitance **C** connected across the port 2 terminals. Find the relationship between i_1 and v_1.

Solution

From the circuit diagram shown, using the alternate model for the ideal transformer, the port 2 variables are constrained by the branch relation for the connected element, namely (noting signs),

$$i_2 = -C\frac{dv_2}{dt}$$

Since, by definition, $v_2 = v_1/n$ and $i_1 = -i_2/n$, i_2 and v_2 are eliminated to yield

$$i_1 = \frac{C}{n^2}\frac{dv_1}{dt}$$

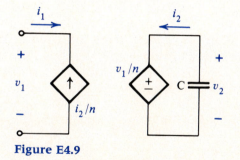

Figure E4.9

This equation implies that, when port 2 is terminated by C, the driving point characteristic at port 1 is that of a branch (one-port) which is equivalent, in the sense used in Chap. 3, to a capacitance of value $C_{eq} = C/n^2$.

Similar calculations made for a resistance termination R and an inductance termination L at port 2 yield $R_{eq} = n^2 R$ and $L_{eq} = n^2 L$, respectively. The ideal transformer thus transforms resistances and inductances as n^2 and conductances and capacitances as $1/n^2$.

4.4.3 The Mutual-Inductance Element

The *mutual-inductance element* is a two-port containing voltage sources controlled by the *derivatives* of the currents. The defining equations for the mutual-inductance element are

$$v_1 = L_1 \frac{di_1}{dt} + M \frac{di_2}{dt} \tag{4.21a}$$

$$v_2 = M \frac{di_1}{dt} + L_2 \frac{di_2}{dt} \tag{4.21b}$$

where L_1, L_2, and M are positive numbers having dimensions of inductance. A symbol and a controlled-source representation for this two-port are shown in Fig. 4.12. The parameters L_1, L_2 are called *self-inductances*, and M is called the *mutual inductance*. The dots shown in Fig. 4.12(a) specify the polarity of the controlled source just as in the ideal transformer. When *both* dots are either at the positive or the negative terminals, $+M$ is used as in Eqs. (4.21a) and (4.21b), otherwise $-M$ is used. If the dots are not shown, $+M$ is assumed.

EXAMPLE 4.10

A mutual inductance defined by Eqs. (4.21a) and (4.21b) is open-circuited at port 2. Find $v_2(t)$ and $v_1(t)$ given $i_1(t) = I_s \cos pt$.

Solution
With port 2 open-circuited, $i_2 = 0$ and $di_2/dt = 0$. From the given i_1, $di_1/dt = -pI_s \sin pt$, hence from Eqs. (4.21a) and (4.21b)

$$v_1(t) = -L_1 pI_s \sin pt$$

and

$$v_2(t) = -MpI_s \sin pt$$

respectively.

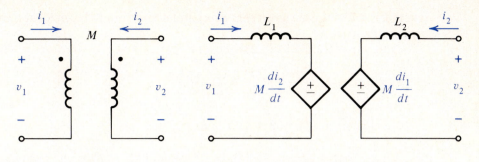

Figure 4.12 The mutual-inductance element: (a) symbol, (b) controlled-source representation.

EXAMPLE 4.11

A mutual inductance is open-circuited at port 2 and driven by a voltage source $v_1(t) = V_1 \cos pt$. Find $v_2(t)$.

Solution
By direct substitution in Eq. (4.21) the condition $i_2 = 0$ means that i_1 must satisfy

$$V_1 \cos pt = L_1 \frac{di_1}{dt}$$

and v_2 must satisfy

$$v_2(t) = M \frac{di_1}{dt}$$

hence

$$v_2(t) = M \frac{V_1}{L_1} \cos pt$$

Because the mutual-inductance element is needed in modeling *practical* transformers it is useful to obtain a network model for it which contains an *ideal transformer*. In order to do this the following two new dimensionless parameters are defined:

$$n = \left(\frac{L_1}{L_2}\right)^{1/2} \tag{4.22}$$

and

$$k = \frac{M}{(L_1 L_2)^{1/2}} \tag{4.23}$$

where k is called the *coefficient of coupling* and n turns out to be an ideal transformer turns ratio.

Using Eqs. (4.22) and (4.23), Eqs. (4.21a) and (4.21b) are now rewritten as

$$v_1 = L_1(1 - k)\frac{di_1}{dt} + \left(M\frac{di_2}{dt} + kL_1\frac{di_1}{dt}\right) \tag{4.24a}$$

$$v_2 = \left(M\frac{di_1}{dt} + kL_2\frac{di_2}{dt}\right) + L_2(1 - k)\frac{di_2}{dt} \tag{4.24b}$$

Since, from Eqs. (4.22) and (4.24), $kL_1 = nM$ and $kL_2 = M/n$, the term in the large parenthesis in the first of the preceding two equations is just n times the term in the large parenthesis in Eq. (4.24b). Consequently a circuit model corresponding to these equations is as shown in Fig. 4.13, where the primed two-port is described by

$$v_1' = M\frac{di_2}{dt} + nM\frac{di_1}{dt} \tag{4.25a}$$

$$v_2' = \frac{M}{n}\frac{di_2}{dt} + M\frac{di_1}{dt} \tag{4.25b}$$

whence

$$v_1' = nv_2' \tag{4.26}$$

From Eq. (4.25a), with nM replaced by kL_1, i_1 is given by

$$i_1 = \frac{1}{kL_1}\int_{-\infty}^{t} v_1' \, d\tau - \frac{i_2}{n} \tag{4.27}$$

Thus the primed two-port of Fig. 4.13 has a network model which consists of an inductance kL_1 and an ideal transformer of turns ratio n as shown in Fig. 4.14. The ideal transformer equation, Eq. (4.26), is seen to be satisfied by the circuit in Fig. 4.14 and the KCL Eq. (4.27) is satisfied at the node labeled A.

It follows that a model for the mutual-inductance element containing an ideal transformer is obtained by combining the models in Figs. 4.13 and 4.14 as shown in Fig. 4.15.

From energy considerations, it can be shown that this two-port element can only store energy; this condition requires that $k \leq 1$.

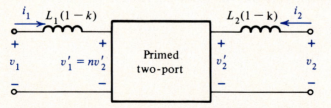

Figure 4.13 Network model for Eqs. (4.24a) and (4.24b).

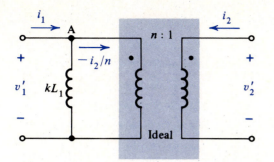

Figure 4.14 Model for primed two-port in Fig. 4.13.

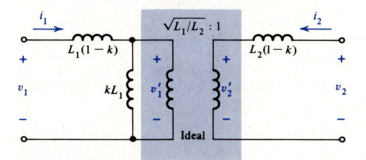

Figure 4.15 A model for the mutual-inductance element based on the ideal transformer.

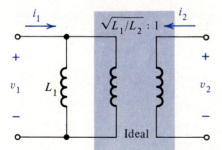

Figure 4.16 Model for unity-coupled mutual-inductance element.

The case $k = 1$ is called the unity-coupled mutual-inductance element for which the model of Fig. 4.15 reduces to that shown in Fig. 4.16. It turns out that this is a reasonably good model for many practical transformers.

4.4.4 The Gyrator

The gyrator is defined by

$$v_1 = -ri_2 \tag{4.28a}$$

$$v_2 = ri_1 \tag{4.28b}$$

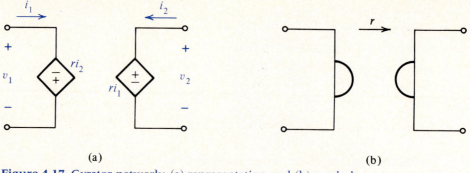

(a)   (b)

Figure 4.17 Gyrator network: (a) representation and (b) symbol.

where r is called the *gyration resistance,* and has the controlled-source representation shown in Fig. 4.17(a). Figure 4.17(b) shows the symbol normally used. The gyrator also does not absorb power since, as in the ideal transformer, $v_1 i_1 + v_2 i_2 = 0$. It has an impedance description [see Table 4.1, Eq. (4.3)] with $z_{12} = -r$ and $z_{21} = r$ as the only nonzero elements.

A gyrator can be viewed as two CCVS ideal amplifiers connected in parallel and back to back with opposite polarities. Obviously, since Eq. (4.28) can also be written as $i_2 = -gv_1$ and $i_1 = gv_2$, there is another way of viewing a gyrator in terms of two voltage-controlled current sources.

EXAMPLE 4.12

Find the i_1, v_1 relationship for the gyrator circuit of Fig. 4.17 when port 2 is terminated by an inductance L.

Solution
The terminal constraint at port 2 is

$$v_2 = -L \frac{di_2}{dt}$$

and, by eliminating v_2 and i_2 using Eq. (4.28), one obtains

$$i_1 = \frac{L}{r^2} \frac{dv_1}{dt}$$

This is the equation for a capacitance having an equivalent value

$$C_{eq} = \frac{L}{r^2}$$

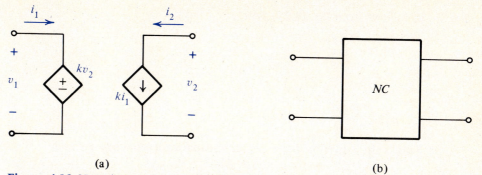

(a)   (b)

Figure 4.18 Negative converter: (a) representation and (b) symbol.

Similar calculations for a resistance termination R and a capacitance termination C yield $R_{eq} = r^2/R$ and $L_{eq} = r^2 C$, respectively. The gyrator effectively interchanges the roles of currents and voltages. This can be seen by showing from Eqs. (4.28) that

$$\frac{v_1}{i_1} = -r^2 \frac{i_2}{v_2}$$

which demonstrates that the port 1 voltage to current ratio is proportional to the port 2 current to voltage ratio.

4.4.5 The Negative Converter (NC)

This two-port is defined by the equations

$$v_1 = kv_2 \tag{4.29a}$$

$$i_2 = ki_1 \tag{4.29b}$$

containing the single parameter k which may be either positive or negative. The NC has the h-parameter description $h_{11} = h_{22} = 0$ and $h_{12} = h_{21} = k$, and the controlled-source representation shown in Fig. 4.18(a). A general symbol is shown in Fig. 4.18(b).

EXAMPLE 4.13

Find the i_1, v_1 relationship for an NC terminated at port 2 by a capacitance C.

Solution
Since

$$i_2 = -C \frac{dv_2}{dt}$$

it follows from Eq. (4.29) that

$$ki_1 = -\frac{C}{k}\frac{dv_1}{dt}$$

or

$$i_1 = \frac{-C}{k^2}\frac{dv_1}{dt}$$

This is the equation for a negative capacitance having a value

$$C_{eq} = \frac{-C}{k^2}$$

Thus at the input terminals the equivalent capacitance is proportional to the "negative" of the terminating capacitance, hence the name NC. Similar calculations for resistive and inductive terminations R and L yield $R_{eq} = -k^2 R$ and $L_{eq} = -k^2 L$, respectively. The NC two-port makes possible the inclusion of negative R, L, and C as network elements.

4.4.6 The Ideal Operational Amplifier

The ideal transvoltage amplifier, described in Sec. 4.4.1, is a four-terminal network which can be viewed as an ideal *differential* amplifier. This point is made clear in Fig. 4.19 which shows the output voltage as the amplified replica of v_I, the difference between v_2 and v_1, with A taken as positive. When $v_2 = 0$ the output $v_O = -Av_1$ has a polarity opposite to v_1, whereas when $v_1 = 0$ the output $v_O = Av_2$ has the same polarity as v_2. All the voltages are referred to a datum node and the circuit shown is a three-port network with two input ports (1 and 2) and one output port (3). As the voltage gain constant A tends to infinity, this circuit becomes an idealization of the *operational amplifier* or op amp, a basic building

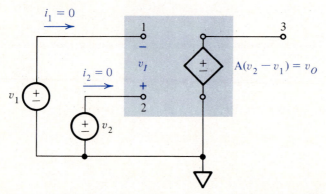

Figure 4.19 Transvoltage element as a differential amplifier.

block in the design of analog electronic circuits, to be discussed in a subsequent chapter.

THE INVERTING AMPLIFIER

The ideal op amp connected as an inverting amplifier ($v_2 = 0$ and $v_I = -v_1$) with *feedback* is shown in Fig. 4.20. The word *feedback* underlines the fact that the element R_f is connected from the output terminal 3 back to the input terminal 1. The amplifying properties of this circuit as $A \to \infty$ are obtained as follows from Fig. 4.20 by KVL

$$v_O = v_1 - R_f i_s \tag{4.30}$$

and, since $v_1 = -v_O/A$, Eq. (4.30) becomes

$$v_O = \frac{-v_O}{A} - R_f i_s$$

or

$$v_O = -R_f i_s \frac{A}{1 + A} \tag{4.31}$$

As $A \to \infty$,

$$v_O \to -R_f i_s \tag{4.32}$$

which, from Eq. (4.30), means that v_1 becomes negligibly small compared to v_O. This condition, that v_1, and hence v_I, approaches zero as A approaches infinity, is referred to as a *virtual short*. Setting $v_1 = 0$ it follows that

$$v_O = -R_f i_s \tag{4.33}$$

and

$$v_S = R i_s \tag{4.34}$$

so that the amplifier input resistance $v_S/i_s = R$. The voltage gain is obtained by

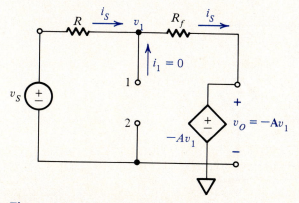

Figure 4.20 Inverting amplifier network.

dividing Eq. (4.33) by Eq. (4.34) to yield

$$A_f = \frac{v_O}{v_S} = \frac{-R_f}{R} \tag{4.35}$$

a value independent of the op amp itself (provided $A \rightarrow \infty$) but completely determined by the external elements R and R_f. Therefore, the gain with feedback A_f is known to an accuracy determined by the tolerance on the external elements R and R_f.

THE NONINVERTING AMPLIFIER

The ideal op amp used as a noninverting amplifier is shown in Fig. 4.21. By inspection of this diagram, since $v_I = 0$ (virtual short), $v_1 = v_2$; since $i_1 = 0$, the current through R is the same as the current through R_f and is given by v_2/R. Therefore,

$$v_O = \frac{(R_f + R)v_2}{R} \tag{4.36}$$

and the voltage gain

$$\frac{v_O}{v_2} = \frac{R_f + R}{R} \tag{4.37}$$

a noninverting gain which is larger in absolute value than that of Eq. (4.35) by the factor of $(R + R_f)/R_f$.

THE DIFFERENTIAL AMPLIFIER

A differential input single-ended output amplifier using an ideal op amp is shown in Fig. 4.22. To get the same absolute value of noninverting gain requires inserting a voltage divider between the source v_B and terminal 2 of the op amp. Such a voltage divider is incorporated in the differential amplifier of Fig. 4.22. From this figure the following relations are obtained by using superposition and setting

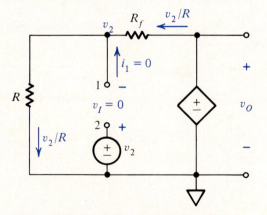

Figure 4.21 Noninverting ideal op amp.

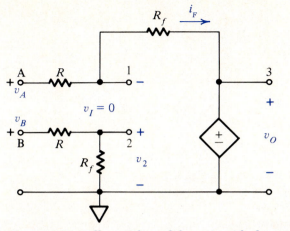

Figure 4.22 Differential amplifier using ideal op amp.

$v_I = 0$ (virtual short property of ideal op amp with feedback). With $v_B = 0$ this is just an inverting amplifier hence Eq. (4.35) holds and

$$v_O = \frac{-R_f}{R} v_A \qquad (4.38)$$

With $v_A = 0$ this is just a noninverting amplifier hence Eq. (4.37) holds and

$$v_O = \frac{(R_f + R)v_2}{R} \qquad (4.39)$$

However, because of the voltage divider inserted between terminal 2 and the source v_B,

$$v_2 = \frac{R_f v_B}{R_f + R} \qquad (4.40)$$

and, therefore,

$$v_O = \frac{R_f}{R} v_B \qquad (4.41)$$

By superposition, when both v_A and v_B are present,

$$v_O = \frac{R_f}{R}(v_B - v_A) \qquad (4.42)$$

a true difference output. From Fig. 4.22 it is also evident that the differential R_{eq} looking into terminals AB is $2R$ since $v_I = 0$.

THE TIME INTEGRATOR
Another application of the op amp, namely, a circuit which performs time integration, is achieved by putting a capacitance as the feedback element in place of R in

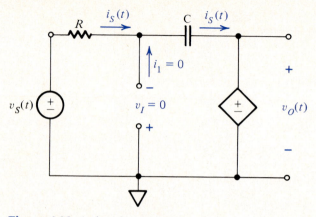

Figure 4.23 A time integrator.

Fig. 4.20. This circuit is shown in Fig. 4.23 where $v_I = 0$, $i_1 = 0$ is the virtual short representation for the op amp input circuit. The voltages are now arbitrary time functions. By KVL

$$v_O(t) = -\frac{1}{C} \int_{-\infty}^{t} i_S(\tau)\, d\tau \qquad (4.43)$$

and since, from Eq. (4.34),

$$i_S(t) = \frac{v_S(t)}{R}$$

one obtains

$$v_O(t) = -\frac{1}{RC} \int_{-\infty}^{t} v_S(\tau)\, d\tau \qquad (4.44)$$

namely, the output voltage is proportional to the time integral of the input voltage. This integrator is a basic element in analog computing as well as in many measurement and singnal processing systems.

DRILL EXERCISES

4.6 Show that at port 1 of a transformer, terminated by an inductance L at port 2, the branch law is equivalent to an inductance of value $n^2 L$.

4.7 Prove that an ideal transformer does not absorb power.

4.8 Show that for a unity-coupled mutual-inductance element that

$$v_2(t) = \left(\frac{L_2}{L_1}\right)^{1/2} v_1(t)$$

4.9   Show that if $v_2 = 0$ in Fig. 4.15 that

$$v_1 = L_1(1 - k^2) \frac{di_1}{dt}$$

4.10 Show that at port 1 of a gyrator, terminated by a capacitance C at port 2, the branch law is equivalent to an inductance of value $r^2 C$.

4.11 Draw the input and output characteristics for the negative converter.

4.12 Show that a time differentiator is obtained by replacing C by L in Fig. 4.23.

4.13 If in Fig. 4.23, the position of R and C are interchanged find $v_o(t)$.

Ans. $v_o = -RC \dfrac{dv_S}{dt}$

4.5   TERMINATED TWO-PORTS

Two-port networks such as filters and amplifiers are invariably terminated, in the sense that one of the ports is driven by a source and the other is terminated by a resistance load. The typical configuration is illustrated in Fig. 4.24. The terminations shown in the figure specify or impose two linear constraints, one at each port. These are KVL

$$v_1 = v_S - i_1 R_S \tag{4.45}$$

at port 1 and the branch relation

$$v_2 = -i_2 R_L \tag{4.46}$$

at port 2. These constraints are imposed by the branches which terminate the two-port, and will be true irrespective of whether the two-port is linear or nonlinear. The two-port is described in terms of its z parameters and the effect of the terminations on the following circuit properties examined:

$$R_{\text{in}} = \frac{v_1}{i_1} \qquad \text{(input resistance)}$$

$$A_v = \frac{v_2}{v_1} \qquad \text{(forward voltage gain)}$$

$$A_i = \frac{i_L}{i_1} = \frac{-i_2}{i_1} \qquad \text{(forward current gain)}$$

$$A_p = \frac{v_2 i_L}{v_1 i_1} = \frac{-v_2 i_2}{v_1 i_1} \qquad \text{(power gain)}$$

In addition to these properties, Thevenin (Norton) equivalent networks as seen at port 2 are calculated. The Thevenin resistance is called the *output resistance R_o*.

Figure 4.24 A terminated two-port.

Although most of these quantities can be obtained using algebraic methods, it is often more efficient to find the answers using circuit transformation methods. The first step when using circuit methods is to draw the appropriate circuit diagram which, for the z-parameter description, is shown in Fig. 4.25. By inspection of the right-hand portion of the diagram it is clear that

$$i_2 = \frac{-z_{21}i_1}{z_{22} + R_L} \tag{4.47}$$

therefore the forward current gain is

$$A_i = \frac{-i_2}{i_1} = \frac{z_{21}}{z_{22} + R_L} \tag{4.48}$$

The network, redrawn as in Fig. 4.26, using $i_2 = -A_i i_1$, has i_1 as the only control variable. The input resistance at port 1 and the forward voltage ratio are now obvious from Fig. 4.26. The input resistance is

$$R_{in} = \frac{v_1}{i_1}$$

$$= \frac{z_{11}i_1 - z_{12}A_i i_1}{i_1}$$

$$= z_{11} - A_i z_{12}$$

$$= z_{11} - \frac{z_{12}z_{21}}{z_{22} + R_L} \tag{4.49}$$

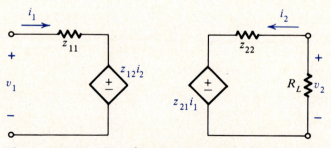

Figure 4.25 Terminated two-port, z parameters.

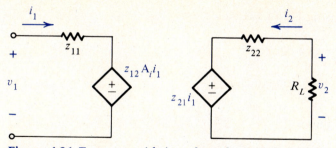

Figure 4.26 Two-port with i_1 as the only control variable.

The forward voltage ratio A_v is found by writing

$$v_2 = -i_2 R_L = A_i R_L i_1 \tag{4.50}$$

and dividing Eq. (4.50) by $v_1 = R_{in} i_1$, as given by Eq. (4.49), to obtain

$$A_v = \frac{v_2}{v_1} = \frac{A_i R_L}{R_{in}} \tag{4.51}$$

This is a very frequently encountered relation which, in the form of Eq. (4.51), is not dependent on the parameter description but only on a knowledge of the current gain A_i and the input resistance R_{in}; the power gain A_p follows from the previous two results as

$$A_p = A_i A_v = \frac{A_i^2 R_L}{R_{in}} \tag{4.52}$$

The Thevenin equivalent network seen at port 2, when port 1 is driven as shown in Fig. 4.24, is most easily obtained by drawing the network as shown in Fig. 4.27. From Fig. 4.27, since i_2 is known,

$$i_1 = \frac{v_S - z_{12} i_2}{R_S + z_{11}} \tag{4.53}$$

and KVL for the right side is just

$$v_2 = z_{21} i_1 + z_{22} i_2 \tag{4.54}$$

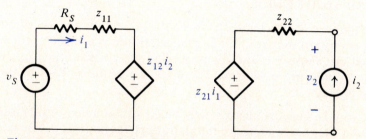

Figure 4.27 Network for Thevenin calculation.

Eliminating i_1 using Eq. (4.53) yields

$$v_2 = \frac{z_{21} v_s}{R_S + z_{11}} + \left(z_{22} - \frac{z_{21} z_{12}}{R_S + z_{11}} \right) i_2 \qquad (4.55)$$

Hence the Thevenin voltage is

$$v_T = \frac{z_{21} v_S}{z_{11} + R_S}$$

and the Thevenin output resistance is

$$R_{\text{out}} = z_{22} - \frac{z_{12} z_{21}}{R_S + z_{11}} \qquad (4.56)$$

There are many two-port networks of practical importance with the property that the 12 parameter is negligible or zero. When this condition applies port 1 has no controlled source term and reduces to a purely resistive branch; the two-port is called *unilateral* since there is no transmission possible from port 2 to port 1.

The input v_1-i_1 characteristic simplifies to only one line going through the origin with slope R_{in}. In terms of the z parameters a terminated two-port with $z_{12} = 0$ has $R_{\text{in}} = z_{11}$, independent of R_L, and has $R_{\text{out}} = z_{22}$, independent of R_S.

DRILL EXERCISES

4.14 Calculate R_{in}, A_v, and R_{out} for a two-port having $y_{11} = 10^{-3}$ S, $y_{12} = 0$, $y_{21} = 0.02$ S, $y_{22} = 10^{-4}$ S when terminated by a 10-kΩ load and driven by a source having $R_S = 50\ \Omega$.
Ans. 1 kΩ, -100, 10 kΩ.

4.15 Show that if $h_{12} = 0$ then z_{12}, y_{12}, g_{12} are also all zero.

4.16 Show that the Thevenin equivalent resistance R_{out} for a two-port driven by a current source is z_{22}.

4.6 SUMMARY AND STUDY GUIDE

The main purpose of this chapter has been the presentation of some basic concepts relating to two-port networks. Their characterizations and their properties when terminated are discussed; and many special two-ports, needed in the study of electronic circuits, are examined. The attention of the student is drawn to the following points.

The two-port network characterization is a consequence of superposition, and two-port parameters are found by applying sources to one port with the other open or shorted.

The graphical representation of two-ports consists of two families of equi-spaced parallel lines.

The various parameters are related; the two ways to convert from one set to another are (a) variable manipulations and (b) applying defining conceptual measurements.

Tee-pi transformations may be written by inspection.

Three terminal two-ports containing no sources with $z_{12} = z_{21}(y_{12} = y_{21})$ have a T(pi) network representation.

The virtual short concept for an ideal op amp with feedback is the condition that for finite output (v_o) the input (v_I) tends to zero as A tends to infinity; this is *not* a short since the current is also zero.

The input resistance of a terminated two-port depends on the load unless the 12 element is zero; the output resistance depends on the source resistance at the input port unless the 12 element is zero.

To calculate Thevenin or Norton equivalents at port 2 easily, drive port 2 with a voltage source (current source) if the input port has a voltage-controlled (current-controlled) source.

PROBLEMS

4.1 Obtain the z parameters for the two-port shown. Calculate v_1/i_1 when $v_2 = -i_2$.

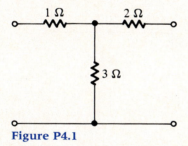

Figure P4.1

4.2 Obtain the y parameters for the linear two-port model of the bipolar transistor shown (the g's are all in siemens).

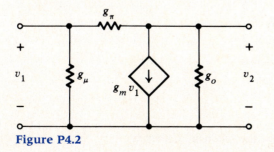

Figure P4.2

4.3 Obtain the h parameters for the network of Prob. 4.1 and hence draw a network representation for it corresponding to Fig. 4.3.

4.4 Given $y_{11} = y_{12} = y_{21} = y_{22} = 1$ S find the h parameters and draw a network representation corresponding to Fig. 4.3.

4.5 A two-port network has $z_{11} = 1$, $z_{12} = 2$, $z_{21} = 3$, and $z_{22} = 4$, all in ohms. Calculate the h parameters for the network.

4.6 Sketch the input and output characteristics for the circuit of Fig. P4.1 using the z description.

4.7 Repeat Prob. 4.6 for the circuit of Fig. P4.2 using the y description assuming $g_\mu = 0$.

4.8 Find the equivalent driving point resistance at ab for the following networks making use of T-pi transformations where appropriate.

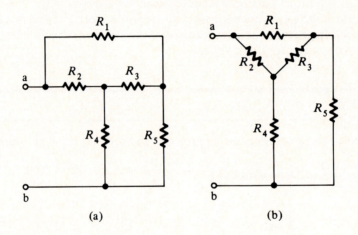

(a)   (b)

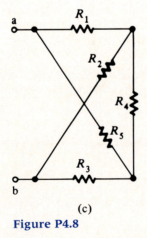

(c)

Figure P4.8

4.9 Repeat Prob. 4.8 for the following networks:

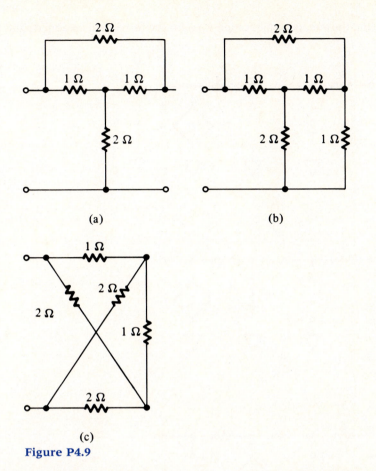

(a)

(b)

(c)

Figure P4.9

4.10 An ideal transformer is terminated at port 2 by a 50-Ω resistance and driven at port 1 by a source having a Thevenin resistance of 600 Ω and a Thevenin voltage of 10 V. If the transformation ratio $n = 4$, what is R_{eq} looking into port 1? How much power is delivered to the load? How much power is supplied by the source?

4.11 Show that the energy stored in a mutual-inductance element can be expressed as

$$E(t) = \frac{1}{2}\left[\left(\sqrt{L_1}\,i_1 \pm \frac{Mi_2}{\sqrt{L_1}}\right)^2 + \left(L_2 - \frac{M^2}{L_1}\right)i_2^2\right]$$

Verify that $E(t)$ is always nonnegative for $M^2 \leq L_1 L_2$.

4.12 A mutual-inductance element has $L_1 = 2$ mH, $L_2 = 8$ mH, and $M = 2$ mH. Find k and, given $i_1 = I_1 \sin pt$, find $v_2(t)$ given $i_2(t) = 0$.

4.13 A unity-coupled mutual-inductance element has $i_1 = i_2 = 1$ A. Calculate the energy stored in the element if the self-inductance values are those given in Prob. 4.12.

4.14 An op amp deviates from the ideal by having nonzero R_{out} and finite R_{in}, A_v. Calculate how each of these deviations separately effects the inverting amplifier voltage gain.

4.15 Find the voltage amplification v_O/v_I for the three ideal op amp circuits in Fig. P4.15.

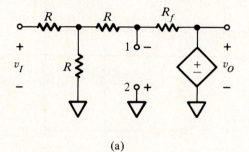

(a)

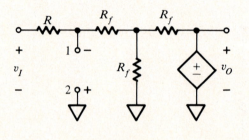

(b)

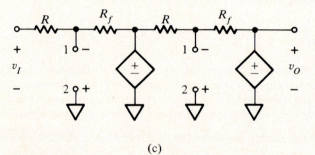

(c)

Figure P4.15

4.16 Choose values of R and R_f in the ideal op amp circuit in Fig. P4.16 such that $v_O = v_B - v_A/2$.

4.17 Two gyrators are connected as shown Fig. P4.17. Obtain an equivalent overall two-port description for the two ports (v_1, i_1) and (v_2, i_2).

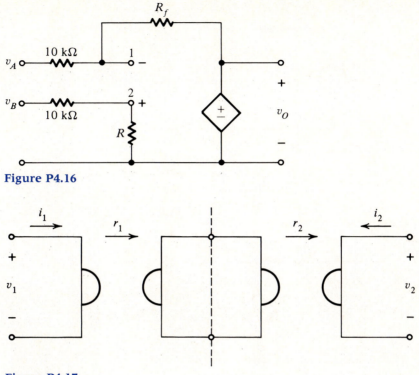

Figure P4.16

Figure P4.17

4.18 For the circuits shown calculate the current amplification i_L/i_1, the driving point resistance v_1/i_1, the voltage amplification, and obtain the Thevenin representation for the circuit to the left of X————X.

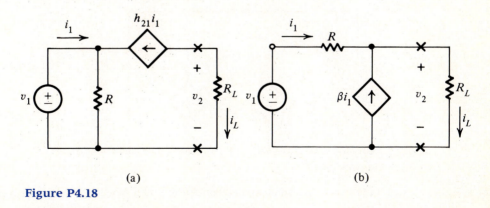

(a)   (b)

Figure P4.18

4.19 Repeat Prob. 4.18 for the circuit of Fig. P4.19.

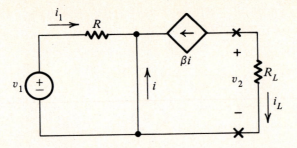

Figure P4.19

4.20 Find v_1/i_1, i_2/i_1, v_2/v_1 for terminated two-ports that have the 12 element zero in terms of the h and y parameters.

4.21 Calculate the y parameters for the circuit in Fig. P4.21.

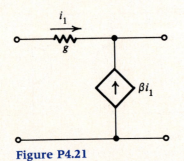

Figure P4.21

4.22 For the circuit of Fig. P4.22 calculate the h parameters, the input resistance at port 1 when port 2 is terminated by 3 Ω, the Thevenin equivalent looking into port 2 when port 1 is driven by a current source.

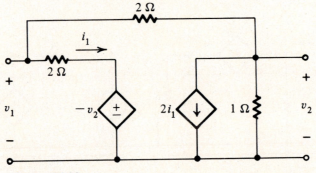

Figure P4.22

4.23 In terms of h parameters, find conditions under which a y-parameter description does not exist.

4.24 Find Thevenin or Norton equivalent networks looking into port 2 of a two-port when driven at port 1 by an ideal voltage source. Do this in terms of the g parameters.

4.25 Repeat Prob. 4.24 in terms of the z parameters.

4.26 Repeat Prob. 4.24 when port 1 is driven by an ideal current source.

4.27 Repeat Prob. 4.25 when port 1 is driven by an ideal current source.

4.28 Calculate the power gain for each of the networks in Fig. P4.18.

4.29 Repeat Prob. 4.28 for Figs. P4.17 and P4.19.

4.30 Calculate the voltage gain for the circuit shown. What is its value as $\beta \rightarrow \infty$?

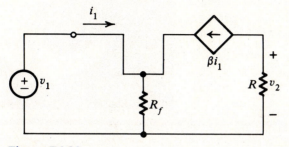

Figure P4.30

4.31 A two-port network has $h_{11} = h_{22} = 0$, and $h_{12} = h_{21} = 2$. Calculate the input resistance, voltage and current gains, and power gain when the network is terminated by a resistance of 100 Ω and driven by a source having a Thevenin resistance of 500 Ω and a Thevenin voltage of 10 V.

4.32 Repeat Prob. 4.31 for a two-port with $y_{11} = 2$ mS, $y_{12} = 0$, $y_{21} = 0.1$ S, $y_{22} = 10$ mS.

4.33 Repeat Prob. 4.31 for a two-port with $z_{11} = 500$ Ω, $z_{12} = 5$ Ω, $z_{22} = 1$ kΩ, $z_{21} = 100$ kΩ.

4.34 Express the Thevenin circuit (v_T, R_O) looking into port 2 of a two-port in terms of the h parameters when port 1 is driven by a source having a Thevenin representation v_S, R_S.

CHAPTER

Nonlinear Circuits

5.1 INTRODUCTION

Nonlinear circuits contain elements with terminal characteristics which cannot be described by linear equations. Such elements are called *nonlinear* and include diodes, transistors, and other devices. The terminal characteristics are either obtained experimentally or derived from the known laws of physics. Models for nonlinear elements tend to be fairly complicated, hence the analysis of nonlinear circuits usually requires the use of digital computers when accurate predictions are required. In many cases, however, a rough analysis is sufficient and much useful insight can be obtained using either very simple graphical techniques or very simple nonlinear models. There are piecewise-linear methods to facilitate large-signal analysis, as well as methods for linearizing circuits for small-signal analysis. In this chapter some of these methods are outlined for circuits containing only resistances and resistive or instantaneous (algebraic) nonlinear elements. Nonlinear energy storage elements are not considered.

5.2 EXAMPLE OF A NONLINEAR CIRCUIT

Figure 5.1 shows the circuit of an inverter containing two transistors Q_1 and Q_2. Because transistors are inherently nonlinear devices and circuits such as the one in Fig. 5.1 are common, it is very important to be able to analyze nonlinear circuits. The circuit of Fig. 5.1 will be analyzed in a later chapter after transistors are introduced. In this section the essential concepts needed for an understanding of nonlinear circuits are introduced using arbitrary nonlinear branches.

The circuit shown in Fig. 5.2(a) consists of two nonlinear branches labeled N1 and N2, a linear branch R_3, and a source represented by the Thevenin circuit v_S, R_S.

Using the current and voltage labels of Fig. 5.2(a) and curly brackets to denote functions, the branch relations for the various elements are

For N1: $v_1 = z_1\{i_1\}$ or $i_1 = y_1\{v_1\}$

For N2: $v_2 = z_2\{i_2\}$ or $i_2 = y_2\{v_2\}$

For R_3: $v_3 = i_3 R_3$ or $i_3 = G_3 v_3$

For the source: $v_4 = v_S - i_1 R_S$ or $i_1 = G_S v_S - G_S v_4$

The KCL equation at node B in terms of mesh currents and i_3 is

$$-j_1 + j_2 + i_3 = 0$$

The mesh KVL equations are

$$-v_4 + v_1 + v_3 = 0$$

$$-v_3 + v_2 = 0$$

It is useful to point out at this point that a nonlinear element which is a branch element or a one-port is just a nonlinear resistance. It can, therefore, be viewed either as a current-controlled voltage source where the control law is nonlinear and of the form $v_1 = z_1\{i_1\}$, or as a voltage-controlled current source where the

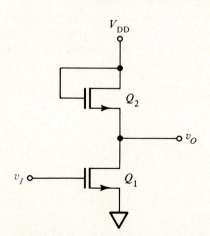

Figure 5.1 A MOSFET inverter circuit.

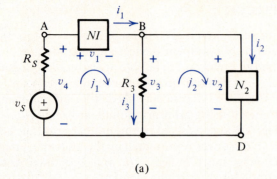

(a)

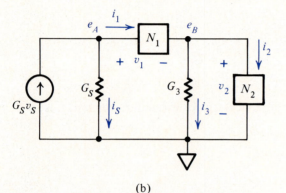

(b)

Figure 5.2 (a) A two-mesh nonlinear circuit. (b) A two-node nonlinear circuit.

control law is nonlinear and of the form $i_1 = y_1\{v_1\}$. When performing mesh analysis it is natural to express nonlinear branches as mesh-current controlled, while for nodal analysis node-voltage control is convenient.

In this example the mesh currents are the same as the nonlinear-branch currents. To carry out mesh analysis for the circuit shown, the KVL equations are, therefore, written for each mesh using the $v = z\{i\}$ form of the nonlinear and linear branch relations, and the branch current i_3 is eliminated using the KCL equation. The result is

$$v_S = z_1\{j_1\} + (R_S + R_3)j_1 - R_3 j_2 \tag{5.1}$$

$$0 = z_2\{j_2\} + R_3(j_2 - j_1) \tag{5.2}$$

Equations (5.1) and (5.2) are the nonlinear mesh equations in the unknown currents j_1 and j_2 and, in general, numerical methods using a computer are necessary for their solution.

To carry out a nodal analysis the network is redrawn using a Thevenin to Norton transformation as shown in Fig. 5.2(b). With the indicated datum node choice, nodal analysis requires formulating the KCL equations at each node using the $i = y\{v\}$ form for the linear and nonlinear branches, i.e,

$$G_S v_S = G_S e_A + y_1\{e_A - e_B\} \tag{5.3}$$

and

$$0 = -y_1\{e_A - e_B\} + G_3 e_B + y_2\{e_B\} \tag{5.4}$$

Equations (5.3) and (5.4) are the nonlinear equations in the unknown node voltages e_A, e_B and, as in the mesh analysis case, numerical methods must be used in general to obtain the solution.

It is important to reiterate at this point that KCL and KVL are circuit laws which always hold and which are established by the circuit topology; the nonlinearity only comes from the branch relations.

EXAMPLE 5.1

If in the circuit of Fig. 5.2(a) $v_S = 1$ V, $R_S = R_3 = 1\ \Omega$, and the branch relations for N1 and N2 are $i_1 = v_1^2$ and $i_2 = v_2 + v_2^2$, write the KCL equations corresponding to Fig. 5.2.

Solution

The equations are just Eqs. (5.3) and (5.4) for the given values and relations, namely

$$1 = e_A + (e_A - e_B)^2$$
$$0 = -(e_A - e_B) + e_B + e_B + e_B^2$$

In summary, for circuit problems containing nonlinear branches, the set of nonlinear network equations to be solved can be formulated using either a mesh or a node approach. In many electronic circuits a simple graphical solution procedure is possible; this is discussed in the next section.

5.3   GRAPHICAL SOLUTION OF NONLINEAR CIRCUITS

Circuits often contain a single nonlinear branch and all the other elements are either sources or linear elements. In this situation either a Thevenin or a Norton equivalent network can replace that part of the network "seen" by the nonlinear branch. This representation is shown in Fig. 5.3 where obviously $i_N = G_{eq} v_T$ or $v_T = R_{eq} i_N$.

When the nonlinear branch relation is given as a CCVS (current-controlled voltage source), $v_A = z\{i_A\}$, it is most convenient to use the Thevenin represen-

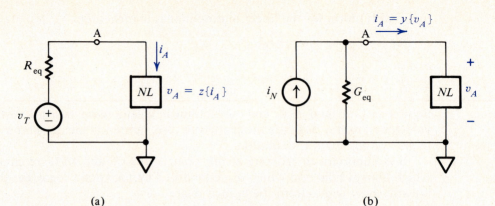

(a)   (b)

Figure 5.3 Network with single nonlinear branch: (a) Thevenin and (b) Norton.

tation shown in Fig. 5.3(a) and write KVL for this one-mesh circuit as the *linear* relation

$$v_A = v_T - R_{eq} i_A \tag{5.5}$$

called the *load line equation*. To solve this nonlinear problem requires satisfying Eq. (5.5) and the branch relation

$$v_A = z\{i_A\} \tag{5.6}$$

simultaneously. Since Eqs. (5.5) and (5.6) are both functions in the v_A-i_A plane (one linear and one nonlinear) the solution is obtained graphically as their intersection point Q, as illustrated in Fig. 5.4.

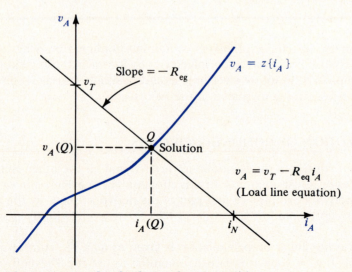

Figure 5.4 Graphical solution for Fig. 5.3(a).

The graphical solution consists simply of drawing the load line Eq. (5.5) on the v-i graph for the nonlinear branch characteristic and finding the intersection point. This line has a slope $-R_{eq}$ (the Thevenin equivalent resistance) and a voltage intercept equal to v_T (the Thevenin open-circuit voltage).

When the nonlinear branch relation is given as a VCCS (voltage-controlled current source) with the voltage as the controlling variable, $i_A = y\{v_A\}$, a dual procedure is followed. Using the Norton representation shown in Fig. 5.3(b), KCL in this one-node network yields the *linear* relation

$$i_A = I_N - G_{eq}v_A \tag{5.7}$$

which is the load line equation for this formulation. To solve this problem requires satisfying Eq. (5.7) and the branch relation

$$i_A = y\{v_A\} \tag{5.8}$$

simultaneously. This is shown graphically in Fig. 5.5. The student should note that the solution shown in Fig. 5.5 is the dual of the solution shown in Fig. 5.4.

To summarize, the graphical solution of a circuit having only one nonlinear branch is first simplified using either Thevenin or Norton's theorem to obtain one of the circuit equivalents shown in Fig. 5.3. The graphical solution is then obtained as follows:

1. Draw the *i-v* (or *v-i*) graph of the nonlinear branch.

2. Draw a straight line with current intercept i_N and voltage intercept v_T (this is called the load line).

3. The solution for *i* and *v* is the intersection of 1 and 2 as illustrated in Figs. 5.4 and 5.5.

In conclusion it should be pointed out that changing v_T or i_N simply moves the load line parallel to itself and changing R_{eq} changes the slope of the load line.

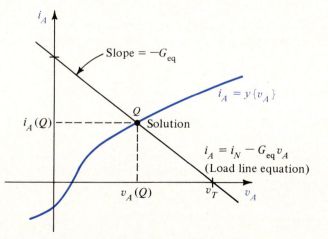

Figure 5.5 Graphical solution for Fig. 5.3(b).

EXAMPLE 5.2

For the nonlinear circuit shown in Fig. E5.2(a) find i and v given $i = 2v^2$ for $v > 0$ and $i = 0$ for $v < 0$.

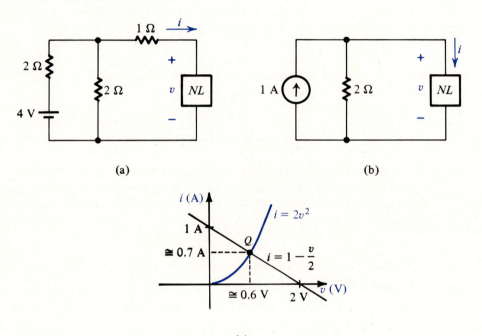

(a)   (b)

(c)

Figure E5.2

Solution

Since for branch NL v is the controlling variable, first convert all elements to the left of NL by the Norton representation as shown in Fig. E5.2(b). The load line equation is, by inspection,

$$i = 1 - \frac{v}{2}$$

Draw the load line and NL branch relation in the i-v plane and find the intersection, to one significant figure, to be at $i \simeq 0.7$ A, $v \simeq 0.6$ V. See Fig. E5.2(c).

For this particular example with a quadratic nonlinearity, an analytic solution is obtained by finding the i which satisfies simultaneously $i = 2v^2$ and $i = 1 - v/2$. This occurs when

$$2v^2 + \frac{v}{2} - 1 = 0$$

hence, to three figures, $v(Q) = 0.593$ V, $i(Q) = 0.703$ A.

DRILL EXERCISE

5.1 If NL in Fig. E5.2 is given by $v = i^2$ for $i > 0$ and $v = 0$ for $i < 0$, show that the solution is at $i(Q) = 0.732$ A and $v(Q) = 0.536$ V.

In some applications there may be two nonlinear branches in a circuit, as shown in Fig. 5.6 where the nonlinear branch relations are given as CCVSs. The KVL equation for the circuit of Fig. 5.6 is the *nonlinear* relation

$$v_2 = v_T - z_1\{i\} \tag{5.9}$$

which, by analogy with Eq. (5.5), is called the *load curve equation*. To solve this nonlinear problem requires satisfying Eq. (5.9) and the NL_2 branch relation

$$v_2 = z_2\{i\} \tag{5.10}$$

simultaneously. Since Eqs. (5.9) and (5.10) are both functions in the v_2-i plane the solution is obtained graphically as their intersection point Q, as illustrated in Fig. 5.7. A practical illustration of a circuit with two nonlinear (transistor) elements is discussed in Chap. 6.

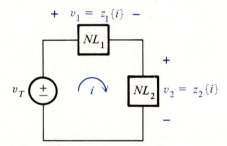

Figure 5.6 Network with two nonlinear branches.

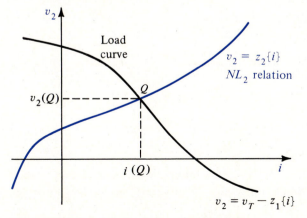

Figure 5.7 Graphical solution for Fig. 5.6.

DRILL EXERCISE

5.2 Show that an alternate graphical solution procedure for the circuit of Fig. 5.6 can be formulated in the v_1-i plane.

EXAMPLE 5.3

Two nonlinear elements having the branch relations $i_1 = 2v_1^2$ and $i_2 = v_2^2$ for v_1, $v_2 > 0$ and $i_1 = i_2 = 0$ for v_1, $v_2 \leq 0$ are connected in parallel and driven by a 1-A current source. Find the Q-point solution values for all the branch variables.

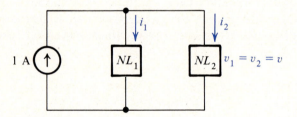

Figure E5.3(a) Circuit for Example 5.3.

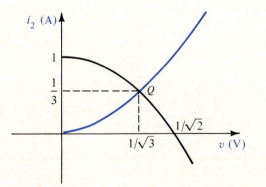

Figure E5.3(b) Solution for Example 5.3.

Solution

The circuit is shown in Fig. E5.3(a) where, since the elements are in parallel, $v_1 = v_2 = v$. This is the dual problem of the single mesh network of Fig. 5.6(a) and therefore it is the KCL equation

$$1 = i_1 + i_2$$

which must be satisfied. A graphical solution in the i_2-v plane is obtained as the intersection between the nonlinear branch relation for network 2

$$i_2 = v^2$$

and the *load curve* equation obtained when i_1 is eliminated from the KCL equation, i.e.,

$$i_2 = 1 - 2v^2$$

The solution is shown in Fig. E5.3(b); it is left as an exercise to show analytically that the solution is

$$i_2(Q) = \frac{1}{3} \text{ A}, \ i_1(Q) = \frac{2}{3} \text{ A}, \ v_1(Q) = v_2(Q) = \frac{1}{\sqrt{3}} \text{V}.$$

5.4 SMALL-SIGNAL MODELS FOR BRANCH ELEMENTS

A large and important class of electronic circuits is used to process information-bearing signals, such as the output of a microphone or the output of a TV antenna or cable. Most of these circuits contain nonlinear elements such as transistors, and include constant independent voltage sources ("power supplies") without which they simply would not work. These constant sources typically have magnitudes of the order of 10 V. The output of a microphone, on the other hand, is measured in millivolts while an antenna delivers microvolt signals. Consequently many of these circuits operate under *small-signal* conditions which means that the varying currents and voltages are small compared to the constant ones. It is therefore useful to have methods for analyzing nonlinear circuits under small-signal conditions. Circuit models which describe the effect of small changes are called *small-signal* or *linear incremental* models and are obtained by using a Taylor series expansion for the nonlinearity and retaining only the linear terms.

5.4.1 Topological Constraints

It is useful to recall at this point that the KCL (KVL) law is not dependent on the kind of elements in a network, but rather on the network graph so that KCL (KVL) will apply separately for the constant and the varying components of current (voltage). For example, KCL at node B, Fig. 5.2(a) is

$$-i_1 + i_2 + i_3 = 0 \tag{5.11}$$

Suppose now that for a given value of v_S the solution values for the currents are labeled $i_1(Q)$, $i_2(Q)$, $i_3(Q)$ where these must satisfy the KCL Eq. (5.11). If v_S is now increased by Δv_S then each of the three currents will also change, i.e.,

$$i_1 = i_1(Q) + \Delta i_1$$

and similarly for i_2 and i_3. These three currents must also satisfy the KCL Eq. (5.11); hence

$$-[i_1(Q) + \Delta i_1] + [i_2(Q) + \Delta i_2] + [i_3(Q) + \Delta i_3] = 0$$

or

$$[-i_1(Q) + i_2(Q) + i_3(Q)] + [-\Delta i_1 + \Delta i_2 + \Delta i_3] = 0$$

Since the Q-point currents satisfy KCL it follows that the increments also satisfy KCL, and

$$-\Delta i_1 + \Delta i_2 + \Delta i_3 = 0$$

the corresponding statement can be made that increments in branch voltages satisfy the network KVL equations.

In summary one can define an *incremental network* in which the branch variables are the Δi's and the Δv's, and which has the same graph as the original network since these incremental variables satisfy the same KCL (KVL) equations as the original nonlinear network.

5.4.2 Constructing Small-Signal Models

In general, a nonlinear branch relation may be written as the function

$$v_A = z\{i_A\} \tag{5.12a}$$

For some combination of values of the independent source(s) there will be a quiescent point solution

$$v_A(Q) = z\{i_A(Q)\} \tag{5.12b}$$

and for some incremental change in the value(s) of the source(s) there will be changes Δv_A and Δi_A. In that case Eq. (5.12a) becomes

$$v_A(Q) + \Delta v_A = z\{i_A(Q) + \Delta i_A\} \tag{5.13a}$$

The essential approximation involved in developing a small-signal model consists in replacing a nonlinear curve, such as the one given by Eq. (5.12a) and shown in Fig. 5.8, by the tangent line at the point Q. In other words the curve is approximated by a straight line passing through Q and having the same slope as the curve at Q. Since, in general, the slope depends on Q it should be emphasized that a dc analysis must precede a small-signal analysis. The Q-point is often called the *quiescent* point.

It is convenient at this stage to introduce a notation which will simplify the equations and which is used in electronic circuits. It consists of indicating quiescent quantities as well as constant sources by *uppercase* letters with *uppercase* subscripts and incremental quantities by *lowercase* letters with *lowercase* subscripts, namely, $V_A = v_A(Q)$, $I_A = i_A(Q)$ and $\Delta v_A = v_a$, $\Delta i_A = i_a$. With this notation the static or quiescent solution is Eq. (5.12a) written as

$$V_A = z\{I_A\} \tag{5.12b}$$

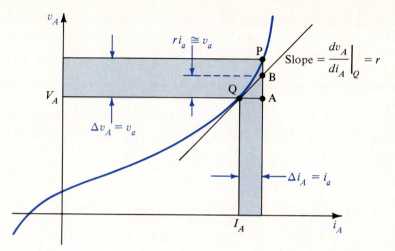

Figure 5.8 Graphical illustration of the small-signal approximation for a nonlinear branch.

and Eq. (5.13a) becomes

$$V_A + v_a = z\{I_A + i_a\} \tag{5.13b}$$

If z is now replaced by its Taylor expansion in Eq. (5.13b) one obtains

$$V_A + v_a = z\{I_A\} + \left.\frac{\partial v_A}{\partial i_A}\right|_Q i_a + \frac{1}{2}\left.\frac{\partial^2 v_A}{\partial i_A^2}\right|_Q i_a^2 + \cdots + \tag{5.14a}$$

where the vertical bar with the Q indicates that these derivatives are evaluated at the quiescent values. Since $V_A = z\{I_A\}$ it follows that the increment v_a is given by the power series

$$v_a = r i_a + \sum_{k=2}^{\infty} a_k i_a^k \tag{5.14b}$$

where r and the a_k denote the derivatives of z evaluated at the Q point.

A *linear incremental* or *small-signal* model for a branch is the approximation to Eq. (5.14b) when only the linear term in the increment i_a is retained, namely,

$$v_a = r i_a \tag{5.15}$$

where r is the *incremental* or *small-signal resistance* of the branch, defined by

$$r = \left.\frac{\partial v_A}{\partial i_A}\right|_Q \tag{5.16}$$

Thus the small-signal or linear incremental network model for the nonlinear branch is a resistance.

A sketch of a possible nonlinear branch characteristic, $v_A = z_A\{i_A\}$, is shown in Fig. 5.8. For an assumed increment i_a in the current from its dc value I_A the resulting increment v_a in the voltage is given by the vertical distance AP. The approximation to AP neglecting all but the linear term in Eq. (5.14b) is ri_a, the vertical distance AB. Hence the error in using the small-signal model is the distance BP which, in general, will depend on Q and i_a.

DRILL EXERCISE

5.3 Derive the relations corresponding to Eqs. (5.15) and (5.16) for a branch defined by $i_A = y\{v_A\}$.

It is now useful to examine how one constructs small-signal models for some of the branch elements encountered in network problems such as the linear resistance, the independent voltage source, the independent current source, and the various kinds of controlled sources.

THE LINEAR RESISTANCE BRANCH
Since $v = Ri$ for all v and i the *incremental resistance* $r = \partial v/\partial i$ is just the resistance R. Thus the small-signal model for a linear resistance is the resistance.

THE INDEPENDENT VOLTAGE SOURCE BRANCH

$$v = v_S = v_S(Q) + \Delta v_S$$

hence

$$\Delta v = \Delta v_S$$

For a *constant* independent source $\Delta v_S = \Delta v = 0$ and in an incremental network a constant independent voltage source is replaced by a short circuit.

THE INDEPENDENT CURRENT SOURCE BRANCH

$$i = i_S = i_S(Q) + \Delta i_S$$

hence

$$\Delta i = \Delta i_S$$

For a *constant* independent source $\Delta i_S = \Delta i = 0$ and in an incremental network a constant independent current source is replaced by an open circuit.

In summary in a linear incremental network

1. The incremental network has the same topology as the nonlinear network.

2. Nonlinear branches are replaced by resistances.

3. Linear branches are unchanged.

4. Constant independent voltage sources are replaced by short circuits.

5. Constant independent current sources are replaced by open circuits.

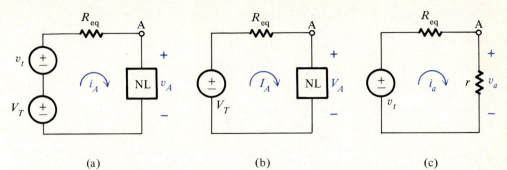

(a)   (b)   (c)

Figure 5.9 Decomposition of nonlinear circuit. (a) The total circuit, (b) the dc circuit, (c) the small-signal circuit.

It is understood that a nonlinear circuit is analyzed in two steps. First the dc or quiescent solution for a given v_T is found, then the small-signal circuit elements are calculated and used to find subsequent small changes. The above decomposition is shown in Fig. 5.9(a) to (c); the lowercase subscript is used to indicate incremental quantities and the small-signal node label. From a small-signal point of view, Fig. 5.9(c) is simply a single loop network with small-signal solutions expressed in terms of an incremental change in voltage v_t,

$$i_a = \frac{v_t}{R_{eq} + r}$$

$$v_a = \frac{r v_t}{R_{eq} + r} \qquad \text{(voltage divider)}$$

EXAMPLE 5.4

In Fig. 5.3(a) given that $v_T = 10$ V, $R_{eq} = 10\ \Omega$ or $G_{eq} = 0.1$ S and $i_A = 0.02 v_A^2$ is the branch relation for NL, find V_A and I_A and the incremental conductance for NL. Calculate the changes in v_A and i_A for a 1 V change in v_T. Also calculate the error in using the small-signal model.

Solution
The graphical solution for these circuit elements is shown in Fig. E5.4(a). Since the current is the dependent variable in the NL branch equation, the Norton form is used with $i_N = 1$ A. From the graph the dc solution is at 0.5 A and 5 V; the small-signal conductance $g = 0.2$ S and therefore the incremental resistance $r = 5\ \Omega$. The incremental model is as shown in Fig. E5.4(b).

The solution corresponding to a change $v_t = 1$ V or $i_n = 0.1$ A is obtained from the model using the current divider rule

$$i_a = \tfrac{1}{15} \text{ A} = 0.0667 \text{ A}$$

$$v_a = \tfrac{1}{3} \text{ V} = 0.3333 \text{ V}$$

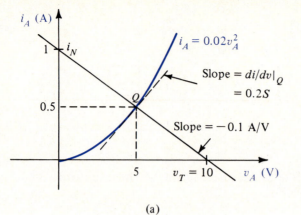

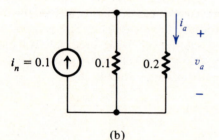

Figure E5.4

As in Example 5.2 an analytical solution for this problem is possible since the NL function is given. The load line equation when $v_T = 11$ V is

$$i_A = 1.1 - \frac{v_A}{10}$$

and the NL function is

$$i_A = 0.02v_A^2$$

hence the two are satisfied when

$$0.02v_A^2 + 0.1v_A - 1.1 = 0$$

This occurs at

$$v_A = 5.326 \text{ V} \quad \text{or} \quad v_a = 0.326 \text{ V}$$

$$i_A = 0.5674 \text{ A} \quad \text{or} \quad i_a = 0.0674 \text{ A}$$

(the other solution to the quadratic lies outside the range of validity of the model) and the error in using the small-signal model is 0.7 mA in the current and 7 mV in the voltage (approximately 1%).

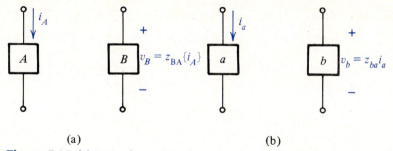

(a)   (b)

Figure 5.10 (a) A nonlinear CCVS element, and its small-signal model (b).

CONTROLLED-SOURCE BRANCHES

For such a branch, say a CCVS, the control relation is

$$v_B = z_{BA}\{i_A\} \tag{5.17}$$

where v_B is the controlled branch voltage and i_A is the controlling branch current. This is represented in the circuit diagram shown in Fig. 5.10(a).

The Taylor expansion for v_B retaining linear terms only is

$$V_B + v_b = z_{BA}\{I_A\} + \left.\frac{\partial v_B}{\partial i_A}\right|_Q i_a \tag{5.18}$$

but since $V_B = z_{BA}\{I_A\}$ the small-signal model is

$$v_b = z_{ba}i_a$$

where z_{ba} is the transimpedance given by the derivative coefficient in Eq. (5.18). Thus the small-signal model for a nonlinear CCVS is a linear CCVS element defined by z_{ba} with the circuit model shown in Fig. 5.10(b). Controlled sources are really two-ports when the controlling branch is included.

DRILL EXERCISE

5.4 Show that the small-signal models for VCCS, CCCS, and VCVS branches are defined by the single parameters y_{ba}, h_{ba}, and g_{ba}, respectively.

The generalization is that the small-signal model for a nonlinear controlled source branch is a linear controlled source branch.

EXAMPLE 5.5

A nonlinear two-port has the transfer voltage relation

$$v_B = 2v_A^2$$

where v_A is the voltage across the controlling input (A) port. Obtain a model which defines the small-signal behavior of the output (B) port and calculate the numerical values for $V_A = 0.5$ V. What is the error in using the small-signal model when $v_a = 0.1$ V?

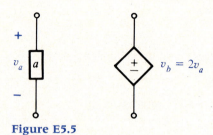

Figure E5.5

Solution

The full Taylor expansion for the given relation is, using the uppercase for dc and lowercase for changes,

$$V_B + v_b = 2V_A^2 + 4V_A v_a + 2v_a^2$$

The dc or quiescent solution ($v_a = 0$) is $V_B = 2V_A^2$, the linear small-signal solution is

$$v_b = 4V_A v_a$$

For $V_A = 0.5$ V, $V_B = 0.5$ V and the corresponding small-signal model is as shown in Fig. E5.5. For $v_a = 0.1$ V the small-signal model gives

$$v_b = 0.2 \text{ V}$$

whereas the full Taylor expansion gives

$$v_b = 0.2 + 2(0.1)^2$$

corresponding to an error of 0.02 V or 10%.

5.5   THE DIODE

5.5.1 The Semiconductor Diode

The most common nonlinear one-port encountered in electronic circuits is the semiconductor diode. The physical configuration of a semiconductor diode consists of two electrically dissimilar types of material (one called positive, the other called negative) as shown in Fig. 5.11(a) forming what is known as a *pn junction*. The labels *p* and *n* indicate the type of material and the device with the geometry of Fig. 5.11(a) is called a *pn* junction diode. Its *i-v* characteristic is obtained by an

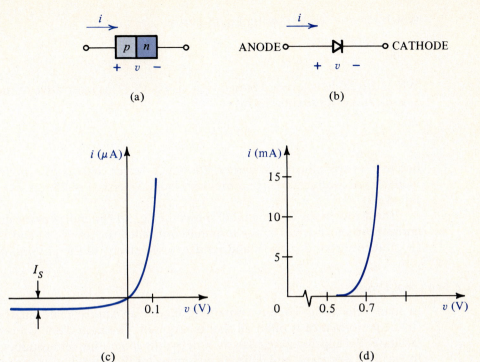

Figure 5.11 The *pn*-junction diode: (a) geometry, (b) its symbol, (c) its *i-v* characteristic, and (d) the same characteristic drawn to a different scale.

analysis of the physical principles governing the flow of charge carriers in semi-conductors or from terminal measurements. It is very closely approximated by the equation

$$i = I_S(e^{v/V_T} - 1) \tag{5.19}$$

The shape of this function, drawn to two different scales, the circuit symbol for a diode, and the polarity conventions are shown in Fig. 5.11. The current I_S is called the *reverse* ($v < 0$) *saturation* current and is typically less than 1 picoampere. V_T is called the thermal voltage and is approximately equal to 0.025 V at room temperature. Because of the exponential in Eq. (5.19) the *forward* current ($i > 0$) rises very rapidly; for example, if $v = 0.25$ V, $\exp(v/V_T) = e^{10} \approx 22000$, whereas if $v = 0.125$ V, $\exp(v/V_T) = e^5 \approx 150$. Thus a 2 to 1 voltage change results in a 150 to 1 change in current. Even more significant is the relative insensitivity of the diode voltage to large changes in current; for example, using $I_S = 10^{-14}$ A, the diode voltage as given by Eq. (5.19) is 0.663 V at 1 mA and 0.691 V at 10 mA. Thus a 10 to 1 current change results in only a 9% change in voltage.

The incremental conductance for a semiconductor diode at a quiescent voltage V is calculated from Eq. (5.19) to be

$$g = \left.\frac{di}{dv}\right|_Q = \frac{I_S}{V_T} e^{V/V_T} \tag{5.20}$$

which may be written, using Eq. (5.19), as

$$g = \frac{I + I_S}{V_T} \tag{5.21}$$

When the diode is forward biased ($v > 0$) Eq. (5.21) is insensitive to the value of I_S and can be approximated by

$$g \simeq \frac{I}{V_T} \tag{5.22}$$

because, as was illustrated earlier, I is much larger than I_S. The corresponding incremental conductance g becomes very large. Thus for $i = 25$ mA, $g = 1$ S from Eq. (5.22).

The arrow in the diode symbol shown in Fig. 5.11(b) indicates the direction of current for a forward bias; the terminals are called the *anode* (+) and the *cathode* (−) and correspond to the p and n sides, respectively, of the physical diode.

5.5.2 The Ideal Diode

In many electronic circuits containing diodes an approximate analysis of the circuit performance is often all that is required. The *ideal* diode concept greatly facilitates the approximate analysis of networks containing semiconductor diodes. In Fig. 5.12 the graph and defining equations for the ideal diode are shown. Note the difference in the symbol used for the ideal diode in Fig. 5.12(b) in contrast to the symbol for a *pn* diode in Fig. 5.11(c). An ideal diode is in one of two possible states; an OFF state when no current flows and the diode voltage is negative, and an ON state when current flows and the voltage is zero.

5.5.3 A Piecewise-Linear Diode Model

The characteristics of Fig. 5.11 may be approximated by the two linear segment characteristics of Fig. 5.13; in this *piecewise-linear* (PL) *model* approximation the current is assumed to be zero for $v < V_B$. R_D is a measure of the slope of the linear line segment which "best" approximates the diode forward characteristic for $v > V_B$. Note the change in scale of the current axis.

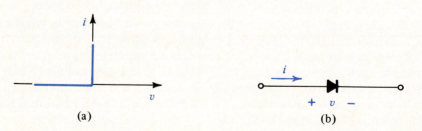

(a)   (b)

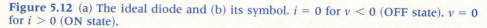

Figure 5.12 (a) The ideal diode and (b) its symbol. $i = 0$ for $v < 0$ (OFF state). $v = 0$ for $i > 0$ (ON state).

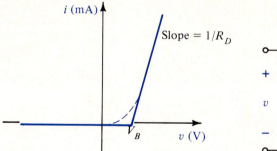

Figure 5.13 A piecewise-linear diode model.

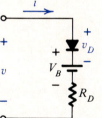

Figure 5.14 Network model for Fig. 5.13 characteristic.

The characteristic shown in Fig. 5.13 is the i-v characteristic for a circuit consisting of three elements: an ideal diode, a resistance, and a voltage source connected as shown in Fig. 5.14. The validity of this network is demonstrated by examining its behavior in each of the two possible diode states. When the diode is ON, then

$$v_D = 0$$

and

$$i = \frac{v - V_B}{R_D}$$

corresponding to the rising linear portion of Fig. 5.13 valid for $i > 0$ or $v > V_B$. When the diode is OFF, $i = 0$ corresponding to the horizontal portion of Fig. 5.13 this occurs if $v < V_B$. The common point to these line segments, known as the *breakpoint*, occurs when the ideal diode current *and* voltage are zero *simultaneously*. The resistance R_D in this model is called the diode forward resistance.

5.5.4 The Diode Rectifier

One of the most common applications of a diode is in the conversion of sinusoidal (as in household) current to a unidirectional current, the simplest circuit is the *half-wave rectifier* shown in Fig. 5.15. For sufficiently large v_I the performance characteristics for this circuit can be obtained with a sufficient degree of accuracy by modeling the diode D by the piecewise-linear circuit as shown in Fig. 5.16(a).

The characteristic of interest is the voltage transfer characteristic $v_O - v_I$. It is obtained by noting that the diode is either ON or OFF. Assume D is OFF, then $i = 0$, $v_O = 0$ and the voltage across D is, $v_I - V_B$. This assumed OFF state can only exist as long as D is reverse biased, i.e., $v_I < V_B$. Assume D is ON then $v_O = R_L i$ with

$$i = \frac{v_I - V_B}{R_L + R_D}$$

This assumed ON state can only exist as long as D is forward biased, i.e., $i > 0$ or

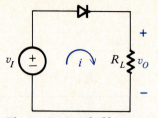

Figure 5.15 A half-wave rectifier.

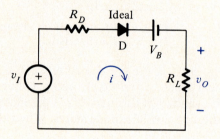

(a)

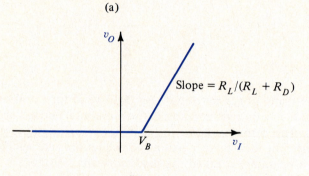

(b)

Figure 5.16 (a) Piecewise-linear model for rectifier circuit and (b) its voltage transfer characteristic.

$v_I > V_B$. Thus the transfer characteristic, shown in Fig. 5.16(b) consists of the two linear segments

$$v_O = 0 \qquad \text{for } v_I < V_B$$

and

$$v_O = \frac{R_L(v_I - V_B)}{R_L + R_D} \qquad \text{for } v_I > V_B$$

By inspection of this circuit when $v_I = V_m \sin \omega t$ with $V_m \gg V_B$ the current flow is adequately approximated by

$$i(t) = \frac{V_m}{R_L + R_D} \sin \omega t$$

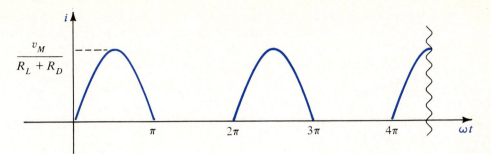

Figure 5.17 Half-wave rectifier current.

for each odd half-cycle when $v_I > 0$ and $i(t) = 0$ for each even half-cycle when $v_I < 0$. The output thus consists of only positive half-cycles of the sine wave as shown in Fig. 5.17, hence the circuit's name. This method of analyzing a piecewise-linear circuit is called the *assumed state method* and will be discussed in the next section.

DRILL EXERCISE

5.5 Show that neglecting V_B the dc or average load current for a half-wave rectifier is given by

$$I_{DC} = \frac{V_m}{(R_L + R_D)\pi}$$

5.6 ANALYSIS OF PIECEWISE-LINEAR (PL) CIRCUITS

In this section some examples of resistive diode circuits with more than one diode are discussed. These circuits perform some wave shaping or logic operations. Such circuits have inputs and outputs and what one needs to determine are the circuit transfer characteristics and the driving point or input characteristics. To keep the models reasonably tractable and to stress basic concepts ideal diodes are assumed in all cases.

5.6.1 The Assumed-State Method

In circuits containing only ideal diodes, the circuit characteristics, be they driving point or transfer, are piecewise-linear (PL). The transition from one linear segment to an adjacent one is called a breakpoint and is the result of a diode in the circuit changing state. In any linear segment of a PL characteristic all the diodes in the circuit remain in a particular state, OFF or ON. Thus, in theory, such circuits can be analyzed by listing all the possible combinations of states of the diodes and obtaining the linear portion of the characteristic for each combination of states.

Since each diode has two states, a circuit containing one diode has two possible states; one with two diodes has at most four possible states, one with three diodes has at most eight, etc. This is the assumed state method used for the rectifier analysis in Sec. 5.5. Only circuits with at most two diodes are considered in this section.

EXAMPLE 5.6

Find the PL i-v characteristics for the ideal diode circuit shown in Fig. E5.6(a).

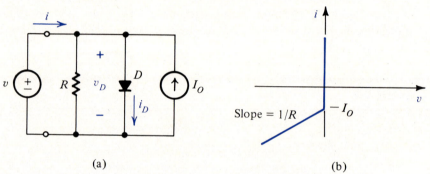

(a)   (b)

Figure E5.6 (a) Circuit for Example 5.6. (b) PL characteristic.

Solution

For the circuit of Fig. E5.6(a) there are only two states.

Diode ON state. In this state

$$i_D \geq 0 \qquad v_D = v = 0$$

and, by KCL

$$i_D = i + I_0$$

The assumption that the diode is ON is true as long as $i_D \geq 0$ or

$$i \geq -I_0$$

thus this state defines the line $v = 0$, i.e., the current axis for $i \geq -I_0$.

Diode OFF state. In this state

$$i_D = 0$$

and

$$v_D \leq 0$$

to guarantee that the diode is OFF. By KCL

$$i = \frac{v}{R} - I_0$$

and the i-v characteristics thus consists of the straight line segment for $v \leq 0$ of slope $1/R$ with current intercept at $-I_0$. The PL segments meet at the breakpoint, which for this circuit occurs at coordinates $(-I_0, 0)$ as shown in Fig. E5.6(b).

EXAMPLE 5.7

Find the PL i-v characteristic for the circuit shown in Fig. E5.7(a).

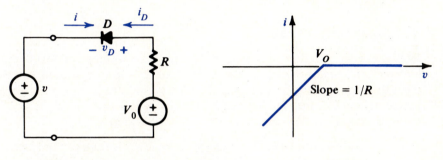

(a)   (b)

Figure E5.7 (a) Circuit for Example 5.7. (b) PL characteristic.

Solution
For the circuit of Fig. E5.7(a) there are two states.
   Diode OFF state. In this state

$$i_D = i = 0$$

and, by KVL

$$v_D = V_0 - v$$

The assumption that the diode is OFF is true as long as $v_D \leq 0$ or

$$v > V_0$$

thus the state defines the $i = 0$ line for $v > V_0$, i.e. the voltage axis for $v \geq V_0$.
   Diode ON state. In this state

$$v_D = 0 \qquad i_D > 0$$

and, by KVL

$$Ri_D = V_0 - v$$

This assumption that the diode is ON is true as long as $v < V_0$. Since $i = -i_D$ the state is characterized by

$$i = \frac{v - V_O}{R}$$

a line with slope $1/R$ and intercept V_O. The PL segments meet at $(0, V_O)$ as shown in Fig. E5.7(b).

DRILL EXERCISE

5.6 Show that changing the polarity of I_o and V_o in Figs. E5.6(a) and E5.7(a) merely shifts the breakpoints to $+I_o$ and $-V_o$, respectively.

5.6.2 Peak Clipper or Limiter Circuit

The nonlinear function which such a circuit is required to perform is to limit the positive and negative excursions of a signal to within specified values without distorting those parts of the signal lying within these limits. A circuit which performs this function is shown in Fig. 5.18. It is assumed that $V_1 \neq V_2$ and both are greater than 0. This circuit has four possible states.

1. D_1 and D_2 both OFF, in which case $e_2 = e_1$. This state can only exist if v_{D1} and v_{D2} are negative. Since, by KVL,

$$v_{D1} = e_2 - V_1$$

and

$$v_{D2} = -V_2 - e_2$$

this state requires that $e_2 < V_1$ and $e_2 > -V_2$. Since $e_1 = e_2$, then $-V_2 < e_1 < V_1$.

2. D_1 ON and D_2 OFF, in which case $e_2 = V_1$. This state exists for $i_{D1} = (e_1 - V_1)/R$ positive, or $e_1 > V_1$.

3. D_1 OFF and D_2 ON, in which case $e_2 = -V_2$. This state exists for $i_{D2} = -(V_2 + e_1)/R$ positive, or $e_1 < -V_2$.

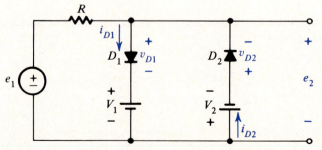

Figure 5.18 Diode clipping circuit.

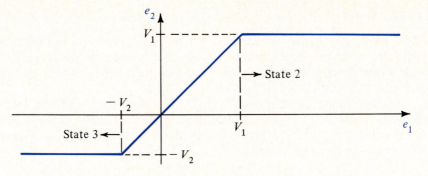

(a)

Figure 5.19(a) Transfer characteristic for circuit of Fig. 5.18.

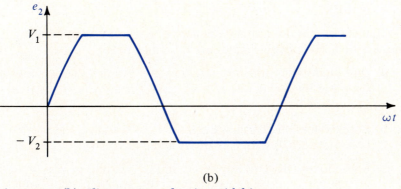

(b)

Figure 5.19(b) Clipper output for sinusoidal input.

Both diodes cannot be simultaneously ON since $V_1 \neq V_2$. The breakpoints occur exactly when $e_1 = V_1$ and $e_1 = -V_2$; hence the transfer characteristic is as shown in Fig. 5.19(a). These results assume that the diodes are ideal and that no current is drawn out of port 2.

When a sinusoidal signal which has a peak amplitude in excess of $|V_1|$ is applied to this clipping circuit, i.e.,

$$e_1 = V_m \sin \omega t$$

then the output voltage e_2 for $|V_2| < |V_1|$ is as shown in Fig. 5.19(b).

DRILL EXERCISE

5.7 Calculate the clipper transfer characteristic for the case when a load R_2 is connected across the port 2 terminals, and show that, instead of unity slope, the central portion will have a slope $R_2/(R + R_2)$.

5.6.3 Diode Logic Gates

Logic gates are circuits which perform logical operations. Because diodes are two-state or binary devices, diode circuits can be designed which perform binary logical operations. For example one can identify a HIGH voltage level with true and a LOW voltage level with false.

The two-diode circuit shown in Fig. 5.20 implements the logical AND function for two input logic variables, and is called a two-input AND gate. Defining a HIGH (true) voltage ≥ 5 V and a LOW (false) voltage $= 0$ V, the circuit must satisfy the following input-output relations:

1. If either of the input variables v_1, v_2 are LOW (v_1, $v_2 = 0$), then the output $v_O = 0$ (LOW);

2. If both the input variables are HIGH (v_1, $v_2 \geq 5$ V) then the output $v_O = 5$ V (HIGH).

That the circuit does satisfy the above relations is verified using the assumed state method. The possible states are described below.

a. One diode OFF and one ON (say D_2). This implies $i_{D2} > 0$ and $v_O = v_2$, $i_{D1} = 0$ and $v_{D1} < 0$, i.e., $v_1 > v_O$. Thus v_2 is LOW, v_1 is HIGH, v_O is LOW and 1 is satisfied.

b. Both diodes OFF. This implies $i_{D1} = i_{D2} = 0$ and v_{D1} and $v_{D2} < 0$. Thus v_1 and $v_2 > 5$ are HIGH and v_O is HIGH and **2** is satisfied.

c. Both diodes ON. This implies i_{D1} and $i_{D2} > 0$ and $v_O = v_1 = v_2$. Thus v_1 and $v_2 < 5$; hence v_1 *and* v_2 must be LOW, and v_O is LOW and 1 is satisfied.

It has been assumed that the diodes are ideal and no tolerances on the voltage levels have been given. These aspects are discussed in a later chapter on logic circuits.

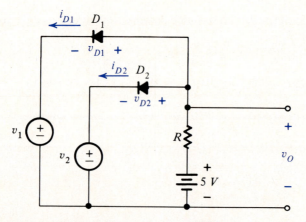

Figure 5.20 Two-input diode AND gate.

DRILL EXERCISES

5.8 Show that the circuit obtained by reversing the diodes in Fig. 5.20 and deleting the 5-V source performs the following operation for the same positive logic levels: (1) with either or both inputs HIGH the output is HIGH; (2) with both inputs LOW the output is LOW. This circuit is known as a two-input OR gate.

5.6.4 A PL Model for a Transistor

As will be seen in the next chapter, a transistor is a three terminal device which behaves very much like a current amplifier. It is quite nonlinear yet its terminal characteristics are reasonably well approximated by the PL circuit shown in Fig. 5.21 which is valid for $i_2 \geq 0$. The circuit contains two ideal diodes and a controlled source and obtaining its v_1-i_1 and v_2-i_2 PL characteristics serves as an example of the use of the assumed state method for a PL network with a controlled source. It also establishes some basis for the subsequent study of transistors.

With two diodes the four possible states are (1) both diodes OFF, (2) both diodes ON, (3) D_1 ON, D_2 OFF, and (4) D_1 OFF, D_2 ON. The PL circuits for these four states are shown in Fig. 5.22.

The assumed state method of analysis can now be used to obtain the input and output characteristics by examining each of the four state circuits.

STATE 1 D_1 AND D_2 OFF

With both diodes assumed OFF as shown in Fig. 5.22(a) it is evident that

$$i_{D1} = 0 \qquad i_{D2} = 0$$

hence

$$i_1 = 0 \qquad i_2 = 0$$

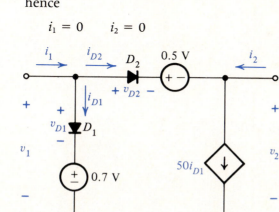

Figure 5.21 A PL model for a transistor.

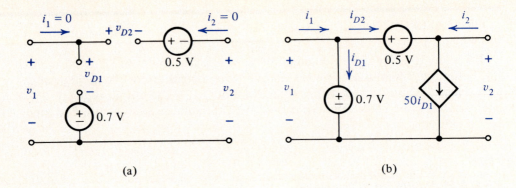

(a)   (b)

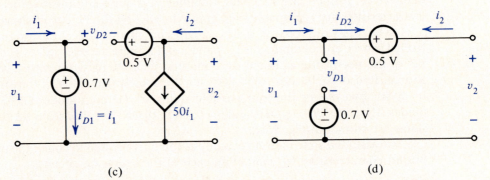

(c)   (d)

Figure 5.22 PL model for the four states of the circuit of Fig. 5.21. (a) State 1,
(b) state 2, (c) state 3, (d) state 4.

The assumption that both diodes are OFF requires that

$$v_{D1} \leq 0 \qquad v_{D2} \leq 0$$

hence, by KVL at port 1

$$v_1 \leq 0.7 \text{ V}$$

and by KVL at port 2, v_2 must satisfy

$$v_2 \geq v_1 - 0.5$$

If this is to hold for all v_1 up to 0.7 V, then

$$v_2 \geq 0.7 - 0.5$$

or

$$v_2 \geq 0.2 \text{ V}$$

Thus the PL segment for state 1 in the i_1-v_1 plane is the v_1 axis ($i_1 = 0$ line) in the
interval $[-\infty, 0.7 \text{ V}]$; and in the i_2-v_2 plane is the v_2 axis ($i_2 = 0$ line) in the interval
$[0.2 \text{ V}, \infty]$.

STATE 2 D_1 AND D_2 ON

With both diodes assumed ON as shown in Fig. 5.22(b) it is evident that

$$v_1 = 0.7 \text{ V} \qquad v_2 = 0.2 \text{ V}$$

The assumption that both diodes are ON requires that

$$i_{D1} \geq 0 \qquad i_{D2} \geq 0$$

Since, by KCL at port 1,

$$i_1 = i_{D2} + i_{D1}$$

it follows that $i_1 \geq 0$ must hold. By KCL at port 2

$$i_{D2} = 50i_{D1} - i_2$$

The current i_2 can be at most $50i_{D1}$ or

$$i_2 \leq 50i_{D1}$$

Since, by KCL, it is also true that

$$i_1 + i_2 = 51i_{D1}$$

the above inequality can be written

$$i_2 \leq \tfrac{50}{51}(i_1 + i_2)$$

or

$$i_2 \leq 50i_1$$

Thus the PL segment for state 2 in the i_1-v_1 plane is the $v_1 = 0.7$ V line for $i_1 \geq 0$; and in the i_2-v_2 plane is the $v_2 = 0.2$ V line for $i_2 \leq 50i_1$.

STATE 3 D_1 ON D_2 OFF

With the assumption D_1 ON and D_2 OFF as shown in Fig. 5.22(c) it is evident that

$$i_{D2} = 0 \qquad v_1 = 0.7 \qquad i_2 = 50i_1$$

The assumptions that D_1 is ON and D_2 is OFF requires that

$$i_{D1} \geq 0 \qquad v_{D2} \leq 0$$

hence, by KVL, since

$$v_{D2} = 0.7 - v_2 - 0.5 = 0.2 - v_2$$

for v_{D2} to remain negative

$$v_2 \geq 0.2 \text{ V}$$

Thus the PL characteristic for state 3 in the i_1-v_1 plane is the $v_1 = 0.7$ V line for $i_1 \geq 0$; and in the i_2-v_2 plane the horizontal family of lines given by

$$i_2 = 50i_1$$

for $v_2 \geq 0.2$ V.

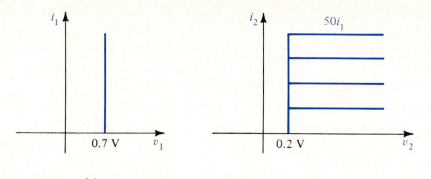

(a)   (b)

Figure 5.23 The (a) i_1-v_1 and (b) i_2-v_2 PL characteristics for the transistor model in Fig. 5.21.

STATE 4 D_1 OFF D_2 ON

With the assumption D_1 OFF and D_2 ON as shown in Fig. 5.22(d) it is evident that i_2 must be negative. Since this model is not valid for $i_2 < 0$ this state is not considered.

The resulting PL characteristics for the transistor are shown in Fig. 5.23.

DRILL EXERCISE

5.9 Show that if state 4 [Fig. 5.22(d)] were allowed the terminal constraints are $i_1 + i_2 = 0$, $v_1 - v_2 = 0.5$ V, $v_1 < 0.7$ V, and $v_2 < 0.2$ V.

5.7 NONLINEAR TWO-PORTS

5.7.1 Characterization of Nonlinear Two-ports

Figure 5.24 shows the general two-port configuration with the same variable designations that were used in Chap. 4. When the two-port is nonlinear it is not

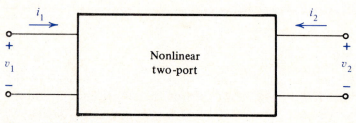

Figure 5.24 Nonlinear two-port.

possible to specify its characteristics by one of the six pairs of linear equations given in Table 4.1; the interrelationships will be one of six pairs of *functional* relations. The corresponding input and output characteristics will, in general, be neither straight nor equispaced. These functional relations, corresponding to the same choice of independent variables as indicated in Table 4.1, are

$$v_1 = h_1\{i_1, v_2\} \tag{5.23a}$$

and

$$i_2 = h_2\{i_1, v_2\} \tag{5.23b}$$

for the nonlinear "hybrid" functional relations, and

$$i_1 = y_1\{v_1, v_2\} \tag{5.24a}$$
$$i_2 = y_2\{v_1, v_2\} \tag{5.24b}$$

for the "admittance" functional relations. In this notation the dependent variables are expressed as functions of the independent variables; h_1, h_2 and y_1, y_2 indicate the four functions with the suffixes 1 and 2 designating input and output plane characteristics, respectively. For example, the set defined by Eq. (5.24) has the graphical representation shown in Fig. 5.25. It is customary to refer to the family in the i_1-v_1 plane as the input characteristics and the family in the i_2-v_2 plane as the output characteristics. With the exception of the last two choices of independent variables listed in Table 4.1, the differences between the h, y, g, and z characterizations lie only in the parameters which identify the members of the families of curves.

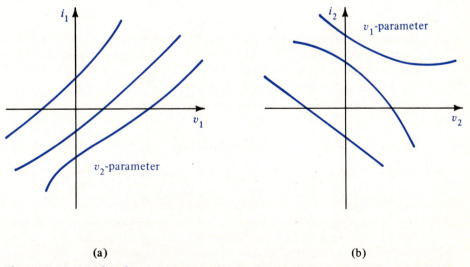

(a) (b)

Figure 5.25 Graphical representation for Eq. (5.24). (a) Input characteristics and (b) output characteristics.

EXAMPLE 5.8

A certain nonlinear two-port is defined by the following equations valid for $v_2 > 0$ only,

$$i_1 = K_1 v_1$$

$$i_2 = K_2 v_2^2 + K_3 v_1$$

Sketch its characteristics.

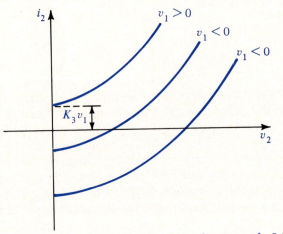

Figure E5.8 Output characteristics for Example 5.8.

Solution

The input equation is linear and independent of v_2; it behaves as a linear resistive branch with $R_i = 1/K_1$; hence the input characteristics is a single straight line. The output characteristics shown in Fig. E5.8 consist of a set of parabolas for $v_2 > 0$ displaced with respect to each other.

5.7.2 Terminated Nonlinear Two-Ports: Load Lines

In Sec. 5.7.1 it was shown that the graphical representation for a nonlinear two-port consisted of two sets or two families of curves. One such representation was given in Fig. 5.25 which shows one family of input characteristics where each member corresponds to a specific value of v_2, and one family of output characteristics where each member corresponds to a specific value of v_1.

What happens when such a two-port is terminated in the manner shown in Fig. 5.26? In this figure the terminations shown impose branch relation con-

Figure 5.26 A terminated two-port.

straints on the network variables. These may be written in terms of the input and output plane variables as

$$i_1 = -G_{S1}v_1 + G_{S1}v_{S1} \tag{5.25a}$$

at port 1, and

$$i_2 = -G_{S2}v_2 + G_{S2}v_{S2} \tag{5.25b}$$

at port 2. These two equations should be recognized as being the *load line equations* for the input and for the output characteristics, respectively (see Sec. 5.3). The constraints of the load lines are shown graphically in Fig. 5.27. The characteristics are given in the form

$$i_1 = y_1\{v_1, v_2\} \qquad \text{and} \qquad i_2 = y_2\{v_1, v_2\} \tag{5.26}$$

Evidently the "solution" for this nonlinear two-port network when terminated as shown in Fig. 5.26 is the set of values $v_1(Q)$, $i_1(Q)$, $i_2(Q)$, $v_2(Q)$ which simultaneously satisfy the nonlinear characteristics *and* the two load line equations. The solution is indicated by the points labeled Q_1 and Q_2 in Fig. 5.27, and is found by trial and error.

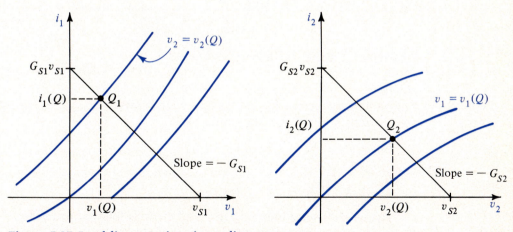

Figure 5.27 Load-line equations in nonlinear two-port.

EXAMPLE 5.9

If $v_{S1} = 4$ V, $G_1 = 3$ S, $v_{S2} = 20$ V, $G_2 = 0.25$ S are connected as in Fig. 5.26 to a two-port with characteristics are shown in Fig. E5.9, find the quiescent operating point. The characteristics are defined for v_1, $v_2 > 0$.

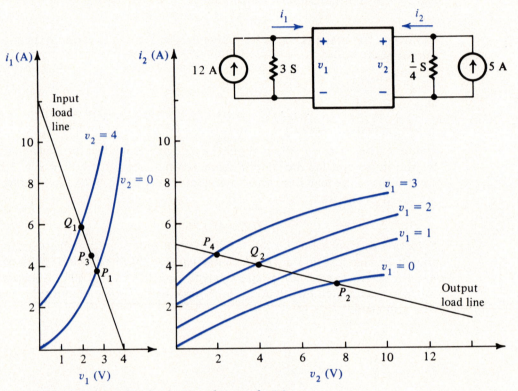

Figure E5.9 Graphical solution of Example 5.9.

Solution

The plotted characteristics and load lines are shown in Fig. E5.9. The solutions for the currents and voltages are found to be

$$i_1 = 6 \text{ A} \qquad v_1 = 2 \text{ V} \qquad \text{for } Q_1$$

and

$$i_2 = 4 \text{ A} \qquad v_2 = 4 \text{ V} \qquad \text{for } Q_2$$

The solutions are found by trial and error as follows: from the graph in Fig. E5.9 with the load lines shown it is clear that v_1 must be less than 4 V and that $v_2 < 20$ V. In fact, from the figure, Q_1 must be on the input load line above the point labeled P_1 therefore v_1 must be less than 2.7 V; Q_2 must be on the output load line above the point labeled P_2; therefore, v_2 must be less than 7.5 V.

Trying $v_1 = 2.5$ V yields P_3 on the input load line, corresponding to $v_2 \simeq 2$ V. The voltage $v_2 = 2$ V defines P_4 on the output load line, which lies on the curve corresponding to $v_1 = 3$ V. Since this is not 2.5 V as assumed, $P_3 P_4$ is not the solution.

Next, try $v_1 = 2$ V. This yields Q_1 on the input load line, corresponding to $v_2 = 4$ V. This voltage defines Q_2 on the output load line, which lies on the curve corresponding to $v_1 = 2$ V, as assumed. Thus, $Q_1 Q_2$ is the solution. This trial-and-error procedure will usually take many iterations.

5.8   SMALL-SIGNAL MODELS FOR TWO-PORTS

In this section the relationships between small changes in terminal variables about the operating point are discussed and methods for finding small-signal models described.

5.8.1 Notation and Terminology

For the two-port circuit shown in Fig. 5.28 a small-signal mathematical model is defined to be the linear relations between small changes in the terminal variables. It is convenient at this point to label the port 1 and 2 variables with the uppercase suffixes A and B, respectively, in order to use lowercase suffixes for small changes or increments, i.e., the notation introduced in Sec. 5.4. The small-signal model corresponding to the hybrid functional form [Eq. (5.23)] is obtained by retaining only the linear terms of the Taylor expansion of v_A and i_B about the operating point Q. Using letter subscripts in Eq. (5.23) repeated here for convenience as

$$v_A = h_1\{i_A, v_B\}$$

and

$$i_B = h_2\{i_A, v_B\}$$

with the notation $\Delta v_A = v_a$, $\Delta i_A = i_a$, $\Delta v_B = v_b$, $\Delta i_B = i_b$ for the changes and $V_A = v_A(Q)$, $I_A = i_A(Q)$, $V_B = v_B(Q)$, $I_B = i_B(Q)$ for the quiescent values the Taylor

Figure 5.28 Basic two-port.

expansions, neglecting higher powers, are written as

$$V_A + v_a = v_A(Q) + \left.\frac{\partial h_1}{\partial i_A}\right|_Q i_a + \left.\frac{\partial h_1}{\partial v_B}\right|_Q v_b \tag{5.27a}$$

$$I_B + i_b = i_B(Q) + \left.\frac{\partial h_2}{\partial i_A}\right|_Q i_a + \left.\frac{\partial h_2}{\partial v_B}\right|_Q v_b \tag{5.28a}$$

Defining, by analogy with linear two-port parameters, a set of four coefficients given by the four first partial derivatives, evaluated at the Q point, which appear in the preceding equations, i.e.,

$$h_{11} = \left.\frac{\partial v_A}{\partial i_A}\right|_Q \tag{5.29a}$$

$$h_{12} = \left.\frac{\partial v_A}{\partial v_B}\right|_Q \tag{5.29b}$$

$$h_{21} = \left.\frac{\partial i_B}{\partial i_A}\right|_Q \tag{5.29c}$$

$$h_{22} = \left.\frac{\partial i_B}{\partial v_B}\right|_Q \tag{5.29d}$$

Equations (5.27a) and (5.28a) can be written in terms of these parameters as

$$v_a = h_{11}i_a + h_{12}v_b \tag{5.27b}$$

and

$$i_b = h_{21}i_a + h_{22}v_b \tag{5.28b}$$

The constant terms drop out since $I_B = i_B(Q)$ and $V_A = v_A(Q)$. The notation is deliberately chosen to coincide with that of Chap. 4 when the two-port is linear. It should be clear, however, that the parameters defined by Eq. (5.29) depend on the quiescent point. A graphical interpretation for these parameters is shown in Fig. 5.29 where two members of the input and output characteristics are drawn to an exaggerated scale to show the details near the Q points.

From the input characteristics the change v_a is seen to contain two components. One corresponds to a change along the $v_B(Q)$ characteristic of vertical component equal to $h_{11}i_a$, the other corresponds to a vertical rise (constant i_A) from A to B of size $h_{12}v_b$. From the output characteristics the change i_b is seen to contain two components, one corresponding to a change along the $i_A(Q)$ characteristic from Q_2 to C having a vertical component $h_{22}v_b$, the other corresponding to a vertical rise (constant v_B) from C to D of size $h_{21}i_a$.

In summary, a Taylor expansion of the two functions h_1 and h_2, in which only the linear part is retained leading to Eqs. (5.27b) and (5.28b), is one of the mathematical small-signal models for a nonlinear two-port. The particular form considered is the hybrid or h-parameter model; others can be derived in an analogous manner for the other five functional forms defined in Chap. 4.

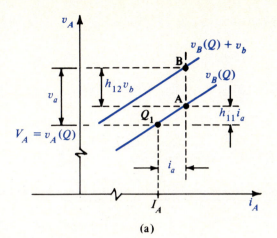

(a)

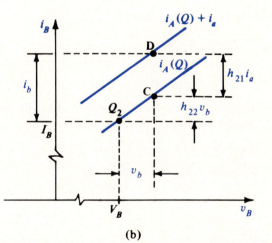

(b)

Figure 5.29 Diagram illustrating small changes in the input (a) and output (b) planes.

It should be stressed that the small-signal model is a linear approximation which is in error by the higher-order terms discarded in the Taylor expansions leading to Eqs. (5.27) and (5.28). It is accuracy considerations which will dictate what is *small* in small signal.

5.8.2 Small-Signal Circuit Representation

Equations (5.27) and (5.28) can be modeled by the circuit shown in Fig. 5.30 which is the same as Fig. 4.3 except for the lowercase letter subscripts. The left side contains a resistance h_{11} in series with a VCVS [Eq. (5.27)]; the right side contains a conductance h_{22} in parallel with a CCCS [Eq. (5.28)]. Each of the four elements has a "physical" interpretation which is evident from the diagram or Eqs. (5.27) and (5.28) or Fig. 5.30.

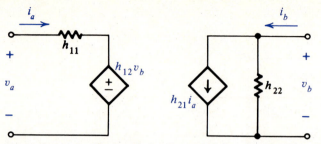

Figure 5.30 Small-signal circuit model (*h* parameters).

EXAMPLE 5.10

A nonlinear two-port is defined by the equations

$$i_A = v_A^2 + \frac{v_B}{2}$$

$$i_B = \sqrt{v_B} + v_A$$

Given that the *Q* point is at $I_A = 6$ A, $V_A = 2$ V, $V_B = 4$ V, obtain a small-signal *y*-parameter model. The model is valid for $v_B \geq 0$.

Solution
First write the generic relations for the *y* parameters:

$$i_a = y_{11}v_a + y_{12}v_b$$

$$i_b = y_{21}v_a + y_{22}v_b$$

where

$$y_{11} = \left. \frac{\partial i_A}{\partial v_A} \right|_Q$$

$$y_{21} = \left. \frac{\partial i_B}{\partial v_A} \right|_Q$$

$$y_{12} = \left. \frac{\partial i_A}{\partial v_B} \right|_Q$$

$$y_{22} = \left. \frac{\partial i_B}{\partial v_B} \right|_Q$$

then evaluate these derivatives at the quiescent point $v_A = 2$ V and $V_B = 4$ V to obtain

$$y_{11} = 4 \text{ S} \qquad y_{21} = 1 \text{ S}$$

$$y_{12} = 0.5 \text{ S} \qquad y_{22} = 0.25 \text{ S}$$

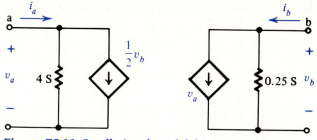

Figure E5.10 Small-signal model for Example 5.10.

Hence the small-signal model is as shown in Fig. E5.10.

EXAMPLE 5.11

For the two-port of Example 5.10 express the y parameters as a function of the Q point.

Solution
From the defining equations, carry out the partial differentiation,

$$y_{11} = 2v_A$$

$$y_{12} = 0.5 \text{ S}$$

$$y_{21} = 1 \text{ S}$$

$$y_{22} = \frac{1}{2\sqrt{v_B}}$$

DRILL EXERCISE

5.10 Find the error in using the small-signal model to calculate i_b and i_a when $v_a = v_b = 0.5$ V are the increments. The quiescent values are given in Example 5.10.
Ans. $i_b = 0.625$ A, $\Delta i_B = 0.6213$ A, error $= 3.7$ mA (.6%); $i_a = 2.25$ A, $\Delta i_A = 2.5$ A, error $= 0.25$ A (10%).

5.9 SUMMARY AND STUDY GUIDE

The main objective of this chapter has been the discussion of the basic concepts necessary for the study of nonlinear networks and the methods used to obtain useful circuit models for nonlinear elements. Formal and graphical methods of solving nonlinear network problems are discussed and the notions of incremental and small-signal models are introduced. The semiconductor diode element and methods of piecewise-linear network analysis are examined and procedures for obtaining small-signal models for nonlinear two-ports are discussed. The attention of students is drawn to the following points.

The solution to a nonlinear network is the result of imposing the KCL, KVL, and the branch laws. A graphical solution is always possible for a network with only one nonlinear branch.

Nonlinear branches are just nonlinear resistances (conductances).

The incremental network for a nonlinear network has the same topology as the nonlinear network and the incremental variables satisfy the same KCL, KVL equations as the nonlinear network.

The linear-incremental or small-signal model for a network is the linear approximation for the incremental network; the small-signal parameters generally depend on the quiescent point.

Small-signal model parameters are obtained by calculating the first derivative term of the Taylor expansion of the branch laws; this may be done analytically or graphically.

A nonlinear network is analyzed in two steps by decomposing it into a dc circuit and a small-signal circuit. The elements of the small-signal circuit, in general, depend on the dc solution.

Nonlinear networks containing diodes and other semiconductor devices can be analyzed approximately using ideal diodes and the method of assumed states. Such piecewise-linear models provide useful insight into the operating characteristics of many electronic devices.

The assumed state method of PL circuit analysis requires two statements. One defines the selected state where ON diodes have zero volts and positive current and OFF diodes have zero current and negative voltage. The other establishes by means of KCL and KVL the range of values of independent source values necessary to ensure that the circuit stays in the selected state.

The Taylor series expansions are used to obtain small-signal models for nonlinear two-port results in the same circuit representations as discussed in Chap. 4. Constant voltage sources in nonlinear networks become short circuits in the small-signal models and constant current sources become open circuits.

The load line concept is a useful concept in understanding the effect of terminations on nonlinear two-ports; in estimating quiescent solutions; and in illustrating graphically that small-signal parameters are quiescent-point dependent.

PROBLEMS

5.1 If in the circuit of Fig. 5.2(a) $v_S = 1$ V, $R_S = 2\ \Omega$, $R_3 = 1\ \Omega$, and the branch relations for N_1 and N_2 are $v_1 = i_1^2$ and $v_2 = i_2 + i_2^2$, write the KVL mesh equations.

5.2 For the nonlinear circuit shown in Fig. 5.6 $v_T = 3$ V, the branch relations are $v_1 = i^2$ and $i = v_2^2$ $(i > 0)$ for NL_1 and NL_2, respectively. Find the quiescent solution for all branch variables.

5.3 Find the small-signal resistance (conductance) model for the two nonlinear elements in Prob. 5.2 at the quiescent value found.

5.4 The output port branch current for a three terminal field effect transistor is approximately given by the voltage-controlled current relation

$$i_B = 16 \left(1 + \frac{v_A}{4}\right)^2 \text{ mA}$$

when v_A is in volts. Obtain a small-signal model for the branch at the quiescent value $V_A = -2$ V. What is the quiescent value I_B?

5.5 Draw the piecewise-linear driving point characteristic for the ideal diode circuits in Fig. P5.5.

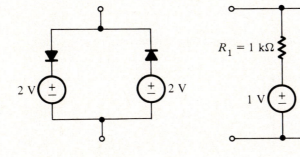

(a)

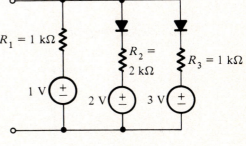

(b)

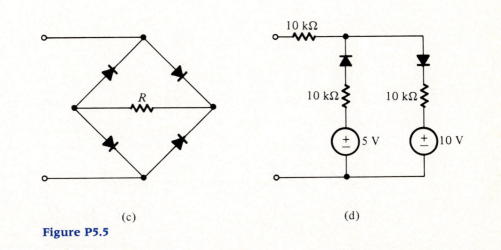

(c)

(d)

Figure P5.5

5.6 For each of the driving point piecewise-linear characteristics given in Fig. P5.6, synthesize a one-port circuit model containing only ideal diodes, independent sources, and resistances.

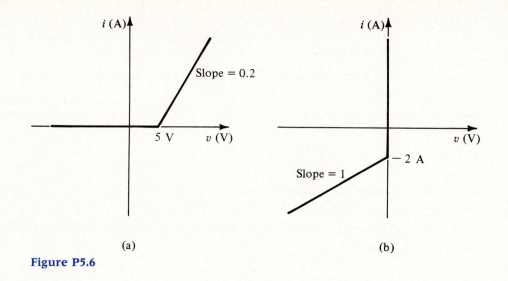

(a)

(b)

Figure P5.6

5.7 Repeat Prob. 5.6 for the driving point characteristic of Fig. P5.7.

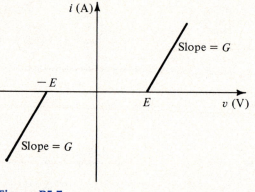

Figure P5.7

5.8 Sketch the v_O-v_I characteristics for the clipping circuits shown in Fig. P5.8.

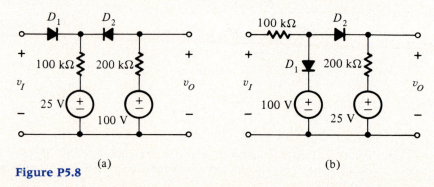

(a)

(b)

Figure P5.8

5.9 If in the two-input diode circuit of Fig. 5.20 HIGH voltage is any value between 5 V and 7 V and LOW voltage is any value between 0 V and 1 V, verify that the circuit behaves as an AND gate by calculating its output for the three conditions
  (a) $v_1 = 0$ V, $v_2 = 6$ V
  (b) $v_1 = 6$ V, $v_2 = 7$ V
  (c) $v_1 = 0$ V, $v_2 = 1$ V

5.10 Show that in the circuit of Fig. P5.10 the diode D_1 is always ON. Sketch v_O vs v_I.

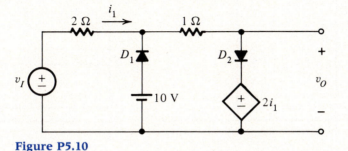

Figure P5.10

5.11 The 10-V battery in series with D_1 in Prob. 5.10 is replaced by a -5-V source. Sketch v_O vs v_I.

5.12 A nonlinear two-port has the following functional description:

$$i_A = 3v_A^2 + v_B \qquad i_B = v_B + v_A^2$$

Sketch the input and output characteristic families and calculate i_A and v_B if a $+1$ voltage source is connected to port A and a $+1$ current source is connected to port B ($v_A = 1$, $i_B = 1$).

5.13 Repeat Prob. 5.12 for $v_A = -1$, $i_B = -1$.

5.14 If the two-port of Prob. 5.12 is terminated as in Example 5.9, estimate the quiescent point by plotting the characteristics and using a graphical trial-and-error procedure.

5.15 Obtain a small-signal y-parameter model for the nonlinear two-port defined in Prob. 5.12 at the quiescent point $V_A = 1$, $V_B = 0$.

5.16 Repeat Prob. 5.15 at the Q point $V_A = 2$, $V_B = 1$.

5.17 For Prob. 5.15 obtain the corresponding h-parameter small-signal model.

5.18 For Prob. 5.16 obtain the z-parameter small-signal model.

5.19 Calculate the incremental input resistance at port A for the model in Prob. 5.15 when port B is terminated by a 1-Ω resistance.

5.20 Obtain a Thevenin and/or Norton representation for the small-signal model looking into port b of Prob. 5.15 when port a is driven by a voltage source v_{as}.

5.21 Repeat Prob. 5.20 when port a is driven by a current source i_{as}.

5.22 For the terminating condition of Prob. 5.19 calculate v_b/v_a and i_b/i_a. Is this a voltage or a current-amplifying circuit?

5.23 Interchange the termination and source of Prob. 5.22 and now calculate v_a/v_b and i_a/i_b. Is this a voltage or a current-amplifying circuit?

5.24 Find the Q point for the two-port characterized by the curves in Fig. P5.24(a) given that the ports are terminated as in Fig. P5.24(b).

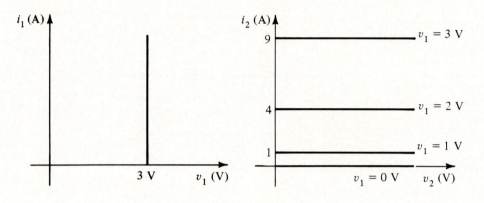

(a)

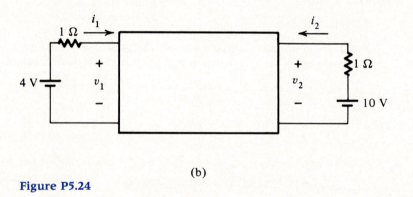

(b)

Figure P5.24

5.25 Find a small-signal model for the two-port characteristics of Fig. P5.24(a).

5.26 Repeat Prob. 5.24 for the two-port characterized by the curves in Fig. P5.26 for the terminations of Fig. P5.24(b).

5.27 Repeat Prob. 5.25 for the two-port of Prob. 5.26.

5.28 If, in the two-port specified in Prob. 5.24, the output port 1 termination resistance is replaced by the nonlinear conductance $i = v^2/2$ find the Q point.

5.29 Sketch the input and output characteristics for the transistor PL model shown in Fig. P5.29.

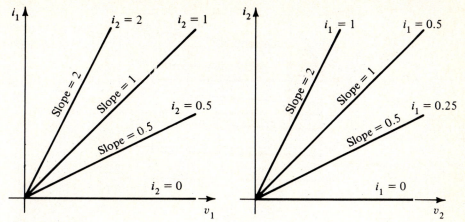

Figure P5.26

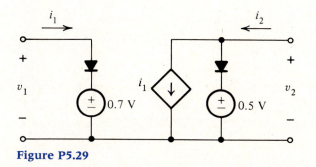

Figure P5.29

5.30 Figure P5.30(a) shows the input characteristic family and Fig. P5.30(b) shows the output family of a certain transistor. When the network is terminated as in Fig. P5.30(c) use the load line method to estimate the quiescent values for I_A, V_A, I_B, V_B.

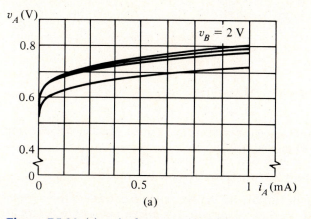

(a)

Figure P5.30 (a) v_A-i_A characteristics; (b) i_B-v_B characteristics; (c) terminations.

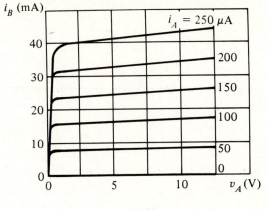

(b)

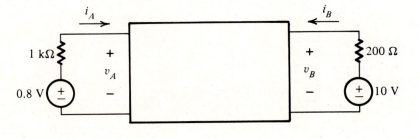

(c)

Figure P5.30 continued

5.31 A nonlinear two-port has the following functional relations:

$$i_A = \begin{cases} \dfrac{v_A}{100} & \text{for } v_A > 0 \\ 0 & \text{for } v_A < 0 \end{cases}$$

and

$$i_B = \begin{cases} \dfrac{v_A}{20} & \text{for } v_B > 0 \\ 0 & \text{for } v_B < 0 \end{cases}$$

Sketch the characteristics and obtain a small-signal model when the quiescent point is
(a) $I_A = 10$ mA, $V_B > 0$
(b) $V_A = -1$ V, $V_B > 0$
Discuss the difference between these two models.

5.32 Calculate the small-signal voltage amplification and current amplification for the two cases of Prob. 5.31 when port a is driven and port b is terminated by a 1000-Ω resistance.

5.33 For the small-signal model of Prob. 5.31 calculate v_b/v_a, i_b/v_a, v_a/v_b, i_a/i_b for cases (a) and (b).

5.34 A nonlinear two-port has the following defining equations:

$$v_A = i_A + 2v_B^2 + 3 \qquad i_B = i_A^2 + \frac{v_B}{2} + 2$$

When terminated as shown in Fig. P5.34, it is found that $I_A = 1$ A.
(a) Find the quiescent values I_A, V_A, I_B, V_B.
(b) Obtain the small-signal h-parameter model at the quiescent values found.

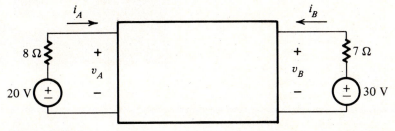

Figure P5.34

5.35 (a) For Fig. P5.34 calculate the small-signal forward current gain i_b/i_a, forward voltage gain v_b/v_a, input resistance at port a seen by the 8-Ω resistance, the output resistance at port b seen by the 7-Ω resistance.
(b) Repeat the calculations for the reverse current and voltage gain.

CHAPTER

Field-Effect Transistors (FET)

6.1 INTRODUCTION

This is the first of two chapters devoted to the study of transistors. These are *three-terminal*, solid-state devices which behave as controlled current sources. They are, therefore, the basic elements used in the design of such important functional building blocks as amplifiers, of which the op amp, which the student encountered in Chap. 4, is probably the most important. Their usefulness extends much further, however, because they are also used as switches; for sufficiently large signal changes at the transistor's controlling port, the current at the output port varies between zero and a relatively large value. Consequently, transistors are also the basic elements used in the design of logic circuits which the student will study in Chap. 8.

Transistors can be divided into two major types: *field-effect transistors* (FET) which are the subject of this chapter, and *bipolar junction transistors* (BJT) which will be treated in Chap. 7. The names reflect important distinguishing features of the physical processes which give rise to their characteristics. Thus the FET behaves as a VCCS while the BJT is more like a CCCS. FETs are further divided into *metal-oxide-semiconductor field-effect transistors* (MOSFET) and *junction field-effect transistors* (JFET). A more general name for the MOSFET is the *insulated-gate* FET (IGFET).

The basic material in all three types of transistor is a semiconductor which, in the earliest devices, was germanium. Today, however, silicon is by far the most important material, although the compound semiconductor gallium arsenide is used in some specialized applications.

Most transistors, and especially MOSFETs, are produced today as elements of *monolithic integrated circuits* (IC) which are commonly known as *chips*. These are complete electronic circuits produced on one single piece of silicon. They range in complexity from one to several hundred thousand transistors. The main IC technologies are the relatively old and mature bipolar and the MOS. The former yields BJTs and JFETs while the latter is evidently based on the MOSFET and can be further subdivided into *p-channel* MOS (PMOS), *n-channel* MOS (NMOS), and *complementary symmetry* MOS (CMOS). Despite its relative youth it is the MOS technology which today is the industry leader in the development of very large-scale integrated (VLSI) circuits containing transistors in numbers approaching a million, on a silicon chip with area dimensions measured in millimeters. This rapid development is driven by such factors as the relative simplicity of the NMOS fabrication process and the lower power consumption of MOSFETs, which permits much greater packing densities than those possible in bipolar ICs.

Reference will be made in this chapter and elsewhere to the use of transistors in *analog* and *digital* circuits. What distinguishes these two classes of circuits is the set of signal values. Whereas in an analog circuit a signal can have *any* value within some practical range, the signal in a digital circuit can only take on a *discrete set* of values. Most digital circuits operate on the basis of the binary system of numbers so that the signals may have only one of two values. The name *analog* derives from the fact that the signal is analogous to the physical signal that it represents. Thus the electrical output of a microphone is analogous to the air pressure signal produced by the vocal tract.

As mentioned earlier, MOSFETs are encountered primarily in ICs, especially VLSI circuits such as microprocessors, memories and complex digital signal-processing ICs for communication systems. Discrete MOSFETs are used in power applications and in amplifiers where a high input resistance is required. Specialized analog ICs such as op amps and very high-frequency circuits are the main areas of application of the JFET.

The MOSFET and the JFET are the subject of this chapter. In each case, the terminal characteristics are presented and simple models for the different modes of the device's operation are developed. These models are then used to analyze simple transistor circuits. The main objective is to develop in the student the ability to carry out the analysis of simple circuits based on these two types of transistor, in preparation for the later study of more complex digital and analog circuits.

6.2   THE METAL-OXIDE-SEMICONDUCTOR FIELD-EFFECT TRANSISTOR (MOSFET)

6.2.1 Structure, Principle of Operation, and Symbols

The starting material in the fabrication of a metal-oxide-semiconductor field-effect transistor (MOSFET) is a thin wafer of crystalline silicon which, depending on its electrical properties, is identified as being either *n*-type or *p*-type. A MOSFET is produced by subjecting the surface of the silicon wafer, called the *substrate*, to a process which, depending on its details, yields either a *p-channel* or an *n-channel* device. Both of these can, in turn, be either an *enhancement* or a *depletion* type.

Figure 6.1 shows the simplified structure of an *n*-channel, depletion MOSFET. In a *p*-channel device the *n* and *p* regions are interchanged. The most important elements which constitute the intrinsic MOSFET, are the *channel* near the substrate surface, the *gate* electrode, the *oxide* layer which insulates the gate from the channel, and the *source* and *drain* regions which terminate the ends of the channel.

Early MOSFETs were *p*-channel and in them the gate was a metal, which is why the device is called a *metal*-oxide-semiconductor transistor. Today, the PMOS technology which was used to produce discrete devices and, later, ICs of this type has been largely superceded by NMOS and CMOS. The former yields *n*-channel MOSFETs which are intrinsically faster than *p*-channel devices, while the latter is characterized by having both *n*-channel and *p*-channel transistors on the same substrate. In both NMOS and CMOS the gates are made of a form of silicon called *polysilicon*. Despite this modification, by tradition, the devices are still called MOSFETs.

The MOSFET behaves like a VCCS, the controlling mechanism being the electric field distribution in the channel where current is constrained to flow. The

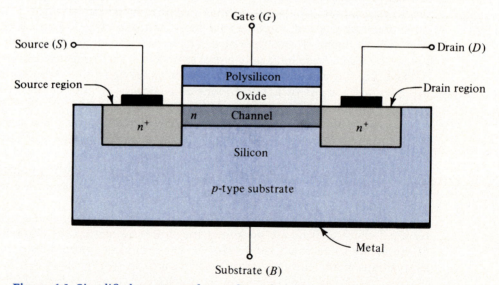

Figure 6.1 Simplified structure of an *n*-channel, depletion MOSFET. The *n* and *p* regions are interchanged in a *p*-channel device.

gate electrode, the silicon surface below it and the intervening insulating oxide behave like a parallel-plate capacitor. When a voltage is applied to the gate, an electric field is set up near the surface of the silicon, and it is this field which controls the flow of current in the narrow channel extending between the source and drain contacts. The presence of the insulating oxide layer also forces the gate current to be zero.

As stated earlier, both the n-channel and the p-channel MOSFET can be of either the enhancement or the depletion type. The main difference between these two types is that a physical channel exists in the latter, as shown in Fig. 6.1, while the channel in the enhancement FET is *created* by the application of a gate voltage of appropriate magnitude and polarity. As a result, the depletion device operates with either polarity of gate voltage, whereas the enhancement MOSFET operates with only one. Despite that, the latter has a much wider range of applications, and will form the basis of the following discussion and examples.

All the important mechanisms which give rise to the MOSFET's useful characteristics occur near the surface of the silicon. However, it is apparent from Fig. 6.1 that the MOS transistor structure gives rise to pn junctions in the bulk of the silicon between the substrate and, respectively, the source and drain regions. These junctions are undesirable, but unavoidable, byproducts of the way the MOSFET is constructed, and have to be always held in a reverse-biased condition. For this reason the *substrate* is connected to a separate terminal B as shown in Fig. 6.1.

The MOSFET is a symmetrical device and the source and drain terminals are identified on the basis of the direction of channel current, which is out of the source terminal for an n-channel FET and in the opposite direction for a p-channel device.

Figure 6.2 shows the symbols for the n-channel and p-channel enhancement MOSFETs, as well as the terminal current and voltage conventions. The corresponding symbols for the n-channel depletion MOSFET appear in Fig. 6.3. For

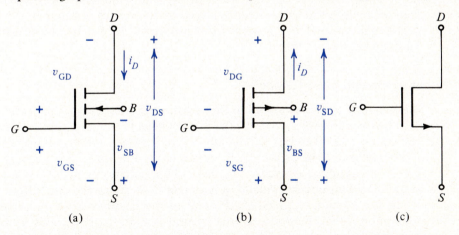

Figure 6.2 Symbols and terminal voltage and current conventions for the (a) n-channel and (b) p-channel enhancement MOSFET. Often the substrate terminal is not shown as in (c) for an n-channel MOSFET.

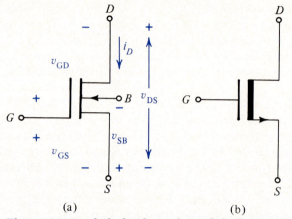

Figure 6.3 Symbols for the n-channel depletion MOSFET: (a) detailed, (b) simplified. For a p-channel device the arrow direction is reversed.

reasons of convenience, voltages are labeled with a double subscript to identify the terminals between which they are defined. The subscripts also define the assumed polarity of the voltage, with the terminal corresponding to the first subscript taken to be the higher potential: thus v_{GS} will be positive if the gate terminal of the FET is positive with respect to the source. This convention is used in the case of the JFET and BJT as well.

The arrowhead on the substrate contact corresponds to the direction in which current *would* flow if the *pn* junction between the substrate and the source or drain were forward biased. In this way one can distinguish between n-channel and p-channel devices. In ICs all the transistors are on a common substrate and, in such a case, for the sake of clarity, a simpler MOSFET symbol is used. Figures 6.2(c) and 6.3(b) show these symbols for, respectively, the enhancement and depletion n-channel devices. The arrowhead is placed on the source terminal in the direction of channel current flow, so that for a p-channel device the arrowhead in Figs. 6.2(c) and 6.3(b) is reversed.

6.2.2 Modes of Operation and Models

The MOSFET can be operated in three distinct modes: *cutoff, triode,* and *pinch-off.* Each mode will be described for the n-channel device, whence the characteristics of the p-channel MOSFET will be inferred by analogy. Because the enhancement and depletion devices have much in common, and because the former is more widely used, the presentation is in terms of the enhancement MOSFET. Reference to the depletion device is made only where their characteristics differ.

In the n-channel MOSFET the end of the channel which is at the higher potential is called the *drain*, so that current flow in the channel is from drain to source. Each mode of operation is identified with reference to a parameter called the *threshold voltage* V_T which, for an n-channel MOSFET, is positive. Thus, if the gate-source voltage v_{GS} satisfies the condition

$$v_{GS} < V_T \qquad\qquad\qquad\qquad\qquad\qquad\qquad (6.1)$$

then the device is cut off, meaning that the drain current $i_D = 0$ for all $v_{DS} \geq 0$.

If $v_{GS} > V_T$ and v_{DS} is increased from zero, the MOSFET enters the triode mode in which it behaves as a nonlinear voltage-controlled resistance. The drain current i_D is found initially to increase linearly with v_{DS} according to the relationship

$$i_D = \beta\,(v_{GS} - V_T)v_{DS} \qquad\qquad\qquad\qquad\qquad (6.2)$$

where the constant β is a function of various process parameters including the MOSFET geometry. Thus the slope of the i_D-v_{DS} characteristic is proportional to v_{GS}. This behavior is depicted in Fig. 6.4. However, as v_{DS} is increased further the slope of the i_D-v_{DS} characteristic decreases and finally tends to approximately zero as v_{DS} approaches the condition

$$v_{GS} - v_{DS} = V_T \qquad\qquad\qquad\qquad\qquad\qquad (6.3)$$

Equation (6.3) defines the boundary between the triode mode and the pinch-off mode.

In the pinch-off mode the MOSFET characteristics are approximated quite well by the expression

$$i_D = \frac{\beta}{2}\,(v_{GS} - V_T)^2 \qquad\qquad\qquad\qquad\qquad (6.4)$$

The device operates in this manner provided that $v_{GS} - v_{DS} < V_T$ or, alternatively, if

$$v_{GD} < V_T \qquad\qquad\qquad\qquad\qquad\qquad\qquad (6.5)$$

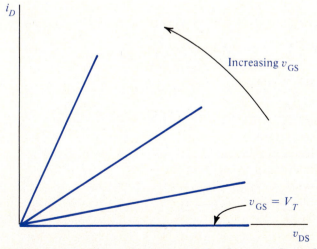

Figure 6.4 Triode mode characteristics for an n-channel MOSFET near $i_D = 0$, $v_{DS} = 0$.

Since the boundary between the two modes of operation is usually defined on the device's i_D-v_{DS} characteristics, it is convenient to express it in the form which results when Eqs. (6.3) and (6.4) are combined, i.e.,

$$i_D = \frac{\beta}{2} v_{DS}^2 \tag{6.6}$$

Figure 6.5 shows the output characteristics of an n-channel enhancement MOSFET based on the simple model considered here. Since they contain the salient features of the device, the model is called a *first-order* model.

A set of equations similar to Eqs. (6.1) to (6.6) can be used to describe the p-channel enhancement MOSFET since its characteristics differ from those of the n-channel device mainly in that the polarity of all currents and voltages is reversed and V_T is negative. Thus, if the polarity convention in Fig. 6.2(b) is used, and the quantities v_{GS}, v_{DS}, v_{GD}, and V_T in Eqs. (6.1) to (6.6) are replaced by, respectively, v_{SG}, v_{SD}, v_{DG}, and $-V_T$, the resulting equations correspond to a first-order model for the p-channel enhancement MOSFET.

The essential difference between the enhancement and depletion MOSFETs is that the latter can be operated with either polarity of gate-to-source voltage. This comes about because, for a depletion MOSFET, V_T is negative in an n-channel device and is positive for a p-channel FET. Otherwise the same equations describe the enhancement and depletion MOSFET. Figure 6.6 shows the simplified characteristics of an n-channel depletion MOSFET. Notice that $i_D > 0$ for all $v_{GS} > 0$ and also for negative v_{GS} down to V_T.

It is in the pinch-off region that the MOSFET is used as an amplifier because, as can be seen from Eq. (6.4), i_D is independent of v_{DS} and the device, therefore, behaves as a VCCS.

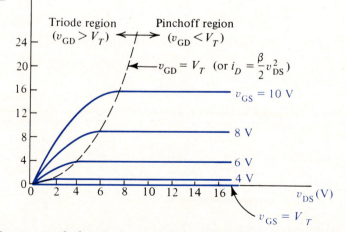

Figure 6.5 Ideal static $i_D - v_{DS}$ characteristics for an n-channel enhancement-type MOSFET with $V_T = 2$ V and $\beta = 0.5$ mA/V².

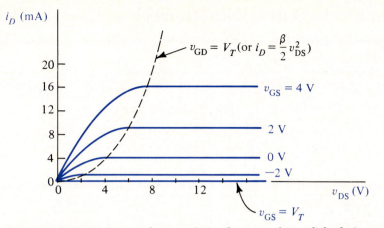

Figure 6.6 Static i_D–v_{DS} characteristics for an *n*-channel depletion MOSFET with the same parameters as the enhancement MOSFET characterized in Fig. 6.5 except for $V_T = -4$ V.

EXAMPLE 6.1

Determine i_D and v_{DS} in the circuit shown in Fig. E6.1 assuming $\beta = 0.5$ mA/V^2 and $V_T = 1$ V.

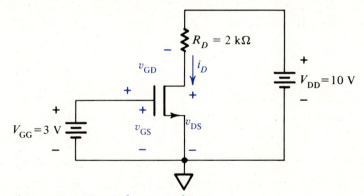

Figure E6.1 Circuit for Example 6.1.

Solution

Assume the MOSFET is in the pinch-off mode. Therefore, from Eq. (6.4),

$$i_D = \frac{\beta}{2}(v_{GS} - V_T)^2 = \frac{0.5}{2}(3 - 1)^2 \text{ mA} = 1 \text{ mA}$$

$$v_{DS} = V_{DD} - i_D R_D = 10 - 1(2) = 8 \text{ V}$$

$$v_{GD} = v_{GS} - v_{DS} = 3 - 8 = -5 \text{ V}$$

Since $v_{GD} < V_T$, the MOSFET is indeed in the pinch-off mode.

DRILL EXERCISE

6.1 The MOSFET in Example 6.1 is replaced by an n-channel depletion device with $\beta = 0.5$ mA/V^2 and $V_T = -2$ V. If $V_{GG} = 0$ compute i_D and v_{DS}.
Ans. 1 mA, 8 V

EXAMPLE 6.2

Determine v_{GS}, i_D and v_{DS} in the circuit in Fig. E6.2(a) using $\beta = 1$ mA/V^2 and $V_T = 2$ V for the MOSFET. Obtain both an analytical and a graphical solution.

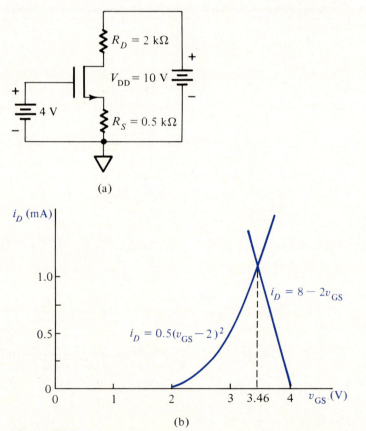

(a)

(b)

Figure E6.2

Solution

1. *Analytical method.* By KVL around the input loop

$$v_{GS} = 4 - i_D R_S = 4 - 0.5 i_D$$

if i_D is in milliamperes. Assuming the transistor to be in the pinch-off mode, i_D is also related to v_{GS} through Eq. (6.4)

$$i_D = \frac{\beta}{2}(v_{GS} - V_T)^2 = 0.5(v_{GS} - 2)^2$$

Eliminating i_D using this pair of equations leads to the following quadratic equation in v_{GS}:

$$v_{GS}^2 - 12 = 0$$

whence $v_{GS} = 3.46$ V. Therefore i_D has a value of

$$i_D = 8 - 2v_{GS} = 1.07 \text{ mA}$$

and v_{DS} is given by

$$\begin{aligned} v_{DS} &= V_{DD} - i_D(R_D + R_S) \\ &= 10 - 1.07(2 + 0.5) \\ &= 7.32 \text{ V} \end{aligned}$$

Since $v_{GS} - v_{DS} = 3.46 - 7.32 = -3.86$ V, which is less than V_T, the FET is indeed in pinch-off.

2. *Graphical method.* The KVL equation relating i_D and v_{GS}

$$i_D = 8 - 2v_{GS}$$

can be plotted on the MOSFET's transfer characteristic corresponding to pinch-off operation. This is done in Fig. E6.2(b). The solution is given by the intersection of the two curves, yielding $v_{GS} = 3.46$ V, $i_D = 1.07$ mA. To compute v_{DS}, it is easier to use the analytical approach.

6.2.3 The MOSFET as an Amplifier

Probably the most important function in analog circuits is signal amplification. Figure 6.7 shows the circuit of an *n*-channel enhancement MOSFET amplifier. The circuit is selected primarily for its conceptual simplicity rather than for its practicality. Because the source is grounded and is common to both the input and output ports, this circuit is called a *grounded-source* or *common-source* amplifier.

DC CONDITIONS
With $v_I = 0$

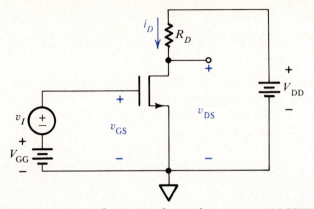

Figure 6.7 Simple circuit of an enhancement MOSFET common-source amplifier.

$$v_{GS} = V_{GS} = V_{GG}$$

$$i_D = I_D$$

and, if $V_{GG} > V_T$, and $V_{GD} < V_T$, the MOSFET will be in pinch-off where it behaves like a voltage-controlled current source. Therefore, from Eq. (6.4)

$$I_D = \frac{\beta}{2}(v_{GS} - V_T)^2 = \frac{\beta}{2}(V_{GG} - V_T)^2$$

whence v_{DS} can be computed from

$$v_{DS} = V_{DS} = V_{DD} - I_D R_D$$

In order to satisfy the pinch-off requirement it is necessary that

$$V_{GD} = V_{GS} - V_{DS} = V_{GG} - V_{DD} + I_D R_D < V_T$$

EXAMPLE 6.3

The MOSFET in Fig. 6.7 is characterized by $\beta = 0.5$ mA/V^2, $V_T = 2$ V. Assuming that $R_D = 2$ kΩ, $V_{DD} = 15$ V and $v_I = 0$, determine both analytically and graphically the range of V_{GG} for which the MOSFET will be in the pinch-off mode with $I_D \geq 0$.

Solution

1. *Analytical method.* If $I_D \geq 0$ then one limit on V_{GG} is

$$V_{GS} = V_{GG} > V_T = 2 \text{ V}$$

The boundary between triode and pinch-off operation is, from Eq. (6.3),

$$V_{GS} - V_{DS} = V_T = 2 \text{ V}$$

According to Eq. (6.6) the drain current I_D, under these conditions, is given by

$$I_D = \frac{\beta}{2} V_{DS}^2 = 0.25 \ V_{DS}^2$$

Combining this expression with the KVL equation which determines V_{DS}, i.e.,

$$V_{DS} = V_{DD} - I_D R_D = 15 - 2I_D$$

one obtains a quadratic equation in one unknown V_{DS}:

$$0.5 V_{DS}^2 + V_{DS} - 15 = 0$$

The positive, and physically meaningful, root is

$$V_{DS} = 4.57 \text{ V}$$

so that, at the boundary between the triode and pinch-off regions,

$$V_{GG} = V_T + V_{DS} = 2 + 4.57 = 6.57 \text{ V}$$

Therefore, for the MOSFET to operate in the pinch-off mode, V_{GG} must be in the range

$$2 \text{ V} < V_{GG} < 6.57 \text{ V}$$

2. *Graphical solution.* The load line equation

$$v_{DS} = 15 - 2i_D$$

is plotted on the MOSFET's output characteristics as shown in Fig. E6.3.

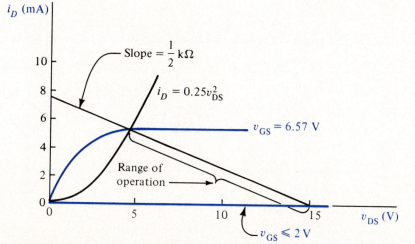

Figure E6.3 Solution of Example 6.3.

The intersection of the load line with the curve

$$i_D = \frac{\beta}{2} v_{DS}^2 = 0.25 \, v_{DS}^2$$

yields the largest value of i_D for pinch-off operation, since the point is on the boundary between the triode and pinch-off modes. This corresponds to $v_{DS} \simeq 4.6$ V and, since here $v_{GS} - v_{DS} = V_T$, it follows that $v_{GS} = 4.6 + 2 = 6.6$ V which, to the accuracy of the graphical solution, agrees with the result obtained analytically.

Evidently the other limit is given by the intersection of the load line with the v_{DS} axis which corresponds to $v_{GS} = V_T = 2$ V.

SMALL-SIGNAL MODEL OF THE MOSFET

Since the gate current is zero it follows that one single nonlinear equation describes the two-port characteristics of the n-channel MOSFET in pinch-off, namely, Eq. (6.4)

$$i_D = \frac{\beta}{2} (v_{GS} - V_T)^2 \tag{6.7}$$

which is valid for $v_{GS} > V_T$. There is, therefore, a single parameter in the first-order small-signal model, namely,

$$g_m = \left. \frac{\partial i_D}{\partial v_{GS}} \right|_Q \tag{6.8}$$

where Q is the quiescent point of the MOSFET determined by V_{GS}, I_D, and V_{DS}. From Eqs. (6.7) and (6.8) g_m, called the *transconductance*, depends on V_{GS} according to

$$g_m = \beta \, (V_{GS} - V_T) \tag{6.9}$$

Figure 6.8(a) shows the resulting first-order small-signal model for the MOSFET in the pinch-off region. Evidently this is a simple y-parameter represen-

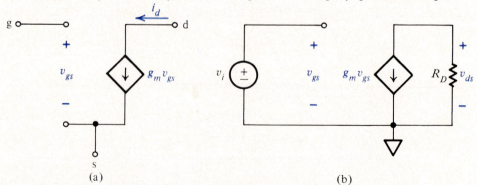

(a)   (b)

Figure 6.8 (a) First-order small-signal model of the MOSFET in pinch-off, (b) small-signal model of the amplifier in Fig. 6.7.

tation with all the y parameters except for $y_{21} = g_m$ equal to zero. Because the model is derived with the MOSFET's source terminal common, it is called a *common-source model*.

The small-signal model of the amplifier in Fig. 6.7 appears in Fig. 6.8(b). By inspection, the small-signal voltage gain v_{ds}/v_i is

$$\frac{v_{ds}}{v_i} = -g_m R_D \tag{6.10}$$

EXAMPLE 6.4

The p-channel MOSFET in the amplifier in Fig. E6.4 has $\beta = 1.0$ mA/V^2, $V_T = -1$ V. Compute the small-signal voltage gain v_o/v_i and compute the limit on the magnitude of v_I for a 10% error in the small-signal approximation.

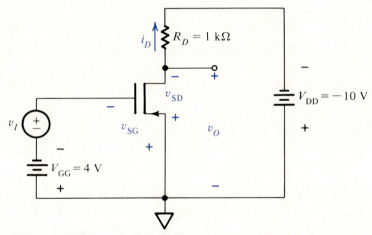

Figure E6.4 Circuit for Example 6.4.

Solution
DC conditions correspond to $v_I = 0$. Noting that the MOSFET is not cut off since

$$V_{SG} = 4 \text{ V} > -V_T$$

and assuming that the device is in pinch-off, Eq. (6.4), with V_{GS} and V_T replaced by, respectively, V_{SG} and $-V_T$, yields the drain current I_D to be

$$I_D = \frac{1.0}{2}(4 - 1)^2 = 4.5 \text{ mA}$$

The dc output voltage is, in that case,

$$V_O = -V_{SD} = I_D R_D + V_{DD} = 4.5 - 10 = -5.5 \text{ V}$$

Therefore, V_{DG} is

$$V_{DG} = v_{SG} - v_{SD} = 4 - 5.5 = -1.5 \text{ V}$$

which is less than $-V_T$, so the MOSFET is, indeed, in pinch-off.

The small-signal model of the amplifier is the same as that in Fig. 6.8(b). The transconductance g_m is computed from Eq. (6.9) with v_{GS} and V_T replaced by, respectively, v_{SG} and $-V_T$, i.e.,

$$g_m = 1.0(4 - 1) = 3 \text{ mS}$$

Therefore, the voltage gain is

$$\frac{v_o}{v_i} = -g_m R_D = -3$$

There remains now the calculation of the limit on v_I for an error or not more than 10% due to the small-signal approximation. This can be done by substituting $v_{GS} = V_{GS} + v_{gs}$ into Eq. (6.4), i.e.,

$$i_D = \frac{\beta}{2}(V_{GS} + v_{gs} - V_T)^2$$

Expanding the expression in brackets and separating dc and small-signal terms results in

$$i_D = I_D + \beta(V_{GS} - V_T)v_{gs} + \frac{\beta}{2}v_{gs}^2$$

Combining this with Eq. (6.9) produces

$$i_D = I_D + g_m v_{gs} + \frac{\beta}{2}v_{gs}^2$$

The same result is obtained if the form of Eq. (6.4) corresponding to a p-channel device is used.

Since the second term is the same as in the small-signal model, the error is given by the last term. Therefore a 10% error means that

$$\frac{(\beta/2)v_{gs}^2}{g_m v_{gs}} = 0.1$$

Substituting in the values of β and g_m leads to the result

$$v_I = v_{gs} = 0.6 \text{ V}$$

So if v_I were a sinusoidal signal

$$v_I = V_i \sin \omega t$$

its amplitude would have to be limited to 0.6 V for a $\pm 10\%$ error.

Before leaving this example it is instructive to note that, for a sinusoidal input, the term in the expression for i_D which is quadratic in v_{gs} and which is excluded in the small-signal approximation gives rise to an output signal component at twice the input frequency ω. This observation follows from

$$\frac{\beta}{2}v_{gs}^2 = \frac{\beta}{2}V_i^2 \sin^2 \omega t = \frac{\beta}{4}V_i^2 (1 - \cos 2\omega t)$$

This *second harmonic* distortion is undesirable in linear signal processing applications such as stereo systems or radio receivers and, therefore, often determines the maximum usable amplitude of the input signal v_I.

The above equation also contains a term which is constant for a given V_i. A little thought should convince the reader that this implies that, as V_i increases, the operating point of the amplifier shifts from its quiescent value ($v_I = 0$) since, for example, the average value of the drain current i_D becomes the sum of I_D and the constant term $(\beta/4)V_i^2$.

6.2.4 The Source Follower

A problem which is frequently encountered in the design of electronic systems is that of coupling a source having a high Thevenin resistance to a low resistance load, with little loss of voltage level. This is normally accomplished with a *buffer* or *isolation* amplifier. The FET, operated with its drain terminal common, exhibits the characteristics that such an amplifier should have. Figure 6.9 shows a simple buffer amplifier based on an *n*-channel, depletion MOSFET. It is usually called a *source follower* for reasons which will become apparent shortly.

As in the case of the common-source amplifier, the MOSFET is operated in the pinch-off mode where it behaves as a voltage-controlled current source. This imposes the constraints

$$v_{GD} = v_G - V_{DD} < V_T$$

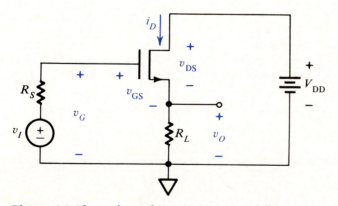

Figure 6.9 The *n*-channel MOSFET source follower.

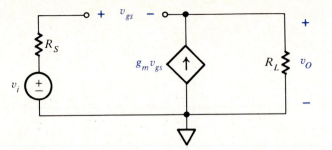

Figure 6.10 Small-signal model of the source follower in Fig. 6.9.

and

$$v_{GS} = v_G - i_D R_L > V_T$$

Under quiescent conditions, because the gate draws negligible current, $v_G = 0$ and the constraints simplify to

$$V_{DD} > -V_T$$

and

$$R_L < \frac{-V_T}{I_D}$$

These two inequalities, together with Eq. (6.4), are used to select a suitable operating point for the amplifier.

Under small-signal conditions the amplifier model is that shown in Fig. 6.10 if the MOSFET is replaced by the small-signal model in Fig. 6.8(a). The voltage gain is obtained by noting that

$$v_o = g_m R_L v_{gs} = g_m R_L (v_i - v_o)$$

whence

$$\frac{v_o}{v_i} = \frac{g_m R_L}{1 + g_m R_L} \tag{6.11}$$

Equation (6.11) shows that if $g_m R_L \gg 1$ the output voltage v_o tends toward the input v_i. The reason for the name *source follower* now becomes apparent if one also notes that v_o, which is taken from the source, has the same polarity as, and therefore follows, v_i. Because, as can be seen from Fig. 6.10, the drain terminal is common to the input and output ports, the source follower is also called a *common-drain* amplifier.

EXAMPLE 6.5

The parameters of the circuit in Fig. 6.9 are $R_S = 100$ kΩ, $R_L = 1$ kΩ, $V_{DD} = 10$ V,

$\beta = 0.5$ mA/V^2 and $V_T = -4$ V. Determine v_o/v_i and compare the result with the case where a source follower is not used and R_L is connected across R_S and v_I.

Solution
Under quiescent conditions

p 241

and, from Eq. (6.4),

$$I_D = \frac{\beta}{2}(V_{GS} - V_T)^2$$

if the transistor is in pinch-off. The solution of these equations with the given parameters is

$$V_{GS} = -1.5 \text{ V} \qquad I_D = 1.5 \text{ mA}$$

Since $V_{GS} > V_T$ and $V_{GD} = V_G - V_{DD} = -10$ V $< V_T$, the MOSFET is indeed in pinch-off. From Eq. (6.9)

$$g_m = \beta(V_{GS} - V_T)$$
$$= 0.5(-1.5 + 4) = 1.25 \text{ mS}$$

so the voltage gain is, from Eq. (6.11),

$$\frac{v_o}{v_i} = \frac{1.25}{1 + 1.25} = 0.56$$

In contrast, if R_L is connected directly across R_S and the voltage source,

$$\frac{v_o}{v_i} = \frac{R_L}{R_L + R_S} = \frac{1}{100 + 1} \simeq 0.01$$

Thus the source follower does act as a buffer amplifier, transferring a signal from a source with a high Thevenin resistance R_S to a comparatively low resistance load R_L with relatively little signal loss or *attenuation*.

6.2.5 The MOSFET Amplifier with Active Load

The simple amplifier circuit in Fig. 6.7 would be impractical in an IC design for several reasons, one of the most important being the fact that the resistor would occupy much more area than the MOSFET. This would limit severely the functional density of the IC. The problem is solved by using MOSFETs as load resistors, albeit nonlinear ones.

Figure 6.11 shows how this is achieved, using an n-channel enhancement device for illustration. Since the boundary between the triode and pinch-off regions is defined by $v_{GS} - v_{DS} = V_T$, it follows that the circuit constraint $v_{GS} = v_{DS}$

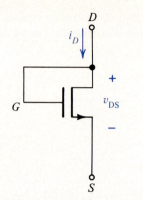

Figure 6.11 The *n*-channel enhancement MOSFET connected as a nonlinear resistance.

forces the MOSFET to operate in the pinch-off region for all $i_D \geq 0$. This is illustrated in Fig. 6.12.

From Eq. (6.4), since $v_{GS} = v_{DS}$, it follows that the circuit in Fig. 6.11 behaves as a nonlinear resistance having a terminal characteristic given by

$$i_D = \frac{\beta}{2}(v_{DS} - V_T)^2 \tag{6.12}$$

Figure 6.13 shows a sketch of the graph corresponding to Eq. (6.12).

Consider now the amplifier circuit in Fig. 6.14 in which the conventional *pull-up* resistor between the drain terminal and the supply voltage V_{DD} is replaced by the arrangement in Fig. 6.11. Details pertaining to the biasing are omitted for the time being. Following a universally accepted convention, the voltage source V_{DD} is not shown explicitly, and the drain of Q_2 is understood to be at a constant

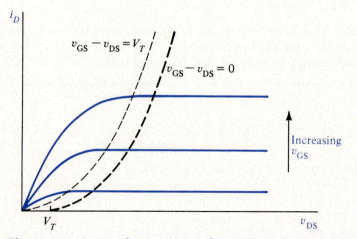

Figure 6.12 Output characteristics of an *n*-channel enhancement MOSFET showing the locus of operating points imposed by the constraint $v_{GS} = v_{DS}$.

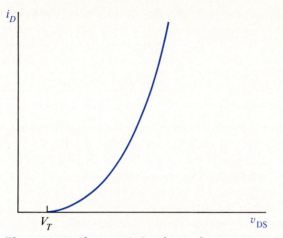

Figure 6.13 Characteristic of an enhancement MOSFET operated as in Fig. 6.11.

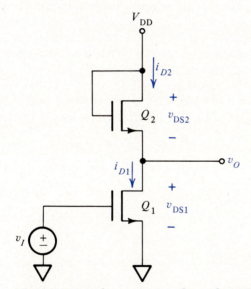

Figure 6.14 Enhancement *n*-channel MOSFET amplifier with the nonlinear load arrangement of Fig. 6.11.

potential of V_{DD} with respect to ground. This convention will be followed henceforth with all constant voltage sources having one terminal grounded. The equation governing the behavior of Q_2 in this circuit is, from Eq. (6.12),

$$i_{D2} = \frac{\beta}{2}(v_{DS2} - V_T)^2$$

where, for the sake of generality, the geometry-dependent parameter β_2 of Q_2 is assumed to differ from the corresponding β_1 of Q_1. The threshold voltages, on the other hand, are assumed to be equal because, in an IC, V_T is approximately the same for, respectively, all enhancement and depletion devices on the same chip.

Since $i_{D2} = i_{D1}$ and $v_{DS2} = V_{DD} - v_{DS1}$, the above equation can be rewritten in the more convenient form

$$i_{D1} = \frac{\beta_2}{2}(V_{DD} - v_{DS1} - V_T)^2 \tag{6.13}$$

The simultaneous solution of Eq. (6.13) with the output characteristics of Q_1 yields the solution of the circuit in Fig. 6.14. This can be done graphically by plotting the *load curve* Eq. (6.13) on the output characteristics of Q_1. Figure 6.15 shows the result of this operation. The solution of the circuit in Fig. 6.14, for a given v_I, is obviously given by the intersection of the load curve, described by Eq. (6.13), with the output characteristic of Q_1 for that v_I.

Figure 6.16 shows a typical voltage transfer characteristic for the circuit in Fig. 6.14. The exact shape depends, among other things, on the relative magnitude of β_2 and β_1. Thus, although the discontinuity in the characteristic at $v_I = v_{GS1} = V_T$ simply depends on V_T, the point at $v_I = V_{IB}$, where Q_1 changes from the pinch-off to the triode mode, is influenced not only by V_T but also by the ratio β_1/β_2.

Since a useful voltage amplifier should have a linear transfer characteristic with a slope that is greater than one, it follows that the central region of the characteristic in Fig. 6.16, corresponding to Q_1 operating in pinch-off, is the most useful one for amplification. The following analysis will show that the voltage transfer characteristic of the circuit in Fig. 6.14 is in fact linear in this region.

Since $v_{GS1} = v_I$ and $i_{D1} = i_{D2} = i_D$ the transfer characteristic of Q_1 can be written as

$$i_D = \frac{\beta_1}{2}(v_I - V_T)^2 \tag{6.14}$$

while for Q_2 the corresponding expression is

$$i_D = \frac{\beta_2}{2}(V_{DD} - v_O - V_T)^2 \tag{6.15}$$

since

$$v_{GS2} = v_{DS2} = V_{DD} - v_O$$

Solving Eqs. (6.14) and (6.15) for v_O in terms of v_I one obtains

$$v_O = \left[V_{DD} - V_T + \left(\frac{\beta_1}{\beta_2}\right)^{1/2} V_T\right] - \left(\frac{\beta_1}{\beta_2}\right)^{1/2} v_I \tag{6.16}$$

which is a linear equation of the form

$$v_O = K_1 - K_2 v_I \tag{6.17}$$

Equation (6.16) shows that the circuit in Fig. 6.14 behaves as a linear amplifier if the input voltage variations are constrained to the pinch-off region of Q_1. The

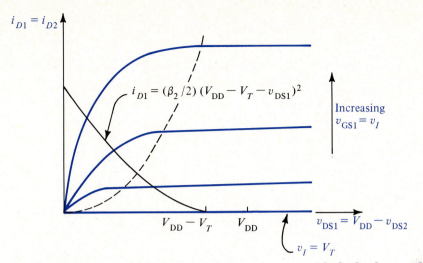

Figure 6.15 Output characteristics of Q_1 in Fig. 6.14 with the load curve due to Q_2 superimposed.

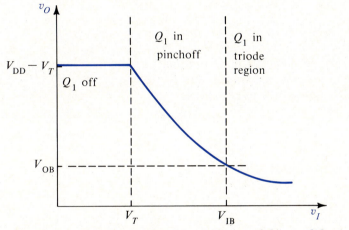

Figure 6.16 Voltage transfer characteristics of the amplifier in Fig. 6.14.

gain of the amplifier is evidently $-(\beta_1/\beta_2)^{1/2}$ which can be controlled by adjusting the relative sizes of Q_1 and Q_2.

As was already noted previously, for this circuit to behave as a linear amplifier, one constraint on the input voltage v_I is

$$v_I > V_T$$

which ensures that Q_1 does not cut off. The other limit is the boundary between the triode and pinch-off regions of Q_1,

$$v_{GS1} - v_{DS1} = V_T$$

or

$$v_I = v_O + V_T$$

Substituting this into Eq. (6.16) yields the result that the point defined by V_{IB} and V_{OB} in Fig. 6.16 is given by

$$V_{IB} = \frac{V_{DD} + (\beta_1/\beta_2)^{1/2}V_T}{1 + (\beta_1/\beta_2)^{1/2}} \tag{6.18}$$

and, since

$$V_{OB} = V_{IB} - V_T$$

therefore,

$$V_{OB} = \frac{V_{DD} - V_T}{1 + (\beta_1/\beta_2)^{1/2}} \tag{6.19}$$

EXAMPLE 6.6

The amplifier in Fig. 6.14 has $V_{DD} = 18$ V, $V_T = 2$ V and $\beta_1/\beta_2 = 9$. If the input voltage v_I consists of a dc voltage source V_{GG} in series with a sinusoidal voltage source v_i, as shown in Fig. E6.6, determine the value of V_{GG} and the amplitude v_i for linear operation with maximum output voltage amplitude. Also compute the corresponding output voltage amplitude.

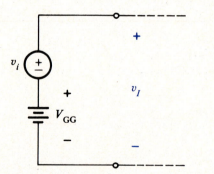

Figure E6.6 Input circuit of the amplifier in Fig. 6.14.

Solution

In order to determine the conditions necessary for maximum output voltage amplitude with linear operation it is necessary first to establish the limits of the linear portion of the transfer characteristic in Fig. 6.16.

One limit corresponds to the point where Q_1 cuts off, and is simply $v_I = V_T = 2$ V. The other limit is given by Eq. (6.18)

$$v_I = V_{IB} = \frac{V_{DD} + (\beta_1/\beta_2)^{1/2} V_T}{1 + (\beta_1/\beta_2)^{1/2}} = 6 \text{ V}$$

Because the transfer characteristic between the above two limits is linear it follows that, since

$$v_I = V_{GG} + V_i \sin \omega t$$

V_{GG} should be midway between the two limits, i.e., $V_{GG} = (2 + 6)/2 = 4$ V, and $V_i = 2$ V.

Since the voltage gain of the amplifier is $-(\beta_1/\beta_2)^{1/2} = -3$ the output voltage amplitude will be 6 V.

6.2.6 The MOSFET as a Switch

The transfer characteristic in Fig. 6.16 indicates that the MOSFET Q_1 in the circuit shown in Fig. 6.14 can be used as a switch. This follows from the observation that for $v_I < V_T$ the switch is open since $i_D = 0$. For sufficiently large v_I, on the other hand, the voltage $v_O = v_{DS1}$ tends to zero and, in view of Eq. (6.4), a large current i_D flows, which corresponds to a closed switch. The voltage across the closed switch can be reduced by using a high β_1/β_2 ratio, as can be seen from Eq. (6.19).

An important application of the circuit in Fig. 6.14 is as an *inverter* in logic circuits. Thus, referring to Fig. 6.16, if the voltage $V(1) = V_{DD} - V_T$ is identified with logical *TRUE*, and some voltage $V(0) < V_T$ is considered to be logical *FALSE*, then, for a sufficiently high β_1/β_2 ratio, applying $v_I = V(1)$ will result in $V(0)$ and vice versa. This means that if the input is TRUE, the output is FALSE and vice versa, which are the conditions for logical inversion.

EXAMPLE 6.7

The circuit in Fig. 6.14, with $V_{DD} = 10$ V, is to be used as a logical inverter. If Q_1 has the characteristics shown in Fig. 6.5 determine β_2 for Q_2 so that the FALSE state $V(0) = V_T/2$.

Solution
Since the LOW voltage state, $V(0) = V_T/2 = 1$ V, has been selected to represent FALSE, the TRUE state will correspond to $V(1) = V_{DD} - V_T = 10 - 2 = 8$ V. If $V(0)$ is to be 1 V then $v_I = 8$ V must result in $v_O = 1$ V. From the characteristics in Fig. 6.5

$$v_I = V_{GS1} = 8 \text{ V}$$

results in

$$v_O = v_{DS1} = 1 \text{ V}$$

if

$$i_D = 4 \text{ mA}$$

Equation (6.15) can now be solved for β_2:

$$\beta_2 = \frac{2i_D}{(V_{DD} - v_O - V_T)^2}$$

$$= \frac{8}{(10 - 1 - 2)^2}$$

$$= \frac{8}{49} \text{ mA/V}^2$$

Since $\beta_1 = 0.5 \text{ mA/V}^2$, therefore,

$$\frac{\beta_1}{\beta_2} = \frac{0.5}{\frac{8}{49}} = 3.06$$

The i_D-v_{DS} characteristics of Q_2 can now be obtained from Fig. 6.5 by simply scaling i_D down by 3.06.

Figure 6.17 shows an alternative form of MOSFET inverter in which a depletion device is used as the load. Because the gate of Q_2 is connected to its source, its drain current i_D and drain-to-source voltage v_{DS} are related through the $v_{GS} = 0$ curve in the output characteristics (see Fig. 6.6). An important advantage of this inverter over the one in Fig. 6.14 is that the output voltage rises all the way to V_{DD} when $v_I < V_T$, the threshold voltage of the enhancement FET Q_1. This circuit is analyzed in Chap. 8.

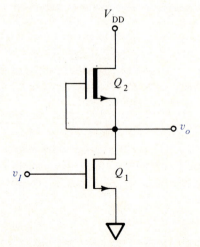

Figure 6.17 *n*-channel MOSFET inverter with depletion load device.

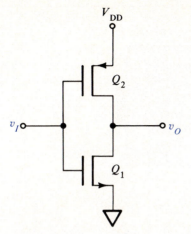

Figure 6.18 A complementary-symmetry MOS (CMOS) inverter.

The fact that n-channel as well as p-channel MOSFETs can be produced on the same IC using CMOS technology makes it possible to exploit some very useful circuits. Figure 6.18 shows a complementary-symmetry MOS (CMOS) inverter. Both transistors are of the enhancement type but, because V_T is positive for an n-channel device and negative for a p-channel, either Q_1 or Q_2 is off, irrespective of whether $v_I = V(1) = V_{DD}$ or $v_I = V(0) = 0$. Since Q_1 and Q_2 are in series, in both states the circuit draws zero current from V_{DD} and, therefore, consumes no power. This makes CMOS circuits very important. A complete analysis of the inverter in Fig. 6.18 is presented in Chap. 8.

6.3 THE JUNCTION FIELD-EFFECT TRANSISTOR (JFET)

6.3.1 Structure, Principles of Operation, and Symbols

The junction field-effect transistor (JFET), as its name implies, is similar to the MOSFET in the sense that, under suitable bias conditions, it too behaves like a VCCS, with the control being via the electric field distribution in a channel where current is constrained to flow. Figure 6.19 shows the simplified structure of an n-channel JFET. In a p-channel device the n and p regions are interchanged.

Just as in the MOSFET, the current in a JFET flows through a channel between source and drain contacts. However, the channel is in the bulk of the silicon rather than near the surface as in the case of the MOS transistor. This current is controlled by the voltage which is applied to the gate terminal and which, together with the potentials at the source and drain terminals, determines the bias conditions at the two pn junctions, and the electric field distribution in the channel. In most applications the pn junctions are reverse biased and, therefore, the gate current is very small and can usually be neglected. Consequently, the input resistance of a JFET is very large, although not as high as that of a MOSFET.

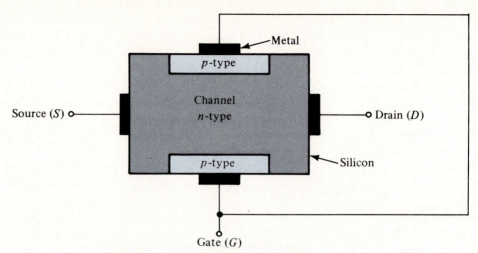

Figure 6.19 Simplified structure of the *n*-channel JFET.

Figure 6.20 shows the symbols and the polarity convention for the *n*-channel and *p*-channel JFETs which, just like the corresponding MOSFETs, are symmetrical. The source is identified by placing the gate close to the source on the symbol. The arrowhead on the gate, by convention, indicates the direction in which current would flow if the *pn* junctions were forward biased and, therefore, identifies the JFET as to whether it is an *n*-channel or *p*-channel device.

6.3.2 Modes of Operation and Models

The JFET, like the MOSFET, can be operated in three distinct modes: cutoff, triode and pinch-off. Each mode will be described for the *n*-channel device, whence the characteristics of the *p*-channel JFET will be inferred by analogy.

In the *n*-channel JFET the end of the channel which is at the higher potential is called the drain, so that current flow in the channel is from drain to source. Each

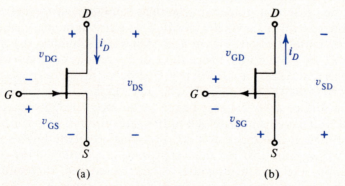

Figure 6.20 Symbols and polarity convention for the (a) *n*-channel and (b) *p*-channel JFET.

mode of operation is identified with reference to a parameter called the *pinch-off voltage* V_P which, for an *n*-channel JFET, is negative. Thus, if the gate-source voltage v_{GS} is sufficiently negative, i.e., if

$$v_{GS} < V_P \tag{6.20}$$

then the device is cut off, meaning that the drain current $i_D = 0$ for $v_{DS} \geq 0$.

If $v_{GS} > V_P$ and v_{DS} is increased from zero, the JFET enters the *triode* mode in which its behavior is the same as that of the MOSFET. The boundary between the triode mode and the pinch-off mode is given by

$$v_{GS} - v_{DS} = V_P \tag{6.21}$$

which can be also expressed as

$$v_{DG} = -V_P \tag{6.22}$$

In the pinch-off mode $v_{DG} > -V_P$ and the JFET behaves like a voltage-controlled current source. The functional relationship between i_D and v_{GS} is approximated quite well by the equation

$$i_D = I_{DSS}\left(1 - \frac{v_{GS}}{V_P}\right)^2 \tag{6.23}$$

which is plotted in Fig. 6.21 for typical values of JFET parameters. This mode is also called the *active region* because the JFET can be used in this region as an amplifier, which is the subject of the next section.

Equation (6.23) shows that i_D is independent of v_{DS} in the pinch-off mode. Therefore, a plot of the JFET's output characteristics, for $v_{DG} > -V_P$, should appear as a set of horizontal lines. These are shown, including the triode region, in Fig. 6.22. The boundary between the two modes of operation is found, by combining Eqs. (6.21) and (6.23), to be defined by

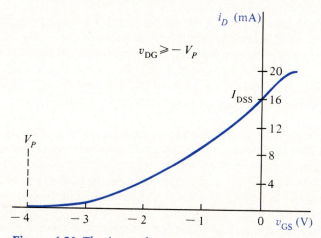

Figure 6.21 The i_D-v_{GS} characteristics of an *n*-channel JFET in pinch-off.

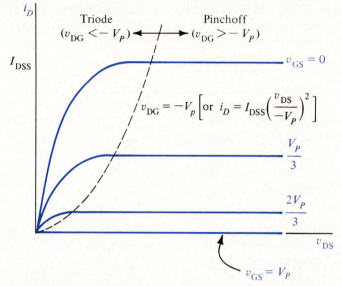

Figure 6.22 n-channel JFET first-order output characteristics.

$$i_D = I_{DSS}\left(\frac{v_{DS}}{-V_P}\right)^2 \tag{6.24}$$

which is a parabola passing through the origin.

The characteristics in Fig. 6.22 are plotted for v_{GS} values not exceeding zero. This is because $v_{GS} = 0$ defines the approximate upper boundary of the JFET's useful region of operation. The main attractions of the JFET in that region are the fact that the gate draws negligible current, and that the device can be used either as a voltage controlled current source or as a voltage controlled resistance. As v_{GS} becomes positive, substantial gate current begins to flow and v_{GS} loses its controlling properties as can be seen in Fig. 6.21.

A set of equations similar to Eqs. (6.20) to (6.24) are used to describe the p-channel JFET since its characteristics differ from those of the n-channel device mainly in that the polarity of all currents and voltages is reversed and V_P is positive. Thus if the polarity convention in Fig. 6.20(b) is used, and the quantities v_{GS}, v_{DS}, v_{DG}, and V_P in Eqs. (6.20) to (6.24) are replaced by, respectively, v_{SG}, v_{SD}, v_{GD}, and $-V_P$, the resulting equations correspond to the first-order model for the p-channel JFET.

For example, the p-channel device is in the cutoff mode if $v_{SG} < -V_P$, and the boundary between the triode and pinch-off regions is defined by $v_{SG} - v_{SD} = -V_P$ or $v_{GD} = V_P$. Note that in the pinch-off mode, where $v_{GD} > V_P$, Eq. (6.23) can be used directly since the signs of v_{GS} and V_P are the same in both types of JFET.

Figure 6.23(a) and (b) shows the approximate shape of, respectively, the output and $i_D - v_{SG}$ characteristic of a p-channel JFET for typical parameter values. As in the n-channel device, the pn junction is normally reverse-biased, i.e., $v_{SG} \leq 0$.

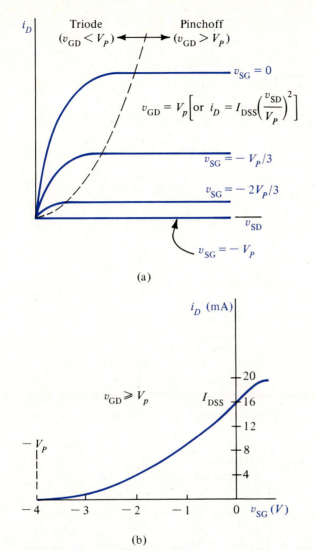

(a)

(b)

Figure 6.23 (a) Output characteristics and (b) transfer characteristic of a *p*-channel JFET.

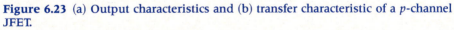

EXAMPLE 6.8

Determine I_D and V_{DS} in the circuit shown in Fig. E6.8 assuming $I_{DSS} = 12$ mA, $V_P = -3$ V.

Solution
Assume the transistor is in the pinch-off mode. Therefore

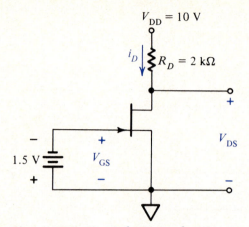

Figure E6.8 Circuit for Example 6.8.

$$I_D = I_{DSS}\left(1 - \frac{V_{GS}}{V_P}\right)^2 = 12\left(1 - \frac{1.5}{3}\right)^2 \text{ mA} = 3 \text{ mA}$$

$$V_{DS} = V_{DD} - I_D R_D = 10 - 3(2) = 4 \text{ V}$$

$$V_{DG} = V_{DS} - V_{GS} = 4 + 1.5 = 5.5 \text{ V}$$

Since $V_{DG} > -V_P = 3$ V, the JFET is indeed in pinch-off.

EXAMPLE 6.9

The p-channel JFET in Fig. E6.9 has $I_{DSS} = 15$ mA, $V_P = 4$ V. Determine the range of R_D for which the device operates in the pinch-off mode.

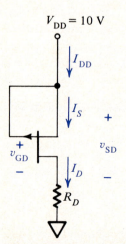

Figure E6.9 Circuit for Example 6.9.

Solution

For pinch-off, $V_{GD} > V_P = 4$ V. Since $V_{SG} = 0$, the JFET will remain in pinch-off as long as

$$V_{GD} = -V_{SG} + V_{SD} = V_{SD} > 4 \text{ V}$$

Also, since $V_{SG} = 0$,

$$I_D = I_S = I_{DSS} = 15 \text{ mA}$$

Therefore, the JFET will be in the pinch-off mode provided that

$$V_{SD} = V_{DD} - I_{DSS} R_D > 4 \text{ V}$$

or

$$R_D < \frac{V_{DD} - 4}{I_{DSS}} = \frac{10 - 4}{15} \text{ k}\Omega = 0.4 \text{ k}\Omega$$

Since the gate draws no current,

$$I_{DD} = I_S = I_{DSS}$$

which points up an interesting application for the JFET with $V_{GS} = 0$, namely, as a constant current source. It will be limited in its range of course by the requirement that $V_{SD} > V_P$ for the p-channel JFET and $V_{DS} > -V_P$ for the n-channel.

6.3.3 The JFET as an Amplifier

Figure 6.24 shows the circuit of an n-channel JFET amplifier. As in the case of the MOSFET, the circuit is selected primarily for its conceptual simplicity rather than for its practicality.

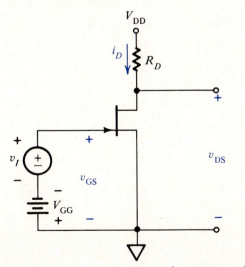

Figure 6.24 Simple circuit of a JFET amplifier.

DC CONDITIONS

With $v_I = 0$

$$v_{GS} = V_{GS} = -V_{GG}$$

$$i_D = I_D$$

and, if $-V_{GG} > V_P$, and $V_{DG} > -V_P$ the JFET will be in pinch-off where it behaves like a voltage-controlled current source. Therefore, from Eq. (6.23)

$$I_D = I_{DSS}\left(1 + \frac{V_{GG}}{V_P}\right)^2$$

whence v_{DS} can be computed from

$$v_{DS} = V_{DS} = V_{DD} - I_D R_D$$

In order to satisfy the pinch-off requirement it is necessary that

$$V_{DG} = V_{DS} - V_{GS} = V_{DD} - I_D R_D + V_{GG} > -V_P$$

SMALL-SIGNAL MODEL OF THE JFET

Comparing Eq. (6.23) with the corresponding Eq. (6.3) for the MOSFET, and noting that, in normal use, just as in the MOSFET, the gate current is zero, the small-signal model for pinch-off operation will be, by direct analogy to the MOSFET, as shown in Fig. 6.25(a). The transconductance g_m is given by

$$g_m = g_{mo}\left(1 - \frac{V_{GS}}{V_P}\right) \tag{6.25}$$

where

$$g_{mo} = 2\frac{I_{DSS}}{|V_P|} \tag{6.26}$$

is the maximum value of g_m, corresponding to $v_{GS} = 0$. These expressions are valid for both n-channel and p-channel JFETs.

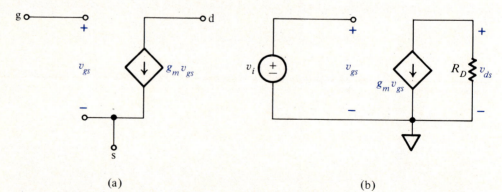

(a)   (b)

Figure 6.25 (a) First-order small-signal model of JFET in pinch-off (b) Small-signal model of amplifier in Fig. 6.24.

The small-signal model of the amplifier in Fig. 6.24 appears in Fig. 6.25(b). By inspection, the small-signal voltage gain v_{ds}/v_i is

$$\frac{v_{ds}}{v_i} = -g_m R_D \tag{6.27}$$

EXAMPLE 6.10

The p-channel JFET in the amplifier in Fig. E6.10 has $I_{DSS} = 16$ mA, $V_P = 4$ V. Compute the small-signal voltage gain v_o/v_i.

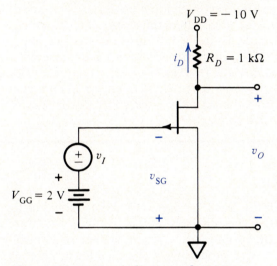

Figure E6.10 Circuit for Example 6.10.

Solution
DC conditions correspond to $v_I = 0$. Noting that the JFET is not cut off since

$$V_{SG} = -2 \text{ V} > -V_P$$

and assuming that the device is in pinch-off, Eq. (6.23) yields the drain current I_D to be

$$I_D = 16(1 - \tfrac{2}{4})^2 = 4 \text{ mA}$$

The dc output voltage is, in that case,

$$V_O = -V_{SD} = I_D R_D + V_D = 4 - 10 = -6 \text{ V}$$

Therefore, V_{GD} is

$$V_{GD} = 2 + 6 = 8 \text{ V}$$

which is greater than V_P, so the JFET is indeed in pinch-off.

The small-signal model of the amplifier is the same as that in Fig. 6.25(b). The transconductance g_m is computed from Eqs. (6.25) and (6.26) to be

$$g_m = \tfrac{32}{4}(1 - \tfrac{2}{4}) = 4 \text{ mS}$$

Therefore, the voltage gain is

$$\frac{v_o}{v_i} = -g_m R_D = -4$$

The applications for JFETs are much more limited than for MOSFETs or for bipolar junction transistors (BJT) which are covered in the next chapter. Because JFETs are superior to BJTs insofar as input resistance is concerned and because they can be merged with BJTs on the same IC, they are used in such bipolar ICs as operational amplifiers where a very high input resistance is required. JFETs also appear in very small-signal amplifiers where the electrical noise which is inherent in all devices has to be carefully controlled. It turns out that the JFET is superior to both the BJT and the MOSFET in this respect. The gallium arsenide technology produces the *metal-semiconductor* FET (MESFET) which is a type of JFET that is used in analog and digital circuits operating at GHz frequencies. Even higher operating frequencies are expected from the *modulation-doped* FET (MODFET) which is currently the subject of intensive research.

6.4 SUMMARY AND STUDY GUIDE

The basic characteristics of the MOSFET and JFET were studied in this chapter. This was done by examining the relatively simple but useful first-order models which are obtained from the terminal current and voltage relationships of these devices. For each transistor type, models were obtained for dc analysis as well as for describing the device's behavior under small-signal conditions. Simple applications, involving not more than two transistors, were considered not only for the purpose of motivating further study of electronic circuits but also to illustrate the techniques which simplify the analysis of transistor circuits.

In later chapters the student will encounter circuits in which features that are neglected in the first-order models become important. These include storage effects which determine the maximum useful frequency of an amplifier or the maximum switching speed of a logic circuit. Important effects also occur when transistors are operated at extremes of voltage or current, and some circuits where these effects cannot be neglected will also be encountered in a later chapter.

The student should pay particular attention to a number of important points which are listed below.

If a consistent polarity convention is used to define terminal currents and voltages, the nonlinear model equations of the p-channel MOSFET or JFET can be easily obtained from their duals, the n-channel MOSFET and JFET, respectively.

The main difference between an enhancement and a depletion MOSFET of the same channel type is the sign of the threshold voltage V_T.

The behavior of transistors is most easily treated on the basis of the modes or states in which they can operate. For the JFET and MOSFET these are the cut-off, pinch-off and triode states.

The analysis of circuits such as small-signal amplifiers is a two step process involving the decomposition of the circuit into a dc model, which is used to obtain the quiescent conditions, and a small-signal model in which some parameters depend on the quiescent currents and voltages.

Small-signal models do not depend on whether the JFET or MOSFET is n-channel or p-channel or on whether the MOSFET is an enhancement or a depletion type.

The characteristics of an amplifier depend on the quiescent (dc) currents and voltages. These determine not only the parameters of the small-signal model and, therefore, the small-signal characteristics of the amplifier, but also the maximum current and voltage variations for large-signal conditions.

Small-signal models cannot be used for large-signal or dc analysis.

The large-signal and dc analysis of practical FET transistor circuits can often be carried out with sufficient accuracy graphically.

The JFET has a very high input impedance.

A very high input impedance is also a characteristic of the MOSFET which, in addition, can be used to design resistorless amplifiers with relatively wide regions of linear operation, and resistorless logic circuits.

PROBLEMS

6.1 An n-channel enhancement MOSFET with $V_T = 2$ V and $\beta = 0.5$ mA/V^2 is to be operated in the pinch-off mode with $I_D = 10$ mA. What is the lowest permissible v_{DS}?

6.2 A gate-to-source voltage v_{GS} of 5 V is applied to an n-channel enhancement MOSFET with $V_T = 1$ V. What will the resulting drain current be if $i_D = 2$ mA when $v_{GS} = 2$ V? Assume pinch-off operation.

6.3 An n-channel MOSFET operated in pinch-off is found to have $i_D = 4$ mA when $v_{GS} = 2V_T$. What v_{GS} must be applied if the device is to operate in pinch-off with $i_D = 2$ mA? What limit must be applied to v_{DS} to ensure constant-current operation?

6.4 A p-channel enhancement MOSFET, with $\beta = 0.8$ mA/V^2 and $V_T = -2$ V, is used in the amplifier in Fig. 6.7. If $V_{GG} = -4$ V, $V_{DD} = -20$ V and $R_D = 4$ kΩ determine:
(a) V_{SG}, I_D, V_{SD} with $v_I = 0$
(b) The small-signal voltage-gain v_{ds}/v_i

6.5 Repeat Prob. 6.4 for a depletion device with $\beta = 0.8$ mA/V^2, $V_T = 2$ V, and $V_{GG} = 0$ V.

6.6 The MOSFET in Fig. 6.7 is replaced by an n-channel depletion device with $\beta = 1.0$ mA/V^2 and $V_T = -3$ V. If $V_{GG} = 0$ V, $V_{DD} = 10$ V and $R_D = 1$ kΩ, compute

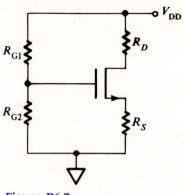

Figure P6.7

(a) I_D, V_{DS}
(b) The small-signal gain v_{ds}/v_i.

6.7 Figure P6.7 shows a popular biasing scheme for discrete enhancement MOSFETs. The MOSFET described in Prob. 6.1 is used with $V_{DD} = 20$ V. Following an analytical approach choose R_{G1}, R_{G2}, and R_S to give a quiescent drain current of 1 mA, and to hold it within 25% of this value if V_T were to increase by as much as 50%. Use resistors of the order of megaohms for R_{G1} and R_{G2}. What is the limit on R_D for pinch-off operation?

6.8 Repeat Prob. 6.7 using graphical techniques to solve for R_S and for the relationship between R_{G1} and R_{G2}.

6.9 The biasing arrangement shown in Fig. P6.9 is sometimes used with enhancement MOSFETs. If $V_T = 2$ V and $\beta = 0.2$ mA/V^2 determine the quiescent I_D and V_{DS} for $V_{DD} = 15$ V.

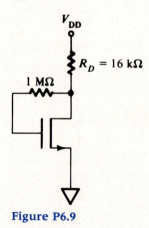

Figure P6.9

6.10 One of the advantages of the biasing arrangement in Fig. P6.9 is that the quiescent point is relatively insensitive to the exact value of V_T which is poorly controlled in discrete MOSFETs. Determine the percentage change in I_D in Prob. 6.9 if V_T changes from 2 to 3 V.

6.11 For the source follower discussed in Example 6.5 determine:
   (a) The maximum and minimum values of v_I for pinch-off operation
   (b) The values of v_O corresponding to the limits on v_I computed in (a).

6.12 The depletion MOSFET in Fig. 6.9 is replaced by an enhancement device with $V_T =$ 2 V and $\beta = 0.5$ mA/V². For proper biasing and additional dc supply voltage V_{SS} must be placed in series with R_L. If $V_{DD} = 10$ V and $R_L = 2$ kΩ determine:
   (a) The magnitude and polarity of V_{SS} for a quiescent current $I_D = 1$ mA in the pinch-off mode
   (b) The voltage gain v_o/v_i corresponding to the conditions in (a).

6.13 A p-channel enhancement MOSFET with $V_T = -2$ V is connected as shown in Fig. 6.11 for an n-channel device. If an applied voltage V_{SD} of 3 V produces a current of 10 mA, what will the voltage be when the current is 1 mA, and 0.1 mA? From your results, suggest what branch element's characteristics are approximated by this connection of an enhancement MOSFET.

6.14 If the voltmeter in Fig. P6.14 has a very high internal resistance, relate the voltmeter's reading, V, to the MOSFET's parameters.

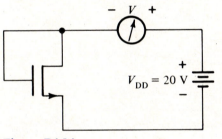

Figure P6.14

6.15 The MOSFETs in Fig. P6.15 are identical with $V_T = 2$ V. When operated individually in pinch-off, they have $i_D = 2$ mA when $v_{GS} = 2V_T$. What is v_O?

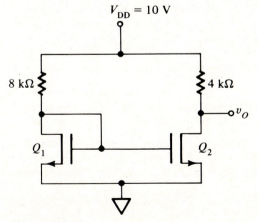

Figure P6.15

6.16 The inverter in Fig. 6.14 uses enhancement MOSFETs with $V_T = 1$ V. The load transistor Q_2 has $\beta = 1$ $\mu A/V^2$, while the switch transistor Q_1, is characterized by $\beta = 64$ $\mu A/V^2$. If $V_{DD} = 10$ V calculate the voltage gain and the maximum output voltage swing for linear operation.

6.17 Obtain the voltage transfer characteristic of the circuit in Fig. 6.14 assuming $\beta_1/\beta_2 = 4$ and $V_{DD} = 10$ V. Use the characteristics in Fig. 6.5 for Q_1. Could this circuit be used as a logic inverter? Explain your answer.

6.18 Repeat Prob. 6.17 using the characteristics in Fig. P6.18 for Q_1 and taking $V_{DD} = 16$ V. These reflect more accurately the behavior of an actual MOSFET.

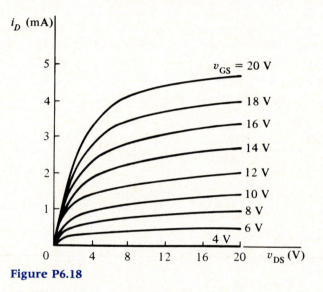

Figure P6.18

6.19 (a) Use the procedure described in Sec. 6.2.5 to compute the limits for linear operation of the inverter circuit described in Prob. 6.18. Determine also the voltage gain from Eq. (6.17). Comment on the validity of the computed limits.

(b) Determine graphically the approximate small-signal voltage gain at a point midway between the limits you found in (a). Compare your result with that obtained in (a) and comment. Estimate more realistic limits for linear operation with maximum voltage gain. Comment.

6.20 Repeat Prob. 6.17 with the gate of Q_2 connected to a separate constant voltage source $V_{GG} = (10 + V_T)$ volts instead of to its drain. Could this circuit be used as a logic inverter? If so, what values would you assign to $V(1)$ and $V(0)$?

6.21 Solve for the voltage transfer characteristic of the inverter in Fig. 6.17 using the MOSFET characteristics in Fig. 6.5. Use $V_{DD} = 10$ V and assume that the $v_{GS} = 6$ V curve in Fig. 6.5 corresponds to $v_{GS} = 0$ V for the depletion device.

6.22 The pinch-off voltage $|V_P|$ of JFETs is typically around 3 or 4 V. Using this fact, determine the JFET parameter which the voltmeter in Fig. P6.22 is measuring, assuming that the voltmeter draws negligible current.

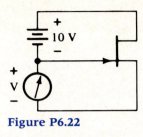

Figure P6.22

6.23 Identify the JFET parameter which the milliammeter in Fig. P6.23 is measuring. Assume the normal range of $|V_P|$ is 2 to 4 V.

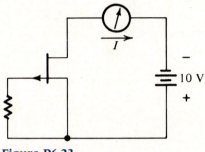

Figure P6.23

6.24 An n-channel JFET has $I_{DSS} = 12$ mA, $V_P = -3$ V. Plot to scale the output characteristics in the pinch-off mode for $v_{GS} = 0$, $V_P/4$, $V_P/2$, $3V_P/4$, and V_P. Clearly identify the coordinates of the points which lie on the boundary defined by Eq. (6.21).

6.25 In what mode will the JFET in Fig. E6.9 operate if $V_{DD} = -10$ V. Explain your answer.

6.26 Determine the voltage V_O in Fig. P6.26 using $V_P = -2$ V and $I_{DSS} = 4$ mA.

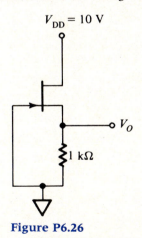

Figure P6.26

6.27 The JFET in Fig. 6.24 is replaced by a P-channel device with $I_{DSS} = 16$ mA and $V_P = 4$ V. If $R_D = 10$ kΩ, $V_{DD} = -20$ V and $V_{GG} = -3$ V, determine the limits on v_I for the JFET to operate in the pinch-off mode.

6.28 Determine g_m for the amplifier described in Prob. 6.27 and, thence, compute the small-signal voltage gain v_{ds}/v_i, and the amplitude of v_{ds} if v_i is a sinusoid of 0.1 V amplitude.

6.29 A JFET, just like a MOSFET, can be operated in the common-drain or source follower configuration (see Sec. 6.2.4). Figure P6.29 shows a somewhat impractical but conceptually simple schematic of a JFET source follower. If the JFET has $I_{DSS} = 15$ mA and $V_P = -3$ V determine the quiescent conditions, i.e., V_{GS}, I_D, V_{DS}, V_O.

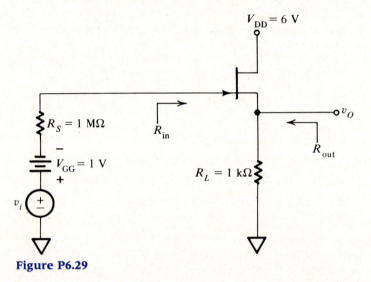

Figure P6.29

6.30 Use the results of Prob. 6.29 to solve for the following quantities:
(a) The small-signal voltage gain v_o/v_i
(b) The small-signal input and output resistances, R_{in} and R_{out}.

CHAPTER

Bipolar Junction Transistors (BJT)

7.1 INTRODUCTION

The bipolar junction transistor (BJT) came into being before either the JFET or the MOSFET, and the first integrated circuits (IC) were also based on the BJT. Thus the silicon bipolar technology is a very mature and stable one. For this reason and for others, such as the higher transconductance and speed of the BJT as compared to either of the silicon FETs, an extremely wide variety of discrete and integrated bipolar devices is available today. The latter include analog ICs such as the op amp, and digital ICs ranging in complexity from small-scale integrated (SSI) circuits such as logic gates to large-scale integrated (LSI) circuits such as the high speed bit-slice microprocessor which contains thousands of transistors.

Because of the maturity of the bipolar technology the BJT's characteristics are very well understood, making it easy to design BJT circuits with predictable performance. A large part of a design can often be carried out using the simple first-order models which are introduced in this chapter. More accurate models are

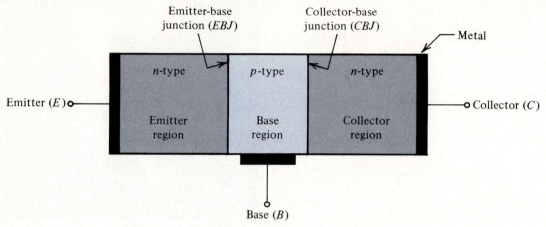

Figure 7.1 Simplified structure of the *npn* transistor.

usually too complex for hand calculations and are normally used during the final computer-aided design (CAD) phases. Thus the objectives of this chapter will have been achieved if, after studying it, the student will be sufficiently familiar with the BJT and its various modes of operation to be able to undertake with confidence the first-order analysis of simple transistor circuits.

7.2 STRUCTURE, MODES OF OPERATION, AND SYMBOLS

The bipolar junction transistor (BJT) essentially consists of a piece of silicon which has been treated in a manner that defines in it three electrically dissimilar regions. These regions are identified, on the basis of their electrical properties, as being either *n*-type or *p*-type. Figure 7.1 shows the simplified structure of one of the two types of BJT that are possible, namely the *npn* transistor. The two *n*-type regions are separated by a *p*-type region. By interchanging the *p*-type and *n*-type regions one obtains the dual, the *pnp* transistor.

Figure 7.1 also shows the names which are used to describe the three regions of the BJT, the boundaries between them, and their corresponding terminals. The different names which are given to the two outside regions reflect the important fact that, unlike the JFET and MOSFET, the BJT is not a symmetrical device, as the simplified picture of Fig. 7.1 might lead one to conclude, and exhibits very different terminal properties with the *emitter* and the *collector* interchanged.

The characteristics of a BJT are due almost entirely to the physical processes which take place in the very narrow *base* region and at the two *pn* junctions, the *emitter-base junction* (EBJ) and the *collector-base junction* (CBJ). Since each of these *pn* junctions behaves like a diode, the BJT can be expected to exhibit four modes of operation. These are identified in Table 7.1 on the basis of whether the EBJ and CBJ are forward or reverse biased.

Table 7.1 BJT MODES OF OPERATIONS

Mode	EBJ	CBJ
Cutoff	Reverse	Reverse
Active	Forward	Reverse
Reverse active	Forward	Forward
Saturation	Forward	Forward

The *active* mode is the most useful one for signal amplification. On the other hand, the *cutoff* and *saturation* modes are exploited in applications where the BJT is used as a switch. One of the rare applications of the *reverse* mode will be encountered when the TTL family of logic circuits is studied in Chap. 8.

Figure 7.2 shows the symbols which are used to represent *npn* and *pnp* BJTs in circuit schematics. The arrowhead on the emitter distinguishes this terminal from the collector, which is important because, as was mentioned earlier, the BJT is not a symmetrical device. The direction of the arrowhead distinguishes between an *npn* and a *pnp* device by defining the direction of current flow in the emitter when the BJT operates in the active mode.

7.2.1 Model for the Active Mode of Operation

Since the BJT is a three-terminal device it can be treated as a two-port and its terminal characteristics described in any one of six possible ways. Figure 7.3(a) shows an *npn* transistor connected as a two-port with the base terminal common. A particularly convenient pair of equations for describing this two-port is

$$v_{BE} = f_1(i_E, v_{CB}) \tag{7.1a}$$

$$i_C = f_2(i_E, v_{CB}) \tag{7.1b}$$

which, referring to Chap. 5, should be recognized as the "*h*" representation except for some discrepancies in assumed voltage polarity and current direction.

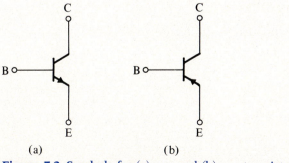

(a)   (b)

Figure 7.2 Symbols for (a) *npn* and (b) *pnp* transistor.

For reasons of convenience, voltages in Fig. 7.3(a) are labeled with a double subscript to identify the terminals between which they are defined. The subscripts also define the assumed polarity of the voltage, with the terminal corresponding to the first subscript taken to be at the higher potential: thus v_{BE} will be positive if the base terminal of the BJT is positive with respect to the emitter. Current directions as well as voltage polarities are chosen so that these quantities are all positive when the BJT is in the active mode.

Starting from an analysis of the physical processes in a BJT or, more directly, from terminal measurements, the active-region input and output characteristics of the *npn* transistor operated with its base terminal common are found to have the approximate form shown in Fig. 7.3(b) and (c).

A simple but useful model results if one neglects the relatively weak dependence on v_{CB} which is apparent from these characteristics. In such a case Eqs. (7.1) are approximated by

$$v_{BE} = f_1(i_E) \tag{7.2a}$$

$$i_C = f_2(i_E) \tag{7.2b}$$

More explicitly, Eq. (7.2a) is described by

$$v_{BE} = V_T \ln\left(\frac{\alpha i_E}{I_S}\right)$$

which is usually written in the form

$$i_E = \frac{I_S}{\alpha} \exp\left(\frac{v_{BE}}{V_T}\right) \tag{7.3a}$$

Equation (7.2b) is approximated by

$$i_C = \alpha i_E \tag{7.3b}$$

where α is relatively constant while I_S and V_T depend on the temperature.

Alternatively, since by KCL, $i_B = i_E - i_C$, and therefore,

$$i_E = \frac{i_B}{1 - \alpha} \tag{7.4}$$

Eq. (7.3) can also be expressed in the form

$$i_B = \frac{I_S}{\beta} \exp\left(\frac{v_{BE}}{V_T}\right) \tag{7.5a}$$

$$i_C = \beta i_B \tag{7.5b}$$

where

$$\beta = \frac{\alpha}{1 - \alpha} \tag{7.6}$$

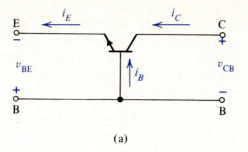

(a)

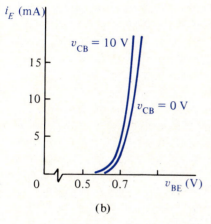

(b)

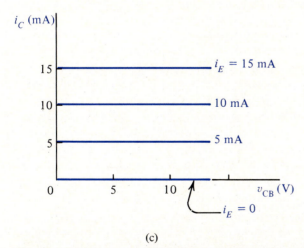

(c)

Figure 7.3 The *npn* transistor in the common-base configuration: (a) circuit with voltage polarities and current directions defined, (b) approximate input characteristics, and (c) output characteristics in the active region (v_{BE}, $v_{CB} > 0$).

Equations (7.3) and (7.5) include the salient features of the *npn* BJT operated in the active mode and, for this reason, they are called *first-order* models. Since both pairs of equations describe a diode and a dependent current source they lead to the circuit interpretations shown in Fig. 7.4. The circuit in Fig. 7.4(a) corresponds to Eq. (7.3) and, since these follow from a two-port treatment of the BJT in which the base is common (see Fig. 7.3), it is referred to as the *common-base* model. Similarly, because α represents the ratio of the output current i_C to the input current i_E in Fig. 7.4(a), it is referred to as the *common-base current gain*.

Figure 7.4(b), on the other hand, is a *common-emitter* model, which follows from Eq. (7.5). The parameter β is called the *common-emitter current gain* since it is the ratio of the output current i_C to the input current i_B.

Since the *p* region of a *pn* junction corresponds to a diode's anode it follows from Table 7.1 that the *pnp* transistor is in the active mode when v_{EB}, $v_{BC} > 0$.

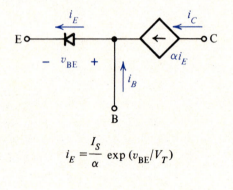

$$i_E = \frac{I_S}{\alpha} \exp(v_{BE}/V_T)$$

(a)

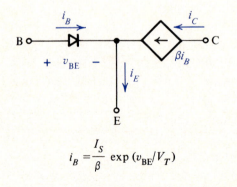

$$i_B = \frac{I_S}{\beta} \exp(v_{BE}/V_T)$$

(b)

Figure 7.4 First-order circuit models of the *npn* BJT in the active mode: (a) common-base, (b) common-emitter.

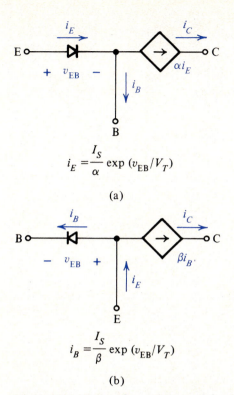

$$i_E = \frac{I_S}{\alpha} \exp (v_{EB}/V_T)$$

(a)

$$i_B = \frac{I_S}{\beta} \exp (v_{EB}/V_T)$$

(b)

Figure 7.5 First-order circuit models of the *pnp* BJT in the active mode: (a) common-base, (b) common-emitter.

Figure 7.5 shows the first-order models for the *pnp* transistor which correspond to the *npn* models in Fig. 7.4. Notice that they are identical except for a reversal of polarity of all currents, voltages and controlled sources.

In a practical BJT α is almost equal to but slightly less than one: typically $\alpha = 0.98$. It follows from Eq. (7.6) that, typically, $\beta = 50$. At a temperature of 20°C, $V_T \simeq 25$ mV, while I_S is so small that the appropriate units for its specification are fA.

7.3   THE BJT AS AN AMPLIFIER

The way a BJT can be used as an amplifier, and some of the necessary conditions for the amplifier to have predictable characteristics can be understood by considering the impractical but conceptually simple and useful circuit in Fig. 7.6(a).

7.3.1 DC Conditions

In Fig. 7.6(b) the BJT has been replaced by its common-emitter model in Fig. 7.4(b). Under dc conditions the signal v_I is set to zero. Using the standard con-

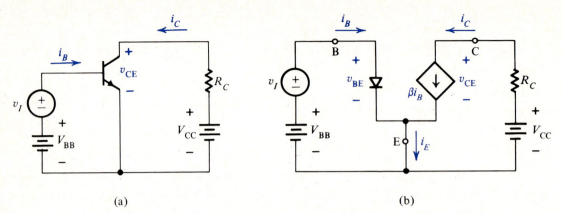

Figure 7.6 The BJT as an amplifier: (a) schematic representation and (b) BJT replaced by common-emitter model.

vention of designating dc quantities by uppercase letters, the dc currents and voltages in Fig. 7.6 are computed beginning with the input loop. Since $V_{BE} = V_{BB}$ the current I_B is computed from Eq. (7.5a) as

$$I_B = \frac{I_S}{\beta} \exp \frac{V_{BB}}{V_T} \tag{7.7}$$

The collector current I_C can now be determined from Eq. (7.5b) to be

$$I_C = \beta I_B \tag{7.8}$$

By KCL,

$$I_E = I_B + I_C = (\beta + 1)I_B$$

while KVL around the output loop yields

$$V_{CE} = V_{CC} - I_C R_C$$

Since the active mode is assumed, the above relationships will only be valid provided that $V_{BB} > 0$ and that $V_{CB} = V_{CE} - V_{BE} > 0$, that is,

$$V_{CE} = V_{CC} - I_C R_C > V_{BB} \tag{7.9}$$

EXAMPLE 7.1

Compute the dc currents and voltages in Fig. 7.6 given that the BJT parameters are $\beta = 50$, $V_T = 25$ mV and $I_S = 10^{-14}$A, and that $V_{CC} = 20$ V, $V_{BB} = 0.69$ V and $R_C = 1$ kΩ.

Solution

From the preceding analysis of this circuit $V_{BE} = V_{BB} = 0.69$ V so

$$I_B = \frac{10^{-14}}{50} \exp\left(\frac{0.69}{0.025}\right) A = 0.194 \text{ mA}$$

Since $I_C = \beta I_B$,

$$I_C = 50(0.194) = 9.69 \text{ mA}$$

$$V_{CE} = V_{CC} - I_C R_C = 20 - 9.69 = 10.3 \text{ V}$$

Because V_{CE} is greater than V_{BE} the BJT is indeed in the active mode.

7.3.2 Small-Signal Model of the BJT

Assuming v_i to be small enough for the currents and voltages in the BJT to undergo small variations about their dc values, the results of Chap. 5 can be used to separate out a small-signal model for the transistor. This model is described by the following equations

$$i_b = \left. \frac{\partial i_B}{\partial v_{BE}} \right|_Q v_{be} \tag{7.10a}$$

$$i_c = \left. \frac{\partial i_C}{\partial i_B} \right|_Q i_b \tag{7.10b}$$

where i_B and i_C are defined by Eq. (7.5) which correspond to the BJT model in Fig. 7.6(b). The subscript Q denotes, as usual, the quiescent (dc) values of the device's currents and voltages.

The partial derivative in Eq. (7.10) is given by

$$\frac{\partial i_B}{\partial v_{BE}} = \frac{I_S}{\beta V_T} \exp\left(\frac{V_{BE}}{V_T}\right)$$

which has a value at the quiescent point of

$$\left. \frac{\partial i_B}{\partial v_{BE}} \right|_Q = \frac{I_B}{V_T} \tag{7.11}$$

where I_B is computed from Eq. (7.7).

The calculation of the partial derivative in Eq. (7.10b) is trivial since

$$i_C = \beta i_B$$

and, therefore,

$$\left. \frac{\partial i_C}{\partial i_B} \right|_Q = \beta \tag{7.12}$$

Equation (7.10) can now be rewritten in the form

$$v_{be} = r_\pi i_b \tag{7.13a}$$

$$i_c = \beta i_b \tag{7.13b}$$

where, from Eq. (7.11),

$$r_\pi = \frac{V_T}{I_B}$$

or, since $I_C = \beta I_B$,

$$r_\pi = \frac{\beta V_T}{I_C} \tag{7.14}$$

This leads to the equivalent circuit interpretation shown in Fig. 7.7(a). Lower-case letters are used to designate the terminals in order to emphasize that the model accounts for only the small-signal behavior of the device.

Comparing the circuit in Fig. 7.7(a) with Fig. 4.3 it is clear that this small-signal BJT model is a hybrid representation with $h_{11} = r_\pi$, $h_{21} = \beta$ and $h_{12} = h_{22} = 0$.

The equivalent circuit in Fig. 7.7(b) is obtained by expressing i_c in terms of v_{be}. Thus, from Eqs. (7.13) and (7.14)

$$i_c = g_m v_{be} \tag{7.15}$$

where $g_m = \beta/r_\pi$ and, therefore,

$$g_m = \frac{I_C}{V_T} \tag{7.16}$$

The quantity g_m, which relates the output collector current to the input base-emitter voltage in the small-signal model, is known as the transconductance of the

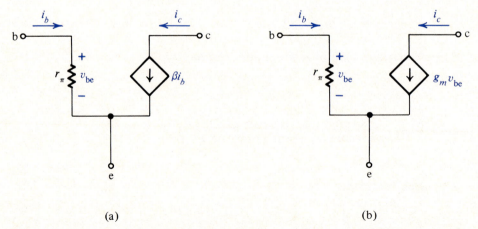

(a)   (b)

Figure 7.7 Two equivalent circuit interpretations of the common-emitter first-order small-signal model of the BJT: (a) hybrid and (b) admittance parameters.

BJT. Referring to Fig. 4.4 it is clear that Fig. 7.7(b) is an admittance representation with $y_{11} = 1/r_\pi$, $y_{21} = g_m$, $y_{12} = y_{22} = 0$.

EXAMPLE 7.2

Using the results of Example 7.1, compute the parameters of the small-signal models in Fig. 7.7.

Solution
From Eq. (7.16)

$$g_m = \frac{I_C}{V_T} = \frac{9.69}{0.025}\text{mS} = 388 \text{ mS}$$

Since, according to Eqs. (7.14) and (7.16),

$$r_\pi = \frac{\beta}{g_m}$$

therefore,

$$r_\pi = \tfrac{50}{388} = 0.13 \ k\Omega$$

DRILL EXERCISE

7.1 Starting with the model in Fig. 7.5(b) show that the corresponding small-signal model of the *pnp* BJT is identical to that shown in Fig. 7.7 and that, therefore, the distinction between *pnp* and *npn* vanishes under small-signal conditions.

By starting with the model represented by Fig. 7.4(a) one obtains a common-base small-signal model for the BJT. However, at this stage it is simpler to compute the common-base parameters directly from the model in Fig. 7.7. If the *h*-parameter model in Fig. 7.7(a) is used, then the circuit to be analyzed is that shown in Fig. 7.8(a).
Computing *h* parameters yields

$$h_{11} = \frac{v_1}{i_1}\bigg|_{v_2 = 0} = \frac{r_\pi}{\beta + 1} \tag{7.17}$$

$$h_{21} = \frac{i_2}{i_1}\bigg|_{v_2 = 0} = \frac{-\beta}{\beta + 1} = -\alpha \tag{7.18}$$

while the remaining two *h* parameters are found to be zero.

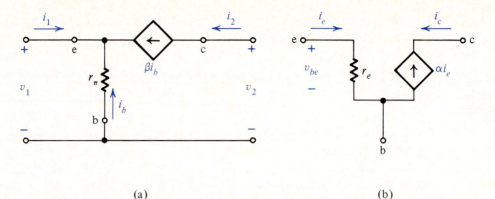

(a)   (b)

Figure 7.8 (a) Figure 7.7(a) redrawn. (b) Resulting equivalent circuit for common-base small-signal model of BJT.

Combining Eqs. (7.14) and (7.17) produces the result that

$$h_{11} = \frac{V_T}{I_C} \frac{\beta}{\beta + 1}$$

which, in view of Eqs. (7.16) and (7.18) as well as the fact that $\alpha \simeq 1$, simplifies to

$$h_{11} = r_e = \frac{1}{g_m} \tag{7.19}$$

The quantity r_e is called the small-signal *emitter resistance*.

Equations (7.18) and (7.19) are the first-order, small-signal, common-base model of the BJT and are represented by the equivalent circuit in Fig. 7.8(b).

Several important points should be noted here. Firstly, the small-signal current gain of the common-emitter transistor is β which is much larger than one, the approximate common-base current gain α. Secondly, the input small-signal driving point resistance r_π of the common-emitter configuration is β times larger than the corresponding driving point resistance r_e of the common-base circuit. Recall that $V_T \simeq 25$ mV at 20°C so, for example, if $\beta = 50$ $r_e = 25$ Ω for $I_C = 1$ mA while the corresponding $r_\pi = 1.25$ kΩ. Finally, because some parameters such as g_m depend on the dc quiescent point, it is necessary to have a stable Q point in order to achieve small-signal amplifiers with stable and predictable characteristics. One of the many reasons why the circuit in Fig. 7.6 is not very practical is that the dc currents depend on V_{BB} and β. Firstly, β is a very poorly controlled parameter and varies from transistor to transistor, and secondly there is the exponential dependence of the currents on V_T and, therefore, on temperature. All of these factors lead to poorly controlled dc and, therefore, small-signal conditions in this amplifier. In Chap. 14 this issue is addressed more fully when the design of biasing networks is considered.

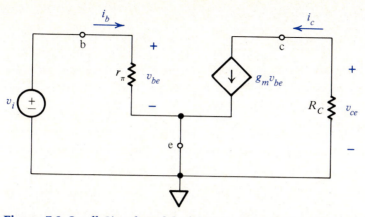

Figure 7.9 Small-Signal model of the amplifier in Fig. 7.6.

7.3.3 Small-Signal Analysis of the Simple BJT Amplifier

The properties of the BJT as a small-signal amplifier using the circuit in Fig. 7.6 can now be examined. Making use of the techniques which were developed in Chap. 5 for obtaining the small-signal circuit, the circuit shown in Fig. 7.9 is obtained. Since the output voltage v_{ce} is given by

$$v_{ce} = -i_c R_C$$

and, because $i_c = g_m v_{be}$, it follows that the small-signal voltage gain of the amplifier is

$$\frac{v_{ce}}{v_i} = -g_m R_C \tag{7.20}$$

EXAMPLE 7.3

Use the results of Examples 7.1 and 7.2 to compute the voltage gain v_{ce}/v_i of the amplifier in Fig. 7.6.

Solution
From Eq. (7.20)

$$\frac{v_{ce}}{v_i} = -g_m R_C = -388$$

7.4 ACTIVE MODE PIECEWISE-LINEAR MODEL

In the analysis of many transistor circuits the simple first-order model which was introduced in Sec. 7.3.2 can be replaced by a still more approximate one, without a significant further loss in accuracy. This comes about because, for example, amplifier circuits are designed to have Q points which are insensitive to the exact value of V_{BE}. This will be illustrated in Example 7.4.

Figure 7.10 shows the common-base input characteristic of a typical general purpose BJT which is normally operated with an emitter current i_E of the order of milliamperes. This characteristic can be approximated by a straight line with slope $1/R_E$ for $i_E > 0$ and by the $i_E = 0$ axis for $v_{BE} < 0.7$ V. Figure 7.11 shows this piecewise-linear approximation and the corresponding circuit model.

In most circuits the external resistors in series with the emitter terminal are much larger than R_E and, consequently, this element is usually ignored resulting in the simple but useful common-base model shown in Fig. 7.12 for both the *npn* and *pnp* transistors.

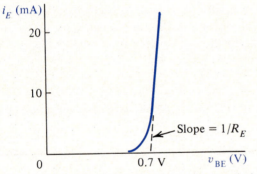

Figure 7.10 Common-base input characteristic of a typical *npn* BJT.

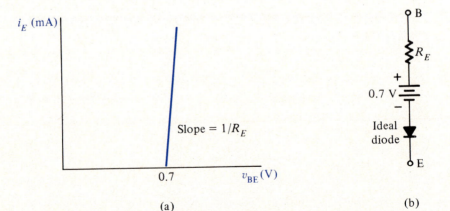

(a)   (b)

Figure 7.11 (a) Piecewise-linear approximation and (b) its circuit model for the common-base input characteristic of the *npn* BJT.

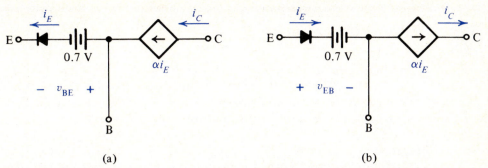

Figure 7.12 Active region, piecewise-linear, common-base models of the (a) *npn* and (b) *pnp* transistor.

A similar piecewise-linear approximation to the diodes in the common-emitter models in Figs. 7.4(b) and 7.5(b) yields immediately the circuits in Fig. 7.13. Henceforth, unless otherwise specified, the voltage source in the piecewise-linear model of the EBJ will be taken to have a value of 0.7 V.

Although the circuits in Fig. 7.12 and 7.13 are called active region models, they account for the BJT's behavior in cutoff as well. Thus, with $v_{BE} < 0.7$ V in Fig. 7.12(a) or Fig. 7.13(a), $i_B = 0$ and, therefore, $i_C = i_E = 0$. Observe that the piecewise-linear approximation results in a modification of the way the boundary between the active and cutoff regions was originally defined. In Sec. 7.2 the active region was defined as corresponding to v_{BE}, $v_{CB} > 0$ for the *npn* BJT. This means that, provided the $v_{CB} > 0$, the device is deemed to be in the active region as soon as i_B rises above I_S/β [see Eq. (7.5a)] which is an extremely small quantity. The piecewise-linear model, on the other hand, effectively sets the boundary between cutoff and the active mode at the point where the current through the EBJ becomes significant on the scale of the device's normal operating currents. This point is approximated by the 0.7 V source in the EBJ's piecewise-linear model. In power transistors, for example, where the device currents are measured in amperes, a value of approximately 1 V for this source is more reasonable.

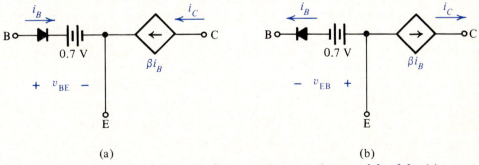

Figure 7.13 Active region, piecewise-linear, common-emitter models of the (a) *npn* and (b) *pnp* BJT.

EXAMPLE 7.4

Determine the dc conditions in the amplifier shown in Fig. E7.4(a), assuming $\beta = 50$, $V_T = 25$ mV, and $I_S = 10^{-14}$A. Hence compute the amplifier's small-signal transimpedance, v_o/i_i, and determine the errors associated with the use of the small-signal model when the output increment $v_o = 1.5$ V.

Solution

DC solution. Figure E7.4(b) shows the circuit under dc conditions, i.e., $i_I = 0$. In view of the polarity of V_{BB} the EBJ is expected to be forward biased. Therefore, solving the input loop by KCL,

$$I_B = \frac{V_{BB} - V_{EB}}{R_B}$$

Since V_{BB} is almost 4 V, whereas V_{EB} is expected to be a fraction of a volt, the above equation suggests that I_B is not very sensitive to the exact value of V_{EB}. It, therefore, seems logical to use the piecewise-linear model in Fig. 7.13(b) for the dc analysis. In that case the base current I_B is

$$I_B = \frac{3.7 - 0.7}{10} = 0.3 \text{ mA}$$

The collector current is determined from

$$I_C = \beta I_B = 50(0.3) = 15 \text{ mA}$$

and, therefore,

$$V_0 = V_{CE} = I_C R_C - V_{CC} = -5 \text{ V}$$

It is instructive at this point to determine the error associated with the use of the piecewise-linear model in this problem. Using $I_S = 10^{-14}$A, $\beta = 50$, $V_T = 25$ mV, and $I_B = 0.3$ mA in the model in Fig. 7.5(b) leads to $V_{EB} = 701$ mV, and the error is indeed negligible. In fact, even the value of I_S is not critical. Thus if $I_S = 10^{-12}$A is used, $V_{EB} = 0.586$ V which leads to a corrected value of I_B

$$I_B = \frac{3.7 - 0.586}{10} = 0.31 \text{ mA}$$

which differs from the previously calculated value by only 3%.

It is important to verify that the BJT is indeed in the active region, which is necessary for the above calculations to be valid.

$$V_{BC} = -I_C R_C + V_{CC} - V_{EB}$$

$$= 5 - 0.7 = 4.3 \text{ V}$$

Since $V_{BC} > 0$ the *pnp* BJT is indeed in the active region. Finally, the emitter

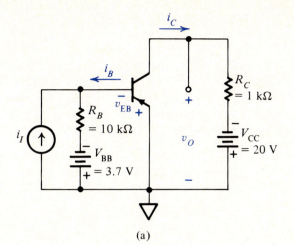

(a)

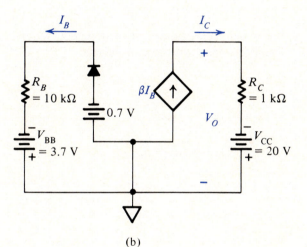

(b)

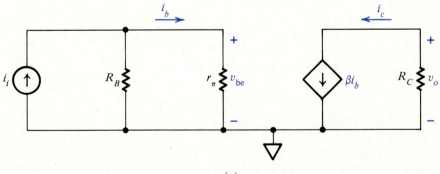

(c)

Figure E7.4 (a) Circuit for Example 7.4 and (b) same circuit under dc conditions with common-emitter, piecewise-linear model shown explicitly. (c) Small-signal model.

current is

$$I_E = I_C + I_B = 15.3 \text{ mA}$$

Small signal solution. Although the piecewise-linear (BJT) model could be used to solve the circuit under small-signal conditions, the procedure is rather awkward and some accuracy is lost because, for example, the dependence of r_π on the quiescent collector current is lost. Consequently the following calculations are based on the small-signal models of the BJT shown in Fig. 7.7.

At 20°C

$$V_T = 25 \text{ mV}$$

$$g_m = \frac{I_C}{V_T} = 600 \text{ mA/V}$$

$$r_\pi = \frac{\beta}{g_m} = 83 \text{ } \Omega$$

All the parameters of the small-signal model, shown in Fig. E7.4(c), are now known. Straightforward analysis yields the output voltage

$$v_o = -\beta R_C i_b = \beta R_C \frac{R_B}{r_\pi + R_B} i_i$$

whence v_o/i_i is

$$\frac{v_o}{i_i} = -50(1)\frac{10}{0.083 + 10} = -49.6 \text{ V/mA}$$

There now remains only the problem of determining the errors which arise due to the use of the small-signal model when the output increment $v_o = 1.5$ V. From the small-signal analysis based on Fig. E7.4(c)

$$\frac{v_o}{i_i} = -49.6 \text{ V/mA}$$

so, for $v_o = 1.5$ V,

$$i_i = \frac{-1.5}{49.6} = -0.0302 \text{ mA}$$

The corresponding i_b and v_{be} are, according to Fig. E7.4(c)

$$i_b = \left[\frac{R_B}{R_B + r_\pi} \right] i_i$$

$$= \left[\frac{-10}{10 + 0.083} \right] i_i$$

$$= -0.992 i_i$$

$$= -0.0299 \text{ mA}$$

and

$$v_{be} = i_b r_\pi$$

$$= -0.0299(83) \text{ mV} = -2.48 \text{ mV}$$

These results must now be compared with those which are obtained by solving the circuit in Fig. E7.4(a) with the BJT replaced by the model in Fig. 7.5(b).

Since the quiescent value of $v_O = v_{CE}$ is -5 V, an output voltage change $v_o = 1.5$ V corresponds to a change in v_O from $v_O = V_O = -5$ V to $v_O = V_O + v_o = -3.5$ V which leaves the transistor in the active region. From KVL around the output loop in Fig. E7.4(a)

$$i_C = \frac{v_O + V_{CC}}{R_C}$$

so $v_O = -3.5$ V corresponds to a collector current i_C of

$$i_C = \frac{-3.5 + 20}{1} = 16.5 \text{ mA}$$

This, in turn, is produced by a base current

$$i_B = \frac{i_C}{\beta} = \frac{16.5}{50} = 0.33 \text{ mA}$$

From Fig. E7.4(a) and (c) $i_B = I_B - i_b$, so the change in i_B which causes the 1.5 V change in v_O is

$$i_b = I_B - i_B = 0.30 - 0.33 = -0.03 \text{ mA}$$

which is in good agreement with the -0.0299 mA result of the small-signal analysis. To solve for the current i_i which produces this i_b, it is necessary to determine the corresponding v_{eb}. This is done by noting that, since according to Fig. 7.5(b)

$$i_B = \frac{I_S}{\beta} \exp\left(\frac{v_{EB}}{V_T}\right)$$

the ratio i_B/I_B is given by

$$\frac{i_B}{I_B} = \frac{\exp(v_{EB}/V_T)}{\exp(V_{EB}/V_T)} = \exp\left(\frac{v_{eb}}{V_T}\right)$$

since $v_{eb} = v_{EB} - V_{EB}$. Therefore,

$$v_{eb} = V_T \ln \frac{i_B}{I_B} = 25 \ln \frac{0.33}{0.30} \text{ mV} = 2.38 \text{ mV}$$

in contrast to 2.48 mV which the small-signal model yields.

Straightforward analysis of the circuit between the base and emitter terminals of the BJT in Fig. E7.4(a) shows that i_I is given by

$$i_I = -i_B + \frac{V_{BB} - v_{EB}}{R_B}$$

Under quiescent conditions $i_I = 0$ and

$$0 = -I_B + \frac{V_{BB} - V_{EB}}{R_B}$$

Since $i_B = I_B + i_b$ and $v_{EB} = V_{EB} + v_{eb}$, it follows from these two equations that

$$i_i = -i_b - \frac{v_{eb}}{R_B}$$

$$= -0.03 - \frac{2.38}{104}\,\text{mA}$$

$$= -0.03024 \text{ mA}$$

which is in excellent agreement with the result of the small-signal analysis.

Comparing the values of i_i, i_b, and v_{be} obtained by the above two methods, it is clear that the error in i_i and i_b is well below 1% while the $v_{eb} = -v_{be}$ obtained from the small-signal model is approximately 4% larger than that obtained using the nonlinear BJT model. The good agreement in i_i and i_b could have been expected because R_B and r_π in Fig. E7.4(c) form a current divider and $R_B \gg r_\pi$. The current i_b is, therefore, relatively insensitive to the exact value of r_π which is the only element in the small-signal model in Fig. 7.7 that is based on the small-signal approximation.

Although the errors in practical amplifier circuits are not always as small as the ones computed in this example, the small-signal analysis is always used where the transistor's current and voltage variations are such that the device remains well within the active mode. If accurate error estimates are needed the circuit is analyzed on the computer by means of a circuit analysis program.

7.5 THE EMITTER FOLLOWER

Although the BJT is most useful as a small-signal voltage amplifier when it is operated in the common-emitter configuration, it also exhibits some very interesting properties in the other two configurations as well. The common-base arrangement is somewhat specialized and its applications will be postponed until Chap. 14. However the *common-collector* amplifier, or *emitter follower* as it is often called, is a very useful configuration which is often used in analog circuits and occasionally in digital circuits.

Figure 7.14 shows the simplified circuit of an emitter follower. Although, generally speaking, this is an impractical arrangement, it is conceptually simple and, in fact, networks which are used to drive the base terminal can often be modeled by the circuit shown between the base and ground (datum) in Fig. 7.14.

DC conditions in the circuit correspond to $v_S = 0$. KVL around the input loop yields

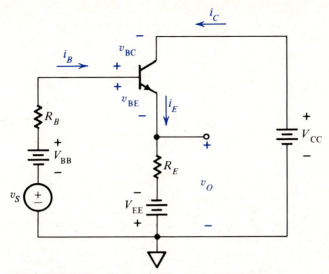

Figure 7.14 Simple circuit of an *npn* BJT in the common-collector (emitter follower) configuration.

$$V_{EE} + V_{BB} = V_{BE} + I_B R_B + I_E R_E \tag{7.21}$$

If the BJT is modeled by means of the piecewise-linear circuit in Fig. 7.12(a) then, for the EBJ to be forward biased, the inequality

$$V_{BB} + V_{EE} > 0.7 \text{ V}$$

must be satisfied. In that case the base current is obtained from Eq. (7.21) as

$$I_B = \frac{V_{EE} + V_{BB} - V_{BE}}{R_B + (\beta + 1)R_E} \tag{7.22}$$

If $V_{BC} < 0$, then

$$I_C = \beta I_B$$

and V_O is given by

$$V_O = I_E R_E - V_{EE} \tag{7.23}$$

For the above analysis to be valid the CBJ must be reverse biased which means that

$$V_{BC} = V_{BE} + I_E R_E - V_{EE} - V_{CC} < 0$$

For the small-signal analysis, all the constant voltage sources in Fig. 7.14 are set to zero and the BJT is replaced by its small-signal model. The resulting circuit is shown in Fig. 7.15, where the first order common-emitter model of Fig. 7.7(a) is used for the BJT. Observe that the collector node of the small-signal model in Fig. 7.15 is connected to ground, which accounts for the use of the term *common-collector*.

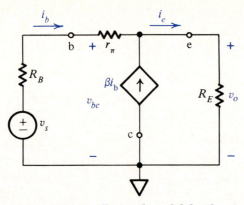

Figure 7.15 Small-signal model for the circuit in Fig. 7.14.

By KVL around the outer loop

$$v_s = i_b(R_B + r_\pi) + i_e R_E \tag{7.24}$$

Since

$$i_e = (\beta + 1)i_b \quad \text{and} \quad v_o = i_e R_E$$

therefore

$$\frac{v_o}{v_s} = \frac{(\beta + 1)R_E}{R_B + r_\pi + (\beta + 1)R_E} \tag{7.25}$$

Clearly the voltage gain v_o/v_s is less than one and tends to one for sufficiently large $(\beta + 1)R_E$. So the emitter follower certainly is not very useful as a voltage amplifier.

Its attributes become apparent from an analysis of its input and output driving point resistances. From Eq. (7.24) the resistance looking into the base terminal of the BJT is

$$R_i = \frac{v_{bc}}{i_b} = r_\pi + (\beta + 1)R_E \tag{7.26}$$

The Thevenin resistance seen by R_E is computed from Fig. 7.16. Since

$$v = -i_b(R_B + r_\pi)$$

and

$$i = -(\beta + 1)i_b$$

it follows that

$$R_o = \frac{v}{i} = \frac{R_B + r_\pi}{\beta + 1} \tag{7.27}$$

The significance of Eqs. (7.26) and (7.27) can be appreciated by considering the problem illustrated in Fig. 7.17. Suppose that R_B and v_s form the small-signal

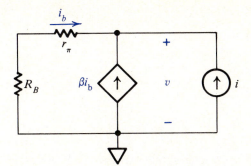

Figure 7.16 Circuit for determining output resistance of circuit in Fig. 7.15.

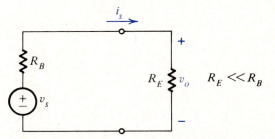

Figure 7.17 Illustration of the need for an emitter follower.

Thevenin equivalent of some electronic circuit's output port, and that R_E is a low-resistance load. If R_E was connected directly to the output port then, because

$$v_o = \frac{v_s R_E}{R_E + R_B} \tag{7.28}$$

v_o would be much smaller than v_s if $R_E \ll R_B$. Moreover, a current i_s, given by

$$i_s = \frac{v_s}{R_B + R_E} \tag{7.29}$$

would be drawn from the source.

If, on the other hand, an emitter follower is interposed between R_B and R_E, as shown in Fig. 7.15, then from Eq. (7.25) the ratio v_o/v_s could be made much closer to unity than in Eq. (7.28) because $\beta \gg 1$. Furthermore, the current drawn from v_s will be determined by $R_B + R_i$ where R_i is given by Eq. (7.26). Again, because $\beta \gg 1$ this current could be much smaller than the value given by Eq. (7.29).

EXAMPLE 7.5

Compare the current drawn from the source and the voltage gain v_o/v_s of the circuits in Figs. 7.15 and 7.17, assuming $R_B = 2$ kΩ, $R_E = 50$ Ω, $\beta = 50$, and $r_\pi = 1$ kΩ.

Solution

In Fig. 7.17

$$i_s = \frac{v_s}{2.05} \text{ mA}$$

and

$$\frac{v_o}{v_s} = \frac{R_E}{R_E + R_B} = 0.024$$

On the other hand in Fig. 7.15 $i_s = i_b$ where, from KVL around the input loop and Eq. (7.26),

$$i_s = \frac{v_s}{R_B + r_\pi + (\beta + 1)R_E} = \frac{v_s}{5.55} \text{ mA}$$

Equation (7.25) yields v_o/v_s as

$$\frac{v_o}{v_s} = 51 \frac{0.05}{2 + 3.55} = 0.46$$

Thus, by using the emitter follower, almost 50% of the available voltage v_s is developed across R_E and this is accompanied by a significant reduction in current drawn from the source.

In summary, the emitter follower acts as a buffer which:

1. Has a voltage gain approximately equal to one

2. Presents a high resistance to the source, therefore drawing a relatively small amount of power from it

3. Presents a low resistance to the load, therefore providing a relatively high output current capacity

The emitter follower is often referred to as an impedance transformer because, as Eqs. (7.26) and (7.27) demonstrate, a resistance connected to the emitter appears to be multiplied by $\beta + 1$ when viewed from the base terminal, while the resistance connected to the base, viewed from the emitter terminal, appears to be divided by $\beta + 1$.

DRILL EXERCISE

7.2 Show that, excluding r_π, the incremental behavior of the emitter follower can be represented by the model shown in Fig. D7.2.

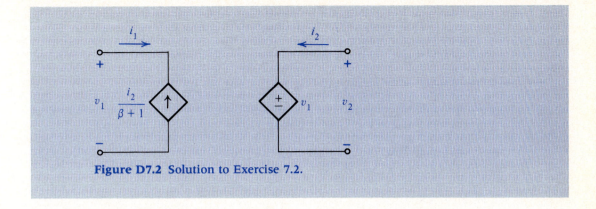

Figure D7.2 Solution to Exercise 7.2.

7.6 LARGE-SIGNAL MODEL OF THE BJT

The piecewise-linear model of the BJT which was introduced in Sec. 7.4 is inadequate if operation of the BJT beyond the limits of the active region has to be considered. In particular, the model does not account for the transistor's terminal characteristics when it is in the saturation mode. In Table 7.1 the saturation mode of the BJT was defined as corresponding to the CBJ as well as the EBJ being forward biased. Since the CBJ is a *pn* junction it can be expected to have a characteristic similar to that of the EBJ shown in Fig. 7.10. This is indeed so with one important difference, namely, that the voltage corresponding to $v_{BE} = 0.7$ V in Fig. 7.10 is, typically, approximately equal to 0.5 V.

If the CBJ characteristics are aproximated by the simple piecewise-linear model which was used for the EBJ in Fig. 7.12, then the BJT model in that figure can be extended to include the saturation mode in the manner shown for the *npn* BJT in Fig. 7.18. The validity of this model is verified using the assumed state method presented in Chap. 5.

CUTOFF
D_E and D_C reverse biased. Therefore,

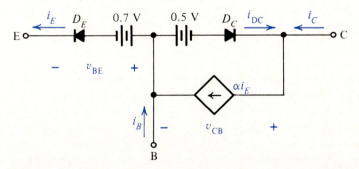

Figure 7.18 Piecewise-linear model of the *npn* BJT which accounts for the saturation mode.

$$i_E = 0 \quad \text{and} \quad v_{BE} < 0.7 \text{ V}$$

and

$$i_{DC} = 0 \quad \text{and} \quad v_{CB} > -0.5 \text{ V}$$

Consequently,

$$i_C = \alpha i_E - i_{DC} = 0$$

$$i_B = i_E - i_C = 0$$

Since all the terminal currents are zero the model in Fig. 7.18 can be replaced by that in Fig. 7.19(a) when the BJT is in the cutoff mode.

ACTIVE MODE

D_E forward biased and D_C reverse biased. Therefore, for diode D_E

$$v_{BE} = 0.7 \text{ V} \quad \text{and} \quad i_E > 0$$

while for D_C

$$v_{CB} > -0.5 \text{ V} \quad \text{and} \quad i_{DC} = 0$$

Hence

$$i_C = \alpha i_E$$

The resulting simplified model, shown in Fig. 7.19(b), corresponds, as expected, to that obtained from Fig. 7.12(a) with the diode replaced by a short circuit.

SATURATION

D_E and D_C forward biased. In that case for D_E

$$v_{BE} = 0.7 \text{ V} \quad \text{and} \quad i_E > 0$$

while for D_C

$$v_{CB} = -0.5 \text{ V}$$

and

$$i_{DC} = \alpha i_E - i_C > 0$$

which means that

$$i_C < \alpha i_E$$

Since both v_{BE} and v_{CB} in the saturation mode are constant voltages, the model in Fig. 7.18 can be replaced by the simpler one in Fig. 7.19(c).

It is left as an exercise for the student to show that in the remaining possible state, namely, D_E reverse biased and D_C forward biased, the model in Fig. 7.18 reduces to an open-circuit between the base and emitter terminals and a voltage source $V_{BC} = 0.5$ V between the base and collector terminals. This state should, according to Table 7.1, correspond to the reverse mode which was declared at the

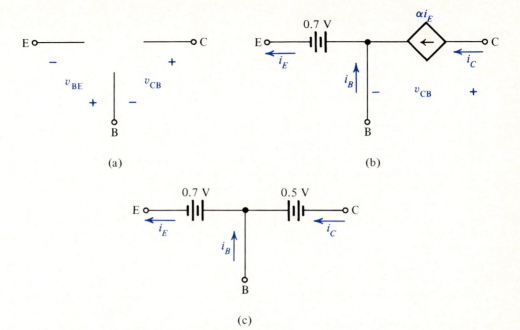

Figure 7.19 Models of the *npn* BJT corresponding to (a) cutoff, (b) active mode, and (c) saturation.

outset to be beyond the scope of this chapter. Although the model in Fig. 7.18 does not, in general, adequately account for the BJT's behavior in this mode, it will be seen in Chap. 8 that in certain cases, such as the TTL family of logic ICs, transistors are designed to behave in a manner which is accounted for quite well by this model.

The preceding analysis leads to the graphical representation shown in Fig. 7.20(b) for the common-base output characteristics. These should be compared to the output curves of a typical *npn* BJT depicted in Fig. 7.20(a). The input characteristics will clearly be the same as those in Fig. 7.11(a) with $R_E = 0$.

A common-emitter, piecewise-linear model of the BJT was analyzed in Chap. 5 and the circuit model of the *npn* transistor is shown in Fig. 5.21. Because the dependent source cannot be expressed in terms of the input current i_B alone, this model is rarely used and the common-base model in Fig. 7.18 is generally preferred. However, the alternative circuit model for the saturation mode shown in Fig. 7.21 is found to be convenient for analyzing BJTs operated in the common-emitter configuration. This model can obviously be derived from that in Fig. 7.19(c) with the collector-to-emitter saturation voltage $V_{CEsat} = 0.7 - 0.5 = 0.2$ V.

Figure 7.22 shows a comparison of the common-emitter output characteristics of a typical *npn* transistor with those which are obtained from the model in Fig. 7.18. For approximate calculations, such as those that are undertaken during the early phases of a design, the accuracy of the piecewise-linear model is quite acceptable.

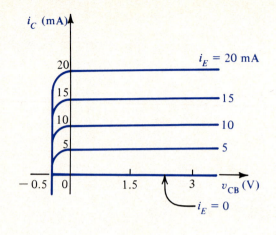

(a)

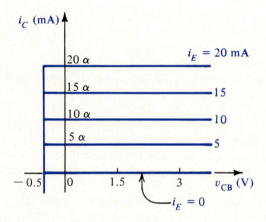

(b)

Figure 7.20 Common-base output characteristics of *npn* BJT: (a) typical, (b) based on model in Fig. 7.18.

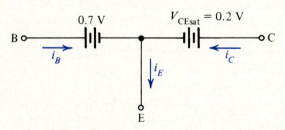

Figure 7.21 Saturation model of the common-emitter *npn* BJT.

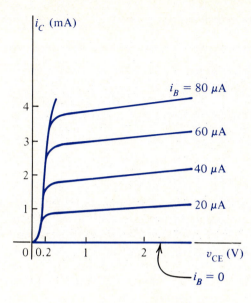

(a)

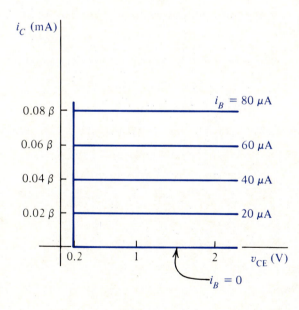

(b)

Figure 7.22 Common-emitter output characteristics of the *npn* BJT: (a) typical, (b) corresponding to model in Fig. 7.18.

DRILL EXERCISE

7.3 Show that if the common-base model of the *pnp* BJT in Fig. 7.5(a) is replaced by a piecewise-linear approximation, and if the saturation mode is included, the resulting model is that in Fig. 7.18 with all diodes, voltage polarities, current sources and current directions reversed.

7.7 THE BJT INVERTER

Figure 7.23 shows the circuit of an *npn* BJT used as an *inverter*. The circuit is a generalization of the common-emitter amplifier with v_I and R_B being the Thevenin equivalent of the actual network connected between the base and emitter terminals of the transistor. In contrast to the preceding analysis of the BJT as an amplifier, the objective here is to analyze the circuit in Fig. 7.23 for variations of the input voltage v_I which cause the transistor to be driven beyond the limits of the active mode.

Operation which extends into cutoff and saturation implies large current and voltage variations, so the appropriate BJT model for analyzing this circuit is the one in Fig. 7.18. However, a graphical approach is easier than working with the circuit equations alone.

Figure 7.24(b) shows the piecewise-linear output characteristics of the common-emitter, *npn* BJT together with the output load line defined by

$$i_C = \frac{V_{CC} - v_{CE}}{R_C} \tag{7.30}$$

The corresponding input characteristics are shown in Fig. 7.24(a) together with

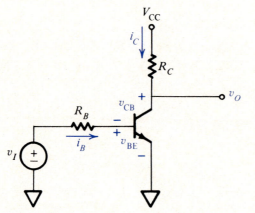

Figure 7.23 The *npn* BJT inverter.

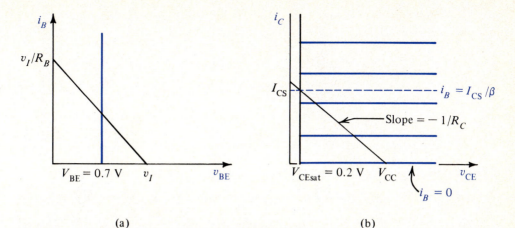

(a)   (b)

Figure 7.24 Graphical solution of the inverter in Fig. 7.23: (a) input plane, (b) output plane.

the input load line given by

$$i_B = \frac{v_I - v_{BE}}{R_B} \tag{7.31}$$

From the preceding analysis of the BJT piecewise-linear model the inverter is expected to have three modes of operation.

CUTOFF
According to Fig. 7.24(a) if

$$v_I < 0.7 \text{ V} \tag{7.32a}$$

then the input load line intersects the BJT characteristic on the v_{BE} axis and, hence,

$$i_B = 0 \tag{7.32b}$$

Therefore, from Fig. 7.24(b),

$$i_C = 0 \tag{7.32c}$$

and

$$v_O = V_{CC} \tag{7.32d}$$

It follows that the transistor can be modelled here by the circuit in Fig. 7.19(a) which results in the inverter cutoff model in Fig. 7.25(a).

ACTIVE MODE
From the input characteristic in Fig. 7.24(a), if

$$v_I > 0.7 \text{ V} \tag{7.33a}$$

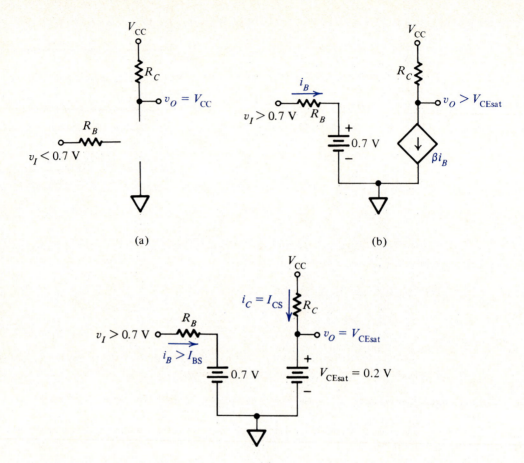

Figure 7.25 Models of the circuit in Fig. 7.23 corresponding to the BJT in cutoff (a), active mode (b), and saturation (c).

then

$$v_{BE} = 0.7 \text{ V} \tag{7.33b}$$

and

$$i_B = \frac{v_I - 0.7}{R_B} > 0 \tag{7.33c}$$

As a result the collector current is, from Fig. 7.24(b),

$$i_C = \beta i_B \tag{7.33d}$$

provided that

$$v_O = v_{CE} > V_{CEsat} \tag{7.33e}$$

Under these conditions the output voltage v_O will be, from Eq. (7.30),

$$v_O = V_{CC} - i_C R_C \tag{7.33f}$$

It follows that the BJT in the inverter can be replaced by the model in Fig. 7.19(b) or, alternatively, by its common-emitter equivalent, namely, Fig. 7.13(a) with the diode replaced by a short circuit. The resulting circuit appears in Fig. 7.25(b).

SATURATION

According to Fig. 7.24(b), when the base current exceeds the value

$$i_B = I_{BS} = \frac{I_{CS}}{\beta} \tag{7.34a}$$

where

$$I_{CS} = \frac{V_{CC} - V_{CEsat}}{R_C} \tag{7.34b}$$

the output load line intersects the BJT characteristics at the point $i_C = I_{CS}$, and v_{CE} is given by

$$v_{CE} = V_{CEsat} = 0.2 \text{ V} \tag{7.34c}$$

The corresponding v_I is, from Eqs. (7.33c) and (7.34a),

$$v_I = I_{BS} R_B + 0.7 \tag{7.34d}$$

Therefore, the transistor in this region of operation can be replaced by the circuit model in Fig. 7.21 leading to the saturated inverter model in Fig. 7.25(c).

The voltage transfer characteristic of the inverter, which is shown in Fig. 7.26, can be obtained either from the models in Fig. 7.25 or directly from the graphical

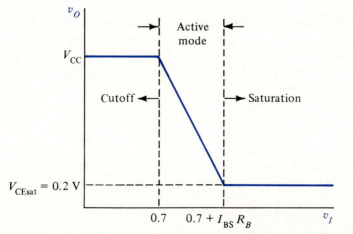

Figure 7.26 Voltage transfer characteristic for the inverter in Fig. 7.23.

solution of the inverter in Fig. 7.24. Using the latter approach, as v_I is increased from zero the first breakpoint occurs where $v_I = 0.7$ V in Fig. 7.24(a). Since $i_B = 0$ here, the corresponding point in the output plane is the intersection of the load line with the v_{CE} axis, namely, $v_O = v_{CE} = V_{CC}$.

The other breakpoint occurs where the load line in the output plane intersects the $v_O = v_{CE} = V_{CEsat}$ line. The intersection point, from Fig. 7.24(b), occurs where $i_B = I_{BS} = I_{CS}/\beta$ which, from Fig. 7.24(a), corresponds to v_I as defined by Eq. (7.34d).

In between the two breakpoints the voltage transfer characteristic is described by the linear circuit in Fig. 7.25(b). The slope of the characteristic is the voltage gain A_v of the inverter for variations in v_I which limit v_O to this linear region. By inspection of Fig. 7.26 this slope is given by

$$A_v = \frac{v_o}{v_i} = \frac{-(V_{CC} - V_{CEsat})}{I_{BS} R_B}$$

Combining this expression with Eqs. (7.34a) and (7.34b) leads to the simple equation

$$A_v = \frac{v_o}{v_i} = -\beta \frac{R_C}{R_B} \tag{7.35}$$

EXAMPLE 7.6

Extend the analysis of the amplifier in Fig. E7.4 by determining the limits on $i_I = I_i \sin \omega t$ for the BJT to remain in the active mode.

Solution

In Example 7.4 linearity considerations constrained the output voltage variation to a value which left the BJT well within the active region. Here the problem is to consider the BJT under large signal conditions which drive it to the edges of the cutoff mode at one end and saturation at the other. Small-signal analysis is thus inappropriate in general.

The circuit in Fig. E7.4(a) is redrawn in Fig. E7.6(a) with the network between the base and emitter of the BJT replaced by its Thevenin equivalent. Instead of solving the circuit directly, the preceding analysis of the inverter is used with appropriate voltage polarity changes to account for the fact that a *pnp* BJT is being considered.

The breakpoint between the cutoff and active modes of the BJT in Fig. E7.6(a) is given by

$$i_B = 0 \qquad v_I = -0.7 \text{ V}$$

in the input plane, and

$$i_C = 0 \qquad v_O = -v_{EC} = -V_{CC} = -20 \text{ V}$$

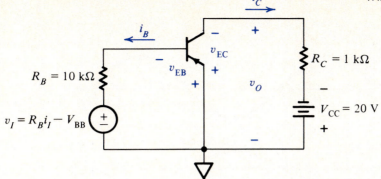

Figure E7.6(a) Amplifier in Fig. E7.4(a) with input circuit replaced by Thevenin equivalent.

in the output plane. From the input loop in Fig. E7.6(a) the conditions $i_B = 0$, $v_I = -0.7$ V corresponds to

$$R_B i_I - V_{BB} = -0.7 \text{ V}$$

Therefore the value of i_I which causes the BJT to cut off is

$$i_I = \frac{V_{BB} - 0.7}{R_B}$$

which, from the dc analysis in Example 7.4, is the value of I_B. Therefore, at the edge of cutoff

$$i_I = 0.3 \text{ mA}$$

The BJT saturates when

$$v_O = -V_{ECsat} = -0.2 \text{ V}$$

From Eq. (7.34b), with V_{CEsat} replaced by V_{ECsat}, this corresponds to a collector current

$$i_C = I_{CS} = \frac{V_{CC} - V_{ECsat}}{R_C}$$

$$= 19.8 \text{ mA}$$

and a base current

$$i_B = I_{BS} = \frac{I_{CS}}{\beta}$$

$$= \frac{19.8}{50} = 0.396 \text{ mA}$$

The corresponding i_I from the input loop in Fig. E7.23(a) is

$$i_I = \frac{V_{BB} - 0.7}{R_B} - i_B = I_B - I_{BS}$$

$$= 0.3 - 0.396 = -0.096 \text{ mA}$$

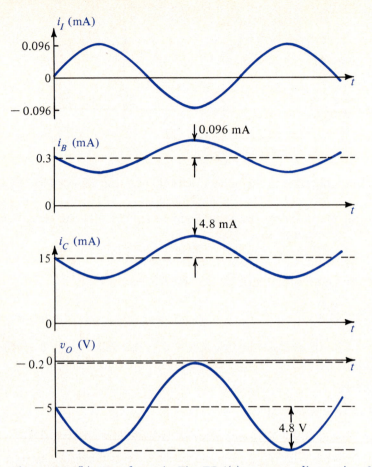

Figure E7.6(b) Waveforms in Fig. E7.6(a) corresponding to $i_I = 0.096 \sin \omega t$ mA. Note that $i_C(t)$ is not to scale.

Since, under quiescent conditions, $i_I = 0$, it follows that, if i_I is a sinusoid, its amplitude must be limited by the smaller of the two values of 0.3 and -0.096 mA corresponding to, respectively, cutoff and saturation. Therefore, the amplitude of i_I is constrained by

$$I_i \leq 0.096 \text{ mA}$$

for the BJT to remain in the active region.

From Example 7.4 the quiescent conditions are:

$$I_B = 0.3 \text{ mA} \qquad I_C = 15 \text{ mA} \qquad V_o = -5 \text{ V}$$

When $i_I = -0.096$ mA the above calculations yield

$$i_B = 0.396 \text{ mA} \qquad i_C = 19.8 \text{ mA}$$

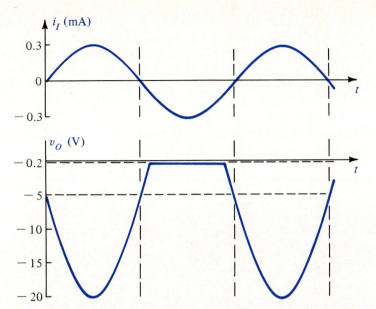

Figure E7.6(c) Waveforms in Fig. E7.6(a) corresponding to $i_I = 0.3 \sin \omega t$ mA.

and, from Fig. 7.26,

$$v_O = -v_{ECsat} = -0.2 \text{ V}$$

Therefore, corresponding to an input current amplitude of $I_i = 0.096$ mA, the other currents and voltages have amplitudes of

$$I_b = 0.396 - 0.3 = 0.096 \text{ mA}$$

$$I_c = 19.8 - 15.0 = 4.8 \text{ mA}$$

and

$$V_o = 5 - 0.2 = 4.8 \text{ V}$$

It is instructive to plot the waveforms associated with these currents and voltages under the above conditions. Figure E7.6(b) shows these waveforms. Notice the polarity reversal between i_I and v_O, in the sense that when i_I goes positive v_O becomes more negative and vice versa. It is very important to realize that this will be the same for both the *pnp* and *npn* BJT.

The current and voltage variations considered in this example are significantly greater than those corresponding to the small-signal conditions calculated in Example 7.4. Nevertheless, note that the amplitude ratio

$$\frac{V_o}{I_i} = 50 \text{ V/mA}$$

is in very close agreement with the small-signal transimpedance computed in Example 7.4. This is not always true and comes about in this case because of the relatively large R_B which makes the transfer function relatively insensitive to v_{EB}.

Before leaving this example, consider what, say, the output voltage waveform would look like if $I_i = 0.3$ mA was chosen instead. Since $i_i = 0.3$ mA corresponds to cutoff, the output voltage v_O would swing to -20 V when $i_I = 0.3$ mA but would remain at -0.2 V for all i_I more negative than -0.096 mA. Consequently $v_O(t)$ would have the approximate shape shown in Fig. E7.6(c). Further increases in I_i would result in clipping of the negative swing in v_O as well.

DRILL EXERCISE

7.4 Replace the BJT in Fig. E7.6(a) by its piecewise-linear model and repeat Example 7.6 using assumed state analysis.

7.8   THE BJT AS A SWITCH

The BJT can be used as a switch because when it is in the cutoff mode of operation the collector current i_C is essentially zero, while in the saturation mode the collector-emitter voltage $v_{CE} = V_{CEsat}$ is approximately zero.

Thus the collector-to-emitter circuit of the inverter in Fig. 7.23 does behave as a switch, albeit a nonideal one due to the nonzero voltage V_{CEsat} across the switch when it is closed. In many practical circuits $V_{CC} \gg V_{CEsat}$ making this voltage error negligible.

The circuit in Fig. 7.23 is called an inverter not only because, when used as an amplifier, the polarity of the output voltage change is opposite to that of the input voltage change, but also because it can perform *logical inversion*.

Suppose that v_I in Fig. 7.23 is allowed to take on only two values: V_{CC} or V_{CEsat}. From the transfer characteristic in Fig. 7.26 it is seen that when v_I has the value

$$v_I = V_{CEsat}$$

the corresponding v_O is

$$v_O = V_{CC}$$

On the other hand when

$$v_I = V_{CC}$$

the output voltage v_O falls to

$$v_O = V_{CEsat}$$

provided that

$$V_{CC} \geq 0.7 + I_{BS} R_B$$

As will be seen in the following example this inequality is easy to satisfy.

Now if V_{CC} is associated with the logical TRUE statement and V_{CEsat} is chosen to correspond to FALSE, then the above circuit conditions mean that

1. If v_I is *TRUE*, then v_O is *FALSE*.

2. If v_I is *FALSE*, then v_O is *TRUE*.

These two statements represent the logical operation of inversion or *complementation* which is normally written in the form

$$Y = \overline{X}$$

where Y and X are two logical variables represented here by v_I and v_O.

EXAMPLE 7.7

Analyze the circuit in Fig. 7.23 with $R_C = 1\ \text{k}\Omega$ and $V_{CC} = 5.2$ V, assuming v_I switches between 0 and 3.7 V. Determine the value of R_B which is required to saturate the BJT if its β can be anywhere between 25 and 100 (a not unrealistic situation).

Solution

$$I_{CS} = \frac{V_{CC} - V_{CEsat}}{R_C} = 5 \text{ mA}$$

$$I_{BS} = \frac{I_{CS}}{\beta} = \frac{5}{\beta} \text{ mA}$$

Therefore the minimum and maximum I_{BS} are

$$I_{BS} \text{ (min)} = \tfrac{5}{100} = 0.05 \text{ mA}$$

and

$$I_{BS} \text{ (max)} = \tfrac{5}{25} = 0.2 \text{ mA}$$

Since

$$R_B = \frac{v_I - 0.7}{I_B}$$

It follows that R_B must be selected on the basis of $I_{BS}(\text{max})$. For the inverter to saturate when v_I has its most positive value R_B must be not greater than

$$R_B = \frac{3.7 - 0.7}{0.2} = 15 \text{ k}\Omega$$

Notice that, with this value of R_B, the *base overdrive factor*, defined as i_B/I_{BS} where

$i_B \geq I_{BS}$, will vary between 1 and I_{BS} (max)$/I_{BS}$ (min) = 4, depending on the value of β.

In Chap. 8 the reader will see how the simple BJT inverter in Fig. 7.23 can be used as a basis for designing circuits which perform more complicated logical operations. Furthermore, the inverter will be shown to be a key element in the design of several logic circuit families.

As a final comment, the reader should note that although the loads in all the circuits considered in this chapter were linear resistances, this does not imply that active loads, such as the MOSFET ones discussed in Chap. 6, cannot be designed using BJTs. Because the associated circuits are more complicated than in the case of the MOSFET, their discussion is postponed until Chap. 14.

7.9 SUMMARY AND STUDY GUIDE

The basic characteristics of the BJT were studied in this chapter. This was done by examining the relatively simple but useful first-order models which are obtained from the terminal current and voltage relationships of these devices. Models were obtained for dc analysis as well as for describing the device's behavior under small-signal conditions. These include a piecewise-linear model which, in contrast to the FET, is useful for the analysis of BJT circuits because of the sharp boundary between the device's characteristics in the active and saturation regions.

Simple applications, involving not more than one transistor, were considered not only for the purpose of motivating further study of electronic circuits but also to illustrate the techniques which simplify the analysis of BJT circuits.

As in the case of FETs, higher order effects in the BJT are left to be considered, where appropriate, in later chapters.

The student should pay particular attention to a number of important points, some of which appear in the Study Guide for Chapter 6 and are repeated below for emphasis.

If a consistent polarity convention is used to define terminal currents and voltages, the nonlinear model equations of the *pnp* BJT can be easily obtained from its dual, the *npn* BJT.

The behavior of transistors is most easily treated on the basis of the modes or states in which they can operate. Thus the BJT is characterized on the basis of its cutoff, active and saturation modes.

The analysis of circuits such as small-signal amplifiers is a two step process involving the decomposition of the circuit into a dc model, which is used to obtain the quiescent conditions, and a small-signal model in which some parameters depend on the quiescent currents and voltages.

Small-signal models do not depend on whether the BJT is *npn* or *pnp*.

The characteristics of an amplifier depend on the quiescent (dc) currents and voltages. These determine not only the parameters of the small-signal model and, therefore, the small-signal characteristics of the amplifier, but also the maximum current and voltage variations for large signal conditions.

Small-signal models cannot be used for large-signal or dc analysis.

The large-signal and dc analysis of practical BJT circuits can usually be carried out with sufficient accuracy either graphically or with the help of a piecewise-linear model.

The strength of the BJT lies both in its potentially high voltage gain and in its good switching characteristics. The high gain properties are associated with the common-emitter configuration.

The common-collector amplifier, or emitter follower, acts as an impedance transformer, such that the resistance connected to the base appears to be divided by $\beta + 1$ when viewed from the emitter terminal, while the resistance connected to the emitter, when viewed from the base terminal, appears to be multiplied by $\beta + 1$.

PROBLEMS

7.1 An *npn* BJT operated in the active mode is found to exhibit a collector current $i_C = 5$ mA when $v_{BE} = 0.68$ V. What is the corresponding v_{BE} when i_C is changed to (a) 50 mA, (b) 0.5 mA, (c) 0.05 mA.

7.2 A process used to manufacture bipolar ICs yields circuits in which the α of the transistors is held within the range of 0.95 to 0.995. What is the corresponding variation in β?

7.3 Measurements on an *npn* BJT operating in the active mode yields the following data: $i_C = 10$ mA, $i_B = 0.5$ mA, $v_{BE} = 0.7$ V. Determine β and I_S.

7.4 The BJT in Fig. 7.6 is replaced by a *pnp* device with $I_S = 1$ fA and $\beta = 30$. If $v_i = 0$, $V_{CC} = -12$ V, $V_{BB} = -0.72$ V and $R_C = 2$ kΩ determine i_C, i_E, and v_{CB}, using the sign convention of Fig. 7.5. Assume the transistor is in the active mode but, finally, verify your assumption.

7.5 The *npn* BJT in Fig. P7.5 is characterized by $\alpha = 0.98$ and $I_S = 10$ fA. Assuming the transistor is in the active mode, compute i_C, i_B and v_{CB}. Verify the above assumption.

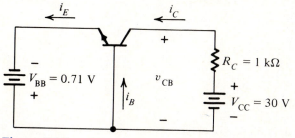

Figure P7.5

7.6 Compute the small-signal transimpedance v_{ce}/i_b for the amplifier described in Prob. 7.4.

7.7 The transistor defined in Prob. 7.2 is used in the amplifier shown in Fig. E7.4. Determine the effect of the specified range for α on
(a) The range of the BJT's quiescent operation, specified in terms of I_C and V_{EC}
(b) The range of the corresponding small-signal ratios v_{ce}/i_b and v_{ce}/i_i

7.8 Using a piecewise-linear model for the BJT in Fig. P7.8 and assuming it is in the active region, compute I_B, I_C, V_{CE}. Verify the assumption.

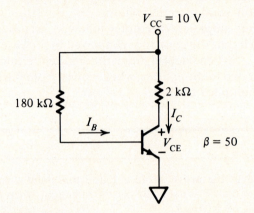

Figure P7.8

7.9 Repeat Prob. 7.8 for the circuit in Fig. P7.9. *Hint.* Replace the network "seen" by the base of the BJT by its Thevenin equivalent circuit.

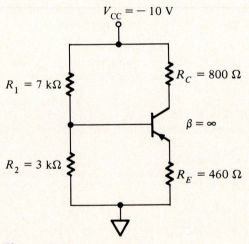

Figure P7.9

7.10 For the circuit in Fig. P7.10 determine I_C and the minimum value of R_C for which the BJT remains in the active region. Assume $\beta \gg 1$.

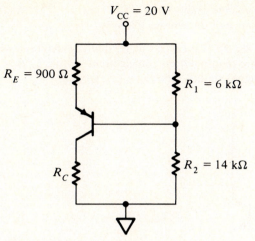

Figure P7.10

7.11 Determine I_C and V_{CE} in the circuit in Fig. P7.11 if the BJT has $\beta = 50$, $V_{BE} = 0.7$ V.

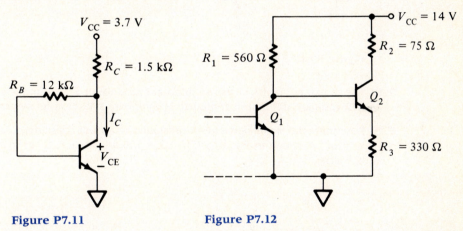

Figure P7.11   **Figure P7.12**

7.12 Transistor Q_1 in Fig. P7.12 is biased in the active mode with $I_{C1} = 5$ mA. Solve for the quiescent conditions in Q_2 assuming it to be in the active region as well. Use $\beta = 50$. Verify the assumption.

7.13 Determine the maximum variation that the collector current of Q_2 in Fig. P7.12 may undergo without the transistor leaving the active mode.

7.14 Derive an incremental model for the circuit in Fig. P7.12 and compute the ratio i_{c2}/i_{c1} where i_{c2} and i_{c1} are, respectively, the small-signal collector currents of Q_2 and Q_1.

7.15 Refer to Fig. 7.14. If $V_{CC} = 12$ V, $V_{BB} = 3.5$ V, $V_{EE} = 0$, $R_B = 2$ kΩ, $R_E = 200$ Ω, and $\beta = 50$, determine I_C, I_B, V_{CE} and V_O, i.e., the quiescent conditions with $v_S = 0$.

7.16 Derive the incremental model of the circuit in Fig. 7.14 with the parameter values specified in Prob. 7.15 and compute:

(a) The small-signal voltage gain v_o/v_s

(b) The output resistance of the emitter follower

7.17 The circuit shown in Fig. P7.17 is called a phase splitter and is used where two outputs of opposite polarity are required from a single input. Assuming $\beta \gg 1$ derive expressions for the small-signal voltage ratios v_c/v_b, v_e/v_b, and v_c/v_e.

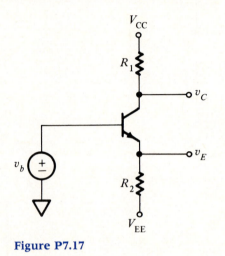

Figure P7.17

7.18 The source in Fig. P7.17 has an internal resistance of R_b. Assuming a finite β for the BJT, compare the output resistance of the phase splitter at its two output ports.

7.19 Determine the small-signal input resistance R_i, output resistance R_o and voltage gain v_o/v_s of the circuit in Fig. P7.19. Use $\beta = 100$ and assume that the average value of v_S is zero.

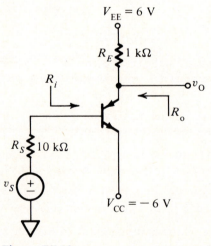

Figure P7.19

7.20 Determine the constraints on V_{CC} and V_{EE} in Fig. P7.17 for the BJT to remain in the active region with $v_c = v_e$. Again assume $\beta \gg 1$.

7.21 Instead of the current source i_t in Fig. E7.4 a voltage source v_t is placed in series with R_B and V_{BB}. Using the piecewise-linear model of the BJT which includes saturation, plot to scale the complete voltage transfer characteristic v_O versus v_I. Clearly dimension all breakpoints and slopes.

7.22 The transistor in Fig. 7.23 is replaced by a *pnp* device. If $V_{CC} = -3$ V, $R_C = 640\ \Omega$, $R_B = 450\ \Omega$ and $\beta = 10$, plot the piecewise-linear voltage transfer characteristic, clearly labelling all breakpoints. Determine the base overdrive factor if v_I varies between -0.2 V and -3 V.

7.23 Determine the range of β for which the transistor in Fig. P7.23 remains saturated.

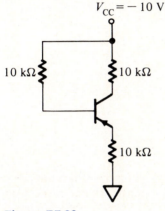

Figure P7.23

7.24 Using the model in Fig. 7.18 determine the currents and voltages indicated on the circuit in Fig. P7.24. In what mode is the BJT operating?

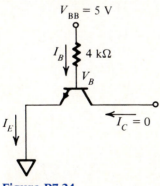

Figure P7.24

7.25 A *pnp* transistor with $\beta = 25$ is used in the inverter circuit in Fig. 7.23 with $R_B = 5\ \text{k}\Omega$ and $R_C = 500\ \Omega$. If v_I varies between -0.2 V and -2.7 V, determine the limits on V_{CC} for the BJT to behave as a switch.

7.26 Determine the slope of the voltage transfer characteristic corresponding to the active region for the transistor in the inverter described in Prob. 7.25.

7.27 The circuit in Fig. P7.27 can be used as a logical NAND gate. What are the minimum values of v_{I1} and v_{I2} for the two BJTs to be saturated and what is the resulting v_O? Use $\beta = 20$, $V_{BE} = 0.7$ V.

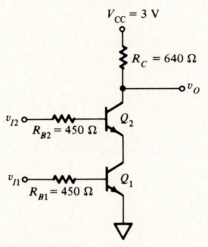

Figure P7.27

7.28 The BJT in Fig. P7.28 has $\beta = 30$.
(a) If $v_I = 12$ V, find the maximum value of R_1 for which the BJT remains saturated.
(b) Using $R_1 = 15$ kΩ, plot the voltage transfer characteristic assuming v_I varies between 0.2 and 12 V. Clearly identify all breakpoints.

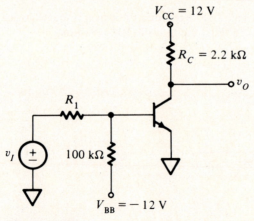

Figure P7.28

CHAPTER

Logic Circuits

8.1 INTRODUCTION

The main objective of this chapter is to study how field-effect and bipolar transistors are used to design logic circuits, in particular, *logic gates* which are the basic elements in *digital* systems such as computers. These systems can be subdivided into *combinatorial* and *sequential* circuits. Both are usually made up of logic gates; however, they differ in that the outputs of the former depend only on the *current* values of the inputs, while the sequential circuits possess memory, so that the outputs depend not only on the present values but also on the *past history* of the inputs.

Logic circuits are available in several logic *families*. The most important MOS families are *n*-channel MOS (NMOS) and complementary-symmetry MOS (CMOS) while the bipolar technology is dominated by *transistor-transistor logic* (TTL), *emitter-coupled logic* (ECL), and, to a lesser extent, *integrated injection logic* (I²L). These logic families differ in such important respects as their speed, power consumption and functional complexity. Thus, for example, ECL is the fastest family, CMOS consumes the least power, while NMOS is used almost exclusively for realizing complex VLSI circuits such as 32-bit microprocessors and 256K dynamic memories.

This chapter is divided into two parts. The first part is a brief introduction to combinatorial and sequential circuits with the logic gate as the basic element. In the second and main part, the characteristics of practical gates are studied and the circuit techniques which are used to design the most important bipolar and MOS logic families are examined.

8.2 BASIC LOGIC OPERATIONS AND LOGIC GATES

Digital systems which are described by means of only two values or states of a variable can be viewed as operating on *binary*-valued signals since the two states can be arbitrarily chosen to be 0 and 1. Such variables are often also called *logic* variables because the two states can be referred to equally well as TRUE and FALSE. Binary systems are analyzed using Boolean algebra which was invented in the nineteenth century by George Boole. The most basic Boolean operations which are performed in Boolean algebra are AND, OR and NOT. In digital circuits these operations are performed by gates. These three gates are described in the following section.

8.2.1 The NOT Gate or Inverter

The NOT gate performs the simplest Boolean operation, that of *complementation*, i.e.,

$$Y = \overline{X}$$

where the bar over the variable indicates the complement. Figure 8.1 shows the symbol used to represent the NOT gate, as well as its *truth table*. The triangle in the symbol identifies the input and output while the circle symbolizes complementation. Figure 8.1(b) shows that the truth table is simply a complete listing of the values of the output variable Y for all possible values of the input variable X.

Because the output is always the complement or inverse of the input this gate is usually called an *inverter*.

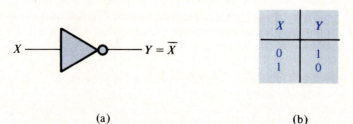

X	Y
0	1
1	0

(a)   (b)

Figure 8.1 NOT gate (a) symbol and (b) truth table.

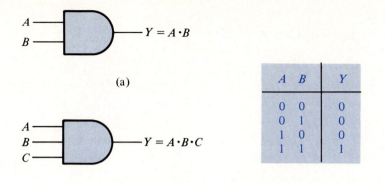

$$A \quad B \quad | \quad Y$$

A	B	Y
0	0	0
0	1	0
1	0	0
1	1	1

(c)   (b)

Figure 8.2 (a) Symbol and (b) truth table for a two-input AND gate. (c) Symbol for three-input AND gate.

8.2.2 The AND Gate

Figure 8.2 shows the symbol and truth table for a two-input AND gate, that is, a gate which performs the Boolean operation

$$Y = A \cdot B$$

where the " $\cdot$ " denotes the AND operation and is often omitted.

The extension of the AND function to three or more input variables is intuitively straightforward. For example, Fig. 8.2(c) shows the symbol for a three-input AND gate.

8.2.3 The OR Gate

A two-input gate which performs the Boolean operation

$$Y = A + B$$

where the " $+$ " denotes the OR operation, is called a two-input OR gate. Its symbol and truth table are shown in Fig. 8.3.

Again, the extension to OR gates with three or more inputs should be obvious. For example, Fig. 8.3(c) shows the symbol of a three-input OR gate.

8.3 SYNTHESIS AND IMPLEMENTATION OF LOGIC FUNCTIONS

The solution of an engineering synthesis problem is always a two-step process: first a function must be found which describes the problem, whereupon one must select the components with which to implement this function. In binary systems the functions are logical expressions that satisfy the laws of Boolean algebra, and the components on the most elementary level are usually gates.

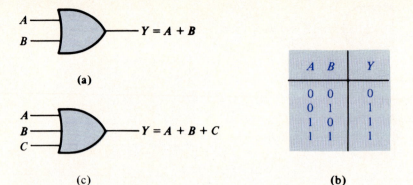

(a)

(c)

A	B	Y
0	0	0
0	1	1
1	0	1
1	1	1

(b)

Figure 8.3 Symbol (a) and truth table (b) for a two-input OR gate. (c) Symbol for three-input OR gate.

8.3.1 Procedures for Obtaining the Logical Function

The most straightforward method for translating a problem into a logical expression is by means of the truth table.

Suppose, for example, the problem is to equip the chairs in a meeting hall with voting buttons connected in such a way that a lamp on the chairman's table lights whenever the majority of those present record a "yes" vote by pressing their buttons. If a small number of seats, such as three, is taken for purposes of illustration, and if complicating issues such as what to do about empty seats are neglected, then this problem can be stated by means of the truth table shown in Fig. 8.4. The variables A, B, C represent the three voting buttons, while Y is the lamp.

The truth table can now be translated into a logical expression. Depending on how one proceeds this expression can take on several different forms. One approach is to note the rows in the truth table which correspond to $Y = 1$. Thus the fourth row in Fig. 8.4 states in effect that $Y = 1$ if $A = 0$ AND $B = 1$ AND $C = 1$. Evidently the $A = 0$ statement can be written as $\overline{A} = 1$. Noting the rows in the truth table which yield $Y = 1$ one can state that $Y = 1$ if the inputs have the values given in row 4 OR row 6 OR row 7 OR row 8. The corresponding algebraic

A	B	C	Y
0	0	0	0
0	0	1	0
0	1	0	0
0	1	1	1
1	0	0	0
1	0	1	1
1	1	0	1
1	1	1	1

Figure 8.4 Truth table for the majority-of-three voting problem.

expression will be

$$Y = \overline{A}BC + A\overline{B}C + AB\overline{C} + ABC \tag{8.1}$$

The general form of logical expression represented by Fig. 8.4 is referred to as the *canonical-sum-of-products* (SOP) form. Its name is derived from the custom in Boolean algebra of calling the OR operation "addition" and the AND operation "multiplication." Each term in the SOP form of the logical expression contains *all* the input variables in either true or complemented form, and is called a *minterm*.

Equation (8.1) is clearly not unique for an equivalent expression could be obtained by considering the input combinations which produce $\overline{Y} = 1$ in Fig. 8.4. This is left as an exercise for the student.

DRILL EXERCISE

8.1 Show that by considering the rows in Fig. 8.4 which correspond to $Y = 0$ the following logical expression is obtained

$$\overline{Y} = \overline{A}\overline{B}\overline{C} + \overline{A}\overline{B}C + \overline{A}B\overline{C} + A\overline{B}\overline{C}$$

Equation (8.1) can be implemented directly using the AND, OR and NOT gates introduced in Sec. 8.2. Three inverters are needed to produce $\overline{A}$, $\overline{B}$, and $\overline{C}$, four 3-input AND gates will yield the minterms, and a single 4-input OR gate will then combine these minterms to produce the output Y. The resulting circuit is shown in Fig. 8.5. The reader is invited to verify that the alternative approach considered in Exercise 8.1 requires the same complement of gates plus one additional inverter to give Y.

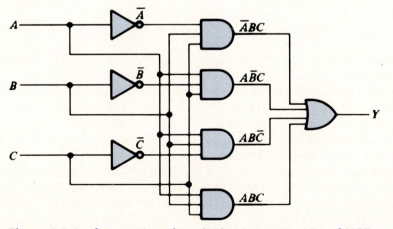

Figure 8.5 Implementation of Eq. (8.1) using AND, OR and NOT gates.

DRILL EXERCISE

8.2 Show that the truth table in Fig. D8.2(a) can be implemented using AND, OR and NOT gates in either of the two forms shown in Fig. D8.2(b) and (c). This is the Exclusive OR, or EXOR, function which is an important building block in digital circuits.

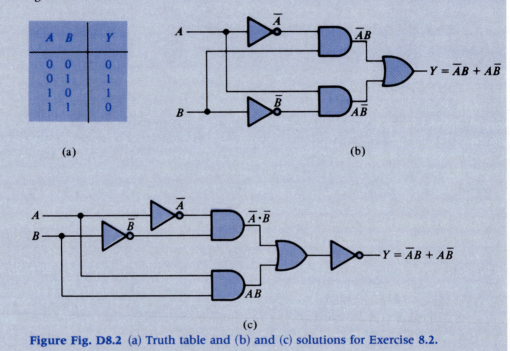

A	B	Y
0	0	0
0	1	1
1	0	1
1	1	0

(a)

(b)

(c)

Figure Fig. D8.2 (a) Truth table and (b) and (c) solutions for Exercise 8.2.

8.3.2   Implementation of Logic Functions

Although the basic Boolean operations are AND, OR, and NOT, the most widely available logic gates are the NAND and NOR, as well as the NOT. Figure 8.6 shows the symbol and truth table for the two-input NAND and NOR gates. A comparison of Figs. 8.6, 8.2, and 8.3 shows that the NAND and NOR gates are simply the complements of the corresponding AND and OR functions. Since the inverter is simply a one-input NOR or NAND gate, it will henceforth be understood to be a member of these two gate types.

Therefore, the function which is to be implemented using logic gates usually must first be expressed in terms of NAND and/or NOR operations. This involves the use of De Morgan's theorems from Boolean algebra, which state that, given logical variables $A, B, C, \ldots$,

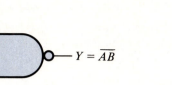

A	B	Y
0	0	1
0	1	1
1	0	1
1	1	0

$Y = \overline{AB}$

(a)

(b)

$Y = \overline{A + B}$

A	B	Y
0	0	1
0	1	0
1	0	0
1	1	0

(c)

(d)

Figure 8.6 Two-input NAND gate (a) symbol and (b) truth table , and two-input NOR gate (c) symbol and (d) truth table.

$$\overline{ABC\cdots} = \overline{A} + \overline{B} + \overline{C} + \cdots \qquad (8.2a)$$

$$\overline{A + B + C + \cdots} = \overline{A}\,\overline{B}\,\overline{C}\cdots \qquad (8.2b)$$

The theorems can be verified by noting that if all variables are 1 then both sides of Eqs. (8.2a) and (8.2b) are zero. However, if one or more inputs are 0 then both sides are equal to 1.

Suppose, for example, that the circuit in Fig. D8.2(b), which is reproduced for convenience in Fig. 8.7(a), is to be implemented using NOR and NAND gates. The conversion procedure is illustrated in Fig. 8.7(b) and (c). In Fig. 8.7(b) every AND gate is replaced by a NAND gate followed by a NOT. Defining the outputs of the AND gates to be

$$X = \overline{A} \cdot B \qquad \text{and} \qquad Y = A \cdot \overline{B}$$

and noting that, according to Eq. (8.2a)

$$\overline{X} + \overline{Y} = \overline{X \cdot Y}$$

one can replace the OR gate together with the inverters connected to its inputs by a single NAND gate.

Equation (8.2b) shows that if two variables A and B are inverted and then ANDed, the resulting pair of inverters and two-input AND gate can be replaced by a two-input NOR gate.

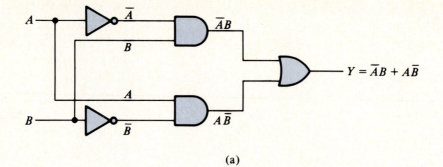

(a)

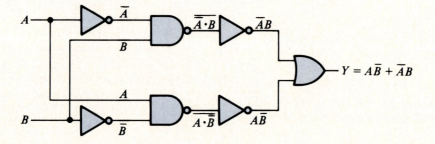

(b)

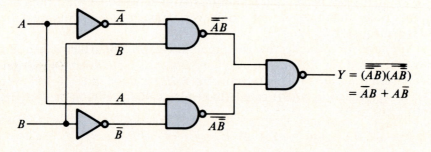

(c)

Figure 8.7 Example of conversion of AND/OR design to NAND/NOR: (a) original circuit, (b) after AND to NAND conversion, (c) after use of De Morgan's theorem.

DRILL EXERCISE

8.3 Show that the circuit in Fig. D8.2(c) can be replaced by the one in Fig. D8.3.

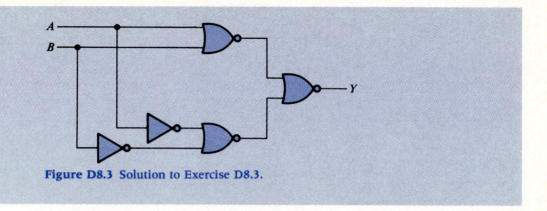

Figure D8.3 Solution to Exercise D8.3.

Notice that the circuit in Fig. 8.7(c) is based entirely on NAND logic whereas the one in Fig. D8.3, which implements the same function, is an all NOR design. This is a very desirable format since, as will be seen later, some logic families, such as NMOS, are basically NOR type whereas others, like TTL, favor the NAND operation.

A quick glance through the catalog of any major integrated circuit (IC) manufacturer will reveal a wide variety of relatively complex combinatorial ICs such as decoders, multiplexers, and arithmetic logic units (ALU), as well as simple circuits such as gates. The basic logical element in all these ICs is the gate.

In ICs based on the MOS technology *switch logic* is mixed with gate logic to great advantage. Figure 8.8 shows the switch implementation of the two-input AND function. The operation is quite clear if one associates the states of A and B with the states of the corresponding switches. Thus, if $A = 1$ and, therefore, $\bar{A} = 0$, the corresponding switches are, respectively, closed and open. The case $A = B = 1$ connects Y to 1, whereas $\bar{A} = 1$ or $\bar{B} = 1$ closes a path between Y and 0 which satisfies the AND truth table.

The advantage of switch logic will become apparent when the MOS family is described later in this chapter. It is interesting to note that computers built before World War II were based entirely on electromechanical switches. There are telephone switching centers to this day which use bulky, slow and failure-prone electromechanical switches (relays), although these are being rapidly phased out.

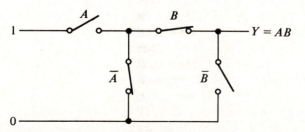

Figure 8.8 Switch implementation of the two-input AND function. Switch settings shown correspond to $A = 0$, $B = 1$.

DRILL EXERCISE

8.4 Show that if A and B in Fig. 8.8 are interchanged with $\overline{A}$ and $\overline{B}$, respectively, the circuit performs the function

$$Y = \overline{AB}$$

8.4 SEQUENTIAL CIRCUITS

Logic gates as well as the logic circuits which were used in the preceding sections to illustrate the principles of logic design are all combinatorial, or *combinational*, circuits. The outputs of such circuits depend only on the *current* values of their inputs, i.e., they do not possess any memory.

There are many applications, however, in which memory elements, capable of storing digital values, are required. Digital computers, for example, require memories for storing programs and data. Data is processed in a computer as a sequence of steps and this requires the temporary storage of intermediate logical values.

Digital circuits which possess memory are called *sequential* circuits. The most basic sequential circuit is the *latch*, which is sometimes also called a *flip-flop*. However, following a widely accepted convention, the latter name will be reserved in this text for a more complex circuit based on the latch.

8.4.1 The Latch

If a logic circuit is to behave as a memory element, it must have inputs through which the following three actions can take place:

1. Change the contents to a 1.

2. Change the contents to a 0.

3. Hold (memorize) the present contents.

In a binary system, at least two input lines are needed since, for example, one line can only do (1) and (2).

Figure 8.9 shows the symbol and *characteristic table* of such a memory element called a *set-reset* (SR) latch. The table contains the logical value of the output Q for all possible combinations of the two inputs, S and R. Q_n denotes the value of Q at a time t_n just before S and R take on a particular set of values, and Q_{n+1} is the value of Q at a time t_{n+1} just after S and R change to these values.

It should be obvious from the table in Fig. 8.9(b) that if the output Q is assumed to display the contents of memory, then $R = 0$, $S = 1$ corresponds to action (1) described above, $R = 1$, $S = 0$ is action (2), and $R = S = 0$ is action (3). The fourth possible input combination, $S = R = 1$, is not needed and, therefore, is not used.

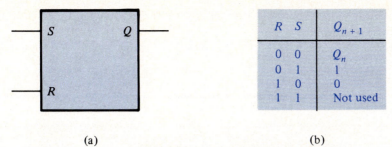

R	S	Q_{n+1}
0	0	Q_n
0	1	1
1	0	0
1	1	Not used

(a) (b)

Figure 8.9 SR latch: (a) symbol and (b) characteristic table.

A NOR gate realization of a *SR* latch, together with the corresponding truth table, are shown in Fig. 8.10. The table can be easily verified with the help of the NOR truth table in Fig. 8.6(d) and the equations

$$Q^* = \overline{S + Q} \tag{8.3a}$$

and

$$Q = \overline{R + Q^*} \tag{8.3b}$$

which describe the relationship between the terminal variables of the two NOR gates in Fig. 8.10(a).

The $S = R = 1$ input combination not only is not required for the operation of a *SR* latch, but also is not allowed. The reason is that if $S = R = 1$ is followed by $S = R = 0$, the next state of Q and Q^* could correspond to either $Q = 1, Q^* = 0$ or $Q = 0, Q^* = 1$, i.e., the next state is not known *a priori*. With this restriction on the valid combinations of S and R, Q^* can obviously be labeled $\overline{Q}$. Comparing now the valid entries in Fig. 8.10(b) with the characteristic table in Fig. 8.9(b) it is clear that the circuit in Fig. 8.10(a) behaves as a *SR* latch with the added bonus that $\overline{Q}$ as well as Q is available.

The *SR* latch in Fig. 8.10(a) illustrates the basic difference between combinatorial and sequential circuits. That difference comes about because of the presence of *feedback* lines in the latter, such as the connection between Q^* and an input of

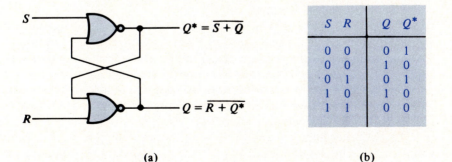

S	R	Q	Q*
0	0	0	1
0	0	1	0
0	1	0	1
1	0	1	0
1	1	0	0

(a) (b)

Figure 8.10 *SR* latch using NOR gates: (a) circuit and (b) truth table. $Q^* = \overline{Q}$ if $S = R = 1$ is forbidden.

the lower NOR gate in Fig. 8.10(a). This causes Q to depend on Q^* which, in turn, depends on Q.

Because the SR latch has only two possible states for all valid inputs, it is often referred to as a *bistable* element.

DRILL EXERCISE

8.5 Verify that the truth table in Fig. D8.5(b) corresponds to the latch in Fig. D8.5(a). Thence show that this is the circuit of a $\overline{SR}$ latch, i.e., that its inputs must be the respective complements of those corresponding to the NOR SR latch in Fig. 8.10 to produce the same values of Q and $\overline{Q}$.

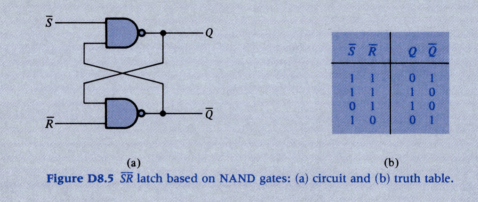

$\overline{S}$	$\overline{R}$	Q	$\overline{Q}$
1	1	0	1
1	1	1	0
0	1	1	0
1	0	0	1

(a)   (b)

Figure D8.5 $\overline{SR}$ latch based on NAND gates: (a) circuit and (b) truth table.

8.4.2 The Latch as a Memory Element

The SR latch is the basic building block in memories, and in the *static random-access memory* (RAM) in particular. This type of memory, which is also known as a read/write memory, is called static because the integrity of the stored data is preserved as long as the memory remains energized. *Dynamic* RAMs, in contrast, must be periodically refreshed.

Figure 8.11 shows a possible design of one RAM cell. Each cell stores one binary digit or *bit*. Because a practical RAM may have to store thousands of bits of data, and because it would be totally impractical to provide separate input and output lines to each cell, a means has to be provided for *addressing* every cell. This is done in Fig. 8.11 by means of the ADDRESS line. Since this line is connected to the inputs of AND gates, the latch will remain in the $Q_{n+1} = Q_n$ state as long as ADDRESS = 0. This will also force DATOUT = 0 irrespective of the value of Q.

On the other hand, if ADDRESS = 1 then what happens to the latch depends on the remaining input lines. If WRITE = 1 as well, then new data can be written into the latch via the DATIN line. Because of the inverter, DATIN = 1 results in $S = 1$, $R = 0$, whereas DATIN = 0 produces $S = 0$, $R = 1$.

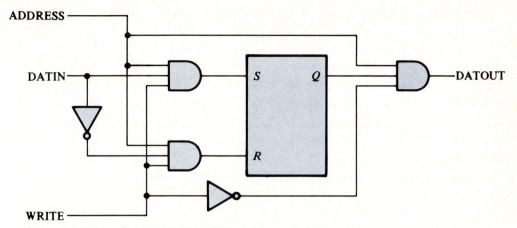

Figure 8.11 Simple static RAM cell based on a SR latch.

Because of the inverter in the WRITE line, WRITE = 1 prevents the contents of the memory from being read since DATOUT = 0 irrespective of Q. When WRITE = 0, i.e., $\overline{\text{WRITE}}$ = 1, the latch is in the "memorize" state, $S = 0$, $R = 0$, and its contents can be read since DATOUT = Q.

8.4.3 The Flip-flop

Most digital systems are designed to operate under the control of a master *clock* which brings order to the system's operation by synchronizing all the latches, thereby ensuring that they change state when their inputs have stable and predictable values. Not only is it relatively easy to design such *synchronous* systems, but predicting limits for their reliable operation is also a comparatively straightforward matter. These synchronized latches will be called, following a widely accepted convention, *flip-flops*.

THE GATED FLIP-FLOP

Figure 8.12(a) shows how a simple flip-flop can be implemented with NAND gates. From the NAND truth table it is clear that, as long as the clock input C is 0, $A = B = 1$ and the $\overline{SR}$ latch, consisting of gates 3 and 4, memorizes its previous state. On the other hand, with $C = 1$, $A = \overline{S}$ and $B = \overline{R}$ and the circuit therefore behaves as a SR latch.

The characteristic table in Fig. 8.12(c) is the same as that in Fig. 8.9(b) except for the meaning of the subscript n. Q_{n+1} is the logical value of Q after the $(n + 1)$th transition of C from 0 to 1 and S_n and R_n are the values of S and R just before this transition. The $S = R = 1$ state is not needed and, as explained earlier, is not allowed.

Thus the clock acts as a simple gate between the flip-flop inputs S and R and the inputs A and B to the $\overline{SR}$ latch. For this reason this type of memory element is referred to as a *gated* or *pulse-triggered* flip-flop.

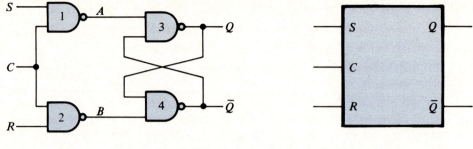

S_n	R_n	Q_{n+1}
0	0	Q_n
0	1	0
1	0	1
1	1	Not allowed

(c)

Figure 8.12 The gated *SR* flip-flop: (a) circuit of NAND-based design, (b) symbol, and (c) characteristic table.

DRILL EXERCISE

8.6 Show that the NOR-based equivalent of the gated *SR* flip-flop in Fig. 8.12(a) is as shown in Fig. D8.6.

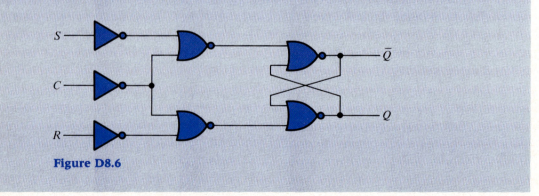

Figure D8.6

Modern flip-flops are much more sophisticated than the simple circuits considered here. A catalog of sequential circuits includes not only flip-flops but such ICs as counters and shift registers, all of which are based on the flip-flop which, in turn, is an assembly of gates.

8.5   PROPERTIES OF PRACTICAL GATES

In the preceding section it was seen that gates are the basic elements in digital circuits since, for example, although memories are based on latches, these can be assembled out of gates. Therefore, practical logic families are compared by comparing the characteristics of the gates on which each family is based. In order to be able to carry out such a comparison it is necessary to define some parameters which reflect, in a compact and simple manner, the important properties of a logic gate. The simplest vehicle for doing this is the BJT inverter that was analyzed in Chap. 7.

8.5.1   The BJT Inverter

The circuit and voltage transfer characteristic of the *npn* BJT inverter which was analyzed in Sec. 7.7 are reproduced for convenience in Fig. 8.13. From Fig. 7.26 and Eqs. (7.32) and (7.34) the quantities V_{OL}, V_{OH}, V_{IL}, and V_{IH} in Fig. 8.13(b) are defined as follows:

$$V_{OL} = V_{CEsat} \tag{8.4}$$

$$V_{OH} = V_{CC} \tag{8.5}$$

$$V_{IL} = V_{BE} \tag{8.6}$$

and

$$V_{IH} = V_{BE} + I_{BS} R_B \tag{8.7}$$

where

$$I_{BS} = \frac{I_{CS}}{\beta} \tag{8.8}$$

and

$$I_{CS} = \frac{V_{CC} - V_{CEsat}}{R_C} \tag{8.9}$$

In Sec. 7.8 it was shown that if V_{OL} and V_{OH} are associated with, respectively, the FALSE and TRUE logical states, then the circuit in Fig. 8.13(a) behaves as a logical inverter or NOT gate. This choice of *signal representation* was clearly an arbitrary one since the TRUE state could have equally well been associated with the lower voltage level V_{OL}. The convention which associates the 1 state with the *more positive* voltage level, is called *positive logic*, while associating logical 1 with the *more negative* signal level leads to *negative logic*. Henceforth, positive logic will be assumed unless the contrary is stated explicitly.

It will be seen further on that the assignment of *distinct* signal levels to the 0 and 1 states is unrealistic; rather one specifies two nonoverlapping signal *ranges*. For example, it is apparent from Fig. 8.13(b) that the BJT inverter still behaves as a NOT gate if the 1 state is associated with a voltage *range* satisfying the condition

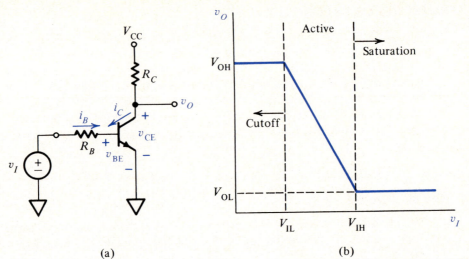

(a)   (b)

Figure 8.13 (a) Circuit and (b) voltage transfer characteristic of the simple *npn* BJT inverter.

$$v \geq V_{IH}$$

and if the 0 state is assumed to correspond to any voltage satisfying the condition

$$v \leq V_{IL}$$

The important question, as will be seen later, is how close do the worst case signal levels lie to V_{IL} and V_{IH}.

8.5.2 Logic Gain: Fanout of BJT Inverter

Logic gates based on *active* elements such as BJTs or FETs exhibit *logic gain*. This simply means that the circuit behaves as a logic gate even when its output is loaded with more than one identical gate. To clarify this important concept the circuit in Fig. 8.14 is considered. This consists of a BJT inverter with its output connected to the inputs of N identical inverters.

If the network in Fig. 8.14 is to exhibit the correct logical behavior then the following two conditions must be satisfied:

1. When Q_0 is saturated, $Q_1, \ldots, Q_N$ are cut off.

2. When Q_0 is cut off, $Q_1, \ldots, Q_N$ are saturated.

The necessary conditions for (1) and (2) to be satisfied are now considered.

Q_0 SATURATED

This corresponds to $v_I > V_{IH}$ and, therefore, $v_O = V_{OL} = 0.2$ V. Since the N load inverters have $V_{IL} = 0.7$ V, it follows that a saturated Q_0 will maintain $Q_1, \ldots, Q_N$ cut off, independent of N.

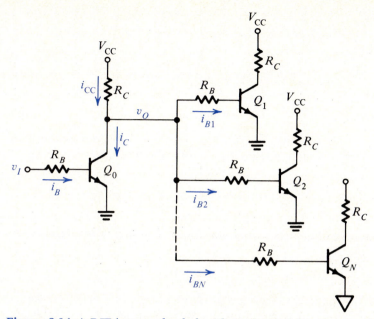

Figure 8.14 A BJT inverter loaded with N identical inverters.

Q_0 CUT OFF

This corresponds to $v_I < V_{IL}$ and, therefore, $i_c = 0$. However, in contrast to the situation in Fig. 8.13, v_O is not equal to V_{CC} because of the current i_{CC} which arises due to the base currents of the N load inverters. It is left as an exercise for the student to show that, if the BJTs in the N load inverters are represented by the piecewise-linear model in Fig. 7.18, then the effect of this load can be represented by the Thevenin circuit shown connected to the collector of Q_0 in Fig. 8.15. The output voltage v_O is given by

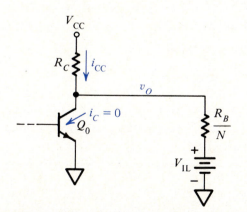

Figure 8.15 Model for computing v_O in Fig. 8.14 when $v_I < V_{IL}$.

$$v_O = \frac{R_B V_{CC} + N R_C V_{IL}}{R_B + N R_C} \tag{8.10}$$

If the N load inverters are, indeed, to perform as inverters then they should be saturated when v_O has the value given by Eq. (8.10). This imposes the requirement that

$$v_O \geq V_{IH} \tag{8.11}$$

Because v_O varies inversely with N, Eq. (8.11) imposes an upper limit on the *fanout* N, which is obtained by solving Eq. (8.10) for $N = N_{max}$ with $v_O = V_{IH}$. The result is

$$N_{max} = \frac{R_B (V_{CC} - V_{IH})}{R_C (V_{IH} - V_{IL})} \tag{8.12}$$

EXAMPLE 8.1

Compute the maximum fanout N_{max} of the inverter in Fig. 8.13 using the following parameter values: $\beta = 25$, $V_{CC} = 5.2$ V, $R_B = 10$ kΩ, $R_C = 1$ kΩ.

Solution
From Eq. (8.9)

$$I_{CS} = \frac{V_{CC} - V_{CEsat}}{R_C} = \frac{5.2 - 0.2}{1} = 5 \text{ mA}$$

Equation (8.8) yields the corresponding value of I_{BS}:

$$I_{BS} = \frac{I_{CS}}{\beta} = \frac{5}{25} = 0.2 \text{ mA}$$

Therefore, V_{IH} has a value, according to Eq. (8.7)

$$V_{IH} = V_{BE} + I_{BS} R_B = 0.7 + 0.2(10) = 2.7 \text{ V}$$

Equation (8.12) can now be used to compute N_{max}:

$$N_{max} = \frac{R_B}{R_C} \frac{V_{CC} - V_{IH}}{V_{IH} - V_{IL}} = \frac{10}{1} \frac{5.2 - 2.7}{2.7 - 0.7} = 12.5$$

Since N_{max} must clearly be an integer the above result has to be rounded off to 12. The conclusion is, therefore, that a BJT inverter with the above parameter values can drive up to 12 identical inverters.

Although, as will be seen shortly, there are other considerations which lead to a more conservative value of N_{max}, nevertheless the example illustrates the important property of logic gain which is quantified by means of the maximum fanout N_{max}. This property is exhibited by all logic families based on active elements, be they BJTs or FETs.

8.5.3 Noise Margin and Logic Swing

Just as electromagnetic interference from a variety of sources can hinder a telephone conversation or radio reception, so it can interfere with the operation of a digital circuit. The immunity of a logic gate to such interference is specified through its *noise margin*. This is defined by considering a chain of identical inverters as shown in Fig. 8.16. The effect of noise can be modeled by the voltage source v_n.

Assume that initially, $v_n = 0$, and that the output voltage $v_O = V_{OH}$ corresponding to the 1 state is greater than V_{IH}. Recall that $V_{OL} = V_{CEsat} < V_{IL}$. Therefore, the two output states, V_{OL} and V_{OH}, of the first inverter are related to the breakpoints in the voltage transfer characteristic of the second inverter as shown in Fig. 8.17.

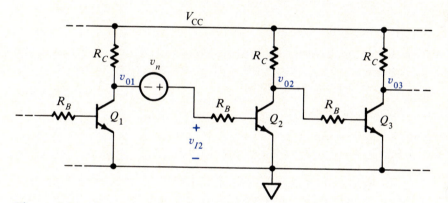

Figure 8.16 Chain of BJT inverters with noise source.

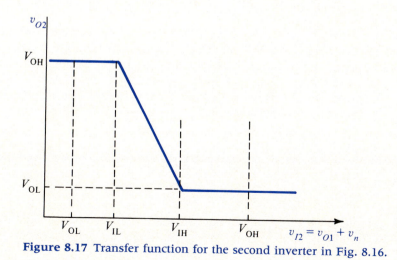

Figure 8.17 Transfer function for the second inverter in Fig. 8.16.

Because, from KVL,

$$v_{I2} = v_{O1} + v_n$$

it follows that, if $v_{O1} = V_{OH}$, then a noise voltage v_n which causes v_{I2} to fall below V_{IH} will drive Q_2 out of saturation. For this not to occur, v_n must satisfy the condition

$$v_{I2} = V_{OH} + v_n \geq V_{IH}$$

or

$$v_n \geq V_{IH} - V_{OH} \tag{8.13}$$

Since $V_{IH} < V_{OH}$, Eq. (8.13) signifies that only negative values of v_n, having a magnitude greater than $V_{OH} - V_{IH}$, would interfere with the normal operation of the circuit.

Following a similar argument, one can show that if

$$v_n \leq V_{IL} - V_{OL} \tag{8.14}$$

then the second inverter will not be driven out of the cutoff state when $v_{O1} = V_{OL}$.

Equations (8.13) and (8.14) lead to the definition of two *noise margins*, NM_H and NM_L, expressed as follows:

$$NM_H = V_{OH} - V_{IH} \tag{8.15a}$$

and

$$NM_L = V_{IL} - V_{OL} \tag{8.15b}$$

Clearly, the greater NM_H and NM_L are, the less susceptible will the logic circuit be to noise. It should be clear that since the noise is random and, therefore, the polarity of v_n is not predictable, the smaller of NM_L and NM_H determines the practical noise margin of the gate. For this reason a single noise margin NM is usually specified.

The above results apply to logic circuits in general, irrespective of whether the elements are inverters or NAND and NOR gates with multiple inputs. One is always faced with the problem of a logic circuit's output being able to maintain its N load gates in their correct states in the presence of noise. As the following example illustrates, this leads to much more conservative values of the maximum fanout N_{max} than Example 8.1 suggested.

EXAMPLE 8.2

Repeat Example 8.1 with the additional constraint that the noise margin NM is to be not less than NM_L.

Solution

From Eq. (8.15b)

$$NM_L = V_{IL} - V_{OL} = 0.7 - 0.2 = 0.5 \text{ V}$$

and, from Eq. (8.15a),

$$V_{OH} = NM_H + V_{IH}$$

Since $NM_H = NM_L$ is required,

$$V_{OH} = 0.5 + V_{IH}$$

In Example 8.1 V_{IH} was found to be 2.7 V, therefore,

$$V_{OH} = 0.5 + 2.7 = 3.2 \text{ V}$$

Equation (8.12) for N_{max} was derived subject to the constraint that the 1 state output voltage V_{OH} be allowed to fall to V_{IH}, i.e., $NM_H = 0$. Therefore, the equation for N_{max}, which is valid for a nonzero noise margin, corresponds to Eq. (8.12) with V_{IH} replaced by V_{OH}, i.e.,

$$N_{max} = \frac{R_B}{R_C} \frac{V_{CC} - V_{OH}}{V_{OH} - V_{IL}}$$

Substituting numerical values

$$N_{max} = \frac{10}{1} \frac{5.2 - 3.2}{3.2 - 0.7} = 8$$

This is still a very reasonable value of N_{max} and is accompanied by what turns out to be a not unreasonable noise margin.

A frequently used figure of merit for logic circuits is *logic swing* which is simply $V_{OH} - V_{OL}$. From Eqs. (8.15)

$$V_{OH} - V_{OL} = NM_H + NM_L + V_{IH} - V_{IL}$$

Although, as was explained earlier, the practical noise margin is determined by the smaller of NM_L and NM_H, nevertheless logic families with a large logic swing often tend to also exhibit a relatively large noise margin, as the above equation suggests.

8.5.4   Propagation Delay Time

Delays occur because of charge storage effects in devices such as transistors. These effects limit the speed with which transistors can respond to input voltage or current changes. Electrical circuits also always contain unwanted, or *parasitic*, capacitances and inductances which give rise to similar energy storage effects. The techniques for analyzing such effects are considered in the following chapters. For the purpose of the present discussion it suffices to define a useful measure of delay time in digital circuits.

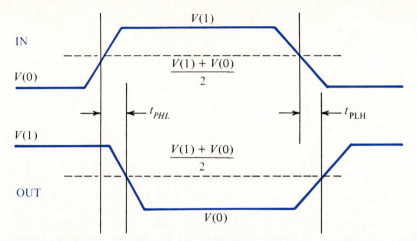

Figure 8.18 Simplified diagram illustrating conventional definition of delay time.

Figure 8.18 shows a very widely used definition of the *propagation delay time* of a digital circuit. As can be seen from the figure, the midpoint level between voltages $V(1)$ and $V(0)$ corresponding, respectively, to the logical 1 and 0 states, is first established. The time interval between the corresponding input and output voltage crossing of this level is then taken as the delay time. Two delays are normally defined, the output voltage low-to-high propagation delay time, t_{PLH}, and the output voltage high-to-low propagation delay time, t_{PHL}. Sometimes a single average propagation delay time t_P is defined as

$$t_P = \frac{t_{PLH} + t_{PHL}}{2}$$

8.6   THE BIPOLAR LOGIC FAMILIES

The most important logic families which use the BJT as the switching element are those based on so-called transistor-transistor logic (TTL), emitter-coupled logic (ECL) and integrated injection logic (I^2L). For each of these, the design techniques that are used to realize the basic gate are studied and, where possible, their characteristics are examined on the basis of the parameters introduced previously.

8.6.1   The Transistor-Transistor Logic (TTL) Family

The transistor-transistor logic (TTL) family is rivaled only by CMOS in the very wide variety of logic functions that are available as ICs. Despite the growing complexity of ICs and the corresponding reduction in the number of ICs needed to assemble a complete digital system, TTL continues to be a very important logic family.

The operating principles of a TTL gate are readily understood by examining how a diode AND gate and a BJT inverter can be combined to produce a diode-transistor logic (DTL) gate, and how the IC technology can then be exploited to evolve the TTL gate from the DTL.

EVOLUTION OF THE TTL GATE

In Sec. 5.6.3 it was shown that the circuit in Fig. 5.20, which is redrawn in a more conventional way in Fig. 8.19, behaves as an AND gate for positive logic. This means that, in particular, as long as the input voltages are less than V_{CC}, the input at the lower potential causes the corresponding diode to conduct, in the process cutting the other diode OFF. Thus if $v_{IB} < v_{IA} < V_{CC}$, D_B conducts and D_A is OFF. Therefore, if a diode D_T is connected to one of the inputs, as shown in Fig. 8.20(a), D_B and D_T will be OFF if $v_{IA} < V_D$, where V_D is the ON voltage in the piecewise-linear model of the diode. Alternatively, with $v_{IA} > V_D$, D_B and D_T conduct while D_A is cut off.

DRILL EXERCISE

8.7 Using assumed state analysis verify that the diode D_T in Fig. 8.20(a) is OFF if $v_{IA} < V_D$, the diode turn on voltage, and, conversely, that it is ON if $v_{IA} > V_D$.

Figure 8.20(b) shows the result of replacing the diode D_T in Fig. 8.20(a) by the base-emitter junction of an *npn* BJT. Since the input characteristic of a BJT is that of a diode, it can be assumed that the conditions determining whether the EBJ is reverse or forward biased are the same as those for diode D_T in Fig. 8.20(a), the sole difference being that the threshold voltage is V_{BE} instead of V_D.

If $v_I < V_{BE}$, D_A is ON and D_B as well as the EBJ are OFF. Therefore,

$$i_B = 0$$

and

$$i_{BB} = i_{DA} = \frac{V_{CC} - V_D - v_I}{R_{BB}} \qquad (8.16)$$

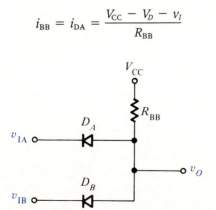

Figure 8.19 A two-input diode logic (DL) AND gate.

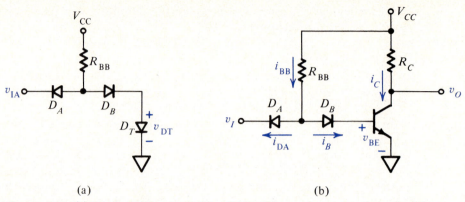

(a)   (b)

Figure 8.20 (a) Diode AND gate in Fig. 8.19 with one input terminated with a diode; (b) simple diode-transistor logic (DTL) inverter.

Since $i_B = 0$ it follows that

$$v_O = V_{CC} \tag{8.17}$$

With $v_I > V_{BE}$, D_B in ON and $v_{BE} = V_{BE}$, while D_A is OFF. Therefore,

$$i_{DA} = 0$$

while

$$i_{BB} = i_B = \frac{V_{CC} - V_D - V_{BE}}{R_{BB}} \tag{8.18}$$

If the base current satisfies the condition

$$i_B \geq I_{BS}$$

the transistor is saturated and

$$v_O = V_{CEsat}$$

The circuit in Fig. 8.20(b), therefore, behaves as an inverter which can be used to build logic circuits because, just like the simple inverter in Fig. 8.13,

$$V_{OH} = V_{CC} > V_{BE}$$

and

$$V_{OL} = V_{CEsat} < V_{BE}$$

Figure 8.20(b) is the simplified circuit of a *diode-transistor logic* (DTL) inverter. The behavior of the diode AND gate in Fig. 8.19 can be generalized to M inputs, with the lowest input voltage causing the diodes associated with the remaining $M - 1$ inputs to be OFF. Therefore, it should be clear that the DTL inverter can be converted to a NAND gate by providing additional inputs as shown in Fig. 8.21 for $M = 2$.

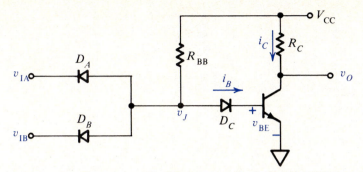

Figure 8.21 Two-input diode-transistor logic (DTL) NAND gate.

Because an *npn* BJT physically appears to consist of two diodes with their anodes connected together,* one would expect that, replacing the two diodes in the DTL inverter by an *npn* BJT, as shown in Fig. 8.22(a), should not alter the behavior of the inverter qualitatively. Furthermore, because the IC technology permits the design of multiemitter transistors, the DTL NAND gate could have its diodes replaced by a BJT with M emitters as shown in Fig. 8.22(b) for $M = 2$. Not only does this transformation yield a significant saving in IC chip area, but the resulting transistor-transistor logic circuit does not suffer from the relatively long delay time of the now obsolete DTL.

*The student should recall, however, that as pointed out in Chap. 7, the narrow base region between the two *pn* junctions is critical to the operation of a transistor.

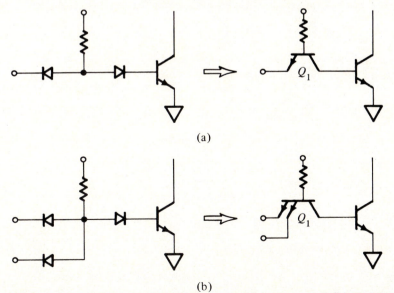

(a)

(b)

Figure 8.22 Evolution of TTL from DTL: (a) inverter and (b) NAND gate.

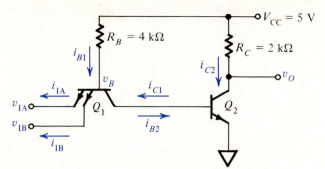

Figure 8.23 Simple two-input TTL NAND gate.

CHARACTERISTICS OF TTL

Figure 8.23 shows the circuit of a simplified 2-input TTL NAND gate. By direct analogy to the DTL gate, from which it was derived, the currents and voltages corresponding to the two logic states are calculated below.

1. $v_{IA} = v_{IB} = V_{CC}$

From the DTL analysis Q_2 is expected to be saturated. Since $V_{CC} \gg V_{BE}$ the EBJs of Q_1 are reverse biased, while the CBJ of Q_1 and EBJ of Q_2 are forward biased. According to Table 7.1 this corresponds to Q_1 operating in the reverse mode. The analysis in Sec. 7.6 of the BJT piecewise-linear model in Fig. 7.18, shows that Q_1 in Fig. 8.23 can be replaced in this state by an open circuit between the base and emitter terminals, and by a 0.5-V source between the base and collector. Therefore, the circuit in Fig. 8.24 can replace that in Fig. 8.23 for the analysis of this state. The base current of Q_2 is

$$i_{B2} = i_{B1} = \frac{5 - 1.2}{4} = 0.95 \text{ mA}$$

For proper operation Q_2 is saturated so

$$i_{C2} = I_{CS} = \frac{V_{CC} - V_{CEsat}}{R_C}$$

$$= \frac{5 - 0.2}{2} = 2.4 \text{ mA}$$

Hence as long as β_2, the common-emitter current gain of Q_2, satisfies the inequality

$$\beta_2 > \frac{i_{C2}}{i_{B2}} = \frac{2.4}{0.95}$$

the output voltage will be

$$v_O = V_{CEsat} = 0.2 \text{ V}$$

The input currents i_{IA} and i_{IB} are evidently both zero in this state.

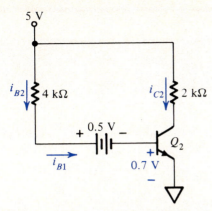

Figure 8.24 Circuit for analyzing Fig. 8.23 with $v_{IA} = v_{IB} = V_{CC}$.

2. $v_{IA} = V_{CC}$, $v_{IB} = V_{CEsat}$

Therefore, one EBJ in Q_1 is forward biased and

$$v_B = V_{BE} + v_{IB} = 0.7 + 0.2 = 0.9 \text{ V}$$

By analogy to the DTL NAND gate it is expected that both the CBJ of Q_1 and EBJ of Q_2 are reverse biased. It is, however, instructive to examine the states of Q_1 and Q_2 a little more closely.

Suppose that the EBJ of Q_2 is assumed to be ON, so that $v_{BE2} = 0.7$ V. Since $v_B = 0.9$ V, then

$$v_{BC1} = v_B - v_{BE2} = 0.2 \text{ V}$$

This is reverse bias for the CBJ of Q_2. But that means Q_1 is in the active mode and, therefore, $i_{C1} > 0$. However,

$$i_{C1} = -i_{B2}$$

and, for Q_2 to be ON, i_{B2} must be positive. Therefore Q_2 cannot be ON.

On the other hand, if Q_1 is assumed to be saturated with $v_{BC1} = 0.5$ V, then

$$v_{BE2} = v_B - v_{BC1} = 0.4 \text{ V}$$

so that $i_{B2} = 0$. Therefore $i_{C1} = -i_{B2} = 0$ which, from Fig. 7.24(b) is a valid condition for a saturated BJT.

So with v_{IB} in Fig. 8.23 at V_{CEsat}, Q_2 is cut off and $v_O = V_{CC}$. The input current i_{IB} is given by

$$i_{IB} = \frac{-(V_{CC} - V_{BE} - v_{IB})}{R_B} = \frac{-(5 - 0.7 - 0.2)}{4} \approx -1 \text{ mA}$$

Evidently the same analysis applies if the input conditions are interchanged or if both inputs are at V_{CEsat}. The only difference in the latter case is that the 1 mA input current divides between the two inputs.

DRILL EXERCISE

8.8 Show that if both inputs in Fig. 8.23 are at V_{OL} the current in each input is approximately -0.5 mA.

There now remains only to establish the values of V_{IL} and V_{IH}. It should be clear from the preceding analysis that, because v_B must be 1.2 V for Q_2 to be ON, if

$$v_B < 1.2 \text{ V}$$

Q_2 will cut off. Since, in that case,

$$v_B = V_{BE} + v_I$$

where v_I is the smaller of v_{IA} and v_{IB}, it follows that

$$v_I = 1.2 - V_{BE} = 0.5 \text{ V}$$

is the threshold voltage corresponding to the transition of Q_2 between cutoff and saturation. Therefore

$$V_{IL} = V_{IH} = 0.5 \text{ V}$$

Thus, the piecewise-linear voltage transfer characteristic of the TTL NAND gate in Fig. 8.23 is as shown in Fig. 8.25. It should be noted that this characteristic is obtained with either the unused input at V_{OH} or with both inputs connected together.

The circuit of a practical TTL NAND gate is shown in Fig. 8.26. The 0 state corresponds to Q_3 saturated so that $v_{OL} = V_{CEsat}$ as before. Q_2 also saturates and, as

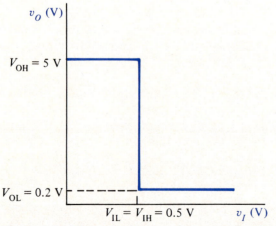

Figure 8.25 Piecewise-linear voltage transfer characteristic of the simple TTL NAND gate in Fig. 8.23. v_I is v_{IA} or v_{IB}.

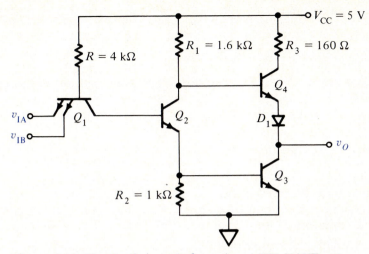

Figure 8.26 Circuit of a practical two-input TTL NAND gate.

a result, Q_4 and D_1 are cut off. On the other hand when $v_O = V_{OH}$, Q_2 and, therefore, Q_3 are cut off, and Q_4 is in the active mode.

Figure 8.27 shows the approximate voltage transfer characteristic of the circuit in Fig. 8.26. The output voltage v_O corresponding to V_{OH} is approximately 3.6 V because Q_4 in this state operates as an emitter follower and D_1 is on. Since a TTL input draws very little current when it is at V_{OH}, the emitter current of Q_4 is very small. Because, moreover, the base current of an emitter follower is $\beta + 1$ times smaller than the emitter current, the voltage drop across R_1 is very small indeed and $V_{OH} \simeq V_{CC} - V_{BE} - V_D = 3.6$ V. A detailed analysis of this circuit is beyond the scope of this text.

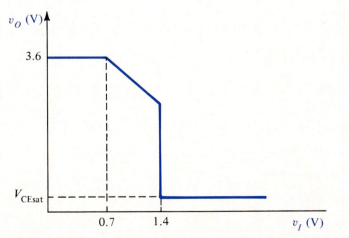

Figure 8.27 Piecewise-linear voltage transfer characteristic of the TTL NAND gate in Fig. 8.26.

The practical noise margin of the TTL gate turns out to be no better than that of the simple DTL circuit in Fig. 8.21. Although NM_H is seen from Fig. 8.27 to be

$$NM_H = V_{OH} - V_{IH} = 3.6 - 1.2 = 2.4 \text{ V}$$

NM_L is much lower. This is because of the first breakpoint at $v_I = 0.7$ V which results in $V_{IL} = 0.7$ V and not the $2V_{BE}$ that may have been anticipated from the circuit. Since $V_{OL} = 0.2$ V, therefore,

$$NM_L = 0.7 - 0.2 = 0.5 \text{ V}$$

Because TTL inputs which are pulled up to V_{OH} draw negligible current, the fanout is very high in this state. However, when $v_o = V_{OL}$, the collector current of the saturated Q_3 in Fig. 8.26 increases with N because of the substantial current which flows out of a TTL input which has $v_I = V_{OL}$. For sufficiently large N, this can cause Q_3 to come out of saturation and, therefore, v_o will rise above V_{OL}. The following example illustrates this effect.

EXAMPLE 8.3

Calculate the maximum fanout N_{max} of the TTL gate in Fig. 8.26. The base current I_{B3} of Q_3 in saturation is found to be 2.6 mA, and $\beta = 30$.

Solution
Since Q_3 in Fig. 8.26 is supposed to be saturated when $v_o = V_{OL}$, the collector-to-emitter circuit can be modeled by a voltage source of $V_{CEsat} = 0.2$ V. Figure E8.3 shows the resulting simplified circuit for computing N_{max}. As explained earlier, Q_4 and D_1 are both off in this state and are, therefore, omitted from the figure. The remaining inputs of each of the N load gates are shown tied to V_{CC} because, as explained earlier, the current in a DTL or a TTL input at V_{OL} is a maximum when the remaining inputs are at V_{OH}.

From Fig. E8.3

$$I_I = \frac{V_{CC} - V_{BE} - V_{OL}}{R} = \frac{5 - 0.7 - 0.2}{4} = 1.03 \text{ mA}$$

Therefore, the collector current of Q_3 is

$$I_{C3} = NI_I = 1.03N \text{ mA}$$

For Q_3 to remain saturated

$$I_{C3} < \beta I_{B3} = 30(2.6) = 78 \text{ mA}$$

Hence N is limited by

$$N_{max} < \frac{78}{1.03} = 76$$

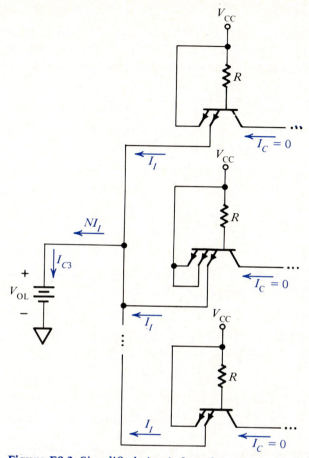

Figure E8.3 Simplified circuit for solving Example 8.3.

The N_{max} for commercial TTL ICs is only 10, but that is due to the inclusion of manufacturing tolerances, ambient temperature variation and a guaranteed maximum delay time which increases with N.

It is primarily the relatively short propagation delay time of the TTL family that established its superiority over DTL. With $N = 10$, $t_P \simeq 14$ ns which is about three times less than the delay time of a commercial DTL gate. Modern TTL circuits are, in fact, a lot faster. The so-called "advanced" TTL series have $t_P < 2$ ns. These speed improvements have occurred primarily because of the use of *Schottky* diodes which are produced between the base and collector of every BJT. Because the ON voltage of these diodes is less than 0.5 V, the CBJ is prevented from turning ON and, therefore, the BJT does not saturate. Since the time taken to bring a BJT out of saturation is the main contributing factor to the delay time of standard TTL circuits, the use of these Schottky diodes leads to a dramatic decrease in t_P.

8.6.2   Emitter-Coupled Logic (ECL)

In emitter-coupled logic (ECL), just like in Schottky TTL, the BJTs also never saturate. This results in a very small delay time although, of course, not without a price, which turns out to be greater circuit complexity and somewhat reduced noise margin.

The basic operating principle of an ECL gate can be understood by considering the diode OR gate in Fig. 8.28. Just as the diode with the *lowest* input voltage conducts in the diode AND gate, in the process cutting the remaining diodes OFF, so in the OR gate the *highest* input voltage dominates.

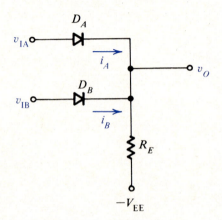

Figure 8.28 Diode OR gate.

DRILL EXERCISE

8.9 Assuming ideal diodes in Fig. 8.28 show that, providing the input voltages are greater than $-V_{EE}$, the more positive input voltage turns on the associated diode and cuts the other one off.

Now if the diodes are replaced by the base-emitter junctions of *npn* transistors, as shown in Fig. 8.29, the collector currents can be controlled by means of the input voltages. In particular, if $v_I > V_R$

$$i_{E2} = 0 \qquad i_{E1} = \frac{V_{EE} + v_I - V_{BE}}{R_E} \tag{8.19}$$

whereas, if $v_I < V_R$,

$$i_{E1} = 0 \qquad i_{E2} = \frac{V_{EE} + V_R - V_{BE}}{R_E} \tag{8.20}$$

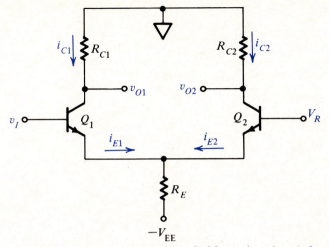

Figure 8.29 Simple emitter-coupled logic (ECL) switch.

Notice that if V_{EE} is made large, then i_{E1} defined by Eq. (8.19) becomes approximately equal to i_{E2} defined by Eq. (8.20). Thus the combination of R_E and V_{EE} behaves as a nonideal current source with a Thevenin resistance of R_E and a short circuit current

$$I_E = \frac{V_{EE}}{R_E} \simeq i_{E1} \simeq i_{E2}$$

If the BJTs are reasonably well matched with respect to V_{BE} then, from the properties of the DL OR gate, it can be seen that this current I_E can be switched between the two transistors with a very small change in v_I about V_R. It is for this reason that this logic family is also referred to as current mode logic (CML) or current switched logic (CSL).

It was stated at the beginning that ECL is nonsaturating logic. By a judicious choice of input voltage range and circuit parameters, the BJTs can be prevented from saturating. This is illustrated in the following example.

EXAMPLE 8.4

Suppose that in Fig. 8.29 $R_{C1} = 270\ \Omega$, $R_{C2} = 290\ \Omega$, $R_E = 1.24\ \text{k}\Omega$, $V_{EE} = 5.2\ \text{V}$, $V_R = -1.15\ \text{V}$, $V_{BE} = 0.75\ \text{V}$, $\beta \gg 1$. Compute the currents and voltages for $v_I = -0.75\ \text{V}$ and $v_I = -1.55\ \text{V}$.

Solution

1. $v_I = -0.75\ \text{V} > V_R$

From Eq. (8.19)

$$i_{E1} = \frac{V_{EE} + v_I - V_{BE}}{R_E} = \frac{5.2 - 0.75 - 0.75}{1.24} = 2.98 \text{ mA}$$

Since $\beta \gg 1$

$$i_{C1} \simeq i_{E1} = 2.98 \text{ mA}$$

and, therefore,

$$v_{O1} = -i_{C1}R_{C1} = -0.805 \text{ V}$$

and

$$v_{CB1} = v_{O1} - v_I = -0.06 \text{ V}$$

Therefore Q_1 is not saturated. Because $v_I > V_R$, therefore $i_{E2} = 0$, and since this results in

$$v_{O2} = 0$$

hence

$$v_{CB2} = v_{O2} - V_R = 1.15 \text{ V}$$

Therefore, Q_2 is not saturated.

2. $v_I = -1.55 \text{ V} < V_R$

Therefore

$$i_{E1} = 0$$

and

$$v_{O1} = 0$$

and again Q_1 is not saturated because $v_{CB1} = -v_I = 1.55$ V. From Eq. (8.20)

$$i_{E2} = \frac{V_{EE} + V_R - V_{BE}}{R_E} = \frac{5.2 - 1.15 - 0.75}{1.24} = 2.66 \text{ mA}$$

Since i_{C2} is almost the same,

$$v_{O2} = -2.66(290) = -0.77 \text{ V}$$

and

$$v_{CB2} = -0.77 + 1.15 = 0.38 \text{ V}$$

So Q_2 is not saturated.

Besides demonstrating that the circuit in Fig. 8.29 can be operated as a switch without saturating the BJTs, the example also confirms that the combination of V_{EE} and R_E does behave as a current source since $i_{E1} \simeq i_{E2}$. The example also highlights another advantage of this circuit, which is that both an inverting and noninverting output are available. Thus, when v_I increases, v_{O1} decreases as Q_1 turns on, while v_{O2} simultaneously increases as Q_2 cuts off. The main problem with the circuit as it stands is the lack of standardization of the logic levels, in the sense that the voltage levels at the input do not correspond to those at the outputs. It is left as an exercise for the reader to show that, in fact, it is not possible to simultaneously standardize the logic levels and prevent the BJTs in Fig. 8.29 from saturating.

Figure 8.30 shows how the standardization problem is solved and how the circuit is modified to perform as an OR/NOR gate. The latter is achieved by adding Q_3 in parallel with Q_1. As a result, by direct analogy to the diode OR gate, if either v_{IA} or v_{IB} rises above V_R, v_{O1} falls and v_{O2} rises as the current in R_E is switched from Q_2 to Q_3 or Q_1. Alternatively, if both v_{IA} and v_{IB} fall below V_R, Q_2 becomes active while Q_1 and Q_3 cut off. Therefore, v_{O1} rises and v_{O2} falls. Thus v_{O1} yields the NOR operation, whereas v_{O2} changes according to the OR function.

Transistors Q_4 and Q_5 operate as emitter followers thereby providing not only a high fanout, N, but also the necessary voltage level shift to standardize the logic levels. This is demonstrated with the right half of the circuit. Using the same input voltage levels as in Example 8.4, if v_{IA} or v_{IB} is set to -0.75 V, Q_2 is cut off and $v_{O2} = 0$. Therefore, the voltage at the emitter of Q_5 is

$$v_O(\text{OR}) = v_{O2} - V_{BE} = -0.75 \text{ V}$$

if $V_{BE} = 0.75$ V.

On the other hand, with v_{IA} and v_{IB} at -1.55 V, Q_2 is ON. If the base current of Q_5 is assumed to be negligible in comparison to the collector current of Q_2, then, as in Example 8.4, $v_{O2} = -0.77$ V and, therefore,

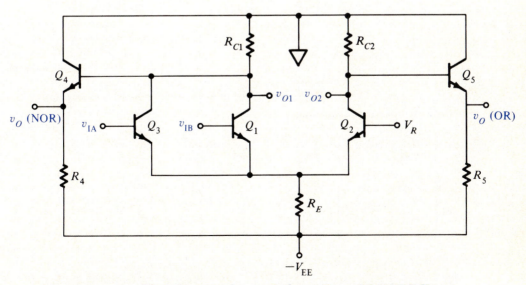

Figure 8.30 Simplified diagram of a practical two-input, OR/NOR ECL gate.

$$v_O(\text{OR}) = v_{O2} - V_{BE} = -0.77 - 0.75 = -1.52 \text{ V}$$

Since the voltage levels at v_{O1} are the same as at v_{O2}, Q_4 will have the same effect as Q_5. Thus the logic levels are standardized to $V(1) = -0.75$ V and $V(0) = -1.52$ V.

DRILL EXERCISE

8.10 Show that the base currents of the emitter follower Q_4 and Q_5 in Fig. 8.30 are negligible in comparison to the collector currents of, respectively, Q_1 and Q_2 when these are on. Assume $\beta = 50$, $R_4 = R_5 = 10$ kΩ, and use the numerical data in Example 8.4.

Figure 8.31 shows the circuit of a complete ECL gate. Transistor Q_6 and the associated diodes and resistors provide a stable voltage reference V_R. The return on as large an investment in circuit size and complexity as Fig. 8.31 represents, is the extremely fast switching speed which makes ECL the fastest silicon logic family. At the time of writing, the state of the art is $t_P = 0.75$ ns. This comes about primarily because BJT saturation is avoided, but also because of the relatively small logic swing: thus, in Fig. 8.30, $V(1) - V(0) = 0.8$ V. The latter is, of course, achieved at the price of a reduced noise margin, $NM_H = NM_L \simeq 0.4$ V.

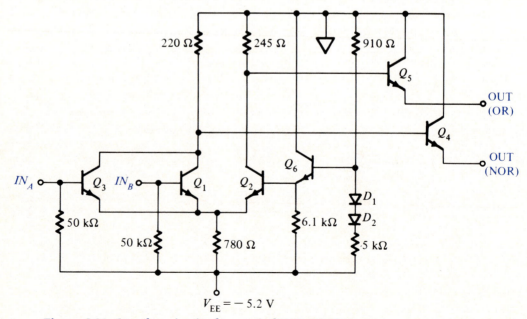

Figure 8.31 Complete circuit of a practical ECL OR/NOR gate.

8.6.3 The Integrated-Injection Logic (I²L) Family

Integrated injection logic (I²L), or *merged transistor logic* (MTL) as it is sometimes called, is a saturating family like standard TTL. However, it is the product of a much more mature technology, which has been exploited in a very clever way, resulting in the elimination of *all* resistors.

THE BASIC GATE

In order to understand the I²L inverter's characteristics it is best to consider it as a member of a chain of inverters as shown in Fig. 8.32. The inverter is seen to consist of a current source and an *npn* BJT. How the current source is achieved is discussed later. Attention is focused on the second inverter, G_2, because it is both loaded at its output and driven at its input by identical inverters.

If Q_1 is saturated, then because $v_{O1} = V_{CEsat}$, Q_2 is cut off. Therefore, the current I flows into the collector of Q_1 and $i_{C1} = I$. On the other hand, if Q_1 is cut off, the current I is *injected* into the base of Q_2 and, since $i_{B2} = I > 0$, $v_{O1} = V_{BE}$. Because $\beta > 1$ and the collector current of Q_2 is limited to $i_{C2} \leq I$, it follows that Q_2 saturates. The resulting $v_{O2} = V_{CEsat}$ cuts off Q_3, and the current I flows entirely into the collector of Q_2.

Thus in a saturated I²L inverter

$$v_O = V_{CEsat}$$

and

$$i_C = i_B = I$$

so that the base overdrive factor is β. On the other hand, when the inverter is cut off,

$$v_O = V_{BE}$$

which is established by its saturated load inverter.

From the above analysis

$$V_{OH} = V_{BE} = 0.7 \text{ V}$$

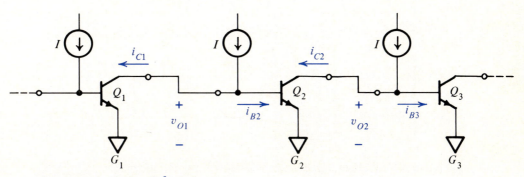

Figure 8.32 Chain of I²L inverters.

and

$$V_{OL} = V_{CEsat} = 0.2 \text{ V}$$

while

$$V_{IL} = V_{IH} = V_{BE}$$

The last equality follows from the piecewise-linear model of the BJT's input characteristic, and leads to the pessimistic conclusion that the noise margin NM is zero because

$$NM_H = V_{OH} - V_{IH} = V_{BE} - V_{BE} = 0$$

An analysis based on a more accurate model of the BJT yields the I²L inverter voltage transfer characteristic shown in Fig. 8.33. The 0.1 V value of V_{OL} reflects the somewhat lower saturation voltage V_{CEsat} of the *npn* BJTs used in I²L ICs.

Despite the fact that the noise margin according to Fig. 8.33 is still only 0.1 V, I²L gates are used with great success in ICs. There are two reasons for this: firstly the concept of a noise margin as defined earlier is too simple for dealing with current injection and, secondly, much lower noise margins can be tolerated when gates form part of a logic array within the same IC. Special circuits are used at the terminal pins of I²L ICs so that their external logic swings are much greater than the 0.6 V which Fig. 8.33 reveals.

The inverter shown in Fig. 8.32 is, in fact, the basic logic gate used in I²L circuits. The only additional feature that has to be added is provision for $N > 1$ since the circuit as it stands turns out to have a practical fanout of 1. If a fanout of, say, 3 is required, then additional BJTs are included as shown in Fig. 8.34(a). Because all the bases and all the emitters are, respectively, connected together, when the circuit is integrated, the *N npn* BJTs are replaced by a single BJT with *N*

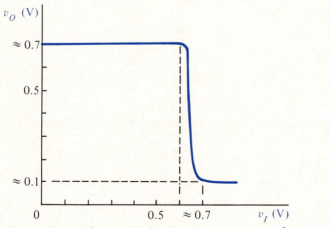

Figure 8.33 Voltage transfer characteristic of the I²L inverter.

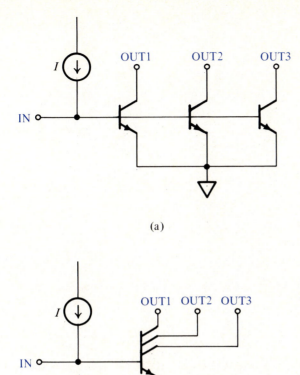

(a)

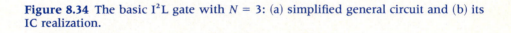

(b)

Figure 8.34 The basic I²L gate with $N = 3$: (a) simplified general circuit and (b) its IC realization.

collectors, as shown in Fig. 8.34(b) for $N = 3$. This results in a significant saving in IC area without any serious degradation in performance.

How the current source in the I²L gate can be achieved should be apparent if one recalls that a BJT operated in the active mode behaves as a current source. Because of the direction of I in the I²L inverter, it turns out to be more practical to use a *pnp* transistor. Figure 8.35(a) shows the resulting circuit for an I²L gate. The emitter of the *pnp* BJT is connected, as shown in Fig. 8.35(b), to a power supply through a resistor R that is external to the IC.

Since V_{EE} is positive, the EBJ of Q_1 is forward biased and

$$I_{EE} = \frac{V_{EE} - 0.7}{R} \tag{8.21}$$

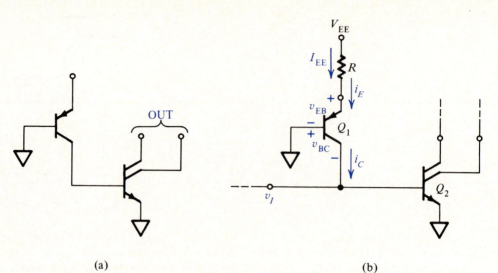

(a)   (b)

Figure 8.35 (a) Actual circuit of an I²L gate; (b) biasing arrangement.

When $v_I = V_{OL}$,

$$v_{BC} = -V_{OL} = -0.1 \text{ V}$$

which is insufficient to forward bias the CBJ of Q_1. Consequently the *pnp* BJT is in the active mode with

$$i_C = \alpha i_E \simeq I_{EE} \tag{8.22}$$

However, when $v_I = V_{OH}$,

$$v_{BC} = -V_{OH} = -0.7 \text{ V}$$

which causes Q_1 to saturate. Recall that, in saturation,

$$i_C < \alpha i_E \tag{8.23}$$

Although the above analysis indicates that the *pnp* BJT does not provide a constant current I, nevertheless the simple model for the I²L gate in Fig. 8.34 is valid for approximate analysis. In fact, one reason why the maximum fanout of the inverter in Fig. 8.32 is limited to one, is the effect expressed by Eq. (8.23) coupled with the relatively low β of the transistors.

The resistor in Fig. 8.35(b) is shared by all the *pnp* current sources in the IC. So, if the IC contains n gates, the current per gate will be I_{EE}/n, where I_{EE} is given by Eq. (8.21). The assumption that the supply current divides equally between the n *pnp* BJTs is valid because, in an IC, transistors are well matched with respect to I_S, and α varies much less than β.

I²L is, therefore, a completely resistorless logic family and permits extremely dense integration. In fact it is only used in LSI and VLSI circuits. An interesting and unique feature of I²L is the fact that the user can determine the current drawn and,

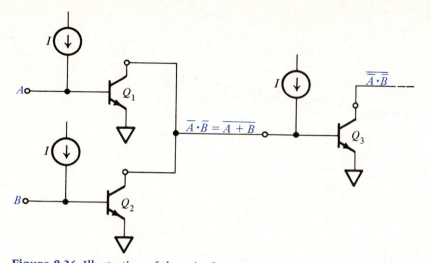

Figure 8.36 Illustration of the wired NOR connection using I²L inverters.

therefore, the power consumed by the IC by controlling the value of the external resistor. Because the propagation delay time is roughly inversely proportional to the current levels at which any logic gate operates, the above feature means that the user can trade power consumption for speed.

DESIGN WITH I²L

Because the I²L family does not include NAND or NOR gates a simple example of how logical expressions are implemented in this technology is appropriate.

Certain logic families such as I²L permit the so-called *wired* NOR connection which amounts to connecting the outputs of gates together. Figure 8.36 shows two I²L inverters with their outputs tied together, driving another inverter. Since it suffices for either inputs, A or B, to be at V_{OH} for the corresponding transistor, Q_1 or Q_2, to be saturated, the logical value of the input of Q_3 is $\bar{A} \cdot \bar{B}$ which, by De Morgan's theorem, is equal to $\overline{A + B}$.

Figure 8.36 also shows that the inverters can be regarded as NAND gates. This follows from the observation that the output of Q_3 is $\overline{\bar{A} \cdot \bar{B}}$, which is the NAND of $\bar{A}$ and $\bar{B}$. Where a NAND gate with $N > 1$ is needed, the fanout is provided by means of circuits such as the one in Fig. 8.34.

The implementation of a logic function with I²L is illustrated by means of the following example.

EXAMPLE 8.5

Use I²L to implement the EXOR circuit in Fig. 8.7(c).

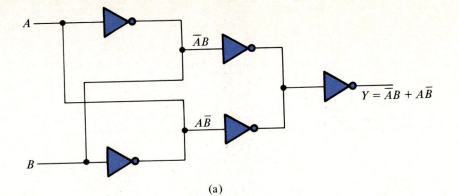

(a)

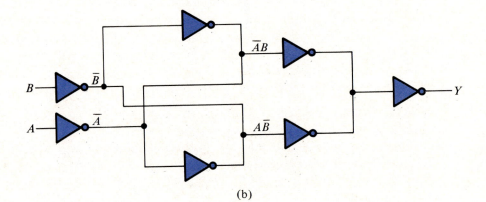

(b)

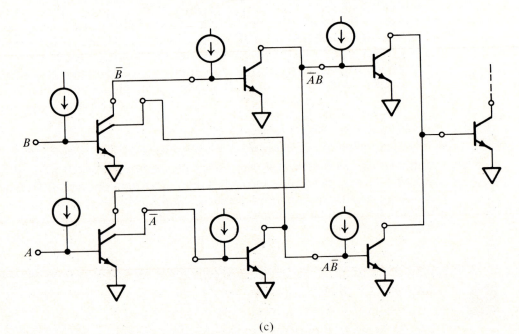

(c)

Figure E8.5

Solution

The first step in translating the logic diagram in Fig. 8.7(c) into an I^2L circuit involves replacing every NAND gate by an inverter with the gate's inputs wired together. The result is shown in Fig. E8.5(a).

All the inverters have a fanout of 1. However, the primary inputs to which A and B are applied have $N = 2$. They must, therefore, be buffered through gates with $N = 2$ to ensure that the I^2L outputs from which A and B are derived are not loaded with a fanout of more than 1. The buffering also ensures that the I^2L inverters in Fig. E8.5(a) which are being used as NAND gates have all their inputs derived from the outputs of other I^2L inverters. The buffering results in the circuit in Fig. E8.5(b).

From Fig. E8.5(b) it is seen that the I^2L circuit requires two gates with $N = 2$ and five with $N = 1$. Figure E8.5(c) shows the final design.

8.7   THE MOS LOGIC FAMILIES

There are three logic families based on the MOS transistor: NMOS and PMOS which, respectively, use the n-channel and p-channel MOSFETs, and CMOS which exploits the complementary properties of the two types of MOSFET. PMOS is the oldest and simplest technology, but because of the intrinsic speed advantage of the n-channel MOSFET over the p-channel device, it is much less frequently encountered than NMOS which is the leading technology in VLSI circuits. CMOS is the principal competitor of TTL, providing a wide variety of logic functions with a much lower power consumption than TTL. However, because of continuing technology improvements, it also appears to be threatening NMOS in the VLSI area.

Because of its lesser importance and its similarity to NMOS, PMOS is not considered here.

8.7.1   The n-Channel MOS (NMOS) Logic Family

The use of the two-transistor, n-channel, enhancement MOSFET circuit in Fig. 6.12 as a logic inverter was discussed in Sec. 6.2.5. One disadvantage of this type of inverter is that its logic swing is limited by the fact that, as can be seen from Fig. 6.14, $V_{OH} = V_{DD} - V_T$. This can be avoided by connecting the gate of Q_2 to a separate supply having a voltage V_{GG} which satisfies the condition $V_{GG} \geq V_{DD} + V_T$ (see Prob. 6.20). However, the main disadvantage of this type of inverter is its relatively slow switching speed which translates into a relatively long delay time t_P. Although ICs are designed using such inverters as well as the associated NOR and NAND gates, nevertheless the more important circuit technique for logic circuits involves the use of an n-channel depletion MOSFET for the nonlinear load.

NMOS INVERTER WITH A DEPLETION MOSFET LOAD
Figure 8.37 shows the circuit of an enhancement-depletion (ED) NMOS inverter

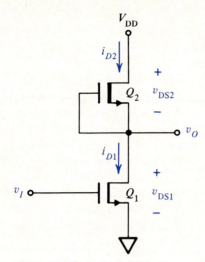

Figure 8.37 An NMOS inverter with a depletion MOSFET load Q_2.

using a depletion MOSFET as the load or *pull-up* transistor Q_2. The *pull-down* transistor Q_1 is, as in Fig. 6.12, an enhancement FET.

The output characteristics of a typical *n*-channel depletion MOSFET are shown in Fig. 8.38. Since the gate and source of Q_2 in Fig. 8.37 are connected together, its operation is constrained to the $v_{GS} = 0$ characteristic in Fig. 8.38, which determines the nonlinear load curve of Q_1. Because in Fig. 8.37

$$i_{D2} = i_{D1}$$

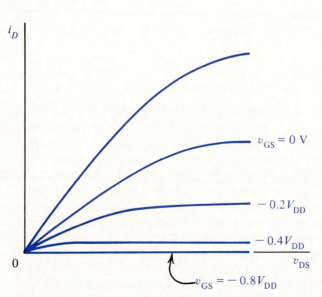
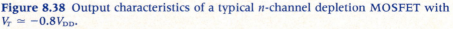

Figure 8.38 Output characteristics of a typical *n*-channel depletion MOSFET with $V_T \simeq -0.8V_{DD}$.

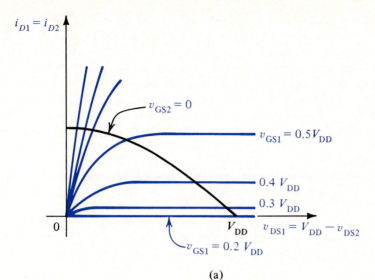

(a)

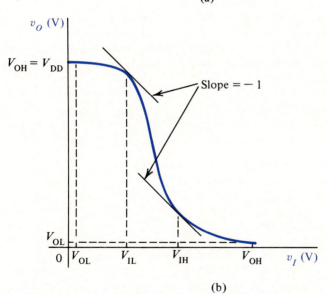

(b)

Figure 8.39 Voltage transfer characteristic of the inverter in Fig. 8.37: (a) graphical solution and (b) resulting typical characteristic showing convention for specifying V_{IL} and V_{IH}.

and

$$v_{DS1} = V_{DD} - v_{DS2}$$

a semigraphical procedure, analogous to the one used in Sec. 6.2.4, can be employed to obtain the voltage transfer characteristic of the circuit in Fig. 8.37.

Figure 8.39(a) shows the output characteristics of an enhancement MOSFET with $V_T \simeq 0.2V_{DD}$. The nonlinear load curve, given by the $v_{GS} = 0$ characteristic of the depletion device in Fig. 8.38 is superimposed on them, subject to the above constraints on the drain currents and drain-to-source voltages. The resulting voltage transfer characteristic, shown in Fig. 8.39(b), is obtained from the intersection of the load curve with the output characteristics in Fig. 8.39(a).

Evidently $V_{OH} = V_{DD}$, and the maximum input voltage for which $v_o = V_{OH}$ is $v_I = V_{T1}$, the threshold voltage of the pull-down transistor Q_1. The output voltage V_{OL} can always be determined from the graph in Fig. 8.39 and, under certain conditions, it can be computed by the method illustrated in the following example.

EXAMPLE 8.6

Determine V_{OL} for the ED inverter in Fig. 8.37 assuming $V_{DD} = 5$ V and that Q_1 is characterized by $\beta_1 = 10$ $\mu A/V^2$ and $V_{T1} = 1$ V while Q_2 has $\beta_2 = 5$ $\mu A/V^2$ and $V_{T2} = -4$ V.

Solution
From Fig. 8.39(a) with $v_{DS1} = v_o = V_{OL}$, Q_1 is likely to be in the triode mode while Q_2 is probably in pinch-off. Substituting V_{T1} and β_1 into Eq. (6.2) one obtains for Q_1

$$i_{D1} = 10(v_{GS1} - 1)v_{DS1}$$

and, since $v_o = V_{OL}$ corresponds to $v_I = v_{GS1} = V_{OH}$, therefore

$$i_{D1} = 10(5 - 1)v_{DS1} = 40v_{DS1}$$

The operation of Q_2 is described by Eq. (6.4) with $V_T = -4$ V, $\beta = 5$ $\mu A/V^2$ and $v_{GS} = 0$. Therefore $i_{D2} = i_{D1}$ is given by

$$i_{D2} = 2.5(0 + 4)^2 = 40 \ \mu A$$

and $v_o = v_{DS1}$ has a value of

$$v_{DS1} = \frac{i_{D1}}{40} = 1 \text{ V}$$

In other words V_{OL} is just small enough to keep the output of an identical load inverter at V_{OH}.

The initial assumptions are valid in this example since $v_{GS1} - v_{DS1} = 5 - 1 = 4$ V, which is greater than V_T, signifying that Q_1 is in the triode mode. Also $v_{GS2} - v_{DS2} = 0 - (V_{DD} - v_{DS1}) = -4$ V $= V_{T2}$, so Q_2 is operating on the boundary between pinch-off and the triode mode.

If V_{OL} has turned out to be greater than 1 V, say, 2 V, the assumption that Q_2 is in pinch-off would not be valid and one would have to fall back to a graphical solution since Eq. (6.2) is not valid for large v_{DS}.

Because of the appreciable curvature in the transfer characteristics of MOS logic circuits, the convention that is used for identifying V_{IL} and V_{IH} and, therefore, noise margins in BJT logic circuits is not convenient. Instead V_{IL} and V_{IH} are frequently defined as the input voltage levels corresponding to the points where the slope of the transfer characteristic is equal to -1, as shown in Fig. 8.39(b). By a judicious choice of β_1, β_2, V_{T1}, and V_{T2} it is possible to yield a nearly symmetrical voltage transfer characteristic with $NM_H \simeq NM_L$.

Figure 8.40(a) shows the circuit of an NMOS two-input NOR gate based on the ED inverter in Fig. 8.37. The circuit behaves as a NOR gate because, if at least one of the inputs is set to V_{OH}, the corresponding enhancement pull-down transistor turns on and its v_{DS} forces the output voltage v_O to fall to V_{OL}. On the other hand if both inputs are held at a voltage V_{OL} which is less than the V_T of the enhancement FET, then both Q_1 and Q_2 are cut off and $v_O = V_{OH} = V_{DD}$.

In the circuit in Fig. 8.40(b) both Q_1 and Q_2 have to be on for v_O to fall below $V_{OH} = V_{DD}$. This means that both v_{IB} and v_{IA} must be high. Consequently the circuit behaves as a NAND gate. However, notice that v_{OL} is the sum of the drain-to-source voltages of Q_1 and Q_2. Therefore, for the same enhancement and depletion MOSFET characteristics, V_{OL} in the AND gate is higher than in the NOR gate. This, in turn, means that the noise margin NM_L of the NAND gate is smaller and is reduced further as the number of inputs is increased.

The degradation of the transfer characteristic which results from the series-connected enhancement FETs in the NAND gate can be minimized but this involves the use of larger FETs. Therefore, NAND logic is avoided in designing MOS integrated circuits. Most current NMOS ICs operate with $V_{DD} = 5$ V as shown in Fig. 8.40. However, as the push to denser integration continues and, therefore, as

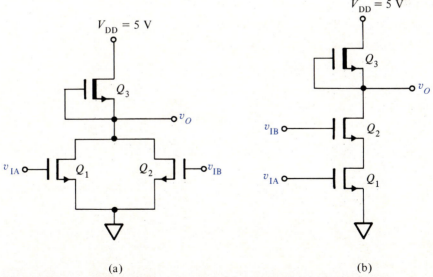

(a)   (b)

Figure 8.40 NMOS two-input positive (a) NOR and (b) NAND gates.

the dimensions of the MOSFETs are reduced, the resulting high electric fields are forcing designers to consider lower operating voltages.

PASS TRANSISTOR (SWITCH) LOGIC

In Sec. 8.3.2 mention was made of the use of switch logic for the implementation of logical functions, and of the fact that this can be realized using the MOS technology. Figure 8.41 illustrates the basis of this design approach.

Observe that, because the source of Q_3 is connected to the gate of Q_1, the drain currents of Q_3 and Q_4 will be zero irrespective of the state of v_I. Since the depletion pull-up Q_4 has its gate and drain connected together, it follows that $v_{DS4} = 0$ and, therefore,

$$v_1 = V_{DD} = V(1) \tag{8.24}$$

It is important to note that, despite the fact that $v_{DS4} = 0$, Q_4 cannot be replaced by a short circuit.

The analysis of the circuit for the two possible values of v_I is now very simple.

1. $v_I = V_{OH} = V_{DD}$

In view of Eq. (8.24)

$$v_I = v_1 = V_{DD}$$

and, therefore,

$$v_{GS3} = v_{DS3}$$

Hence Q_3 is operating like the enhancement pull-up in Fig. 6.12 and, therefore, since $i_{D3} = 0$,

$$v_{GS3} = v_{DS3} = V_T$$

Consequently

$$v_G = v_I - v_{GS3} = V_{DD} - V_T \tag{8.25}$$

It can be shown that this is an acceptable value for V_{OH} in the sense that the inverter

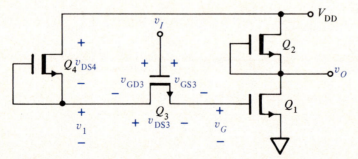

Figure 8.41 Circuit illustrating the use of an enhancement MOSFET as a contact switch.

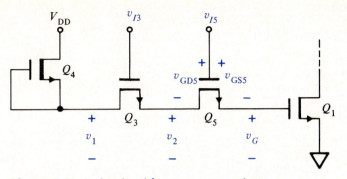

Figure 8.42 A circuit with two pass transistors.

consisting of Q_2 and Q_1 can be designed to operate properly with this lower value of V_{OH}.

2. $v_I = V_{OL} \simeq 0$

Because $v_1 = V_{DD}$, therefore

$$v_{GD3} = v_I - v_1 \simeq 0 - V_{DD} < V_T$$

so Q_3 is either in pinch-off or, if $v_{GS3} < V_T$, it is cut off. Observe that v_G in Fig. 8.41 must lie somewhere between 0 and V_{DD}. Since

$$v_{GS3} = v_I - v_G = 0 - v_G$$

it follows that Q_3 is cut off.

The preceding analysis demonstrates that Q_3 in the circuit in Fig. 8.41 behaves like the contact switch which is used in switch logic. When v_I is at logical 1, the switch is closed and the output (source) terminal of Q_3 is connected to logical 1. On the other hand, when v_I corresponds to logical 0, the switch is open. MOSFETs which are used in this manner are often referred to as *pass* transistors.

Observe that two or more pass transistors can be connected in series as shown in Fig. 8.42. The preceding analysis applies to Q_3 here as well. Because $v_2 = V_{DD} - V_T$ when $v_{I3} = V_{DD}$, setting $v_{I5} = V_{DD}$ results in $v_{GD5} = V_T$. Since the drain current is zero, Fig. 6.5 shows that $v_{DS5} = 0$ and, therefore, $v_{GS5} = V_T$. So, $v_G = V_{DD} - V_T$ as in the case of one pass transistor.

As in Fig. 8.41, setting v_{I5} or v_{I3} to V_{OL} cuts off the corresponding FET.

DRILL EXERCISE

8.11 Refering to Fig. 8.8 show that the circuit in Fig. D8.11 behaves as a two-input NAND gate.

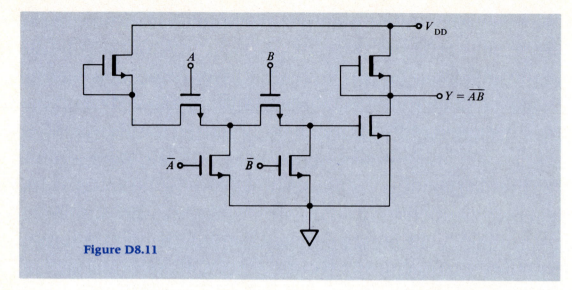

Figure D8.11

The circuit in Fig. D8.11 is not impressive when compared to Fig. 8.40(b). However, consider the circuit in Fig. 8.43, which is the contact switch implementation of a *selector* circuit described by

$$Y = S_0 \overline{A}\overline{B} + S_1 \overline{A}B + S_2 A\overline{B} + S_3 AB$$

If S_0, S_1, S_2, S_3 as well as A, B, $\overline{A}$, $\overline{B}$ were the outputs of NMOS gates and Y was developed across the input of a gate, such as v_G in Fig. 8.41, then a pass transistor implementation of Fig. 8.43 would require eight MOSFETs. A gate implementation on the other hand would involve 23 transistors. Moreover, the pass transistor circuit would occupy much less area on the IC because all the MOSFETs are enhancement devices which can be made much smaller than the depletion pull-ups in NOR gates. Finally, since the currents in the pass transistors are zero in both states, their use leads to a reduced power dissipation.

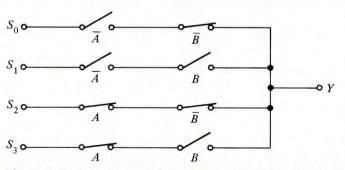

Figure 8.43 A contact switch implementation of a selector circuit. Switches are set for $A = 1$, $B = 0$; therefore $Y = S_2$.

NMOS logic is encountered in memories, microprocessors and other LSI and VLSI ICs where the high circuit density which this technology affords can be exploited fully.

8.7.2 The Complementary-Symmetry MOS (CMOS) Logic Family

As its name implies, complementary-symmetry MOS (CMOS) logic uses both p-channel and n-channel MOSFETs. The circuit of a CMOS inverter is shown in Fig. 8.44(a). Because of the topology, the circuit operates under the following constraints:

$$i_{D1} = i_{D2} \tag{8.26a}$$

$$v_I = v_{GS1} = V_{DD} - v_{SG2} \tag{8.26b}$$

$$v_O = v_{DS1} = V_{DD} - v_{SD2} \tag{8.26c}$$

As the following analysis demonstrates the inverter switches between $V_{OH} = V_{DD}$ and $V_{OL} = 0$ and draws essentially zero current in both states.

1. $v_I = V_{DD}$

From Eq. (8.26b), $v_{SG2} = 0$ and, therefore, $i_{D2} = 0$, which can be seen from the characteristic of the p-channel Q_2 in Fig. 8.44(c). Since $v_I = v_{GS1} = V_{DD}$ and $i_{D1} = i_{D2} = 0$, it follows from the output characteristic of the n-channel Q_1 in Fig. 8.44(b) that $v_{DS1} = v_O = 0$.

2. $v_I = 0$

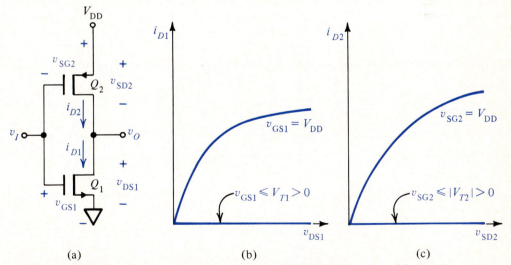

(a) (b) (c)

Figure 8.44 The complementary-symmetry MOS (CMOS) inverter: (a) circuit, (b) output characteristic of n-channel MOSFET Q_1 for $v_{GS1} = v_I = V_{DD}$, and (c) output characteristic of p-channel MOSFET Q_2 for $v_{SG2} = V_{DD} - v_I = V_{DD}$.

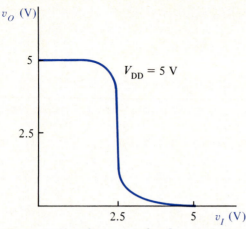

Figure 8.45 Voltage transfer characteristic of CMOS inverter with $V_{DD} = 5$ V.

From Eq. (8.26b) $v_{GS1} = 0$ and, therefore, $i_{D1} = i_{D2} = 0$. The same equation yields $v_{SG2} = V_{DD}$ and, as can be seen from Fig. 8.44(c), since $i_{D2} = 0$, $v_{SD2} = 0$. Therefore, from Eq. (8.26c) $v_O = V_{DD}$.

So the logic swing of the CMOS inverter is the maximum possible, that is V_{DD}, and the current drawn from the supply is negligibly small in both states. The large logic swing is, moreover, accompanied by a noise margin approaching $V_{DD}/2$ as can be seen from the voltage transfer characteristic for $V_{DD} = 5$ V shown in Fig. 8.45.

Figure 8.46 shows the circuits of a CMOS NOR gate and a CMOS NAND gate. In contrast to all the other logic families there is no preferred logic in CMOS.

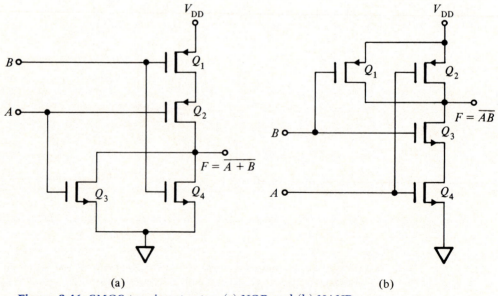

(a)   (b)

Figure 8.46 CMOS two-input gates: (a) NOR and (b) NAND.

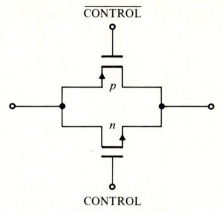

CONTROL

Figure 8.47 A CMOS transmission gate.

Moreover, like NMOS, CMOS provides attractive design techniques which complement gate logic. One of these is the use of the *transmission* gate which, despite its name, serves the same function as the pass transistor in NMOS. Figure 8.47 shows the circuit of a transmission gate which can be demonstrated to have characteristics that are superior to those of the single pass transistor.

Although CMOS cannot compete with TTL in speed, its low power consumption and high noise margin, which can be increased further by raising V_{DD} to as much as 15 V, make it a very attractive logic family in the many applications where speed is not of paramount importance. As a result, CMOS, just like TTL, is available in a wide variety of functions. But because, unlike TTL, it is a resistorless logic family it has become an important competitor in the field of VLSI circuits, with many applications such as microprocessors.

8.8   SUMMARY AND STUDY GUIDE

Logic circuits were studied in this chapter on two levels of abstraction: the level of ideal logic elements having terminal relationships which satisfy the laws of Boolean algebra, and the level of transistor circuits which have terminal characteristics that approximate those of ideal logic elements. The gate, which is the most important ideal logic element, was used to introduce the reader to the subject of the analysis and design of combinatorial circuits using NOR and/or NAND gates. Switch logic, which has become an attractive alternative to gate logic in VLSI IC design, was also mentioned. The basic binary memory element, the *SR* latch, was used to introduce the subject of sequential circuits. This was followed by an examination of the clocked *SR* latch, or flip-flop.

The main purpose of this chapter was to study the characteristics of the more important bipolar and MOS logic families. This was done by examining the circuit design techniques which went into creating the basic gates in each family. Where possible these techniques were related to the relevant IC technologies. Although the scope of this text restricted the analysis to dc conditions corresponding to the two logic states, the reader will encounter in the following chapters simple examples of some of the more tractable charge and flux storage effects which give rise to delays in logic circuits.

Several important points which must be noted and clearly understood are listed below.

Problems which can be expressed in terms of TRUE and FALSE statements can be solved using binary logic. The resulting logical functions are interpreted in terms of networks of interconnected AND, OR, and NOT gates. Simple procedures exist for translating these networks into corresponding ones based on the practically more important NAND and NOR gates.

The states of combinatorial circuits depend only on the current values of their inputs. In contrast, sequential circuits possess memory. Truth tables are adequate to describe the former while characteristic tables are needed to deal with sequential circuits such as the latch and flip-flop.

The NAND and NOR gates are the most important basic elements used in all logic circuits, be they combinatorial or sequential.

A practical logic circuit is characterized on the basis of a set of parameters which includes logical gain, or maximum fanout N_{max}, noise margin, NM, propagation delay time, and power dissipation.

Three simple circuits are keys to the understanding of the operation of the bipolar logic families: these are the diode OR gate, diode AND gate and the BJT inverter. The very popular TTL family is based on the latter two while the very fast ECL logic circuits operate on the principle of the former.

The TTL family is based on the NAND gate and, in the standard form, is relatively slow because some of the BJTs saturate.

ECL circuits are designed on the basis of the OR/NOR functions. Like Schottky TTL, ECL is a high-speed family because the transistors are not permitted to saturate. However, this is achieved at the expense of greater circuit complexity and power consumption.

I^2L is a resistorless, saturating, bipolar family. The basic element is the inverter, and logical circuits are designed using wired logic.

Because of the very sharp knees in the large signal characteristics of the BJT, piecewise-linear models are used to simplify the analysis of the bipolar logic families. In contrast, this method is inappropriate for the analysis of MOS logic circuits where a semigraphical approach is often the most straightforward for simple circuits.

The logical primitive, the inverter, is a resistorless, two transistor circuit in both NMOS and CMOS. It is based on the use of a transistor as a nonlinear load resistor in the inverter.

Switch logic is an attractive alternative to gates in the design of MOS integrated circuits.

NMOS is a NOR family and, because it is resistorless, is used almost exclusively in VLSI circuits.

CMOS is a very attractive family because it permits the use of both NOR and NAND logic, it consumes negligible power in both states and, like NMOS, it is resistorless.

PROBLEMS

8.1 A desk lamp L will not light unless the plug P on the end of the lamp cord is plugged into a wall mains socket and the switch S on the lamp is turned on. Express this statement in the form of a truth table.

8.2 The starting circuit of a car is wired in such a way that the car will not start when the ignition key K is turned to the *start* position if the safety belt of either the driver D or the other front seat passenger P is not buckled. However, the driver must be able to start the engine if the other front seat is not occupied. This is achieved by means of a pressure-sensitive switch S mounted in the seat. Express the above problem in the form of a truth table letting $C = 1$ correspond to the car starting. From the truth table derive a logical function relating C to K, D, P, and S.

8.3 Verify De Morgan's theorems for three variables by constructing truth tables for both sides of each equation and verifying that the outputs of the truth tables are the same.

8.4 By applying De Morgan's theorem to Fig. 8.10(a) show that the NAND-based version of the SR latch is as shown in Fig. P8.4.

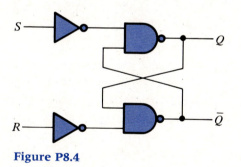

Figure P8.4

8.5 Using only NAND gates and inverters implement the following functions:
(**a**) $F = A\overline{B} + \overline{A}B$
(**b**) $F = \overline{A} + B(\overline{C} + DE)$
(**c**) $F = WXZ + X\overline{Z} + WY\overline{Z}$

8.6 Repeat Prob. 8.5 using NOR gates. Assume that the complement of every input is also available.

8.7 The SR flip-flop in Fig. 8.12(a) has the sequence of HIGH and LOW states shown in Fig. P8.7 applied to its inputs. Sketch the corresponding sequence of logical states at Q and $\overline{Q}$. Include the clock (C) sequence in your sketch.

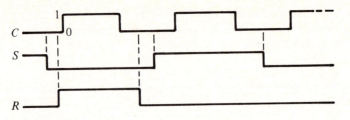

Figure P8.7

8.8 Show that if an inverter is connected between the S and R inputs in Fig. 8.12(a), i.e., if $R = \bar{S}$, then the resulting flip-flop has the characteristic table shown in Fig. P8.8. This is the frequently used *data* (D) flip-flop.

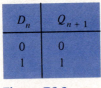

D_n	Q_{n+1}
0	0
1	1

Figure P8.8

8.9 Verify Eqs. (8.10) and (8.12).

8.10 The inverter in Fig. 8.13 is operated with $V_{CC} = 3.2$ V. If the BJT is characterized by $\beta = 20$, $V_{BE} = 0.7$ V, $V_{CEsat} = 0.2$ V, determine the values of R_B and R_C that are required to give an $N_{max} = 10$ if the collector current must not exceed 5 mA.

8.11 Refer to Example 8.2. Compute the base overdrive factors (defined in Example 7.7) corresponding to an N_{max} of 12, 8, and 1.

8.12 Repeat Example 8.1 with $R_B = 0$. Explain why elimination of the base resistor R_B would be impractical if the BJT parameter I_S could not be well controlled.

8.13 The fanout limitation for a practical TTL gate which was illustrated in Example 8.3 also applies to the simple TTL circuit in Fig. 8.23. Determine N_{max} for this circuit assuming $\beta = 30$.

8.14 Refer to Fig. E8.3. In a practical situation the other inputs of the load gates, shown in the figure tied to V_{CC}, would be connected to the outputs of other TTL gates. Determine the conditions at those other outputs for NI_l to be a maximum.

8.15 The objective of this problem is to account for the slope in the TTL transfer characteristic in Fig. 8.27 for v_l between 0.7 and 1.4 V. Neglecting the base current of Q_4 in Fig. 8.26, determine the relationship between the collector voltage of Q_2 and the circuit's input voltage, in the range where Q_2 is in the active mode and Q_3 is cut off. Assume $\beta \gg 1$.

8.16 When the input of a practical TTL gate is high, it draws approximately 50 μA from the source, i.e., the input current as defined in Fig. 8.23 is negative. Noting that in a practical gate, such as the one shown in Fig. 8.26, Q_2 and Q_3 are cut off when $v_o = V_{OH}$:
(a) Explain why there will be a maximum fanout, N_{max}, for the condition $v_o = V_{OH}$.

(b) Compute V_{OH} for a fanout $N = 10$ using $\beta = 30$, $I_S = 10^{-14}$A and $V_T = 25$ mV for Q_4 and D_1. Refer to Eq. (5.19) and Fig. 7.4 for the respective diode and BJT models.

8.17 On the same graph plot the voltage transfer characteristics (i.e., v_{o1} and v_{o2} as a function of v_I) of the ECL switch in Fig. 8.29 using the element values specified in Example 8.4. Choose a wide enough input voltage range to include the saturation of Q_1. Clearly label all breakpoints on the characteristics.

8.18 The reference voltage V_R on the base of Q_2 in Fig. 8.31 can be shown to be equal to -1.29 V. Neglecting the 50 kΩ resistors and assuming β to be sufficiently large to permit neglecting base currents, show that $V_{OH} \approx -0.88$ V, $V_{OL} \approx -1.77$ V.

8.19 Starting with the results of Prob. 8.18, determine the approximate shifts that can occur in V_{OL} and V_{OH} if Q_4 and Q_5 in Fig. 8.31 can have β as low as 50. In this case do not neglect the loading of the 50 kΩ resistor on the output of the driving gate.

8.20 The β of a multicollector BJT such as the one in Fig. 8.34(b) is defined as the ratio of the sum of the collector currents to the base current. Given $\beta = 5$, what is N_{max} if the base overdrive factor (see Example 7.7) is to be not less than 1.25?

8.21 Design an I^{2}L two-input NAND gate which is to function with external inputs, i.e., the inputs may not be derived from the outputs of I^{2}L circuits.

8.22 The inputs to a decoder are A and B. Draw an I^{2}L circuit for the decoder.

8.23 Because of the importance of TTL most logic families are made "TTL-compatible." Figure P8.23 shows how the input of an I^{2}L gate could be made TTL-compatible. If the current source in the I^{2}L gate delivers a current I = 0.1 mA, and assuming $\beta = 5$, $V_{BE} = 0.7$ V, $V_{CEsat} = 0.2$ V:
(a) Show that Q_1 is saturated with $v_I = V_{CEsat}$, and compute its collector current.
(b) Show that, with $v_I = V_{CC}$, Q_1 is in the reverse mode, and determine the limit on R_1 for Q_2 to be saturated.
(c) If a pull-up resistor were connected between the collector of Q_3 and V_{CC}, could the output of this circuit drive the input of a standard TTL inverter? Explain.

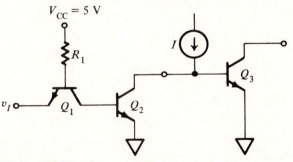

Figure P8.23

8.24 The output characteristics of an enhancement and a depletion n-channel MOSFET are plotted in Fig. P8.24. Plot to scale the voltage transfer characteristic of an ED inverter, such as the one in Fig. 8.37, using these transistors and $V_{DD} = 5$ V. Is the characteristic acceptable for a logical inverter? Explain.

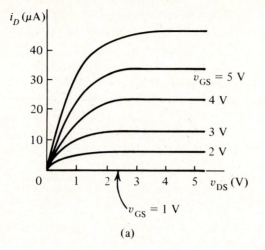

(a)

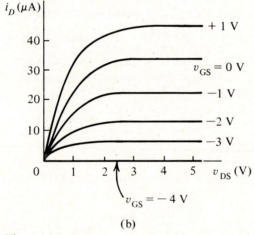

(b)

Figure P8.24

8.25 Repeat Prob. 8.24 after reducing by a factor of 4 the β of the depletion device described by the characteristics in Fig. P8.24(b). Determine the noise margins of the inverter.

8.26 Compute V_{OL} for the inverter in Prob. 8.25 using the method of Example 8.6. Compare your answer with the results of Prob. 8.25.

8.27 Determine the logical transfer function of the circuit in Fig. P8.27. Compare the number of transistors in this circuit with the number that would be involved in a design based on NMOS NOR gate logic.

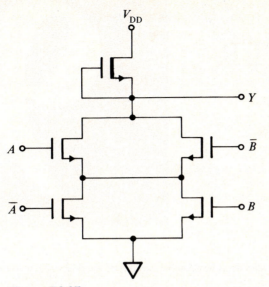

Figure P8.27

8.28 Sketch to scale the approximate voltage transfer characteristic of a CMOS inverter using the enhancement MOSFET characteristics in Fig. P8.24 for the n-channel and, after appropriate polarity changes, for the p-channel MOSFET as well.

8.29 A CMOS inverter uses n-channel and p-channel MOSFETs with $|V_T| = 2$ V. If $V_{DD} = 5$ V sketch the channel current i_D as a function of the input voltage v_I. Only a qualitative solution is required; however, the points where i_D goes to zero should be clearly labeled.

8.30 Show that the circuits in Fig. 8.46(a) and (b) do perform, respectively, the two-input NOR and NAND operations.

8.31 Draw the circuit of an *SR* latch using CMOS NOR gates.

8.32 (a) Two NMOS inverters have their outputs connected together. If the input to one inverter is A and to the other is B, what is the output Y?
(b) Could the wired logic method of (a) be used with CMOS inverters? Explain.

CHAPTER

First-Order Circuits

9.1 INTRODUCTION

In this chapter circuits with a single inductance or capacitance are examined. Recalling from Chap. 1 that the branch relation for any one of these two energy-storage elements has either a first-order differential equation form [Eqs. (1.4) and (1.7)] or an integral equation form [Eqs. (1.5) and (1.8)], it is evident that networks with one L or one C are described by first-order differential equations. Such circuits are called first-order circuits. It follows then that the circuit variables are, in general, time functions, and solving a circuit problem means finding these time functions.

This chapter begins with a brief review of the properties of the two energy-storage elements, after which the unit step and impulse functions are introduced. This is followed by a discussion of the solutions to first-order differential equations which leads to the development of inspection methods for their solution. A number of networks containing switches and diodes are next solved using these inspection methods. It is also shown here that, by restricting the class of time functions permitted, the solution to the circuit equations is reduced to solving a problem in complex-number arithmetic. Only linear, lumped, time-invariant elements are considered.

9.1.1 Circuits with Inductance and Capacitance

In Fig. 9.1(a) and (b) the symbol and variable definitions are shown for the L and C branch elements, respectively. The branch relation for the inductance element is either the differential equation Eq. (1.7),

$$v_L = L\frac{di_L}{dt} \tag{9.1a}$$

which is valid at any time, or the integral equation Eq. (1.8) with the initial time $t_0 = 0-$,

$$i_L(t) = \frac{1}{L}\int_{0-}^{t} v_L(\tau)\,d\tau + i_L(0-) \tag{9.1b}$$

which is valid for $t \geq 0$. The choice $t_0 = 0-$ facilitates the handling of situations with discontinuities at the time origin.

It is important to remember that Eq. (9.1b) is only valid for $t \geq 0$; all information about $i_L(t)$ *prior* to $t = 0-$ is "lost," and the entire negative time history of $v_L(t)$ is contained in the *initial condition* $i_L(0-)$.

The branch relation for the capacitance element is either the differential equation Eq. (1.4),

$$i_C = C\frac{dv_C}{dt} \tag{9.2a}$$

which is valid at any time, or the integral equation Eq. (1.5) with $t_0 = 0-$,

$$v_C(t) = \frac{1}{C}\int_{0-}^{t} i_C(\tau)\,d\tau + v_C(0-) \tag{9.2b}$$

which is valid for $t \geq 0$. The initial condition $v_C(0-)$ contains the entire negative time history of $i_C(t)$.

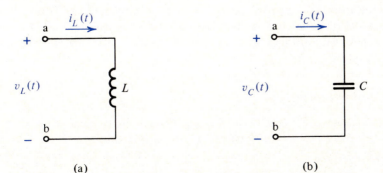

(a)   (b)

Figure 9.1 (a) The inductance and (b) the capacitance elements.

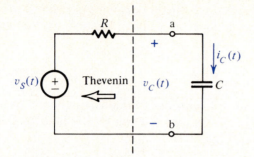

Figure 9.2 A single capacitance circuit.

For circuits containing a single energy-storage element, that portion of the circuit connected to the element terminals *ab* of Fig. 9.1 can be represented by either a Thevenin or a Norton equivalent circuit. Thus a circuit containing a single capacitance can always be reduced to the form shown in Fig. 9.2 where $v_S(t)$ is a known forcing function.

This single mesh circuit has the KVL equation

$$Ri_C + v_C = v_S \tag{9.3}$$

and one can either solve for $v_C(t)$ by eliminating $i_C(t)$ using Eq. (9.2a), or one can solve for $i_C(t)$ by eliminating $v_C(t)$ using Eq. (9.2b).

Solving for $v_C(t)$ one obtains the first-order differential equation

$$RC\frac{dv_C}{dt} + v_C = v_S \tag{9.4}$$

Solving for $i_C(t)$ one obtains the integral equation

$$Ri_C(t) + \frac{1}{C}\int_{0-}^{t} i_C(\tau)\,d\tau + v_C(0-) = v_S \tag{9.5}$$

which, by differentiation, yields the first-order differential equation

$$RC\frac{di_C}{dt} + i_C = C\frac{dv_S}{dt}$$

A similar examination of a circuit containing a single inductance also leads to first-order differential equations. Thus single C or single L circuits are called *first-order* circuits. Finding the solutions to first-order circuit problems is the subject of this chapter.

9.1.2 Energy and Power Considerations

In Chap. 1 it was shown that the L and C branch elements store energy, and that the energy stored in a capacitance C at time t is given by

$$E_C(t) = \tfrac{1}{2}Cv_C^2(t) \tag{1.22}$$

Correspondingly, the energy stored in an inductance L is given by

$$E_L(t) = \tfrac{1}{2}Li_L^2(t) \tag{1.25}$$

The instantaneous energy state of the capacitance [inductance] element is therefore completely specified by $v_C(t)[i_L(t)]$ and these variables are called state variables. Thus the requirement that thet energy stored be finite or bounded is equivalent to $v_C(t)[i_L(t)]$ being bounded.

It is also important to realize that when the current $i_C(t)$ through a capacitance is bounded then the voltage, $v_C(t)$, across the capacitance is continuous. This is shown by calculating the voltage difference between two adjacent time points t and $t + \Delta t$ using the integral form of the branch Eq. (9.2b), i.e.,

$$v_C(t + \Delta t) - v_C(t) = \frac{1}{C}\int_t^{t+\Delta t} i_C(\tau)\,d\tau \tag{9.6}$$

Clearly if i_C is bounded then as $\Delta t \to 0$, the right hand side of Eq. (9.6) vanishes,

$$v_C(t + \Delta t) \to v_C(t) \qquad \text{as } \Delta t \to 0$$

and v_C is continuous.

What happens if the capacitance voltage is forced to be discontinuous? Such a situation clearly applies in the "thought experiment" of connecting an ideal voltage source of V_0 volts across an uncharged capacitance, say at $t = 0$. This connection is represented in the circuit of Fig. 9.3 by a switch S closing at $t = 0$. The mathematical answer is that the capacitance current $i_C(t)$ must be unbounded (infinite)! However, using the notation $v_C(0-)$ and $v_C(0+)$ to designate, respectively, the capacitance voltage just before and just after the connection is made, Eq. (9.6) can be evaluated. Since

$$v_C(0+) = V_0 \qquad \text{and} \qquad v_C(0-) = 0$$

then, by direct substitution,

$$\int_{0-}^{0+} i_C(t)\,dt = CV_0 \tag{9.7}$$

is finite and precisely equal to the charge supplied instantaneously at $t = 0$ by the source.

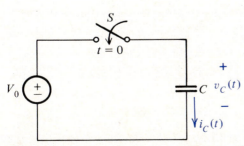

Figure 9.3 Connecting a source to a capacitance instantaneously.

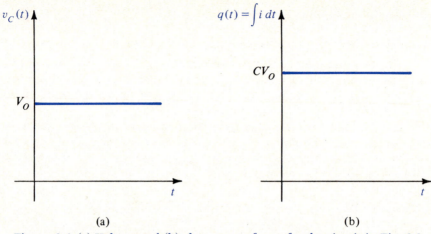

(a)   (b)

Figure 9.4 (a) Voltage and (b) charge waveforms for the circuit in Fig. 9.3.

It should also be realized that accepting that energy has been instantaneously transferred to the capacitance means accepting *infinite power*. In Fig. 9.4 the voltage and charge waveforms for this thought experiment are shown; these waveforms are called step functions. What can be said about the current $i_C(t)$ waveform? Since the charging is instantaneous, $i(0-)$ and $i(0+)$ must be zero. From Eq. (9.7) $i_C(t)$ must be unbounded (infinite) at $t = 0$, yet the area under its waveform is finite. The term *impulse* is used to designate an unbounded waveform of zero duration which has finite area; this notion is formally discussed in a later section.

A similar procedure carried out for an inductance leads to the conclusion that the current through an inductance, $i_L(t)$, is a continuous function when the voltage, $v_L(t)$, across it is bounded. For a jump discontinuity in $i_L(t)$ the voltage $v_L(t)$ is impulsive (unbounded).

In summary, a capacitance voltage (inductance current) is finite and continuous when the capacitance current (inductance voltage) is finite. A jump discontinuity in capacitance voltage (inductance current) is associated with an impulsive capacitance current (inductance voltage).

DRILL EXERCISE

9.1 Show that if a constant current source of value I_0 is suddenly connected to an inductance L, the source supplies an instantaneous flux LI_0.

9.2 STEPS AND IMPULSES

In this section the step and the impulse time functions are introduced and the solutions to some simple circuits excited by these functions are examined.

9.2.1 The Unit Impulse and the Unit Step

The *unit area pulse* $U(t, T)$, illustrated in Fig. 9.5(a), is defined by

$$U(t, T) = \begin{cases} \dfrac{1}{T} & \text{for } 0 \le t \le T \\ 0 & \text{for } t < 0 \end{cases} \tag{9.8}$$

so that, as its name suggests, it has unit area. Its time integral is shown in Fig. 9.5(b).

The *unit impulse* $u_0(t)$ is formally defined as the limit of the unit area function as $T \rightarrow 0$, i.e.,

$$u_0(t) = \lim_{T \to 0} U(t, T) \tag{9.9}$$

even though this limit does not in fact exist. Conceptually, visualize the pulse U shrinking in width as it gets taller, its area remaining unity as in Fig. 9.5(c). Justification for the use of the impulse rests on the mathematical Theory of Distributions. The need for such a formal concept was motivated in the previous section in the discussion of a jump change in capacitance voltage.

The same limit, when applied to the integral of $U(t, T)$, generates the *unit step* $u_{-1}(t)$ as the limit

$$u_{-1}(t) = \lim_{T \to 0} \int_{-\infty}^{t} U(\tau, T)\, d\tau \tag{9.10}$$

An alternate definition is

$$u_{-1}(t) = \begin{cases} 1 & \text{for } t \ge 0 \\ 0 & \text{for } t < 0 \end{cases} \tag{9.11}$$

One can also consider the impulse to be the derivative of the step, namely,

$$\frac{du_{-1}(t)}{dt} = u_0(t) \tag{9.12}$$

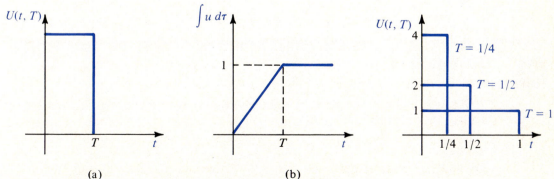

(a) (b)

Figure 9.5 (a) The unit area function and (b) its integral.

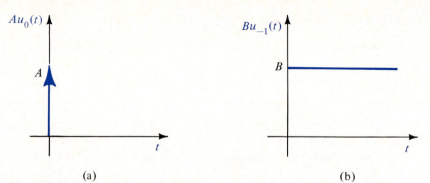

(a)   (b)

Figure 9.6 Graphical representations for (a) an impulse and (b) a step.

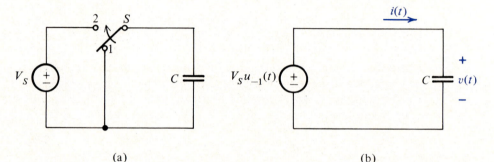

(a)   (b)

Figure 9.7 (a) Circuit with switch and (b) equivalent representation with step function.

These two *singularity* functions are usually represented by the diagrams shown in Fig. 9.6. In Fig. 9.6(a) an impulse at $t = 0$ of area A is denoted by an arrow of height A. In Fig. 9.6(b) a step of height B is shown. The step is a convenient way of representing the closing or opening of a switch as illustrated in Fig. 9.7. Figure 9.7(a) depicts a circuit containing a switch S which is changed from position 1 to position 2 at $t = 0$. Figure 9.7(b) shows an equivalent way of representing this circuit problem using a step function. This circuit of course can only provide solutions for $t \geq 0$.

EXAMPLE 9.1

In Fig. 9.7(b) find the current $i(t)$ and hence the charge supplied by the source. The capacitance is initially uncharged.

Solution
By KVL

$$v(t) = V_S u_{-1}(t)$$

and, by definition,

$$i = C\frac{dv}{dt}$$

Thus, one obtains

$$i = CV_S\frac{du_{-1}}{dt}$$

or, since the derivative of a step is an impulse,

$$i = CV_S u_0(t)$$

This is the formal result that a voltage step across a capacitance is accompanied by a current impulse. The charge is given by

$$q(t) = CV_S\int_{-\infty}^{t} u_0(\tau)\, d\tau = CV_S u_{-1}(t)$$

in other words, the charge is also a step function. These are the formal statements corresponding to the waveforms of Fig. 9.4.

EXAMPLE 9.2

Calculate the current $i(t)$ and the voltages $v_C(t)$ and $v(t)$ in the circuit shown in Fig. E9.2. The capacitance is initially uncharged.

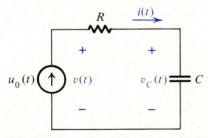

Figure E9.2 Circuit for Example 9.2.

Solution
From the diagram

$$i(t) = u_0(t) = C\frac{dv_C}{dt}$$

hence $v_C(t)$ is a step (as one expects) given by

$$v_C(t) = \frac{1}{C}\int_{-\infty}^{t} u_0(\tau)\, d\tau = \frac{1}{C}u_{-1}(t)$$

By KVL

$$v(t) = Ri + v_C(t)$$

or

$$v(t) = Ru_0(t) + \frac{1}{C}u_{-1}(t)$$

Hence $v(t)$ has an impulse of size R at $t = 0$ and is a constant of size $1/C$ for $t > 0$.

DRILL EXERCISE

9.2 Show that for a unit step current excitation in Fig. E9.2 the voltage v is given by

$$v = Ru_{-1}(t) + \frac{1}{C}t \qquad \text{for } t \geq 0$$

A unit impulse occurring at $t = T$ as shown in Fig. 9.8(a) is written $u_0(t - T)$ and has the useful property that when it multiplies any continuous function, say $g(t)$, the result is an impulse of strength (area) $g(T)$. Although the two product forms shown in Eq. (9.13) are identical, it is the right-hand side which *must* be used since $u_0(t - T)$ is zero except at $t = T$:

$$g(t)u_0(t - T) = g(T)u_0(t - T) \tag{9.13}$$

Integrating Eq. (9.13) gives rise to the *sampling* property of the impulse:

$$\int_{-\infty}^{\infty} g(t)u_0(t - T)\, dt = \int_{-\infty}^{\infty} g(T)u_0(t - T)\, dt = g(T) \tag{9.14}$$

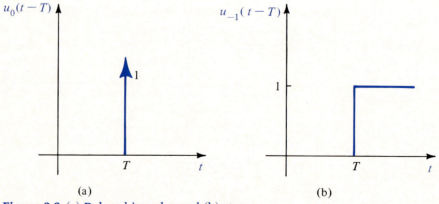

(a)   (b)

Figure 9.8 (a) Delayed impulse and (b) step.

One further result from the theory of distributions is that an impulse of zero area is equal to zero.

Another special function, used occasionally, is the *unit ramp* defined as the integral of the unit step, or

$$u_{-2}(t) = \begin{cases} t & \text{for } t \geq 0 \\ 0 & \text{for } t < 0 \end{cases}$$

The unit step and ramp are shown in Fig. 9.9. Note that one can also consider the step as the formal derivative of the ramp, namely,

$$\frac{du_{-2}(t)}{dt} = u_{-1}(t) \tag{9.15}$$

Steps and ramps are used to construct functions composed of linear segments. For example, the pulse waveform of Fig. 9.10 is written as the sum of two steps

$$p(t) = Au_{-1}(t) - Au_{-1}(t - T) \tag{9.16}$$

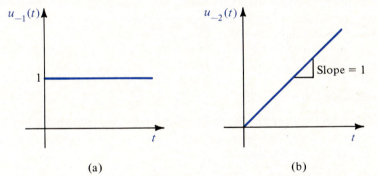

(a) (b)

Figure 9.9 (a) A unit step and (b) its integral, the unit ramp.

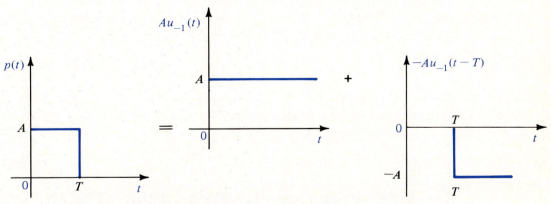

Figure 9.10 Pulse waveform as sum of two steps.

9.3 SOLUTION OF FIRST-ORDER CIRCUITS

9.3.1 First-Order Differential Equations

Since first-order circuits are described by first-order differential equations such as Eq. (9.4), it is necessary briefly to review their solution.

The general linear, first-order, constant coefficient, differential equation can be written as

$$\frac{dy}{dt} + a_0 y = f(t) \tag{9.17}$$

where the right-hand side, $f(t)$, is a known forcing function and the unknown y has a known initial value $y(0+)$. The function $y(t)$ which is to be determined is called the *response* or *output*. It is useful to take $t = 0+$ as the initial time since, as was seen earlier, jump changes can occur.

From the theory of elementary differential equations, it is well known that a *general* solution to Eq. (9.17) is given by the sum of the solution of the *homogeneous equation*, $y_h(t)$, and any *particular* solution, $y_p(t)$, for a given $f(t)$.

Equation (9.17) with the right-hand side set to zero is the *homogeneous equation*, and its solution is denoted by $y_h(t)$. When a trial solution $y_h(t) = Ae^{st}$ is inserted in the homogeneous equation, one obtains

$$sAe^{st} + a_0 Ae^{st} = 0$$

or

$$(s + a_0)Ae^{st} = 0$$

Hence, since $Ae^{st} \neq 0$, Ae^{st} is a solution to the homogeneous equation if, and only if, $s = -a_0 = s_1$, say. The value s_1 is called the *natural frequency* and the solution to the homogeneous equation is

$$y_h(t) = Ae^{-a_0 t} = Ae^{s_1 t} \tag{9.18}$$

a decaying exponential for a_0 positive.

The general solution to Eq. (9.17) is therefore Eq. (9.18) plus any *particular* solution to Eq. (9.17) for a given $f(t)$. For the special case where $f(t) = F$ is a *constant*, an evident particular solution to Eq. (9.17) is

$$y_p(t) = \frac{F}{a_0} \tag{9.19}$$

as may be easily verified by direct substitution into Eq. (9.17). This particular solution is a constant and is usually called the *dc steady-state* solution; it is designated by the notation y_{ss}, i.e.,

$$y_{ss} = \frac{F}{a_0} \tag{9.20}$$

The complete solution $y = y_h + y_p$ is therefore written

$$y = Ae^{s_1 t} + y_{ss} \tag{9.21}$$

The constant A is now found by evaluating Eq. (9.21) at $t = 0+$, yielding

$$A = y(0+) - y_{ss} \tag{9.22}$$

Therefore, Eq. (9.21) is, for $t \geq 0$,

$$y(t) = y_{ss} + [y(0+) - y_{ss}]e^{s_1 t} \tag{9.23}$$

Equation (9.23) can also be written in the form

$$y(t) = y_{ss} + [y(0+) - y_{ss}]e^{-t/T} \tag{9.24}$$

where T is called the *time constant* and is given by

$$T = \frac{1}{a_0} = \frac{-1}{s_1} \tag{9.25}$$

Thus, the solution to a first-order differential equation for a constant excitation, as given by Eq. (9.24), contains a decaying exponential component called the *transient* solution and a constant component called the *dc steady-state* solution. First-order circuits are often called *single time-constant* (STC) circuits since the transient solution can always be written in the form $e^{-t/T}$.

EXAMPLE 9.3

The circuit shown in Fig. E9.3 is the first-order circuit containing a single capacitance with the remainder of the circuit replaced by a Thevenin equivalent at terminals ab. Find $v_C(t)$ for $v_s = V_s$, a constant, given $v_C(0+)$. Identify the transient and steady-state components of v_C. What is the time constant?

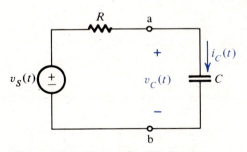

Figure E9.3 First-order RC circuit.

Solution
The KVL equation for the circuit shown is

$$Ri_C + v_C = V_s$$

and, eliminating i_C using Eq. (9.2a), one obtains

$$RC\frac{dv_C}{dt} + v_C = V_S$$

as the differential equation for v_C, with v_C having a known initial value $v_C(0+)$.

The solution to the homogeneous equation is given by the single exponential function

$$v_{Ch}(t) = Ae^{-t/RC}$$

and the particular solution is the steady-state value

$$v_{Cp}(t) = V_S$$

Thus, the complete solution $v_C = v_{Ch} + v_{Cp}$ is

$$v_C = V_S + Ae^{-t/RC}$$

The constant A is now found by evaluating this equation at $t = 0+$, yielding

$$A = v_C(0+) - V_S$$

so the complete solution is, for $t \geq 0$,

$$v_C(t) = V_s + [v_C(0+) - V_S]e^{-t/RC} \tag{9.26a}$$

The dc steady-state solution is the first term in Eq. (9.26)

$$v_{Css} = V_S$$

the transient solution has an initial amplitude $v_C(0+) - V_S$ and decays exponentially with time constant $T = RC$. In a single capacitance (inductance) network the time constant is thus the product of C (L) times the Thevenin equivalent resistance (conductance).

Because $v_C(0-) = v_C(0+)$ in this example, Eq. (9.26) can be rewritten as

$$v_C(t) = V_S(1 - e^{-t/RC}) + v_C(0-)e^{-t/RC} \tag{9.26}$$

a two-term decomposition where the first term is proportional to the input V_S and the second is proportional to the *initial state* $v_C(0-)$. The term initial state is normally used for initial conditions at $t = 0-$. These components are called the *zero-state*, $v_{Czs}(t)$, and the *zero-input*, $v_{Czi}(t)$, solutions, respectively.

9.3.2 The Zero-State Response

The solution for the circuit of Fig. E9.3 when $v_C(0-) = 0$ and $v_s = V_S$ is the zero-state response and is given by the first component of Eq. (9.26)

$$v_{Czs}(t) = V_S(1 - e^{-t/RC}) \qquad \text{for } t \geq 0 \tag{9.27}$$

Equation (9.27) is also called the zero-state step response since it corresponds to V_S being suddenly applied at $t = 0$. The waveform $v_{Czs}(t)$ is a rising exponential showing that the capacitance starts at zero volts and charges up exponentially to a dc steady-state value equal to the applied voltage V_s. The exponential term is characterized by the circuit time constant RC.

9.3.3 The Zero-Input Response

For the circuit of Fig. E9.3 with no excitation the zero-input circuit is shown in Fig. 9.11. The initial state is the capacitance voltage $v_C(0-)$.

The zero-input solution for the capacitance voltage is the second component of Eq. (9.26),

$$v_{Czi}(t) = v_C(0-)e^{-t/RC} \qquad\qquad (9.28)$$

for $t \geq 0$. The waveform $v_{Czi}(t)$, drawn in Fig. 9.12, is seen to have an initial tangent line or slope (shown dashed) which intersects the time axis at $t = RC$, i.e., one time constant. As can be shown from Eq. (9.28), in one time constant the exponential decays by $1/e$ or to approximately 37% of its starting value. Equation (9.28) is the zero-input response for this STC circuit.

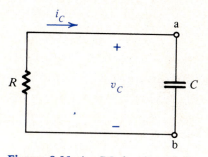

Figure 9.11 An RC circuit with known initial state.

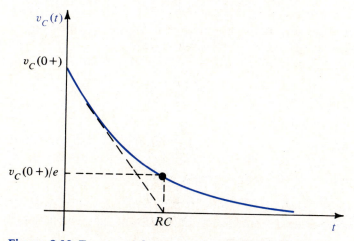

Figure 9.12 Exponential response for circuit of Fig. 9.11.

DRILL EXERCISES

9.3 Show that after 2, 3, 4 time constants an exponential decays to 13.5%, 4.9%, 1.8% of its initial value, respectively.

9.4 Show that the dual circuit to that of Fig. E9.3 is a current source $i_s(t)$ driving a parallel GL circuit.

9.5 Show that the time constant of the circuit in Exercise 9.4 is GL.

EXAMPLE 9.4

A 3-μF capacitance with a charge q of 6 μC is connected to a 2-kΩ resistance at $t = 0$. Find the capacitance current and voltage time evolution. [The given exercise is equivalent to finding $i_C(t)$ and $v_C(t)$ in the circuit of Fig. 9.11 with $R = 2$ kΩ, $C = 3\mu$F and $v_C(0+) = q/C = 2$ V.] Verify that the energy initially stored in C is eventually dissipated in R.

Solution
The KVL equation, using milliamperes for current and kiloohms for resistance, is

$$2i_C + v_C = 0$$

The branch relation using milliamperes for current, microfarads for capacitance, and milliseconds for time is

$$i_C = 3\frac{dv_C}{dt}$$

Eliminating i_C, one obtains the homogeneous first-order equation

$$6\frac{dv_C}{dt} + v_C = 0$$

with zero-input solution

$$v_C = 2e^{-t/6} \qquad \text{for } t \geq 0$$

Hence the time constant RC for this STC circuit is 6 ms. Using the branch relation, the current is calculated to be, for $t \geq 0$,

$$i_C = -e^{-t/6}$$

Thus the capacitance voltage decays exponentially to zero from its 2-V initial value, the capacitance current jumps to -1 mA at the instant the connection is made and then also decays to zero exponentially.

The energy dissipated in the resistance, in microjoules for $k\Omega$, mA, ms units, is given by Eq. (1.16) as

$$E_R = \int_0^\infty R i_C^2 \, dt$$

$$= \int_0^\infty 2e^{-t/3} \, dt = 6 \ \mu J$$

The initial energy stored in C is

$$E_C = \tfrac{1}{2}(3 \ \mu F)(2)^2 = 6 \ \mu J$$

thus verifying the conservation of energy.

These examples serve to illustrate how the solution to a first-order or STC circuit is obtained using the standard methods of elementary differential equations. However, since the complete solution to a first-order problem is completely specified by three quantities only, namely, y_{ss}, $y(0+)$, and T (or s_1) in Eq. (9.24), inspection methods, which permit a rapid solution to first-order problems, will be developed in a later section.

9.4   USE OF SOURCES TO REPLACE INITIAL CONDITIONS

When a switching event occurs at $t = 0$ the voltage across a capacitance may be written, using Eq. (9.2b) as

$$v_C(t) = v_C(0-) + \frac{1}{C} \int_{0-}^{t} i_C(\tau) \, d\tau \qquad (9.29)$$

valid for $t \geq 0$. Using the step function this equation can just as well be written, for $t \geq 0$, as

$$v_C(t) = v_C(0-)u_{-1}(t) + \frac{1}{C} \int_{0-}^{t} i_C(\tau) \, d\tau \qquad (9.29b)$$

which can be interpreted as the equation for a Thevenin network consisting of a voltage source in series with an uncharged capacitance. Thus Eq. (9.29) is represented by the circuit equivalence shown in Fig. 9.13. It is important to note that the capacitance terminals are ab and that the portions of the circuit enclosed by the screened boundary is an equivalence with respect to those terminals ab.

Similarly, for an inductance, Eq. (9.1b) is written with a step function as

$$i_L(t) = i_L(0-)u_{-1}(t) + \frac{1}{L} \int_{0-}^{t} v_L(\tau) \, d\tau \qquad (9.30)$$

an expression valid for $t \geq 0$. This has the Norton circuit representation shown in Fig. 9.14.

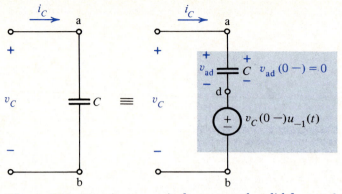

Figure 9.13 Capacitance equivalent network valid for $t \geq 0$.

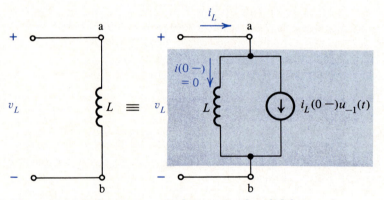

Figure 9.14 Inductance equivalent network valid for $t \geq 0$.

These two results mean that a problem with nonzero initial conditions may be replaced by an equivalent zero-state problem; the initial conditions appear as *independent* step sources in the equivalent circuit.

Usually, there are no impulsive capacitance currents (inductance voltages) and the capacitance voltages (inductance currents) are continuous so that $v_C(0+) = v_C(0-)$ $[i_L(0+) = i_L(0-)]$ and the initial conditions which are required in the solution to a circuit problem [see Eq. (9.26) for example] are therefore known. The presence of impulsive components is flagged whenever KVL (or KCL) and continuity cannot *both* be satisfied at $t = 0+$; an example of such a situation was encountered in the circuit problem defined in Fig. 9.3. The best procedure for testing that there are no impulses is to *assume* continuity and to verify that KVL (or KCL) are also satisfied at $t = 0+$.

EXAMPLE 9.5

Use the equivalent circuit of Fig. 9.13 to obtain the solution to Example 9.4 for $v_C(t)$.

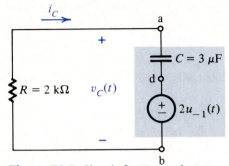

Figure E9.5 Circuit for Example 9.5.

Solution

The initial voltage $v_C(0-)$ is 2 V and, assuming there is no impulsive current, $v_C(0-) = v_C(0+)$. The equivalent circuit problem is as shown in Fig. E9.5. This circuit is just an RC circuit with a step excitation and can be solved by inspection for $v_{ad}(t)$ by expressing it in the form of Eq. (9.27) with $V_s = -2$ V, i.e., for $t \geq 0$

$$v_{ad}(t) = -2(1 - e^{-t/6})u_{-1}(t)$$

Hence, by KVL, for $t \geq 0$

$$v_C(t) = v_{ad}(t) + 2u_{-1}(t) = 2e^{-t/6}u_{-1}(t)$$

Using the branch law for a resistance implies that $i_C(0+) = -v_C(0+)/R = -1$ mA; KVL and continuity of v_C are thus *both* satisfied at t = 0+ verifying that i_C has no impulse.

DRILL EXERCISES

9.6 Show that the Norton representation for the circuit of Fig. 9.13 is as shown in Fig. D9.6.

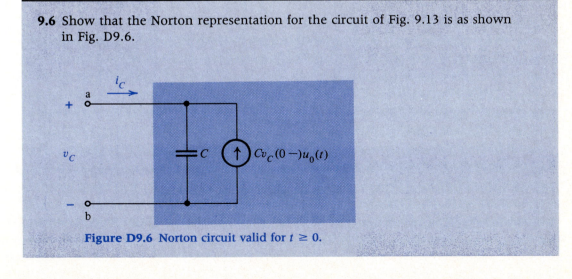

Figure D9.6 Norton circuit valid for $t \geq 0$.

9.7 Show that the Thevenin representation for the circuit of Fig. 9.14 is as shown in Fig. D9.7.

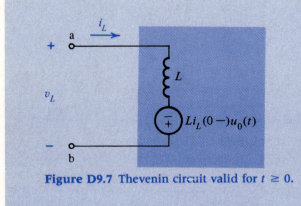

Figure D9.7 Thevenin circuit valid for $t \geq 0$.

9.5 INSPECTION METHODS OF SOLUTION

In this section a number of examples of first-order or STC circuits are discussed. The objective is to give the reader a basis for solving such network problems by inspection for step or constant excitations. There are three parameters which are needed in order to write solutions of the form Eq. (9.24) by inspection, namely:

1. The time constant T, given by RC (GL) for a single capacitance (inductance) network

2. The dc steady state

3. The initial condition $v_C(0+)$ $[i_L(0+)]$

The first is obvious from the network and the third is a given of the problem or must be found from the initial conditions known at $t = 0-$. The second, the dc steady state, is easily found since, under dc conditions there are no time variations and thus capacitance currents and inductance voltages are zero by definition.

9.5.1 The DC Steady State

This is easily found by writing

$$i_C = C \frac{dv_C}{dt}$$

for the capacitance current, and noting that, if a dc steady state exists, then $v_C(t) \to v_{Css}$ as $t \to \infty$, and

$$i_{Css} = \lim_{t \to \infty} i_C(t) = C \lim_{t \to \infty} \frac{dv_C(t)}{dt}$$

or

$$i_{Css} = C\frac{dv_{Css}}{dt}$$

assuming the limit and differentiation can be interchanged. Since v_{Css} is a constant, its derivative vanishes and

$$i_{Css} = 0$$

For an inductance similar reasoning leads to

$$v_{Lss} = 0$$

By the substitution theorem these elements can therefore be replaced by an open circuit and a short circuit, respectively.

EXAMPLE 9.6

Find the dc steady-state values for the network variables in Fig. E9.6(a).

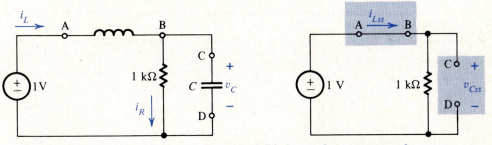

Figure E9.6 (a) Circuit for Example 9.6 and (b) dc steady-state network.

Solution
In Fig. E9.6(b), the inductance is shown as a short circuit and the capacitance as an open circuit. The dc steady-state solution for the variables can be written by inspection as

$$i_{Lss} = 1 \text{ mA} = i_{Rss} \qquad v_{Css} = 1 \text{ V}$$

EXAMPLE 9.7

Find the complete solution for the inductance current in the STC circuit shown, given $i_L(0+)$. Sketch its waveform for $I_S > i_L(0+)$.

Solution

By inspection the time constant $T = GL$, the dc steady-state current

$$i_{Lss} = I_S$$

since $v_{Lss} = 0$, and thus the complete solution is, for $t \geq 0$,

$$i_L(t) = I_S + [i_L(0+) - I_S]e^{-t/GL}$$

The waveform is as shown in Fig. E9.7(b).

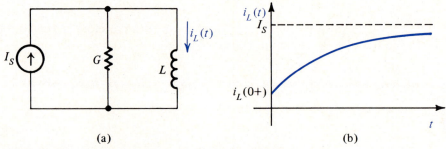

(a)   (b)

Figure E9.7 (a) Circuit for Example 9.7. (b) Solution.

DRILL EXERCISES

9.8 If in Fig. E9.7(a) $G = 1$ mS, $L = 1$ mH, $i_L(0+) = 5$ mA, $I_s = 10$ mA, show that $i_L = 10 - 5e^{-t}$ mA for t in microseconds.

9.5.2 Initial Conditions at $t = 0+$

Normally the known data for a network problem are the values of capacitance voltages and inductance currents at $t = 0-$ prior to any step excitation or switching event. It is therefore important to distinguish between variable values before a switching event ($t = 0-$) or after a switching event ($t = 0+$) since, some variables may have step discontinuities.

Given an initially uncharged capacitance $v_C(0-) = 0$, if $i_C(t)$ has no impulse at $t = 0$ then, according to Eq. (9.6), the voltage is continuous and

$$v_C(0+) = v_C(0-) = 0$$

By the substitution theorem, the capacitance may thus be replaced, for calculation purposes, by a short circuit at $t = 0+$.

Similarly, for an inductance with zero initial energy, $i_L(0-) = 0$, and if $v_L(t)$ has no impulse at $t = 0$, then the current is continuous,

$$i_L(0+) = i_L(0-) = 0$$

and, by the substitution theorem, the inductance can be replaced, for calculations, by an open circuit at $t = 0+$.

Remembering that in some cases jump discontinuities may occur in capacitance voltages and inductance currents, it is usually wise to check on the finiteness of capacitance currents and inductance voltages. Note that these are the only variables "allowed" to have impulsive components.

EXAMPLE 9.8

In Fig. E9.8 the step clearly forces the capacitance voltages to be discontinuous and hence i_C will be an impulse. Thus C_1 and C_2 cannot both be replaced by short circuits even though $v_{C1}(0-) = v_{C2}(0-) = 0$, since KVL imposes

$$v_{C1}(0+) + v_{C2}(0+) = V_0$$

Under those conditions find the impulsive current i_C and the capacitance voltages v_{C1} and v_{C2}.

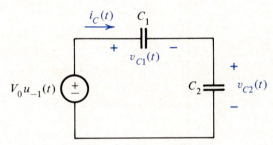

Figure E9.8 Circuit for Example 9.8.

Solution

Because the equivalent capacitance of two capacitances in series is, in this case,

$$C_{eq} = \frac{C_1 C_2}{C_1 + C_2}$$

the charge delivered at $t = 0$ is given by

$$\int_{0-}^{0+} i_C \, dt = C_{eq} V_0$$

Hence the current impulse is

$$i_C = C_{eq} V_0 u_0(t)$$

The voltage step amplitudes are now obtained using Eq. (9.2b) as

$$v_{C1}(0+) = \frac{1}{C_1} \int_{0-}^{0+} i_C \, dt = \frac{C_2 V_0}{C_1 + C_2}$$

and

$$v_{C2}(0+) = \frac{C_1 V_0}{C_1 + C_2}$$

which sum to V_0 as required.

DRILL EXERCISES

9.9 The network of Fig. D9.9(a) is initially at rest $[v_C(0-) = 0, i_L(0-) = 0]$. Show that the circuit of Fig. D9.9(b) correctly describes the situation at $t = 0+$ and find $i_C(0+)$, $v_L(0+)$, $i_R(0+)$. Are there any impulses?
Ans. $i_C(0+) = i_R(0+) = 1$ mA, $v_L(0+) = 1$; no

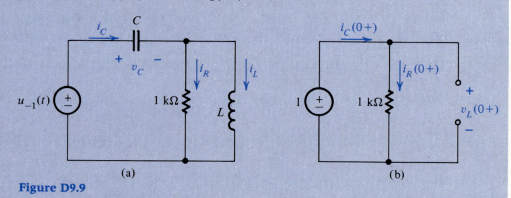

(a)   **(b)**

Figure D9.9

9.10 The network of Fig. D9.10(a) has $v_C(0-) = 1$ V, $i_L(0-) = \frac{1}{2}$A. Show that the circuit of Fig. 9.10(b) correctly describes the situation at $t = 0+$ and find $i_C(0+)$, and $i_R(0+)$. Are there any impulses?
Ans. $i_C(0+) = \frac{1}{2}$A, $i_R(0+) = 0$, $v_L(0+) = 0$; no

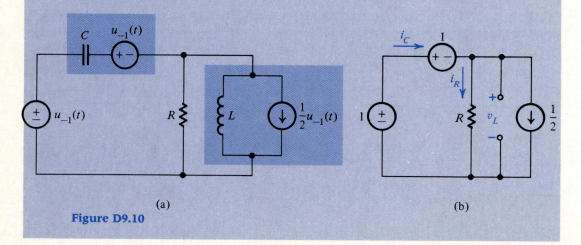

(a)   **(b)**

Figure D9.10

9.11 For the network of Fig. D9.11 solve for $i(t)$ and $v_C(t)$ by inspection, given $v_C(0-) = 1$ V.
Ans. $v_C = 1$ V, $i = 0.5$ mA

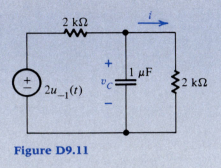

Figure D9.11

9.5.3 Step Response Examples

 EXAMPLE 9.9

For the circuit of Fig. E9.9(a) with an inductance connected to ab, find the inductance current and voltage for $v_S = u_{-1}(t)$. There is no energy stored in the inductance at $t = 0$ (zero state).

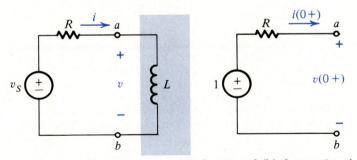

Figure E9.9 (a) Circuit for Example 9.9 and (b) the $t = 0+$ circuit.

Solution
This is just finding the step response. As before the three quantities needed are: the time constant T, the dc steady-state value i_{Lss}, and the initial condition $i_L(0+)$. These are obtained by inspection of the circuit and from the given as

$$T = \frac{L}{R} \qquad i_{Lss} = \frac{1}{R}$$

and, from Fig. E9.9(b),

$$i_L(0+) = 0.$$

The solution is therefore

$$i_L(t) = \frac{1}{R} + \left(0 - \frac{1}{R}\right)e^{-tR/L} \qquad \text{for } t \geq 0$$

or

$$i_L(t) = \frac{1 - e^{-tR/L}}{R} u_{-1}(t)$$

The inductance voltage is given by either the branch relation $L di_L/dt$ or by KVL. By KVL

$$v_L = u_{-1} - Ri_L$$

hence

$$v_L(t) = e^{-tR/L} u_{-1}(t)$$

Use of the branch relation requires some care since it is necessary to use the derivative of a product rule, i.e.,

$$v_L(t) = \frac{L di_L}{dt} = L(1 - e^{-tR/L})u_0(t) + e^{-tR/L} u_{-1}(t)$$

hence

$$v_L(t) = e^{-tR/L} u_{-1}(t)$$

Since the square bracket vanishes at $t = 0$, no impulse appears in i_L as required by finite energy constraints.

EXAMPLE 9.10

The small-signal model of an emitter-follower amplifier used to drive a capacitive load is shown in Fig. E9.10. When driven by a unit voltage step calculate how long it will take for the capacitance voltage to reach 50% of its dc steady-state value. Assume the transistor remains active and that $v_o(0-) = 0$.

Solution
By inspection this is a zero-state problem with $v_o(0+) = v_o(0-) = 0$ and, since by KVL

$$51i_{bss} = 1 - i_{bss}$$

$$i_{bss} = \frac{1}{52} \text{ mA}$$

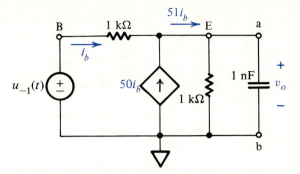

Figure E9.10 Circuit for Example 9.10.

and

$$V_{oss} = \frac{51}{52} \text{ V}$$

The only other quantity needed to obtain the solution is the circuit time constant; this requires finding R_{eq} to the left of ab. This is given by Eq. (7.27) which, when evaluated using the circuit values of Fig. E9.10, yields

$$R_{eq} = \frac{1}{51} \text{ k}\Omega = 19.6 \text{ }\Omega$$

$$T = CR_{eq} \approx 20 \text{ ns}$$

The 1 kΩ across the capacitance clearly has a negligible effect on R_{eq}. The step response is, with t in nanoseconds given by

$$v_o(t) = \frac{51}{52}(1 - e^{-t/20}) \text{ V} \qquad \text{for } t \geq 0$$

This reaches 50% of its dc steady-state value when $e^{-t/20} = 0.5$ which occurs at $t \approx 14$ ns (about 0.7 time constants).

DRILL EXERCISE

9.12 For the problem in Example 9.10 calculate the time taken for the output to go from 10 to 90% of its final value. How many time constants is this?
Ans. 44 ns, 2.2

EXAMPLE 9.11

A common-emitter amplifier is used to drive a resistive load R_L which can influence the amplifier's Q point. The load is therefore "dc blocked" by a capaci-

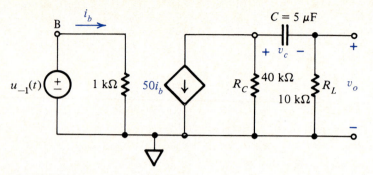

Figure E9.11 Circuit model for Example 9.11.

tance; the small-signal model for the circuit is shown in Fig. E9.11. Find the response for v_o to a 1-mV input step assuming the transistor remains active and $v_c(0-) = 0$. After how long will v_o drop to 90% of its initial $(0+)$ value?

Solution
By inspection this is a zero-state problem with $i_b(0-) = 0$, $v_c(0-) = v_c(0+) = 0$. At $t = 0+$, i_b jumps to $i_b(0+) = 1$ μA; hence, since $v_c(0+) = 0$, $50i_b(0+)$ flows through R_C in parallel with R_L, i.e.,

$$v_o(0+) = (-50 \ \mu A)(8 \ k\Omega) = -400 \ mV$$

The dc steady state for v_o is zero since C is replaced by an open circuit for dc. The time constant, found by setting the independent source to zero, is

$$T = C(R_L + R_C) = 250 \ ms$$

The complete solution is now written, with t in ms, as

$$v_o(t) = -0.4e^{-t/250} \text{ volts}$$

The 90% value occurs at t_1, where t_1 is given by

$$t_1 = -250 \ln 0.9 = 26.3 \ ms$$

9.5.4 Summary

For STC circuits it has been shown that the step response for a network variable can be found by inspection by calculating three quantities as follows:

1. The circuit time constant CR_{eq} or LG_{eq} obtained by inspection,

2. The dc steady-state value of the variable obtained by replacing inductances by short circuits and capacitances by open circuits

3. The $t = 0+$ value of the variable obtained by replacing a capacitance with $v_C(0-) \neq 0$ by a step voltage source of value $v_C(0-)$, and by replacing an

inductance with $i_L(0-) \neq 0$ by a step current source of value $i_L(0-)$. This presumes no impulses in capacitance (inductance) currents (voltages).

9.6 NETWORKS WITH SWITCHES

In this section inspection methods for solving network problems with switching events, not necessarily occurring at $t = 0$, are outlined. Switching events can occur either because there are switches in the circuit which open or close at predetermined times, or because there are diodes in the circuit which change state or switch according to network conditions.

In Fig. 9.15 a network with a switch S is shown. The switch has been in position 1 for a long time before $t = 0$, when it is thrown to position 2. The question is to solve for $v_L(t)$ with V_S a constant voltage.

In this problem there are two states to the circuit. Prior to $t = 0$ the circuit is in a dc steady state, $v_L = 0$, and

$$i_L(0-) = i_{Lss} = \frac{V_S}{R_1} \tag{9.31}$$

At $t = 0$, the switch is thrown to 2 and the network problem is that of an STC LR circuit with a known initial value as shown in Fig. 9.16. The relevant networks

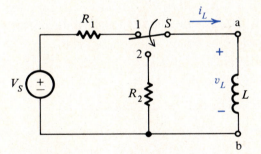

Figure 9.15 Network with switch.

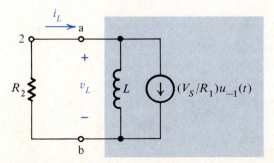

Figure 9.16 Equivalent network of Fig. 9.15 valid for $t \geq 0$.

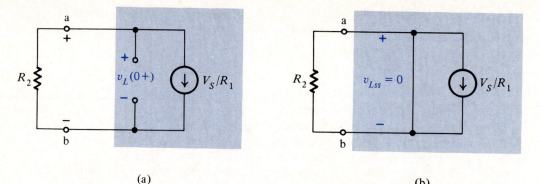

(a)   (b)

Figure 9.17 (a) The 0+ network and (b) the dc steady-state network corresponding to the circuit in Fig. 9.16.

for the 0+ and dc steady-state calculations are shown in Fig. 9.17(a) and (b), respectively, and yield

$$v_L(0+) = -V_S\frac{R_2}{R_1}$$

and

$$v_{Lss} = 0$$

Hence, with a time constant $T_2 = L/R_2$, the solution for the inductance voltage for $t \geq 0$ is, by inspection,

$$v_L(t) = -\left(\frac{V_S R_2}{R_1}\right)e^{-t/T_2} \tag{9.32}$$

The inductance current is calculated, using Ohm's law, as

$$i_L(t) = \frac{-v_L}{R_2} = \frac{V_S}{R_1}e^{-t/T_2} \tag{9.33}$$

If, after a long time $t_1 \gg T_2$ such that $v_L \simeq 0$, the switch S is moved back to position 1, the voltage across the inductance is found using the circuit of Fig. 9.18.

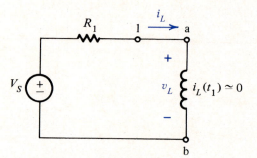

Figure 9.18 Circuit in Fig. 9.15 after switch moved to position 1.

Since $i_L(t_1) \simeq 0$, the "initial" condition, taking t_1 as a new time origin, i.e., letting $\tau = t - t_1$, is

$$i_L(\tau = 0+) = 0 \tag{9.34}$$

Hence

$$v_L(0+) = V_S - i_L(0+)R_1 = V_S \tag{9.35}$$

and, the steady-state voltage

$$v_{Lss} = 0 \tag{9.36}$$

The time solution is now written by inspection as

$$v_L(\tau) = V_S e^{-\tau/T_1} \tag{9.37}$$

for $t \geq t_1$, where $T_1 = L/R_1$ is the time constant of the STC circuit of Fig. 9.18. Equation (9.37) can be written in terms of t and a delayed unit step as

$$v_L(t) = V_S e^{[-(t-t_1)/T_1]} u_{-1}(t - t_1) \tag{9.38}$$

Figure 9.19 illustrates $v_L(t)$, for the case $R_2 > R_1$. Actually, this network illustrates a method of providing the spark in an internal combustion engine. If $R_2 \gg R_1$, the large negative voltage spike appearing across the inductance (the spark coil) causes ionization, i.e., an arc across the sparkplug gap. The coil is then charged, at a much slower rate, then discharged, and so on. For example, with $V_S = 12$ V, $R_2 = 10$ kΩ, and $R_1 = 5$ Ω the spike amplitude is 24 kV.

This network calculation illustrates the use of inspection methods for situations where a switching event establishes two different STC circuits, i.e., Figs. 9.16 and 9.18. When such a situation occurs there are two related STC problems which must be solved, each with a different initial condition and time constant. The initial condition after a switching event is, for the LR problem just discussed, the value of i_L just prior to the switching time.

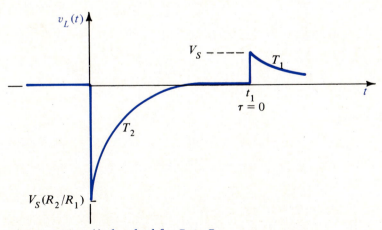

Figure 9.19 $v_L(t)$ sketched for $R_2 > R_1$.

For circuits containing diodes, switching events are determined by changes of state which depend on circuit dynamics. An illustration of the type of problem is shown in Fig. 9.20(a) which contains an ideal diode. It is desired to find the step response of this ideal diode circuit given $v_C(0+) = 0$ and to sketch the i_R, i_C, and v_C waveforms.

The diode D is obviously OFF ($i_D = 0$) prior to $t = 0$ and will remain OFF as long as $v_D < 0$. The capacitance voltage starts to rise exponentially from $v_C(0+) = 0$ toward a dc steady-state value of 1 V and at some point D will switch to the ON state ($v_D = 0$) clamping v_C to 0.7 V and v_R to 0.3 V.

Analytically, since $i_C = I_R$ when D is OFF, v_C is given by

$$v_C = 1 - e^{-t}$$

with t in milliseconds since the time constant is 1 ms. By KVL this OFF state requires $v_D < 0$ or $v_C - 0.7 \leq 0$. Thus D changes state when the equality is satisfied at t_1. Hence t_1 is given by

$$0.3 = e^{-t_1} \qquad \text{or} \qquad t_1 = 1.2 \text{ ms}$$

The current i_R is given by KVL

$$i_R = \frac{1 - v_C(t)}{R}$$

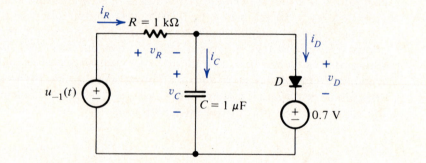

(a)

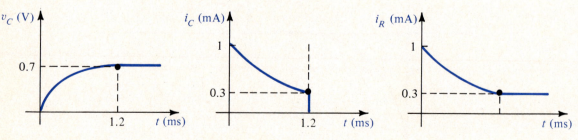

(b)

Figure 9.20 (a) An ideal diode circuit. (b) Waveforms.

and the current i_C is given by the branch law

$$i_C = C\frac{dv_C}{dt}$$

The waveforms, not drawn to scale, are as shown in Fig. 9.20(b).

9.6.1 Switching the BJT Inverter

A very important application of a switching calculation is the study of the effect of a capacitance load on the dynamics of a logic gate such as the BJT inverter circuit in Fig. 7.23. The circuit of the inverter with the capacitive load C is shown in Fig. 9.21(a). The input waveform is shown in Fig. 9.21(b). From the analysis of Chap. 7, such an input signal will cause the BJT to go from cutoff, through the active mode to saturation, and back to cutoff. There are thus three piecewise-linear states with three corresponding STC equivalent circuits as shown in Fig. 9.22. These circuits are the models in Fig. 7.25, augmented by C, with the simplification that v_{BE} and v_{CEsat} are both made zero. The circuit of Fig. 9.22(a) establishes the conditions at $t = 0-$ showing $v_0(0-) = V_{CC}$. That of Fig. 9.22(b) shows the circuit for the active region with

$$v_0(0+) = v_0(0-) = V_{CC}$$

The steady-state value of v_o is just the Thevenin voltage seen to the left of x—x, namely,

$$v_{Oss} = V_{CC} - \beta i_B R_C$$

where $i_B = V_{CC}/R_B$.

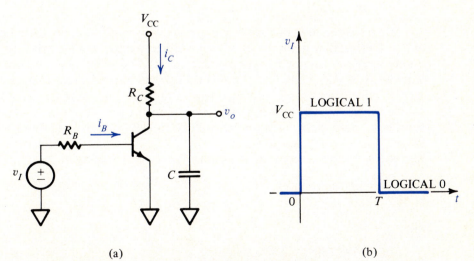

(a) (b)

Figure 9.21 (a) BJT inverter and capacitance load and (b) input waveform.

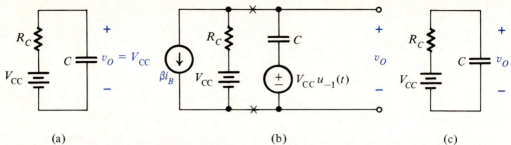

(a)   (b)   (c)

Figure 9.22 The three STC circuits for BJT inverter dynamics: (a) $t = 0-$, cutoff; (b) $t \geq 0$, active; (c) $t \geq T$, return to cutoff with $v_o(T-) = 0$.

The dynamic equation for the voltage v_o is, for $t \geq 0$,

$$v_o(t) = [v_o(0+) - v_{Oss}]e^{-t/(R_C C)} + v_{Oss}$$

which is valid as long as the transistor is active or, equivalently, as long as $v_o \geq 0$. Thus the voltage v_o decays exponentially from V_{CC} to $V_{CC} - \beta V_{CC} R_C/R_B$ which is normally designed to be a large negative number to ensure that the transistor is eventually well saturated. Hence $v_o(t)$ becomes zero at a time t_f which is a small fraction of the time constant $C R_C$. At this time the transistor saturates and the BJT output v_o clamps to zero. If $t_f \ll R_C C$ the exponential solution can be accurately approximated by the linear expansion

$$v_o(t) \simeq [v_o(0+) - v_{Oss}]\left(1 - \frac{t}{R_C C}\right) + v_{Oss} = V_{CC} - \beta V_{CC}\frac{t}{R_B}\frac{t}{C}$$

which equals zero at

$$t_f \simeq \frac{C R_B}{\beta}$$

The transistor remains saturated with $v_o = 0$ until the input switches to zero at $t = T$. From that point on, the circuit of Fig. 9.22(c) applies, with $v_o(T-) = 0$, and v_o increases from zero exponentially toward V_{CC} with time constant $C R_C$. The waveform of $v_o(t)$, shown in Fig. 9.23 illustrates these switching dynamics. The BJT is active from 0 to t_f, saturated from t_f to T, and rises toward cutoff for $t > T$.

In contrast to the instantaneous transitions in the v_I waveform, the v_o waveform has a finite delay in going from HIGH to LOW and from LOW to HIGH. The delays, t_{PLH} and t_{PHL}, are defined in Fig. 9.23 following the convention used in Sec. 8.5.4. These are computed from the preceding analysis to be approximately

$$t_{PHL} \simeq \frac{C R_B}{2\beta}$$

and

$$t_{PLH} \simeq 0.69\, C R_C$$

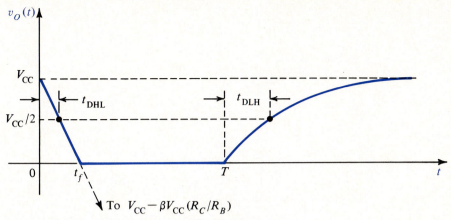

Figure 9.23 Switching dynamics for BJT inverter.

To illustrate the orders of magnitude involved, the typical values $R_B = 450 \; \Omega$, $R_C = 640 \; \Omega$, $\beta = 30$ can be used to estimate these delays for a 20-pF load. The results are

$$t_{\text{PHL}} = 0.02 \frac{450}{60} = 0.15 \text{ ns}$$

$$t_{\text{PLH}} = 0.7(0.02)(640) \simeq 9.0 \text{ ns}$$

EXAMPLE 9.12

Use the model of the BJT in Fig. 7.18 to calculate the delay times t_{PLH} and t_{PHL} of the circuit shown in Fig. E9.12(a) for the input waveform shown in Fig. E9.12(b). The BJT does not saturate and R_E is comparable to R_S. Assume T is long enough for the circuit to reach steady state.

Solution
At $t = 0-$ the transistor is cut off $i_E = 0$, $v_0 = 0$, and $v_{\text{BE}} < 0.7$ V. As the input jumps to V_{CC} the transistor moves into the active region and the capacitance C_E charges according to the circuit model shown in Fig. E9.12(c). The Thevenin equivalent circuit seen at terminals EG is found by noting that the open-circuit voltage

$$v_T = V_{\text{CC}} - 0.7$$

and the short-circuit current $i_E(sc)$ must satisfy the KVL equation

$$(1 - \alpha)i_E(sc)R_S = V_{\text{CC}} - 0.7$$

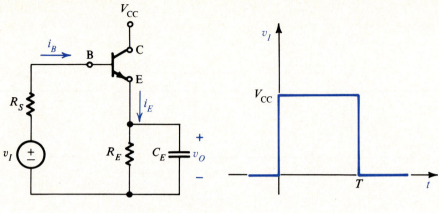

(a)

(b)

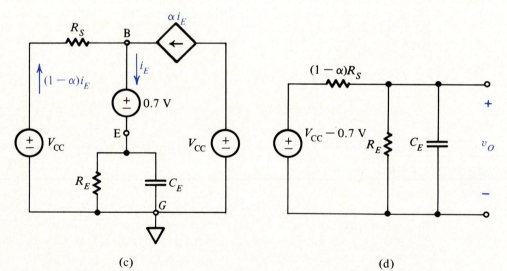

(c)

(d)

Figure E9.12 (a) Circuit and (b) input voltage waveform for Example 9.12.

hence

$$R_{eq} = (1 - \alpha)R_s$$

The Thevenin circuit obtained is shown in Fig. E9.12(d) from which, if one assumes $R_E \gg (1 - \alpha)R_s$, the output will just be the exponential rise

$$v_O(t) = (V_{CC} - 0.7)(1 - e^{-t/T_{ON}})$$

with the turn-on time constant

$$T_{ON} = C_E R_s (1 - \alpha)$$

At $t = T$ the input voltage drops to zero. Since $v_O(T) > V_{BE}$ the transistor cuts off, and C_E discharges through R_E toward zero, with an "initial" value $v_O(T) = V_{CC} - 0.7$, according to

$$v_O(t) = (V_{CC} - 0.7)e^{-t/T_{OFF}}$$

where the turn-off time constant T_{OFF} is

$$T_{OFF} = C_E R_E$$

Since R_E and R_S are assumed to be comparable,

$$T_{OFF} \gg T_{ON}$$

and the time delays are very different. They are, using the delay definitions in Sec. 8.5.4,

$$t_{PLH} = 0.69 C_E R_S (1 - \alpha)$$

and

$$t_{PHL} = 0.69 C_E R_E$$

In contrast to the common-emitter circuit discussed in Sec. 9.6.1, here t_{PLH} is the smaller time delay.

9.7   RESPONSE TO EXPONENTIALS AND SINUSOIDS*

Exponentials will be extensively used in this book. The general form of such an excitation will be

$$x(t) = Xe^{s_p t} \tag{9.39a}$$

or

$$x(t) = Xe^{s_p t}u_{-1}(t) \tag{9.39b}$$

where Eq. (9.39b) differs from Eq. (9.39a) only by virtue of the unit step which acts as a "switch," turning on the exponential at $t = 0$.

Here X is a constant *complex number*, called the *complex amplitude*; s_p is another complex number called the *complex frequency*.

Writing

$$X = Ae^{j\theta} = A \cos \theta + jA \sin \theta \tag{9.40}$$

and

$$s_p = \sigma_p + j\omega_p \tag{9.41}$$

*The student should make sure that the material on complex numbers in the appendix has been thoroughly mastered.

Eq. (9.39b) becomes

$$x(t) = Ae^{\sigma_p t}e^{j(\omega_p t + \theta)}u_{-1}(t) \tag{9.42}$$

Equations (9.39a) or (9.39b) provide a means of defining a modest number of functions in terms of the two complex numbers X and s_p. In Fig. 9.24 some of the various possible functions are sketched for the case of real X ($\theta = 0$). For $\omega_p = 0$ Eq. (9.42) represents a real exponential when $\theta = 0$. If $\sigma_p = 0$ and $\omega_p = 0$, $x(t)$ is a step. If $\omega_p \neq 0$, then

$$x(t) = Ae^{\sigma_p t}[\cos{(\omega_p t + \theta)} + j\sin{(\omega_p t + \theta)}] \tag{9.43}$$

and to represent real signals, one uses the real or imaginary part of Eq. (9.43). The symbol ω is usually used to specify the *angular frequency* in radians per second of a sinusoid. It is related to the *frequency* f in hertz, where 1 Hz equals one cycle per second, by $\omega = 2\pi f$. The magnitude of X, A, determines the amplitude of the signal, while the angle of X, θ, determines the phase. Note that neither the impulse nor the ramp have a complex exponential representation.

The general linear first-order, constant coefficient, differential equation may be written as

$$\frac{dy}{dt} + a_0 y = b_1\frac{dx}{dt} + b_0 x \tag{9.44}$$

when the right-hand side has at most one first derivative. $x(t)$ is the known excitation, and $y(t)$ is the response or output to be found. The homogeneous solution is of the form

$$v_h = Ae^{-a_0 t} = Ae^{s_1 t} \tag{9.45}$$

Consider now an excitation given By (9.39), and let a complex exponential solution candidate $y_p(t) = Ye^{s_p t}$ be inserted into Eq. (9.44). $y_p(t)$ is a solution if, and only if,

$$Ys_p e^{s_p t} + Ya_0 e^{s_p t} = Xb_1 s_p e^{s_p t} + Xb_0 e^{s_p t}$$

or, collecting terms and dividing out by $e^{s_p t}$,

$$(s_p + a_0)Y = (b_1 s_p + b_0)X \tag{9.46}$$

Thus, provided that the excitation complex frequency s_p is not the natural frequency $-a_0$, the complex number Y is given by

$$Y = \frac{b_1 s_p + b_0}{s_p + a_0}X \tag{9.47}$$

The *system function* corresponding to the first-order differential equation (9.44) is defined, for an arbitrary value of s, as

$$H(s) = \frac{b_1 s + b_0}{s + a_0} \tag{9.48}$$

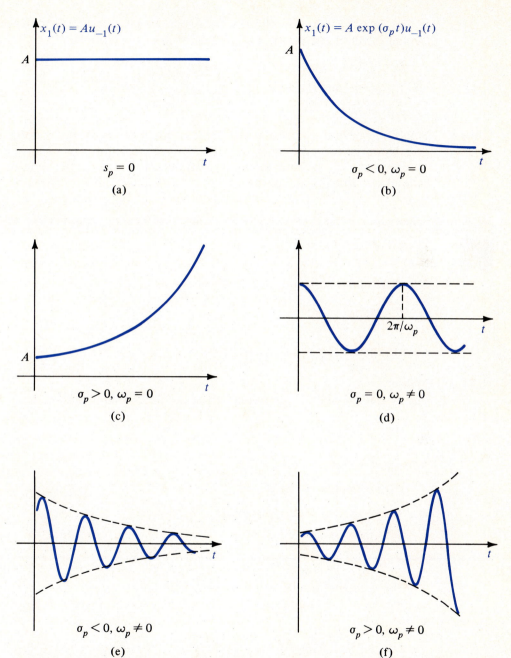

Figure 9.24 Some of the waveforms that can be expressed in terms of complex frequencies. (a) A step, (b) decaying exponential, (c) growing exponential, (d) sinusoid, (e) decaying sinusoid, (f) growing sinusoid.

so that Eq. (9.47) may be written as the system function evaluated at the excitation frequency times the complex amplitude of the excitation, i.e.,

$$Y = H(s_p)X \tag{9.49}$$

and the *particular solution* is

$$y_p(t) = H(s_p)Xe^{s_p t} \tag{9.50}$$

Thus one can find a particular solution to a complex excitation by doing a simple calculation in complex arithmetic. The system function can be written by inspection from the given differential equation and, for first-order systems, it is the ratio of two linear functions of *s*.

EXAMPLE 9.13

The differential equation for a first-order system is

$$\frac{dy}{dt} + \frac{y}{2} = \frac{x}{2}$$

Find the particular solution for $x = 3e^t$.

Solution

The system function is, by inspection of the differential equation,

$$H(s) = \frac{\frac{1}{2}}{s + \frac{1}{2}}$$

With $X = 3$ and $s_p = 1$,

$$H(s_p) = H(1) = \tfrac{1}{3}$$

and hence the particular solution is

$$y_p = H(1)3e^t = e^t$$

The complete solution to Eq. (9.44) with an excitation $x(t) = Xe^{s_p t}$ is written as the sum of the homogeneous and particular solution, namely, Eqs. (9.45) and (9.50), i.e.,

$$y(t) = Ae^{s_1 t} + H(s_p)Xe^{s_p t} \tag{9.51}$$

where $s_1 = -a_0$ is the system natural frequency and where $s_p \neq s_1$. This solution contains (1) an exponential determined by the natural frequency, and (2) a term containing the *same exponential* as the excitation with complex amplitude given by $XH(s_p)$.

In order to determine A, it is necessary to know the initial condition or initial value of y. Since $t = 0$ is usually taken as the time when an excitation is turned on, it will be assumed that $y(0+)$ is known.

EXAMPLE 9.14

Find the general solution to the preceding example given that $y(0+) = 2$.

Solution
The complete solution is given by Eq. (9.51) with $s_p = 1$, $X = 3$, $H(s_p) = \frac{1}{3}$, and $s_1 = -\frac{1}{2}$, i.e.,

$$y(t) = Ae^{-t/2} + e^t \quad \text{for } t \geq 0$$

Inserting the initial condition one finds $A = 1$; hence for $t = 0$,

$$y(t) = e^{-t/2} + e^t$$

EXAMPLE 9.15

Find the particular solution to the differential equation in Example 9.13 for the complex excitation

$$x(t) = e^{(-1+2j)t}$$

Solution
Here $X = 1$, $s_p = -1 + 2j$, the system function is

$$H(s_p) = \frac{1}{2s_p + 1} = \frac{1}{4j - 1}$$

Hence the complex response is

$$y_p = \frac{1}{4j - 1} e^{(-1+2j)t}$$

EXAMPLE 9.16

Repeat Example 9.15 for the conjugate excitation

$$x^*(t) = e^{(-1-2j)t}$$

Solution

Here $X = 1$, $s_p = -1 - 2j$, the system function is

$$H(s_p) = \frac{1}{-2 - 4j + 1} = \frac{1}{-4j - 1}$$

and the complex response is

$$y_p = \frac{1}{-4j - 1} e^{(-1-2j)t}$$

This particular solution is the complex conjugate of the solution to Example 9.15.

Example 9.16 illustrates the property, which arises from linearity, that if the particular solution, $y_p(t)$, to a complex excitation, $x(t)$, is known, then the particular solution to the conjugate excitation, $x^*(t)$, is simply the complex conjugate function $y_p^*(t)$.

As a corollary to this property if a real excitation can be written in the form

$$x(t) = Xe^{s_p t} + X^* e^{s_p^* t} = 2\text{Re}(Xe^{s_p t})$$

then the particular solution to this excitation, by superposition, will be

$$y_p(t) = Ye^{s_p t} + Y^* e^{s_p^* t} = 2\text{Re}\{Ye^{s_p t}\}$$
$$= 2\text{Re}\{H(s_p)Xe^{s_p t}\}$$

since $Y = H(s_p)X$ and where $Y^* = H(s_p^*)X^* = H^*(s_p)X^*$. The student should test this result by direct substitution in the differential equation (9.44).

DRILL EXERCISES

9.13 Express the excitation of Example 9.15 in trigonometric form.
Ans. $e^{-t}(\cos 2t + j \sin 2t)$

9.14 Repeat Drill Exercise 9.13 for the excitation of Example 9.16.
Ans. $e^{-t}(\cos 2t - j \sin 2t)$

9.15 Show that the sum of the answers to the preceding exercises is real and equals $2\text{Re}\{e^{(-1\pm 2j)t}\}$.

9.16 Express $\cos pt$ as the sum of two complex exponentials.
Ans. $\cos pt = \dfrac{e^{jpt}}{2} + \dfrac{e^{-jpt}}{2}$

9.17 Using the result that $y_p(t) = 2\text{Re}\{H(s_p)Xe^{s_p t}\}$ is the particular solution to $x(t) = 2 \text{Re}\{Xe^{s_p t}\}$, find the particular solution to $x = \cos pt$.
Ans. $y_p = \text{Re}\{H(jp)e^{jpt}\}$

9.18 Express the answer to Exercise 9.17 in trigonometric form.

Ans. $y_p = |H(jp)| \cos [pt + \sphericalangle H(jp)]$

EXAMPLE 9.17

Repeat Example 9.13 for the excitation $x = e^{-t} \cos 2t$.

Solution

Method 1. This excitation is just the real part of $e^{(-1+2j)t}$ as was seen in Drill Exercise 9.13. The particular solution to this excitation is therefore the real part of the solution to Example 9.15, i.e.,

$$y_p(t) = \text{Re}\left\{ \frac{1}{4j - 1} e^{(-1+2j)t} \right\}$$

$$= \text{Re}\{0.2425 e^{-t} e^{-j104°} e^{2jt}\}$$

$$= 0.2425 e^{-t} \cos (2t - 104°)$$

Method 2. Since this excitation can also be written as

$$x = e^{-t}\left(\frac{e^{j2t}}{2} + \frac{e^{-j2t}}{2} \right)$$

$$= \tfrac{1}{2} e^{(-1+2j)t} + \tfrac{1}{2} e^{(-1-2j)t}$$

which is half the sum of the excitations of Examples 9.15 and 9.16, respectively; the solution is, by superposition, just half of the sum of the two solutions, i.e.,

$$y_p(t) = \frac{1}{2}\left(\frac{1}{4j - 1} e^{(-1+2j)t} + \frac{1}{-4j - 1} e^{(-1-2j)t} \right)$$

which reduces to the same answer as given above.

The response to a real sinusoidal excitation is obtained by using the results of Exercises 9.17 and 9.18.

EXAMPLE 9.18

Repeat Example 9.14 for the excitation $\cos 2t$.

Solution

First find the solution to $x = e^{j2t}$ as

$$y_p = H(j2)e^{j2t} = \frac{1}{4j+1}e^{j2t}$$

$$= \frac{1}{(17)^{1/2}}e^{-j76°}e^{j2t}$$

$$= \frac{1}{(17)^{1/2}}e^{j(2t-76°)}$$

Now the real part of this last result is the particular solution to the cosine excitation, i.e.,

$$y_p = \frac{1}{(17)^{1/2}}\cos(2t - 76°)$$

The output amplitude is 0.2425 and the output phase is $-76°$. The engineering convention of writing phase in degrees is used.

9.8 SUMMARY AND STUDY GUIDE

The purpose of this chapter has been to present methods of solving first-order circuit problems. The solutions to first-order differential equations were reviewed and some simple STC circuits subjected to steps and impulses were examined. Inspection methods were developed for solving switching problems, and the network function method was introduced for finding particular solutions to complex exponential excitations. The attention of the student is drawn to the following points.

A capacitance voltage (inductance current) is continuous unless the capacitance current (inductance voltage) is impulsive, in which case the voltage (current) has a step discontinuity.

The step response for any variable in an STC circuit contains a dc steady-state component and a single exponential transient component.

Replacing initial conditions with step sources reduces network transient problems to zero-state problems.

Under dc steady-state conditions capacitances act as open circuits and inductances as short circuits.

Initial conditions at $t = 0+$ are found by inspection since capacitance voltages (inductance currents) are continuous unless there are capacitance current (inductance voltage) impulses.

The absence of impulses in inductance voltages (capacitance currents) is checked by verifying that continuity of inductance currents (capacitance voltages) *and* the Kirchhoff laws are satisfied at $t = 0+$.

In problems with switches, the capacitance voltages and inductance currents at the end of an interval are the initial conditions for the next interval.

The difference between problems with switches and diodes is that switches close and open at predetermined times, whereas diodes switch according to network conditions.

Particular solutions to excitations that can be represented by complex exponentials are easily found using the system function method and some complex arithmetic.

PROBLEMS

9.1 Find the differential equation for i_L in the circuit of Fig. P9.1.

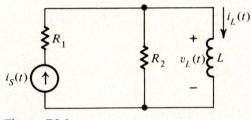

Figure P9.1

9.2 What is the homogeneous solution for i_L in the circuit of Fig. P9.1? Given $i_L(0+)$ what is the general solution for $i_S(t) = I_S$?

9.3 Identify, in the general solution obtained in Prob. 9.2, the steady-state and the transient solutions.

9.4 Identify, in the general solution to Prob. 9.2, the zero-input and the zero-state solutions.

9.5 Solve for $i(t)$ and $v(t)$ for $t \geq 0$ in the networks of Fig. P9.5 given $v(0+) = 1$ V and $i(0+) = 0.5$ mA.

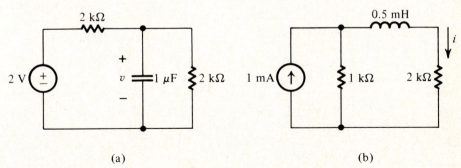

(a)   (b)

Figure P9.5

9.6 Express each of the pulse waveforms of Fig. P9.6 in terms of steps and ramps.

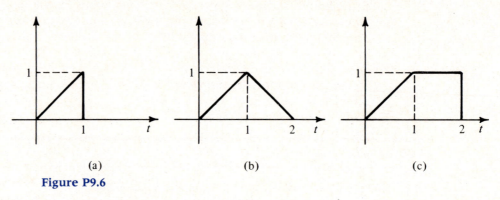

(a)   (b)   (c)

Figure P9.6

9.7 Find the unit step response for the zero-state circuits shown in Fig. P9.7 for each of the variables labeled.

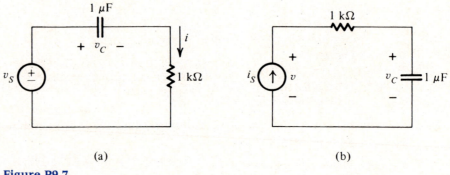

(a)   (b)

Figure P9.7

9.8 Solve for v, i_{L1} and i_{L2} for the zero-state circuit in Fig. P9.8.

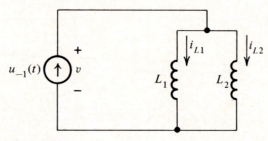

Figure P9.8

9.9   Find the dc steady-state values for the variables labeled in Fig. P9.9(a) and (b).

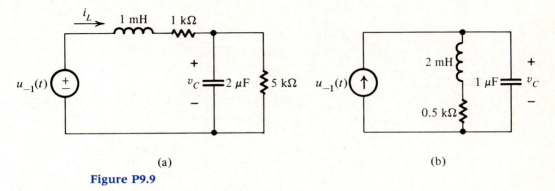

(a)   (b)

Figure P9.9

9.10   Find $v(0+)$ and $i(0+)$ for the zero-state network of Fig. P9.10.

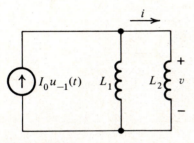

Figure P9.10

9.11   For the networks of Fig. P9.11, find $v(0+)$ and $i(0+)$ given $v_C(0-) = 1$ V and $i_L(0-) = -1$ A.

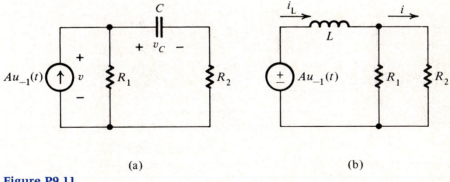

(a)   (b)

Figure P9.11

9.12   Find the zero-state response for v and i in Fig. P9.10.

9.13   Find the unit step response for v_C for the circuit of Fig. P9.11(a) and for i in the circuit of Fig. P9.11(b). $v_C(0-) = 1$ V and $i_L(0-) = -1$ A.

9.14 Find the time constant for each of the following STC circuits.

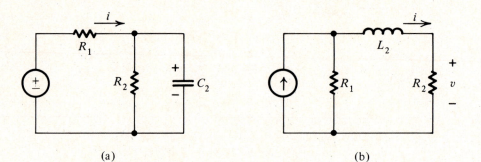

(a)   (b)

(c)

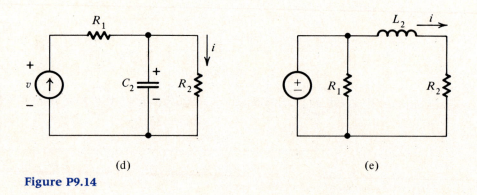

(d)   (e)

Figure P9.14

9.15 Find the unit-step zero-state response for the variables indicated in Fig. P9.14.

9.16 Find the step response for v in the small-signal model of a common-emitter amplifier shown in Fig. P9.14(c).

9.17 For the circuit in Fig. P9.17 find the differential equation relating v_1 and i_1.

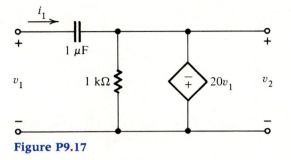

Figure P9.17

9.18 If, in the zero-state circuit of Fig. P9.17, i_1 is a 1-mA step, calculate $v_1(t)$ and $v_2(t)$ for $t \geq 0$. Repeat if the 1-μF capacitance voltage is 1 V at $t = 0-$.

9.19 Calculate the $v_1(t)$, $i_1(t)$ and $v_2(t)$ if the zero-state network in Fig. P9.17 is driven by the Thevenin equivalent circuit: $v_T = u_{-1}(t)$, $R_{eq} = 1$ kΩ.

9.20 Solve for the variables listed in Fig. P9.14 problems for a unit step excitation given all capacitance voltages at $t = 0-$ are 1 V and all inductance currents are 1 A. (*Hint*: Use the method of replacing initial conditions by generators and solve as zero-state problems by inspection.)

9.21 The network of Fig. P9.21 is in the zero-state at $t = 0-$. The switch is open for $t < 2$ ms, closed for $t > 2$ ms. Calculate and sketch $v(t)$, $0 < t < \infty$.

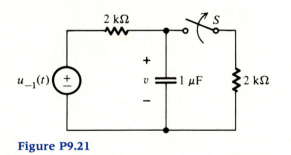

Figure P9.21

9.22 Switch S in Fig. P9.22 has been closed for a long time prior to $t = 0$. It opens at $t = 0$, closes again at $t = 0.5$ μs and remains closed thereafter. Find and sketch $v(t)$, $0 < t < \infty$.

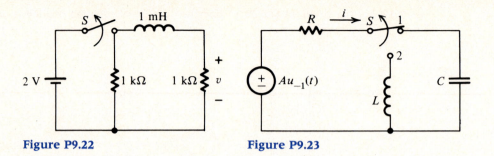

Figure P9.22   **Figure P9.23**

9.23 The network of Fig. P9.23 is in the zero state at $t = 0-$. Switch S is in position 1 for $t < T$, and in position 2 for $t \geq T$. Calculate and sketch $i(t)$.

9.24 The input $v_s(t)$ in Fig. P9.24(a) is a rectangular pulse, i.e.,

$$v_s(t) = \begin{cases} 0 & \text{for } t < 0 \\ A & \text{for } 0 \leq t < T \\ 0 & \text{for } t \geq T \end{cases}$$

This excitation is equivalent to the circuit in Fig. P9.24(b), with S in position 2 for $t < 0$, in position 1 for $0 \leq t < T$, and in position 2 for $t \geq T$. It is not difficult to see that this produces an excitation identical to $v_s(t)$.
(a) Solve for $v(t)$ in Fig. P9.24(b) from the zero state.
(b) Sketch $v(t)$ for $T \ll RC$, $T \simeq RC$, and $T \gg RC$.

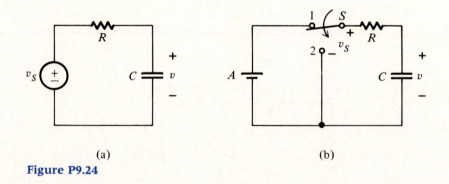

(a)   (b)

Figure P9.24

9.25 The circuit in Fig. P9.25(a) is used as a dynamic memory in NMOS logic circuits. Figure P9.25(b) shows a simple model of this circuit which is adequate to estimate the relative magnitudes of some important timing parameters.
(a) Determine the time it takes for v_2 to reach the midpoint between V_{OH} and V_{OL} if Q_3 is OFF before Q_2 is turned ON at $t = 0$ and C_1 is initially uncharged.
(b) Repeat part (a) with Q_3 ON and $v_2(0-) = 5$ V.
(c) Assuming $V_{TH} = 2.5$ V for Q_1 [see Fig. 8.39(b)], determine how long Q_2 can be OFF if $v_2(0-) = 5$ V.

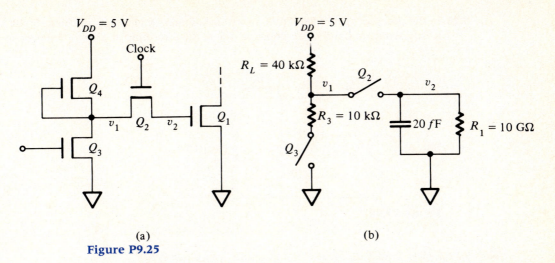

(a)

(b)

Figure P9.25

9.26 Find $v_o(t)$ in the circuit of Fig. P9.26(a) for the $v_i(t)$ excitation shown in Fig. P9.26(b).

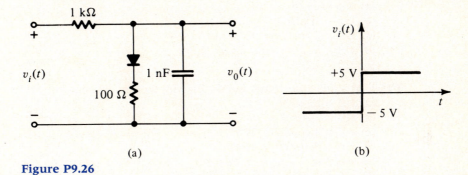

(a)

(b)

Figure P9.26

9.27 In Fig. P9.27(a) a circuit called a diode clamp is shown. Find and sketch $v_o(t)$ for the $v_i(t)$ excitation shown in Fig. P9.27(b) and hence show that the $v_o(t)$ waveform is approximately the $v_i(t)$ waveform shifted down by 5 V (or having its positive peaks clamped to zero).

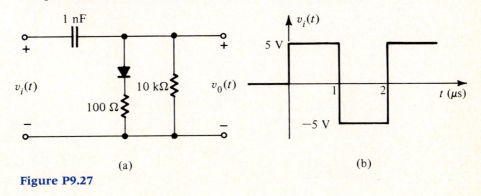

(a)

(b)

Figure P9.27

9.28 In the circuit of Fig. P9.28 $i_L(0-) = -10$ mA. Find $i_L(t)$ for $t \geq 0$ and identify the dc steady-state and transient response as well as the zero-input and zero-state response.

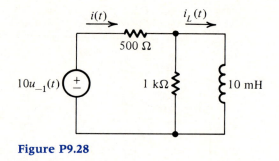

Figure P9.28

9.29 Repeat Prob. 9.28 for the input current $i(t)$ for $t \geq 0$.

9.30 Given $H(s) = (s + 1)/(s + 4)$, find particular solutions for the following inputs.
(a) $x(t) = u_{-1}(t)$
(b) $x(t) = tu_{-1}(t)$
(c) $x(t) = 2 \cos (3t + 45°)u_{-1}(t)$
(d) $x(t) = u_{-1}(t) + \cos t \, u_{-1}(t)$
(e) $x(t) = u_0(t)$

9.31 Find X and s_p such that $\text{Re}\{Xe^{s_p t}\} = x(t)$, for the following $x(t)$:
(a) $x(t) = 3 \cos 3t$
(b) $x(t) = 2 \sin 3t$
(c) $x(t) = 3 \cos (2t + 30°)$
(d) $x(t) = 3 \sin (2t + 30°)$
(e) $x(t) = 4$
(f) $x(t) = -e^{-t}$
(g) $x(t) = 3e^{-t} \cos (2t - 30°)$

9.32 In the circuit of Fig. P9.32 $v_C(0-) = 1$ V. Find $i(t)$ for $t \geq 0$ given $v_i(t) = 5 \cos (2\pi \times 10^3 t)u_{-1}(t)$. If the phase, θ, of the v_i generator can be adjusted, is there a θ such that $i(t)$ has no transient component? If so, what is it?

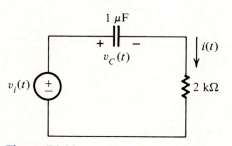

Figure P9.32

9.33 Show that the solution to Prob. 9.32 becomes only sinusoidal as t → ∞, and show that this is identical to the particular solution obtained using the network function approach. This is called the *sinusoidal steady state*.

9.34 Find the step response $v_o(t)$ for the inverting ideal op amp ($A \to \infty$) circuit shown in Fig. P9.34. What is the transient response? What is the dc steady-state response?

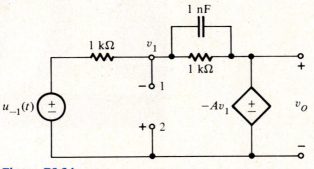

Figure P9.34

CHAPTER 10

Transient Response and Impedance

10.1 INTRODUCTION

The transient response of a network of order greater than one is studied in this chapter. The accent is placed on the second-order case, which has particular practical importance. The natural frequencies of second-order networks are characterized by two parameters, Q and ω_0, and give rise to a wide variety of responses. In particular, second-order networks account for the oscillatory behavior encountered in many practical situations.

The system function has already been introduced in the preceding chapter. Its relationship to the differential equation is explored in this chapter, and turns out to be simple and direct. The system function is used to find the particular solution, which is that part of the solution attributable to the input.

Impedance is seen to be a special system function. It is a generalization of resistance to networks with inductance and capacitance and makes possible the use of the techniques of linear resistive networks to obtain the functions needed to calculate system responses.

10.2 RESPONSE OF NETWORKS OF ORDER GREATER THAN ONE

10.2.1 Introduction

Second-order networks are described by a second-order differential equation. The network of Fig. 10.1 is a member of this class.

A single KVL equation is written to describe the network:

$$v_L + v_R + v_C = v_I$$

or, using the branch laws,

$$L\frac{di}{dt} + Ri + \frac{1}{C}\int_{-\infty}^{t} i\,d\tau = v_I$$

Differentiating each side yields

$$L\frac{d^2i}{dt^2} + R\frac{di}{dt} + \frac{i}{C} = \frac{dv_I}{dt}$$

or

$$\frac{d^2i}{dt^2} + \frac{R}{L}\frac{di}{dt} + \frac{1}{LC}i = \frac{1}{L}\frac{dv_I}{dt} \qquad (10.1)$$

10.2.2 Solution of the Homogeneous Equation

The homogeneous equation corresponding to Eq. (10.1) is

$$\frac{d^2i}{dt^2} + \frac{R}{L}\frac{di}{dt} + \frac{1}{LC}i = 0 \qquad (10.2)$$

The trial solution Ae^{st} satisfies Eq. (10.2) if

$$As^2e^{st} + A\frac{R}{L}se^{st} + A\frac{1}{LC}e^{st} = 0$$

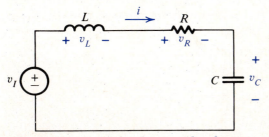

Figure 10.1 Example of a second-order network.

which, since $Ae^{st} \neq 0$, is true if and only if

$$s^2 + \frac{R}{L}s + \frac{1}{LC} = 0 \qquad (10.3)$$

Equation (10.3) is the *characteristic equation* corresponding to the differential equation, and the left-hand side of Eq. (10.3) is the *characteristic polynomial*. The solution of Eq. (10.2) is

$$i_h(t) = A_1 e^{s_1 t} + A_2 e^{s_2 t} \qquad (10.4)$$

where A_1, A_2 are constants and where s_1, s_2 are the roots of the characteristic polynomial. In general, these roots are complex numbers and are called the *natural frequencies*. The constants are found from the known initial values of i and its derivative.

It is customary to write a second-order characteristic equation in terms of two parameters, ω_0 and Q as

$$s^2 + \frac{\omega_0}{Q}s + \omega_0^2 = 0 \qquad (10.5)$$

where Q is dimensionless and ω_0 has dimension of inverse time.

For the example at hand, ω_0 and Q are calculated by matching the coefficients of Eqs. (10.3) and (10.5). This yields

$$\omega_0 = \frac{1}{\sqrt{LC}} \qquad (10.6)$$

and

$$Q = \frac{\omega_0 L}{R} = \frac{1}{\omega_0 CR} = \frac{1}{R}\sqrt{\frac{L}{C}} \qquad (10.7)$$

The natural frequencies are the solutions of Eq. (10.5), given by

$$s_1, s_2 = -\frac{\omega_0}{2Q} \pm \left[\left(\frac{\omega_0}{2Q}\right)^2 - \omega_0^2\right]^{1/2} = \left(-\frac{1}{2Q} \pm \sqrt{\frac{1}{4Q^2} - 1}\right)\omega_0 \qquad (10.8)$$

They are real if $Q \leq \frac{1}{2}$ and complex if $Q > \frac{1}{2}$. The network is said to be *underdamped* if the natural frequencies are complex, *overdamped* if they are real and distinct and *critically damped* if they are coincident.

In the complex case, which is of greater practical interest, the natural frequencies are the complex conjugate pair

$$s_1, s_1^* = \left(-\frac{1}{2Q} \pm j\sqrt{1 - \frac{1}{4Q^2}}\right)\omega_0 \qquad (10.9)$$

$$= -\sigma \pm j\omega \qquad (10.10)$$

where

$$\sigma = \frac{\omega_0}{2Q} \tag{10.11}$$

$$\omega = \left(1 - \frac{1}{4Q^2}\right)^{1/2} \omega_0 \tag{10.12}$$

The homogeneous solution is, from Eq. (10.4),

$$i_h(t) = A_1 e^{(-\sigma+j\omega)t} + A_2 e^{(-\sigma-j\omega)t} \tag{10.13}$$

Since $i_h(t)$ must be real, it can be shown (Example 9.9) that $A_2 = A_1^*$, so that

$$i_h(t) = A_1 e^{(-\sigma+j\omega)t} + A_1^* e^{(-\sigma-j\omega)t} = 2\,\text{Re}(A_1 e^{(-\sigma+j\omega)t}) \tag{10.14}$$

Now, A_1 is a complex number, written as $|A_1| e^{j\theta}$, and, therefore,

$$
\begin{aligned}
i_h(t) &= 2\text{Re}(|A_1| e^{j\theta} e^{(-\sigma+j\omega)t}) \\
&= 2|A_1| e^{-\sigma t}\,\text{Re}(e^{j(\omega t+\theta)}) \\
&= A e^{-\sigma t}\cos(\omega t + \theta)
\end{aligned} \tag{10.15}
$$

where $A = 2|A_1|$.

The constants A and θ are evaluated from the initial conditions at $t = 0+$ given by Eq. (10.15) and its derivative

$$\frac{di_h}{dt} = -\sigma A e^{-\sigma t}\cos(\omega t + \theta) - \omega A e^{-\sigma t}\sin(\omega t + \theta) \tag{10.16}$$

Thus, from Eq. (10.16),

$$\frac{di_h}{dt}(0+) = -\sigma A \cos\theta - \omega A \sin\theta$$

or

$$\omega A \sin\theta = -\sigma A \cos\theta - \frac{di_h}{dt}(0+)$$

Combining this equation with Eq. (10.15), evaluated at $t = 0+$, namely,

$$A \cos\theta = i_h(0+) \tag{10.17}$$

yields

$$A \sin\theta = -\frac{\sigma i_h(0+) + di_h/dt(0+)}{\omega} \tag{10.18}$$

Since $\cos^2\theta + \sin^2\theta = 1$, it follows that

$$A^2 = i_h^2(0+) + \frac{[\sigma i_h(0+) + di_h/dt(0+)]^2}{\omega^2} \tag{10.19}$$

Equations (10.17) and (10.18) are used to find θ once A is known.

EXAMPLE 10.1

In the network of Fig. 10.1, the inductance and capacitance values are $L = 4$ H and $C = 1$ μF. Given $v_i(t) = 0$, $i(0+) = 2$ mA, $di/dt(0+) = 0$, find $i(t)$ for $R = 1$ kΩ, 100 Ω, and 10 Ω.

Solution
From Chap. 1, the units in the problem statement form a consistent set, if time is in milliseconds. From Eqs. (10.6) and (10.7),

$$\omega_0 = 0.5 \text{ rad/ms} = 0.5 \text{ krad/s}$$

$$Q = \begin{cases} 2 & \text{for } R = 1 \text{ k}\Omega \\ 20 & \text{for } R = 0.1 \text{ k}\Omega \\ 200 & \text{for } R = 0.01 \text{ k}\Omega \end{cases}$$

For $R = 1$ kΩ, Eqs. (10.11) and (10.12) yield $\sigma = 0.125$ ms^{-1} and $\omega = 0.5(1 - \frac{1}{16})^{1/2} = 0.484$ ms^{-1}.

From Eq. (10.19),

$$A^2 = 4 + \left(\frac{1}{0.484}\right)^2 0.125^2(2^2)$$

or

$$A = 2.0656$$

From Eq. (10.17)

$$\cos \theta = \frac{2}{2.0656}$$

hence

$$\theta = \pm 14.48°$$

From Eq. (10.18), $A \sin \theta$ is seen to be negative, so θ lies in the fourth quadrant, i.e., $\theta = -14.48°$. The solution is

$$i(t) = 2.0656 e^{-0.125t} \cos(0.484t - 14.48°) \text{ mA}$$

where t is in milliseconds.

The same calculation is performed for the other two values of R. There results

$$i(t) = 2.000 e^{-0.0125t} \cos(0.5t - 1.479°) \text{ mA} \qquad \text{for } R = 0.1 \text{ k}\Omega$$

$$i(t) = 2.000 e^{-0.00125t} \cos(0.5t - 0.1480°) \text{ mA} \qquad \text{for } R = 0.01 \text{ k}\Omega$$

The three responses are sketched in Fig. E10.1.

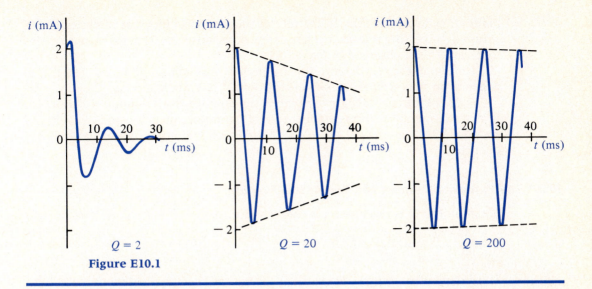

Figure E10.1

The homogeneous solution is a damped sinusoid, i.e., a sinusoid with an exponentially decreasing envelope, as shown in Fig. 10.2. The cosine has a period $2\pi/\omega$. During that time, according to Eq. (10.15), the envelope's amplitude changes by a factor $e^{-2\pi\sigma/\omega}$. From Eqs. (10.11) and (10.12), the ratio σ/ω depends only on Q. The Q values of 2, 20, and 200 of Example 10.1 yield values of 0.197, 0.854, and 0.984, respectively, for $e^{-2\pi\sigma/\omega}$; for example, if $Q = 2$, the envelope magnitude shrinks by a factor of 0.197 during one period of the cosine. In this

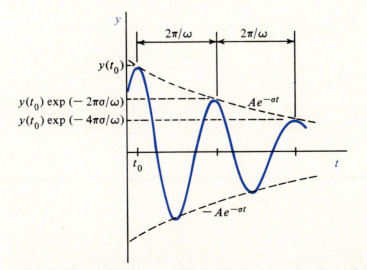

Figure 10.2 Homogeneous solution for the second-order underdamped case.

manner, Q determines the decay rate of the response, in terms of cycles of the sinusoid: increasing Q leads to a less damped response, one that takes more cycles to decay. This is observed in Fig. E10.1.

The parameter ω_0 determines the period, through ω. From Eq. (10.12), ω depends on Q as well as ω_0; however, if $1/4Q^2 \ll 1$, then $\omega \simeq \omega_0$.

For networks of order greater than 2, the homogeneous solution will be a sum of real exponentials, corresponding to real natural frequencies, and of sines and cosines with exponential envelopes, corresponding to complex natural frequencies.

10.2.3 The Particular Solution and the System Function

Equation (10.1) is rewritten, for convenience, as

$$\frac{d^2i}{dt^2} + \frac{R}{L}\frac{di}{dt} + \frac{1}{LC}i = \frac{1}{L}\frac{dv_I}{dt}$$

The independent source voltage $v_i(t)$ is the excitation. Only excitations of the form $v_I(t) = V_I e^{s_p t}$ are considered.

A solution $i_P(t) = Ie^{s_p t}$ is postulated. Insertion in the differential equation yields

$$\left(s_p^2 + \frac{R}{L}s_p + \frac{1}{LC}\right)Ie^{s_p t} = \frac{1}{L}s_p V_I e^{s_p t}$$

or, since $e^{s_p t} \neq 0$,

$$\left(s_p^2 + \frac{R}{L}s_p + \frac{1}{LC}\right)I = \frac{1}{L}s_p V_I$$

Provided s_p is not a natural frequency, the quantity in parentheses on the left is not zero, and

$$\frac{I}{V_I} = H(s_p) = \frac{(1/L)s_p}{s_p^2 + (R/L)s_p + 1/LC} \tag{10.20}$$

The constants I and V_I are known as the complex amplitudes of the particular solution and of the excitation, respectively. Their ratio $H(s_p)$ is the system function $H(s)$, evaluated at $s = s_p$.

EXAMPLE 10.2

For the network of Fig. 10.1, find the particular solution for (1) $v_I(t) = 2e^{-t}$ volts, (2) $v_I(t) = 1$ V.

Solution

For $v_I(t) = 2e^{-t}$, $V_I = 2$ V and $s_p = -1$. Using this in the system function Eq. (10.20),

$$I = H(-1)2$$

$$I = \frac{2(-1/L)}{1 - R/L + 1/LC}$$

and

$$i_p(t) = Ie^{-t} = \frac{(-2/L)e^{-t}}{1 - R/L + 1/LC}$$

For $v_I(t) = 1$, $V_I = 1$ and $s_p = 0$, in which case, from Eq. (10.20),

$$H(0) = 0$$

and

$$i_p(t) = 0$$

Using Eqs. (10.1) and (10.20) as an example, it is not difficult to see how the system function is generated from the differential equation in the general case. The system function $H(s)$ is a ratio of polynomials. The numerator is constructed by replacing each operation d^k/dt^k on the right-hand side of the equation by s^k, and the denominator is constructed in a similar fashion from the left-hand side of the equation.

EXAMPLE 10.3

Write the system function corresponding to the differential equation

$$\frac{d^3y}{dt^3} + \frac{2d^2y}{dt^2} + \frac{dy}{dt} = \frac{-2dx}{dt} + x$$

Solution

By inspection

$$\frac{Y}{X} = H(s) = \frac{-2s + 1}{s^3 + 2s^2 + s}$$

EXAMPLE 10.4

Given

$$\frac{Y}{X} = H(s) = \frac{s^2 + 2s}{s^3 + s^2 + s + 1}$$

write the differential equation relating $y(t)$ to $x(t)$.

Solution
By inspection,

$$\frac{d^3y}{dt^3} + \frac{d^2y}{dt^2} + \frac{dy}{dt} + y = \frac{d^2x}{dt^2} + \frac{2dx}{dt}$$

10.3   THE LAPLACE TRANSFORM

The *Laplace transformation* is a mathematical operation that replaces a function of time by its Laplace *transform*, a function of a complex variable. A function $f(t)$ and its transform $F(s)$ form a *transform pair*, as indicated by the notations

$$F(s) = \mathscr{L}[f(t)]$$

or

$$f(t) \leftrightarrow F(s)$$

Capital letters are used for the transforms, lowercase letters for the time functions. If the time function is subscripted, the transform will have the same subscript; for example, $\mathscr{L}[v_S(t)] = V_S(s)$ and $\mathscr{L}[v_s(t)] = V_s(s)$. Possible confusion with constants, denoted by capital letters with capital subscripts, is avoided by showing the s dependence explicitly in the transform when necessary.

For systems described by linear, constant-coefficient differential equations, it turns out that $Y(s)$, the transform of the output, is just the product of the system function times the transform of the input. It will be seen that $y(t)$ is obtained from $Y(s)$ by algebraic steps, and the solution of differential equations is thus reduced to algebra. Further, the calculation of the system function is also reduced to a problem in algebra. The Laplace transformation therefore provides an elegant method to solve networks described by linear, constant-coefficient differential equations.

10.3.1 Definition

The Laplace transform of $f(t)$ is defined as

$$F(s) = \int_{0-}^{\infty} f(t)e^{-st}\, dt \tag{10.21}$$

where s is a complex variable written as

$$s = \sigma + j\omega \tag{10.22}$$

The notation $t = 0-$ is actually shorthand for $t = -\epsilon$, where ϵ is an infinitesimally small positive quantity. The use of $0-$ is favored in engineering, so that impulses occurring at $t = 0$ are included in calculating $F(s)$.

Note that the Laplace transform does not depend on $f(t)$ for $t < 0$; in other words, two time functions that are different for $t < 0$ but identical for $t \geq 0$ have identical Laplace transforms.

The integral of Eq. (10.21) must exist, of course; in general, existence is guaranteed for Re(s) greater than some number σ_o. The set of values of s for which the integral exists, called the *region of convergence*, is of great importance for a more general class of transforms but is of little consequence in the present context.

EXAMPLE 10.5

Calculate the Laplace transform of

$$f(t) = e^{at}u_{-1}(t)$$

Solution

From Eq. (10.21),

$$F(s) = \int_{0-}^{\infty} e^{at}e^{-st}\, dt = \left(\frac{e^{(a-s)t}}{a-s}\right)\Bigg|_{0-}^{\infty}$$

$$= \frac{1}{s-a} + \lim_{t\to\infty}\frac{e^{(a-s)t}}{a-s}$$

The second term of the right-hand side is zero if Re(s) $> a$ [if Re(s) $< a$, the second term "blows up" and the transform does not exist]. Thus,

$$e^{at}u_{-1}(t) \leftrightarrow \frac{1}{s-a}$$

Note that this is also the Laplace transform of e^{at}, since $e^{at} = e^{at}u_{-1}(t)$ for $t \geq 0$. If $a = 0$, then $e^{at}u_{-1}(t) = u_{-1}(t)$, and therefore

$$u_{-1}(t) \leftrightarrow \frac{1}{s}$$

10.3.2 Transform Theorems

Certain theorems prove useful in the manipulation of transforms. A limited number of them are now given.

THE LINEARITY THEOREM

If $f_1(t)$, $f_2(t)$ are time functions and a_1, a_2 are constants, then

$$a_1 f_1(t) + a_2 f_2(t) \leftrightarrow a_1 F_1(s) + a_2 F_2(s) \tag{10.23}$$

It is straightforward to verify this result by inserting the left-hand side of Eq. (10.23) into Eq. (10.21).

THE REAL DIFFERENTIATION THEOREM

The Laplace transform of df/dt is

$$\mathcal{L}\frac{df}{dt} = sF(s) - f(0-) \tag{10.24}$$

To show this, write

$$\mathcal{L}\frac{df}{dt} = \int_{0-}^{\infty} \frac{df}{dt} e^{-st}\, dt$$

and integrate by parts to yield

$$\mathcal{L}\frac{df}{dt} = f(t)e^{-st}\Big|_{0-}^{\infty} + s\int_{0-}^{\infty} f(t)e^{-st}\, dt$$

$$= -f(0-) + sF(s)$$

for $\mathrm{Re}(s)$ greater than some real number σ_o.

The extension to higher derivatives is

$$\mathcal{L}[d^n f/dt^n] = s^n F(s) - s^{n-1}f(0-) - s^{n-2}\left[\frac{df}{dt}(0-)\right] - \cdots - \left[\frac{d^{n-1}f}{dt^{n-1}}(0-)\right] \tag{10.25}$$

THE REAL INTEGRATION THEOREM

The Laplace transform of $\int_{-\infty}^{t} f(\tau)\, d\tau$ is

$$\mathcal{L}\left[\int_{-\infty}^{t} f(\tau)\, d\tau\right] = \frac{F(s)}{s} + \frac{\displaystyle\int_{-\infty}^{0-} f(\tau)\, d\tau}{s} \tag{10.26}$$

To prove this, use the real differentiation theorem to write

$$\mathcal{L}\left[\frac{d}{dt}\int_{-\infty}^{t} f(\tau)\, d\tau\right] = s\mathcal{L}\left[\int_{-\infty}^{t} f(\tau)\, d\tau\right] - \int_{-\infty}^{0-} f(\tau)\, d\tau$$

Since $\dfrac{d}{dt}\displaystyle\int_{-\infty}^{t} f(\tau)\, d\tau = f(t)$, Eq. (10.26) follows.

THE COMPLEX DIFFERENTIATION THEOREM

If $f(t)$ and $F(s)$ are a transform pair, then

$$\frac{d^k F(s)}{ds^k} = (-1)^k \mathcal{L}[t^k f(t)] \tag{10.27}$$

This is established by successively differentiating Eq. (10.21), as follows:

$$\frac{dF}{ds} = \int_{0-}^{\infty} f(t)(-t)e^{-st}\, dt$$

$$\frac{d^2 F}{ds^2} = \int_{0-}^{\infty} f(t)(-t)^2 e^{-st}\, dt$$

$$\cdot\ \cdot\ \cdot\ \cdot\ \cdot\ \cdot\ \cdot\ \cdot\ \cdot\ \cdot\ \cdot\ \cdot\ \cdot\ \cdot$$

$$\frac{d^k F}{ds^k} = (-1)^k \int_{0-}^{\infty} t^k f(t)e^{-st}\, dt$$

$$= (-1)^k \mathcal{L}[t^k f(t)]$$

THE TIME DELAY THEOREM

The Laplace transform of $f(t - t_d)u_{-1}(t - t_d)$, $t_d > 0$, is

$$\mathcal{L}[f(t - t_d)u_{-1}(t - t_d)] = F(s)e^{-st_d} \tag{10.28}$$

To show this, write

$$\mathcal{L}[f(t - t_d)u_{-1}(t - t_d)] = \int_{0-}^{\infty} f(t - t_d)u_{-1}(t - t_d)e^{-st}\, dt$$

$$= \int_{t_d-}^{\infty} f(t - t_d)e^{-s(t-t_d)}e^{-st_d}\, dt$$

Since e^{-st_d} is not a function of t,

$$\mathcal{L}[f(t - t_d)u_{-1}(t - t_d)] = e^{-st_d} \int_{t_d-}^{\infty} f(t - t_d)e^{-s(t-t_d)}\, dt$$

Now, let $t' = t - t_d$. Then,

$$\mathcal{L}[f(t - t_d)u_{-1}(t - t_d)] = e^{-st_d} \int_{0-}^{\infty} f(t')e^{-st'}\, dt'$$

$$= e^{-st_d} F(s)$$

EXAMPLE 10.6

Calculate the transform of $\cos \omega t$.

Solution

Since $\cos \omega t = e^{j\omega t}/2 + e^{-j\omega t}/2$, then by the linearity theorem and the result of

Example 10.5,

$$\mathcal{L}[\cos \omega t] = \frac{1}{2(s - j\omega)} + \frac{1}{2(s + j\omega)}$$

$$= \frac{s}{s^2 + \omega^2}$$

EXAMPLE 10.7

Calculate the transform of the unit impulse $u_0(t)$.

Solution
Since $u_0(t) = du_{-1}(t)/dt$, the real differentiation theorem yields

$$\mathcal{L}[u_0(t)] = s\mathcal{L}[u_{-1}(t)] - u_{-1}(0-)$$

$$= s\frac{1}{s} = 1$$

It is worth pointing out that the function $f(t) = 1$ has the same transform as the unit step. However, the transform of its derivative is

$$\mathcal{L}\left[\frac{df}{dt}\right] = s\frac{1}{s} - f(0-)$$

$$= 1 - 1 = 0$$

which is the correct answer since $df/dt = 0$.

EXAMPLE 10.8

Calculate the transform of $tu_{-1}(t)$.

Solution
Since $tu_{-1}(t) = \int_{-\infty}^{t} u_{-1}(\tau)\, d\tau$, the real integration theorem yields

$$\mathcal{L}[tu_{-1}(t)] = \frac{\mathcal{L}[u_{-1}(t)]}{s} + \frac{\int_{-\infty}^{0-} u_{-1}(\tau)\, d\tau}{s}$$

$$= \frac{1}{s^2}$$

EXAMPLE 10.9

Calculate the transform of the rectangular pulse of Fig. E10.9.

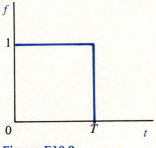

Figure E10.9

Solution
Since $f(t) = u_{-1}(t) - u_{-1}(t - T)$, the application of the delay theorem results in

$$\mathcal{L}[f(t)] = \frac{1}{s} - \frac{1}{s}e^{-sT} = \frac{1 - e^{-sT}}{s}$$

EXAMPLE 10.10

Calculate the transform of $t^k e^{at}$.

Solution
By the complex differentiation theorem,

$$\mathcal{L}[t^k e^{at}] = (-1)^k \frac{d^k[1/(s - a)]}{ds^k}$$

Now,

$$\frac{d(s - a)^{-1}}{ds} = -1(s - a)^{-2}$$

$$\frac{d^2(s - a)^{-1}}{ds^2} = 2(s - a)^{-3}$$

$$\frac{d^3(s - a)^{-1}}{ds^3} = -3[2(s - a)^{-4}]$$

Table 10.1 LAPLACE TRANSFORMS

$f(t)$	$F(s)$
$u_0(t)$	1
$u_{-1}(t)$	$\dfrac{1}{s}$
e^{at}	$\dfrac{1}{s-a}$
$\dfrac{t^k}{k!}e^{at}$	$\dfrac{1}{(s-a)^{k+1}}$
$\cos \omega t$	$\dfrac{s}{s^2+\omega^2}$
$\sin \omega t$	$\dfrac{\omega}{s^2+\omega^2}$
$Ae^{-at}\cos(\omega t+\theta)$	$\dfrac{A[(s+a)\cos\theta-\omega\sin\theta]}{(s+a)^2+\omega^2}$
$Ae^{-at}\sin(\omega t+\theta)$	$\dfrac{A[(s+a)\sin\theta-\omega\cos\theta]}{(s+a)^2+\omega^2}$

. . . .

$$\frac{d^k(s-a)^{-1}}{ds^k}=(-1)^k k(k-1)\cdots 2(s-a)^{-(k+1)}$$

Therefore,

$$\mathscr{L}[t^k e^{at}]=\frac{k!}{(s-a)^{k+1}}$$

The theorems are used to construct the table of Laplace transforms (Table 10.1).

DRILL EXERCISES

10.1 Derive the last two entries in Table 10.1.

10.2 Compute the Laplace transform of $\sin \omega t$, using the transform of $\cos \omega t$ and the differentiation theorem.

10.3 Using Laplace transforms, show that

$$\frac{d}{dt}[\cos\omega t\, u_{-1}(t)]=u_0(t)-\omega\sin\omega t \qquad \text{for } t\geq 0$$

10.4 Calculate the Laplace transform of the waveform $f(t)$ of Fig. D10.4.

Ans. $F(s) = \dfrac{A(1 - e^{-sT})^2}{s}$

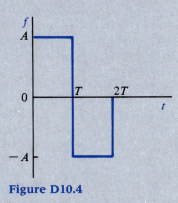

Figure D10.4

10.5 Calculate the Laplace transform of $t \cos \omega t$.

Ans. $F(s) = \dfrac{s^2 - \omega^2}{(s^2 + \omega^2)^2}$

10.3.3 Inversion of Laplace Transforms

The operation whereby the time function is recovered from the Laplace transform is called *inversion*. The inversion formula is

$$f(t) = \frac{1}{2\pi j} \int_{\sigma - j\infty}^{\sigma + j\infty} F(s)e^{st} \, ds \qquad (10.29)$$

This is a line integral in the complex plane, along a vertical line in the region for which the integral of Eq. (10.21) exists. The inversion integral is not used if $F(s)$ is a rational function of s (or a rational function multiplied by e^{-st_d}). In such cases a *partial fraction expansion* is used to reduce $F(s)$ to a sum of terms of the form of those in Table 10.1. Linearity is then invoked to obtain $f(t)$. A rational function $F(s)$ is expressed as

$$F(s) = \frac{b_m s^m + b_{m-1} s^{m-1} + \cdots + b_1 s + b_0}{s^n + a_{n-1} s^{n-1} + \cdots + a_1 s + a_0} \qquad (10.30)$$

$$= b_m \frac{(s - z_1)(s - z_2) \cdots (s - z_m)}{(s - p_1)(s - p_2) \cdots (s - p_n)} \qquad (10.31)$$

The complex numbers $z_1, z_2, \ldots, z_m$ are the roots of the numerator polynomial and are known as the *zeros* of $F(s)$. The roots $p_1, p_2, \ldots, p_n$ of the denominator are called the *poles* of $F(s)$. If the coefficients in Eq. (10.30) are real, then the poles and zeros are either real or occur in complex conjugate pairs.

For the time being, $F(s)$ is assumed to be *strictly proper*, which means that there are more poles than zeros. Two cases must be considered, depending on whether or not the poles are distinct.

THE CASE OF DISTINCT ROOTS
As an example, consider

$$F(s) = \frac{P(s)}{(s - s_1)(s - s_2)(s - s_3)}$$

where $P(s)$ is a polynomial of order 2 or less and where s_1, s_2, and s_3 are distinct.

It is known that $F(s)$ can be written as

$$F(s) = \frac{A_1}{(s - s_1)} + \frac{A_2}{(s - s_2)} + \frac{A_3}{(s - s_3)} \tag{10.32}$$

where the right-hand side is known as the partial-fraction expansion of $F(s)$.

One may evaluate A_1, A_2, and A_3 by putting the right-hand side sum over one common denominator and equating the resulting numerator to $P(s)$. Another method is suggested by the observation that

$$(s - s_1)F(s) = A_1 + \frac{(s - s_1)A_2}{s - s_2} + \frac{(s - s_1)A_3}{s - s_3}$$

Since s_2, $s_3 \neq s_1$, it is seen that

$$(s - s_1)F(s)\big|_{s=s_1} = A_1 \tag{10.33}$$

with similar formulas for A_2 and A_3.

EXAMPLE 10.11

Invert the transform $F(s) = (s^2 + 1)/[s(s^2 + 2s + 2)]$.

Solution
Factoring out the denominator yields

$$F(s) = \frac{s^2 + 1}{s(s + 1 + j)(s + 1 - j)}$$

$$= \frac{A_1}{s} + \frac{A_2}{s + 1 + j} + \frac{A_3}{(s + 1 - j)}$$

Applying Eq. (10.33),

$$A_1 = sF(s)\big|_{s=0} = \tfrac{1}{2}$$

$$A_2 = (s + 1 + j)F(s)\big|_{s=-1-j}$$

$$= \frac{(-1-j)^2 + 1}{(-1-j)(-1-j+1-j)}$$

$$= \frac{1 + 2j}{2j(1+j)}$$

$$= 0.791 \;\angle{-71.6°}$$

$$A_3 = (s + 1 - j)F(s)\big|_{s=-1+j} = A_2^*$$

$$= 0.791 \;\angle{71.6°}$$

From Table 10.1 and linearity,

$$f(t) = \tfrac{1}{2} + 0.791e^{-j71.6°}e^{(-1-j)t} + 0.791e^{j71.6°}e^{(-1+j)t}$$

$$= \tfrac{1}{2} + 2\,\mathrm{Re}(0.791e^{j71.6°}e^{(-1+j)t})$$

$$= \tfrac{1}{2} + 2\,\mathrm{Re}(0.791e^{-t}e^{j(t+71.6°)})$$

$$= \tfrac{1}{2} + 1.58e^{-t}\cos(t + 71.6°) \qquad t \geq 0$$

THE CASE OF MULTIPLE ROOTS

If a root s_1 of the denominator of $F(s)$ has multiplicity p, then its contribution to the partial fraction expansion is

$$\frac{A_{11}}{(s-s_1)^p} + \frac{A_{12}}{(s-s_1)^{p-1}} + \cdots + \frac{A_{1p}}{s-s_1}$$

The coefficients are given by

$$A_{11} = (s-s_1)^p F(s)\big|_{s=s_1}$$

$$A_{12} = \frac{d}{ds}[s-s_1)^p F(s)]\big|_{s=s_1}$$

$$A_{13} = \frac{1}{2!}\frac{d^2}{ds^2}[(s-s_1)^p F(s)]\big|_{s=s_1} \qquad\qquad (10.34)$$

$$A_{1p} = \frac{1}{(p-1)!}\frac{d^{p-1}}{ds^{p-1}}[(s-s_1)^p F(s)]\big|_{s=s_1}$$

EXAMPLE 10.12

Invert the transform $F(s) = 1/[s(s+1)^3]$.

Solution

The partial fraction expansion is

$$F(s) = \frac{A_1}{s} + \frac{A_{21}}{(s+1)^3} + \frac{A_{22}}{(s+1)^2} + \frac{A_{23}}{s+1}$$

The coefficients are given by

$$A_1 = sF(s)|_{s=0} = 1$$

$$A_{21} = (s + 1)^3 F(s)|_{s=-1} = -1$$

$$A_{22} = \frac{d}{ds}[(s + 1)^3 F(s)]|_{s=-1}$$

$$= \frac{d}{ds}\frac{1}{s}\bigg|_{s=-1} = -1$$

$$A_{23} = \frac{1}{2}\frac{d^2}{ds^2}\frac{1}{s}\bigg|_{s=-1} = -1$$

Thus, $F(s)$ is given by

$$F(s) = \frac{1}{s} - \frac{1}{(s + 1)^3} - \frac{1}{(s + 1)^2} - \frac{1}{s + 1}$$

and, using Table 10.1,

$$f(t) = 1 - \frac{t^2}{2}e^{-t} - te^{-t} - e^{-t} \qquad t \geq 0$$

THE CASE OF $F(s)$ NOT STRICTLY PROPER

Up to this point, $F(s)$ has been assumed strictly proper. If the order of the denominator is not greater than that of the numerator, long division is required as a preliminary step, to obtain a polynomial in s plus a strictly proper rational function. This is illustrated in the following example.

EXAMPLE 10.13

Invert the transform $F(s) = (s^2 + s + 1)/[s(s + 2)]$.

Solution

By long division with remainder,

$$F(s) = 1 + \frac{-s + 1}{s(s + 2)} = 1 + \frac{\frac{1}{2}}{s} - \frac{\frac{3}{2}}{s + 2}$$

Using Table 10.1,

$$f(t) = u_0(t) + \tfrac{1}{2} - \tfrac{3}{2}e^{-2t} \qquad t \geq 0$$

FUNCTIONS WITH DELAYS

Partial fraction expansion and the delay theorem are combined to invert terms of the form $e^{-st_d}F(s)$, as shown in the following example.

EXAMPLE 10.14

Invert the transform $F(s) = (1 - e^{-2s})/[s(s + 1)]$.

Solution

By partial fraction expansion,

$$F'(s) = \frac{1}{s(s + 1)} = \frac{1}{s} - \frac{1}{(s + 1)}$$

and the inverse of $F'(s)$ is

$$f'(t) = 1 - e^{-t} \qquad t \geq 0$$

By the delay theorem, the inverse transform of $e^{-2s}F'(s)$ is $f'(t - 2)u_{-1}(t - 2)$. Therefore,

$$f(t) = f'(t) - f'(t - 2)u_{-1}(t - 2)$$
$$= 1 - e^{-t} - (1 - e^{-(t-2)})u_{-1}(t - 2) \qquad t \geq 0$$

This can also be expressed as

$$f(t) = \begin{cases} 1 - e^{-t} & 0 \leq t < 2 \\ (e^2 - 1)e^{-t} & t > 2 \end{cases}$$

DRILL EXERCISES

10.6   Invert $F(s) = 1/(s + 1)(s + 2)$.
   Ans. $f(t) = e^{-t} - e^{-2t} \qquad t \geq 0$

10.7   Invert $F(s) = (2s + 1)/[(s^2 + s + 1)(s^2 + 2s + 2)]$.
   Ans. $f(t) = 2e^{-t/2}\cos[(\frac{3}{4})^{1/2}t - 60°]$
   $-\sqrt{5}e^{-t}\cos(t - 63°) \qquad t \geq 0$

10.8   Invert $F(s) = 1/s^2(s + 2)^2$.
   Ans. $f(t) = (t - 1 + te^{-2t} + e^{-2t})/4 \qquad t \geq 0$

10.9   Invert $F(s) = (s + 1)/(s + 2)$.
   Ans. $f(t) = u_0(t) - e^{-2t} \qquad t \geq 0$

10.10 Invert $F(s) = (1 - e^{-s})/s^2$.
   Ans. $f(t) = t - (t - 1)u_{-1}(t - 1) \qquad t \geq 0$

10.3.4 Solution of Differential Equations by Laplace Transforms

The principal use of Laplace transforms is in the solution of linear, constant-coefficient differential equations. Consider, as an example, the second-order equation

$$\ddot{y} + a_1\dot{y} + a_0 y = b_1\dot{x} + b_0 x \tag{10.35}$$

The dot notation introduced here is commonly used to denote differentiation with respect to time. For example, $\ddot{y} = d^2y/dt^2$. Taking Laplace transforms and using the real differentiation theorem yields

$$s^2 Y(s) - sy(0-) - \dot{y}(0-) + a_1[sY(s) - y(0-) + a_0 Y(s)$$

$$= b_1[sX(s) - x(0-)] + b_0 X(s)$$

or

$$(s^2 + a_1 s + a_0)Y(s) = (b_1 s + b_0)X(s) + sy(0-) + \dot{y}(0-) + a_1 y(0-) - b_1 x(0-)$$

Solving for $Y(s)$, one obtains

$$Y(s) = \frac{b_1 s + b_0}{s^2 + a_1 s + a_0}X(s) + \frac{sy(0-) + \dot{y}(0-) + a_1 y(0-) - b_1 x(0-)}{s^2 + a_1 s + a_0} \tag{10.36}$$

The first term on the right-hand side of Eq. (10.36) is the transform of the zero-state solution obtained when all initial conditions at $t = 0-$ are zero. The second term is the transform of the solution obtained when $x(t) = 0$ [$X(s) = 0$], i.e., the zero-input solution. Invoking the linearity theorem, the solution is seen to be the sum of the zero-state and zero-input solutions.

The function multiplying $X(s)$ in Eq. (10.36) is seen to be the system function $H(s)$. Therefore, the transform of the zero-state solution $y_{zs}(t)$ is

$$Y_{zs}(s) = H(s)X(s) \tag{10.37}$$

Equation (10.37) provides yet another interpretation of $H(s)$: it is the ratio of the transforms of the zero-state solution and of the input.

Since the denominator of $H(s)$ is the characteristic polynomial of the differential equation, the poles of the system function are the natural frequencies. The break-down of a solution into zero-input and zero-state components is not the same as the split into homogeneous and particular solutions. The two are related as follows.

If $x(t) = Ae^{s_p t}$ for $t \geq 0$, then

$$X(s) = \frac{A}{s - s_p}$$

and

$$Y_{zs}(s) = H(s)\frac{A}{s - s_p}$$

Assuming p_1, p_2 to be the poles of $H(s)$ in Eq. (10.36), and assuming $s_p \neq p_1$, p_2, the partial-fraction expansion yields

$$Y_{zs}(s) = \frac{AH(s_p)}{s - s_p} + \frac{B_1}{s - p_1} + \frac{B_2}{s - p_2}$$

and thus

$$y_{zs}(t) = AH(s_p)e^{s_p t} + B_1 e^{p_1 t} + B_2 e^{p_2 t} \tag{10.38}$$

The zero-state solution is seen to contain the particular solution, as well as some terms belonging to the homogeneous solution, since p_1, p_2 are the natural frequencies.

From Eq. (10.36) the transform of the zero-input solution is

$$Y_{zi}(s) = \frac{sy(0-) + \dot{y}(0-) + a_1 y(0-) - b_1 x(0-)}{s^2 + a_1 s + a_0}$$

The solution $y_{zi}(t)$ Eq. (10.36), has only p_1 and p_2 as its poles, which leads to terms of the form $e^{p_1 t}$ and $e^{p_2 t}$ in $y_{zi}(t)$. Therefore, $y_{zi}(t)$ contributes only to the homogeneous solution. Note that the matching of initial conditions is taken into account automatically if the Laplace transform is used.

EXAMPLE 10.15

Solve the differential equation

$$\ddot{y} + 3\dot{y} + 2y = \dot{x} - x$$

given that $x(t) = u_{-1}(t)$ and $\dot{y}(0-) = 1$, $y(0-) = 0$. Identify the zero state and zero-input solutions, and also the particular and homogeneous solutions.

Solution
Transforming each side yields

$$s^2 Y(s) - sy(0-) - \dot{y}(0-) + 3sY(s) - 3y(0-) + 2Y(s) = sX(s) - x(0-) - X(s)$$

Since $x(t) = u_{-1}(t)$, $x(0-) = 0$, so that

$$(s^2 + 3s + 2)Y(s) = (s - 1)X(s) + 1$$

or

$$Y(s) = (s - 1)\frac{X(s)}{s^2 + 3s + 2} + \frac{1}{s^2 + 3s + 2}$$

Since $X(s) = 1/s$, the transform of the zero-state solution is

$$Y_{zs}(s) = \frac{s - 1}{s(s + 1)(s + 2)}$$

$$= \frac{-\frac{1}{2}}{s} + \frac{2}{s + 1} - \frac{\frac{3}{2}}{s + 2}$$

and, therefore,

$$y_{zs}(t) = -\frac{1}{2} + 2e^{-t} - \frac{3e^{-2t}}{2} \qquad t \geq 0$$

The transform of the zero-input solution is

$$Y_{zi}(s) = \frac{1}{(s + 1)(s + 2)} = \frac{1}{s + 1} - \frac{1}{s + 2}$$

so that

$$y_{zi}(t) = e^{-t} - e^{-2t} \qquad t \geq 0$$

The full solution is

$$y(t) = y_{zs}(t) + y_{zi}(t) = -\frac{1}{2} + 3e^{-t} - \frac{5e^{-2t}}{2} \qquad t \geq 0$$

The particular solution is that term which corresponds to e^{0t}, i.e.,

$$y_p(t) = -\tfrac{1}{2} \qquad t \geq 0$$

The homogeneous solution is

$$y_h(t) = 3e^{-t} - \frac{5e^{-2t}}{2} \qquad t \geq 0$$

DRILL EXERCISES

10.11 Obtain the zero-input solution of Eq. (10.1), using the Laplace transform. Use the element values in Example 10.1, with $R = 1$ kΩ, and the initial condition $i(0-) = 3$ mA, $di/dt(0-) = 0$, $v_l(0-) = 0$.
Ans. $i(t) = 3.098e^{-.125t}\cos(.484t - 14.48°)$

10.12 Calculate the zero-state solution for the equation of Exercise 10.11, with $v_l(t) = u_{-1}(t)$.
Ans. $i(t) = 0.516e^{-0.125t}\sin 0.484t$

10.4   THE TRANSFORM NETWORK

10.4.1 The Kirchhoff Laws for Transforms

Corresponding to a network composed of sources and linear elements there is a *transform network* which has the same graph, but where the branch voltages and

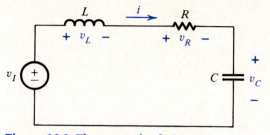

Figure 10.3 The network of Fig. 10.1.

currents are replaced by their transforms. The transform network is solved by algebraic steps involving transforms: the time response is obtained as the final step, by an inversion operation.

Consider, for example, the network of Fig. 10.1, redrawn for convenience as Fig. 10.3.

By KVL,

$$v_L(t) + v_R(t) + v_C(t) = v_I(t) \tag{10.39}$$

Taking Laplace transforms on each side results in

$$V_L(s) + V_R(s) + V_C(s) = V_I(s) \tag{10.40}$$

Thus, it is seen that the transforms of the voltages satisfy the same KVL equation as the voltages themselves. This will be true for all KVL equations, and a similar statement holds for the KCL equations.

The Kirchhoff-law constraints between the transforms are the interconnection laws of a network identical to the original one, but where the branch voltages and currents are replaced by their transforms. This new network, shown in Fig. 10.4, is the transform network.

10.4.2 The Branch Relationships

There remains to define the branch relationships in this new transform network, i.e., the equations relating the voltage and current transforms in each branch.

For a resistance,

$$v_R(t) = Ri_R(t)$$

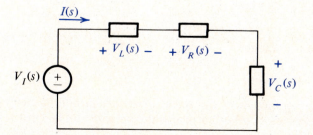

Figure 10.4 The transform network corresponding to the network of Fig. 10.3.

Taking transforms yields

$$V_R(s) = RI_R(s) \tag{10.41}$$

For an inductance,

$$v_L(t) = L\frac{di_L(t)}{dt}$$

For the time being, zero initial conditions are assumed at $t = 0-$. Taking transforms yields

$$V_L(s) = sLI_L(s) \tag{10.42}$$

Similarly, for a capacitance,

$$I_C(s) = sCV_C(s) \tag{10.43}$$

In each case, the voltage and current transforms are related by a linear, algebraic equation similar to Ohm's law. A ratio of voltage to current transforms is called an *impedance*, denoted by $Z(s)$ and expressed in ohms. Thus, for an R, L, or C,

$$V(s) = Z(s)I(s) \tag{10.44}$$

where

$Z(s) = R$ for a resistance

$Z(s) = sL$ for an inductance

$Z(s) = \dfrac{1}{sC}$ for a capacitance

If the voltage and current refer to the same port, as is the case here, the impedance is called a *driving-point* impedance; otherwise, it is known as a *trans-impedance*.

A ratio of current to voltage transforms is called an *admittance*, denoted by $Y(s)$ and expressed in siemens. Thus,

$$I(s) = Y(s)V(s) \tag{10.45}$$

where

$Y(s) = \dfrac{1}{R} = G$ for a resistance

$Y(s) = \dfrac{1}{sL}$ for an inductance

$Y(s) = sC$ for a capacitance

As in the case of impedance, the terms *driving-point* admittance and *trans-admittance* are used to denote whether or not the voltage and current refer to the same port.

Ratios of current transforms or voltage transforms are called *transfer ratios*, and are dimensionless.

Driving-point impedances or admittances have the special property that their inverses are also system functions. If a port is connected to a current source $i_i(t)$, the port voltage is calculated from $V(s) = Z(s)I_i(s)$, where $Z(s)$ is the driving-point impedance; conversely, if a voltage source $v_i(t)$ is used, the transform of the current is $I(s) = [1/Z(s)]V_i(s)$. This property does not apply to other system functions. To give a simple example, if an input $v_1(t)$ applied to port 1 of a voltage divider produces an output $v_2(t) = [R_1/(R_1 + R_2)]v_1(t)$, at port 2, it is *not* true that excitation $v_2(t)$ at port 2 produces a voltage $[(R_1 + R_2)/R_1]v_2(t)$ at port 1.

In the transform network, the branch relationship for an R, L, or C is given by Ohm's law, but with resistance replaced by impedance (conductance by admittance). A controlled source need no longer relate the controlling and controlled variables by a constant. For example, if

$$\frac{di(t)}{dt} + ai(t) = \beta i_c(t)$$

then the controlled-source relationship is written as

$$I(s) = \frac{\beta}{s + a} I_c(s) \tag{10.46}$$

in the transform network. Thus, transfer ratios, transimpedances or transadmittances are used to write branch laws for controlled sources.

The transform network corresponding to the network of Fig. 10.3 is shown in Fig. 10.5.

In summary, the transform network is constructed as follows:

1. Replace all voltages and currents, including independent sources, by their transforms.

2. Replace all inductances and capacitances by their impedances (or admittances).

3. If a controlled source is described by a differential equation between the controlled variables and the controlling variable, express the transform of the controlled variable as a system function times the transform of the controlling variable.

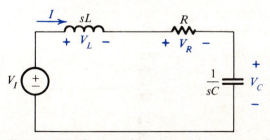

Figure 10.5 The transform network corresponding to Fig. 10.3, with impedance labeling.

Note that networks with nonlinear elements, such as diodes, are excluded from this development because such networks are not described by linear differential equations.

10.4.3 Calculation of the System Function

The transform network is described by two sets of equations: a set of KCL and KVL equations, and a set of branch laws. Because the branches are described either by Ohm's law (with impedance replacing resistance) or by the relationships describing independent or controlled sources, the mathematical structure of this problem is the same as that of resistive networks. The only differences are that (1) the branch variables are complex and (2) the ohmic relationships are functions of s, rather than real constants.

The mathematical similarity of the two problems allows a straightforward transfer of results from the resistive circuits to the present case, simply by replacing resistance by impedance. For example:

1. Mesh and nodal analysis are valid.

2. Superposition holds.

3. All equivalent networks of Chap. 3 can be used.

4. Linear two-port theory is applicable.

EXAMPLE 10.16

Calculate the driving point impedance at terminals *ab* for the network of Fig. E10.16(a). Obtain the differential equation relating $i(t)$ to $v_i(t)$.

Solution

The transform network is shown in Fig. E10.16(b). Using the series rule, the

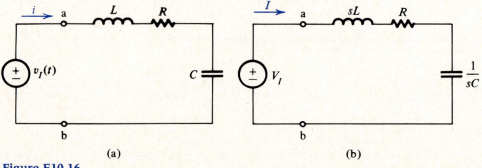

(a)   (b)

Figure E10.16

driving point impedance at *ab* is

$$\frac{V_I}{I} = Z(s) = sL + R + \frac{1}{sC}$$

$$= \frac{L[s^2 + (R/L)s + 1/LC]}{s}$$

Since *I* is the output for this problem, the system function is

$$\frac{I}{V_I} = \frac{s/L}{s^2 + (R/L)s + 1/LC}$$

and the differential equation is, by inspection,

$$\frac{d^2i}{dt^2} + \frac{R}{L}\frac{di}{dt} + \frac{1}{LC}i = \frac{1}{L}\frac{dv_I}{dt}$$

EXAMPLE 10.17

Find the Thevenin equivalent for the network of Fig. E10.17(a) to the left of *ab* and calculate the system function V_O/V_S.

Solution

With *ab* open-circuited as in Fig. E10.17(b), the voltage divider rule yields

$$\frac{V_t}{V_S} = \frac{1/sC}{R_1 + 1/sC} = \frac{1}{sCR_1 + 1}$$

With $V_S = 0$, the impedance looking into *ab* is the parallel combination of R_1 and $1/sC$, i.e.,

$$Z_{eq}(s) = [R_1/sC]/[R_1 + 1/sC] = R_1/[sCR_1 + 1]$$

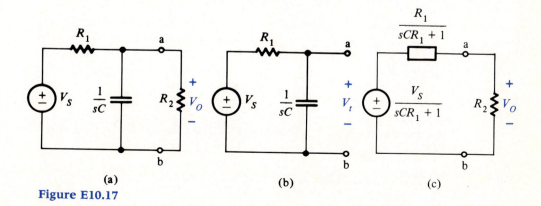

(a) (b) (c)

Figure E10.17

The Thevenin equivalent is shown in Fig. E.10.17(c). From Fig. E10.17(c), by the voltage divider rule,

$$V_O = \frac{R_2}{R_2 + R_1/(sCR_1 + 1)} \frac{V_S}{sCR_1 + 1}$$

or

$$\frac{V_O}{V_S} = \frac{R_2}{sCR_1 R_2 + R_1 + R_2}$$

EXAMPLE 10.18

The combination of R_2 and C_2 in Fig. E10.18(a) is an equivalent circuit of the input port of an oscilloscope. The combination of R_1 and C_1 represents a probe, which is to be designed so as to make $v_O(t)$ as close in shape to $v_S(t)$ as possible. Given zero initial conditions at $t = 0-$ and $v_S(t) = V_S u_{-1}(t)$, calculate $v_O(t)$. For what values of the probe parameters R_1 and C_1 is $v_O(t)$ a step?

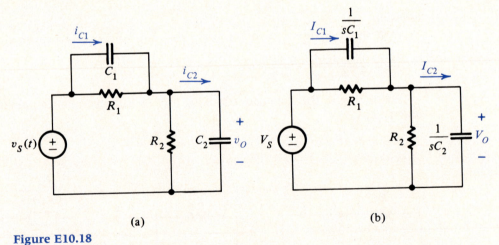

(a)   (b)

Figure E10.18

Solution
The transform network is shown in Fig. E10.18(b). By the voltage divider rule, the transfer function V_O/V_S is

$$\frac{V_O}{V_S} = \frac{1/(G_2 + sC_2)}{1/(G_2 + sC_2) + 1/(G_1 + sC_1)}$$

$$= \frac{G_1 + sC_1}{G_1 + G_2 + s(C_1 + C_2)}$$

$$= \frac{C_1}{C_1 + C_2} \frac{s + 1/R_1 C_1}{s + 1/\tau}$$

where $\tau = (C_1 + C_2)R_1 R_2/(R_1 + R_2)$.

If $v_S(t) = u_{-1}(t)$, then

$$V_0(s) = \frac{C_1}{C_1 + C_2} \frac{s + 1/R_1 C_1}{s + 1/\tau} \frac{1}{s}$$

$$= \frac{R_2/(R_1 + R_2)}{s} + \left(\frac{C_1}{C_1 + C_2} - \frac{R_2}{R_1 + R_2} \right) \frac{1}{s + 1/\tau}$$

and the time response is

$$v_0(t) = \frac{R_2}{R_1 + R_2} + \left(\frac{C_1}{C_1 + C_2} - \frac{R_2}{R_1 + R_2} \right) e^{-t/\tau} \qquad t \geq 0$$

If

$$\frac{C_1}{C_1 + C_2} = \frac{R_2}{R_1 + R_2}$$

then

$$v_0(t) = \frac{R_2}{R_1 + R_2} \qquad t \geq 0$$

or

$$v_0(t) = \frac{R_2}{R_1 + R_2} u_{-1}(t)$$

since the network was initially at rest. The condition is rewritten as

$$\frac{C_1 + C_2}{C_1} = \frac{R_1 + R_2}{R_2}$$

or

$$R_1 C_1 = R_2 C_2$$

Figure E10.18(c) shows $v_0(t)$ for the three cases $R_1 C_1 < R_2 C_2$, $R_1 C_1 = R_2 C_2$, $R_1 C_1 > R_2 C_2$.

For $R_1 C_1 = R_2 C_2$, it is easy to show that $\tau = R_1 C_1$, so that for *any* input,

$$V_0(s) = \frac{C_1}{C_1 + C_2} V_S(s)$$

Thus, the output is seen to be a replica of the input not only for a step, but for any input as well.

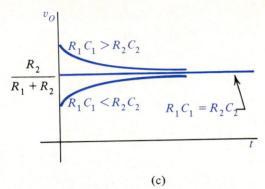

(c)

Figure E10.18 (continued)

The fact that $v_o(t)$ is discontinuous means that $i_{C2}(t)$ contains an impulse. For a unit step input and with $R_1 C_1 = R_2 C_2$,

$$i_{C2}(t) = C_2 \frac{dv_o}{dt} = C_2 \frac{R_2}{R_1 + R_2} \frac{du_{-1}(t)}{dt}$$

$$= \frac{C_1 C_2}{C_1 + C_2} u_0(t)$$

EXAMPLE 10.19

The circuit of Fig. E10.19(a) is a small-signal model of a common-emitter BJT amplifier, where capacitances are included to improve the BJT model. Calculate the voltage gain V_o/V_s, and the zero-state response to a unit step.

Solution

It can be shown, using the methods of Chap. 1, that the units mA, V, mS, pF, and ns are consistent. The transform diagram is redrawn, in terms of admittance units, in Fig. E10.19(b). The analysis is simplified by replacing the network to the left of ab by its Norton equivalent, as shown in Fig. E10.19(c).

Applying nodal analysis yields

$$\begin{bmatrix} 12 + 103s & -3s \\ -3s & 1 + 3s \end{bmatrix} \begin{bmatrix} V_b \\ V_o \end{bmatrix} = \begin{bmatrix} 10V_s \\ -40V_b \end{bmatrix}$$

or

$$\begin{bmatrix} 12 + 103s & -3s \\ 40 - 3s & 1 + 3s \end{bmatrix} \begin{bmatrix} V_b \\ V_o \end{bmatrix} = \begin{bmatrix} 10V_s \\ 0 \end{bmatrix}$$

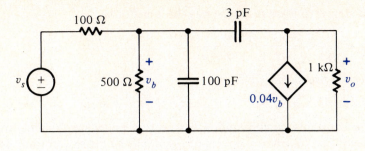

(a)

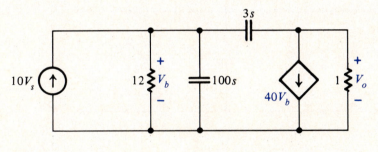

(b)

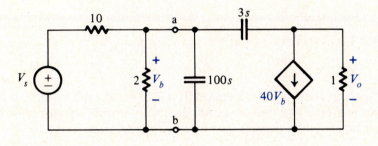

(c)

Figure E10.19

By Cramer's rule,

$$\frac{V_o}{V_s} = \frac{10(3s - 40)}{300s^2 + 259s + 12}$$

or

$$\frac{V_o}{V_s} = \frac{0.100(s - 13.3)}{s^2 + 0.863s + 0.04001}$$

The unit step zero-state response has the transform

$$V_o(s) = \frac{0.100(s - 13.3)}{s(s + 0.0492)(s + 0.814)}$$

Inverting the transform yields

$$v_o(t) = -33.3 + 35.5e^{-0.0492t} - 2.26e^{-0.814t} \qquad t \geq 0$$

The time constants corresponding to the two poles are 20.3 ns and 1.23 ns, respectively. The transient is governed essentially by the slower time constant, and the steady state is reached, for practical purposes, in about 100 ns (5 time constants).

EXAMPLE 10.20

The transistor model used in Example 10.19 is reproduced in Fig. E10.20(a). Find the y-parameter description, and calculate the input admittance when the transistor is terminated at its output port by a 1-kΩ resistance.

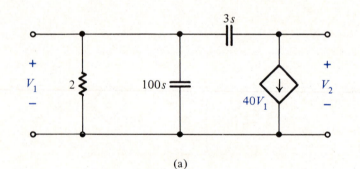

(a)

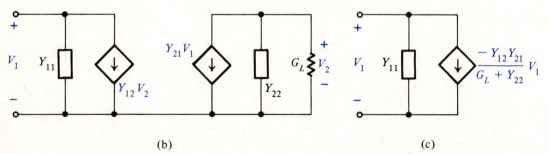

(b)   (c)

Figure E10.20

Solution

By inspection, using mA, V, mS, pF, and ns,

$$Y_{11}(s) = \left. \frac{I_1}{V_1} \right|_{V_2=0} = 2 + 103s$$

$$Y_{21}(s) = \left. \frac{I_2}{V_1} \right|_{V_2=0} = -3s + 40$$

$$Y_{12}(s) = \left. \frac{I_1}{V_2} \right|_{V_1=0} = -3s$$

$$Y_{22}(s) = \left. \frac{I_2}{V_2} \right|_{V_1=0} = 3s$$

To calculate the input admittance, use the network of Fig. E10.20(b).
Clearly,

$$V_2 = \frac{-Y_{21} V_1}{G_L + Y_{22}}$$

and the input side of the network is as in Fig. E10.20(c). The controlled source is, in fact, an admittance, because the current of the controlled source is proportional to the voltage across the source. Therefore, the input admittance Y_{in} (in mS) is

$$Y_{in}(s) = Y_{11} - \frac{Y_{12} Y_{21}}{G_L + Y_{22}}$$

$$= \frac{300s^2 + 229s + 2}{3s + 1} \, mS$$

for $G_L = 1$ mS.

EXAMPLE 10.21

Use the node method to solve the network of Fig. E10.21(a) for the system functions $E_1/I_{S1}|_{I_{S2}=0}$ and $E_1/I_{S2}|_{I_{S1}=0}$.

Solution

By the rules of construction of node equations for the transform network of Fig. E10.21(b),

$$\begin{bmatrix} sC + G & -G \\ -G & G + \dfrac{1}{sL} \end{bmatrix} \begin{bmatrix} E_1 \\ E_2 \end{bmatrix} = \begin{bmatrix} I_{S1} \\ I_{S2} \end{bmatrix} \tag{10.47}$$

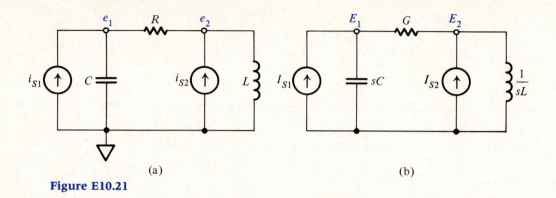

Figure E10.21

The matrix on the left-hand side of Eq. (10.47) is called the *node admittance matrix,* by extension of the term node conductance matrix used in Chap. 2. By Cramer's rule,

$$E_1 = \frac{\det\begin{bmatrix} I_{S1} & -G \\ I_{S2} & G + 1/sL \end{bmatrix}}{\det\begin{bmatrix} sC + G & -G \\ -G & G + 1/sL \end{bmatrix}} \tag{10.48}$$

$$E_1 = \frac{(G + 1/sL)I_{S1} + GI_{S2}}{sCG + C/L + G/sL}$$

$$= \frac{1}{C}\frac{(s + R/L)I_{S1} + sI_{S2}}{s^2 + (R/L)s + 1/LC}$$

The two system functions are

$$\left.\frac{E_1}{I_{S1}}\right|_{I_{S2}=0} = \frac{1}{C}\frac{s + R/L}{s^2 + (R/L)s + 1/LC}$$

and

$$\left.\frac{E_1}{I_{S2}}\right|_{I_{S1}=0} = \frac{1}{C}\frac{s}{s^2 + (R/L)s + 1/LC}$$

10.4.4 Natural Frequencies and Stability of Networks

Example 10.21 brings out an important point. Both system functions have the same denominators, hence the natural frequencies are identical for the differential equations relating $e_1(t)$ to $i_{S1}(t)$ and $e_1(t)$ to $i_{S2}(t)$. This fact is traced back to Eq. (10.48), where the denominator is seen to be the determinant of the left-hand side of the node equation. In fact, this denominator is the same if one solves for E_2.

In general, the denominator is the same for all node voltages (or mesh currents) hence the natural frequencies are also the same. Since all branch variables can be expressed in terms of node voltages (mesh currents), it follows that all branch variables have the same natural frequencies. Thus, it is justified to speak of the natural frequencies of a *network*, as opposed to the natural frequencies of a particular input and output pair.

The natural frequencies of a network are obtained by setting the determinant of the node admittance matrix (mesh impedance matrix) to zero and solving for s. The natural frequencies do not depend on the independent sources, since such sources do not appear in the node admittance (mesh impedance) matrix. They do, however, depend on the controlled sources.

Given a natural frequency $s_i = \sigma_i + j\omega_i$, a typical term of the homogeneous solution has the form $A_i t^k e^{\sigma_i t} e^{j\omega_i t}$. This term decays if $\sigma_i < 0$, grows if $\sigma_i > 0$, and has constant magnitude if $\sigma_i = 0$. A network is said to be *stable* if *all* its natural frequencies have negative real parts; the homogeneous solution for a stable network decays with time. Conversely, if at least one natural frequency has a real part that is positive or zero, the network is said to be *unstable*.

A network composed exclusively of L's, R's, C's, and independent sources cannot have a natural frequency with a real part greater than zero, because this would mean that, with the sources set to zero, some branch currents or voltages grow indefinitely. This is not possible, because a network composed only of L's, R's, and C's cannot generate energy, as required by such growth. Such networks are called *passive*, and are either stable or, at most, have natural frequencies with zero real parts. On the other hand, controlled sources can supply energy and networks with controlled sources can be unstable, as the next example shows.

EXAMPLE 10.22

The network of Fig. E10.22(a) is known as a phase-shift oscillator. An oscillator is an example of a network that is unstable by design. It is intended to produce a sinusoid output without any excitation, so that the homogeneous solution of its differential equation must contain the sinusoid. Since a sinusoid of frequency ω is the sum of two exponentials $e^{\pm j\omega t}$, the oscillator must have $\pm j\omega$ among its natural frequencies. The other natural frequencies must have negative real parts, because their contribution must die out, leaving only the sinusoid.

For the network of Fig. E10.22, calculate the characteristic polynomial, find the value of k for which two of the natural frequencies are imaginary and give these natural frequencies.

Solution
Figure 10.22(a) is redrawn to give the transform network shown in Fig. E10.22(b).

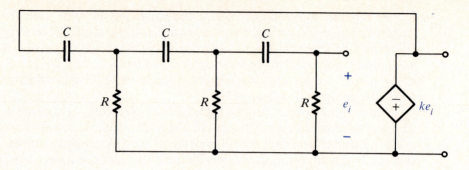

(a)

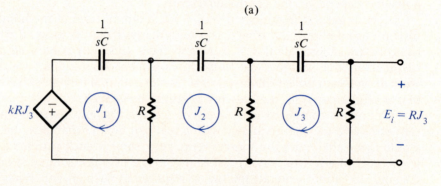

(b)

Figure E10.22

The mesh equations are written by inspection,

$$
\begin{bmatrix}
R + \dfrac{1}{sC} & -R & 0 \\
-R & 2R + \dfrac{1}{sC} & -R \\
0 & -R & 2R + \dfrac{1}{sC}
\end{bmatrix}
\begin{bmatrix}
J_1 \\ J_2 \\ J_3
\end{bmatrix}
=
\begin{bmatrix}
-kRJ_3 \\ 0 \\ 0
\end{bmatrix}
$$

or

$$
\begin{bmatrix}
R + \dfrac{1}{sC} & -R & kR \\
-R & 2R + \dfrac{1}{sC} & -R \\
0 & -R & 2R + \dfrac{1}{sC}
\end{bmatrix}
\begin{bmatrix}
J_1 \\ J_2 \\ J_3
\end{bmatrix}
=
\begin{bmatrix}
0 \\ 0 \\ 0
\end{bmatrix}
$$

The right-hand side is zero because there is no input to this network. The determinant of the mesh impedance matrix is calculated to be

$$
\frac{(k + 1)T^3 s^3 + 6T^2 s^2 + 5Ts + 1}{C^3 s^3}
$$

where $T = RC$. The natural frequencies are the roots of the characteristic equation

$$
(k + 1)T^3 s^3 + 6T^2 s^2 + 5Ts + 1 = 0
$$

Solving for the roots of a cubic is usually difficult. In this case, however, two of the roots are known to be of the form $s = \pm j\omega$ by design. Inserting this in the equation yields

$$-j(k + 1)T^3\omega^3 - 6T^2\omega^2 + j5T\omega + 1 = 0$$

This complex equation actually contains two real equations, one for the real part and one for the imaginary part. They are

$$-6T^2\omega^2 + 1 = 0$$

and

$$-(k + 1)T^3\omega^3 + 5T\omega = 0$$

The solution for ω and k is

$$\omega = \frac{1}{\sqrt{6}\,T} = \frac{0.408}{T}$$

and

$$k = 29$$

For $k = 29$, there are natural frequencies at $s = \pm j0.408/T$, leading to sinusoidal terms in the homogeneous solution of frequency $f = 0.0649/T$.

Once two roots of the characteristic polynomial are known, it is not difficult to solve for the other. For $k = 29$, this turns out to be $-1/5T$, corresponding to a decaying exponential.

The oscillator is "turned on" by almost any initial condition. In practice, k is made slightly greater than 29, so that the two complex natural frequencies have a small positive real part. The growth of the oscillation is actually limited by circuit nonlinearities.

DRILL EXERCISES

10.13 Find the transfer function V/I_s for the network of Fig. D10.13 and calculate the zero-state response, given $i_s(t) = u_{-1}(t)$.
Ans. $v(t) = 1.005e^{-0.1t}\cos(0.995t - 90°)$ volts, $t \geq 0$

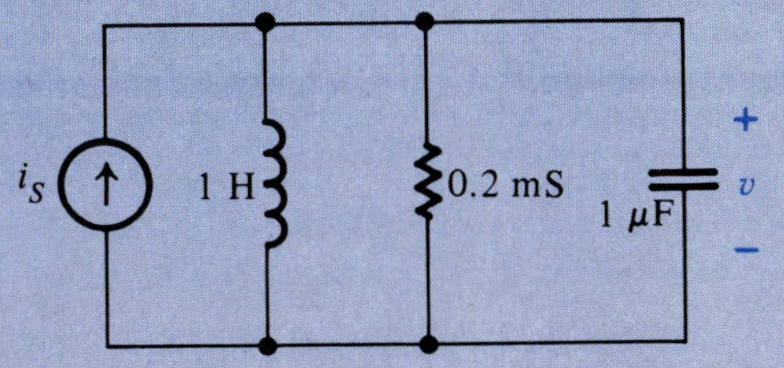

Figure D10.13

10.14 Calculate the transfer function I/I_S for the network of Fig. D10.14.

Ans. $\dfrac{I}{I_S} = \dfrac{sL_1 + R_1}{s(L_1 + L_2) + R_1}$

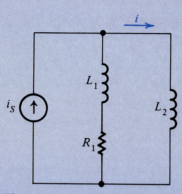

Figure D10.14

10.15 Find the Norton equivalent at port 2 for the network of Fig. E10.20(a), if port 1 is driven by a voltage source $v_S(t)$ in series with a 100 kΩ resistance.

Ans. $I_n = \dfrac{0.291(s - 13.3)V_S}{s + 0.116}$

$Y_{eq} = \dfrac{2.91\, s(s + 0.520)}{s + 0.116}$

10.16 Show that the natural frequencies of the network of Fig. D10.16 are $-G_2/C_2$ and $(g_m - G_1)/C_1$. For what values of g_m is the network stable?
Ans. $g_m < G_1$

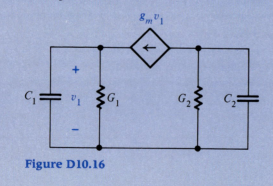

Figure D10.16

10.5 NETWORKS WITH NONZERO INITIAL STATE

In a network, the initial state is not specified by the values of the input, the response and their derivatives at $t = 0-$, but is given by the capacitance voltages and inductance currents at that time.

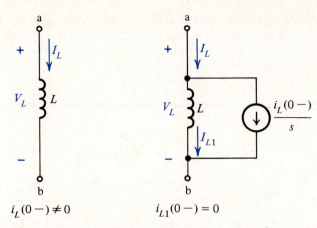

Figure 10.6 (a) Inductance with nonzero initial current. (b) Equivalent network using an inductance with zero initial current.

For an inductance,

$$v_L(t) = \frac{L \, di_L(t)}{dt}$$

If $i_L(0-) \neq 0$, then transforming yields

$$V_L(s) = sLI_L(s) - Li_L(0-) \tag{10.49}$$

Equation (10.49) is also written as

$$I_L(s) = \frac{1}{sL} V_L(s) + \frac{i_L(0-)}{s} \tag{10.50}$$

Equation (10.50) is represented by the network of Fig. 10.6(b), where $I_L = i_{L1} + i_L(0-)/s$. Since $i_L(0-) = 0$, it follows that $I_{L1}(s) = (1/sL)V_L(s)$, so that the two networks of Fig. 10.6 are equivalent. These are the transform networks corresponding to Fig. 9.14.

A similar argument is used to show the equivalence of the two networks of Fig. 10.7, which are the transform networks for Fig. 9.13.

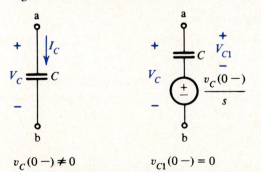

Figure 10.7 (a) Initially charged capacitance. (b) Equivalent network using initially uncharged capacitance.

In each case, an element that is not initially at rest is replaced in the transform network by an element at rest and a source. In this manner, any problem involving a network that is not in the zero state is transformed into a zero-state problem at the cost of introducing additional sources.

EXAMPLE 10.23

In the network of Fig. E10.23(a), the element values are $L = 4$ H, $C = 1\mu$F, and $R = 1$ kΩ. Solve for $i(t)$, given that $v_C(0-) = 1$ V and $i(0-) = -2$ mA.

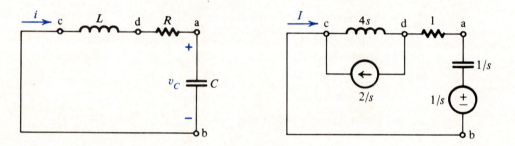

(a)   (b)

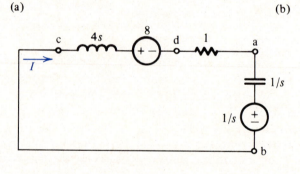

(c)

Figure E10.23

Solution

With V, mA, kΩ, H, μF, and ms as units, the transform network is given in Fig. E10.23(b), with impedance labeling. Applying Thevenin's theorem leads to Fig. E10.23(c). By KVL,

$$I(s) = \frac{-8 - 1/s}{4s + 1 + 1/s}$$

$$= \frac{-(8s + 1)}{4s^2 + s + 1}$$

or

$$I(s) = \frac{-2(s + 0.125)}{s^2 + 0.25s + 0.25}$$

Inverting, one obtains

$$i(t) = -2e^{-0.135t} \cos 0.484t \text{ mA} \qquad t \geq 0$$

with t in milliseconds.

EXAMPLE 10.24

Solve for $v_0(t)$ in the network of Fig. E10.24(a), given that $v_{C1}(0-)$ and $v_{C2}(0-)$ are not zero. The switch is closed at $t = 0$.

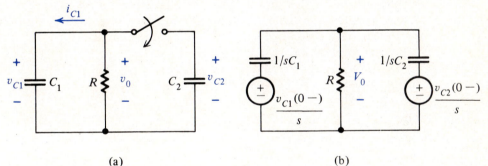

(a)   (b)

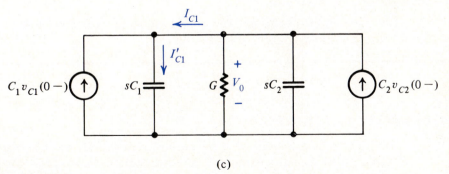

(c)

Figure E10.24

Solution

Norton's theorem is used to transform the diagram of Fig. E10.24(b) to that of Fig. E10.24(c), where the labeling is in terms of admittances. By inspection,

$$V_0(s) = \frac{C_1 v_{C1}(0-) + C_2 v_{C2}(0-)}{s(C_1 + C_2) + G}$$

$$= \frac{C_1 v_{C1}(0-) + C_2 v_{C2}(0-)}{(C_1 + C_2)[s + 1/R(C_1 + C_2)]}$$

Inverting yields

$$v_0(t) = \frac{C_1 v_{C1}(0-)}{C_1 + C_2} + \frac{C_2 v_{C2}(0-)}{C_1 + C_2} e^{-t/[R(C_1+C_2)]} \qquad t \geq 0$$

Since $v_{C1}(t) = v_{C2}(t) = v_0(t)$, $t > 0$, the capacitance voltages at $t = 0+$ are both equal to

$$v_0(0+) = \frac{C_1 v_{C1}(0-) + C_2 v_{C2}(0-)}{C_1 + C_2}$$

In general, $v_{C1}(0+) \neq v_{C1}(0-)$ and $v_{C2}(0+) \neq v_{C2}(0-)$; i.e., the capacitance voltages are discontinuous. This implies the presence of current impulses at $t = 0$. This can be verified by calculating $I_{C1}(s)$ in Fig. E10.24(c); however, a short cut is available if one notes that an impulse of current cannot flow through the resistance, as it would dissipate infinite energy. Therefore, to study the network near $t = 0$, the resistance can be replaced by an open circuit.

By current divider,

$$I'_{C1} = \frac{sC_1}{s(C_1 + C_2)}[C_1 v_{C1}(0-) + C_2 v_{C2}(0-)]$$

and

$$I_{C1} = I'_{C1} - C_1 v_{C1}(0-)$$

$$= \frac{C_1 C_2}{C_1 + C_2}[v_{C2}(0-) - v_{C1}(0-)]$$

This expression is valid near $t = 0$ only. It shows that there is a current impulse if the two initial voltages are different.

10.6 SUMMARY AND STUDY GUIDE

The homogeneous solution of second-order networks was the first topic of this chapter, with special attention given to the underdamped case. The system function was introduced from the point of view of classical differential equations, and reappeared in the next section on Laplace transforms. It was then shown that the transforms of the branch variables are themselves branch variables for another network, the transform network. For linear branch elements, the transform network is solved by the same techniques as resistive networks. Finally Laplace transform networks were introduced to analyze networks with nonzero initial states.

The student should bear in mind the following points.

The parameters Q and ω_0 determine the homogeneous solution of a second-order network; Q dictates the general shape and ω_0 sets the time scale.

The system function is written by inspection from the differential equation, and vice versa.

The system function, evaluated at the complex frequency of the excitation, yields a complex number that multiplies the exponential excitation to yield the particular solution.

The system function poles are the natural frequencies.

For rational functions of s, the partial fraction expansion is used to express the transform as a sum of easily invertible terms.

The techniques of linear resistive circuits are applicable to networks with L's and C's by using impedance or admittance.

The natural frequencies of a network are the roots of the determinant of the mesh impedance or node admittance matrices. They depend on the controlled-source parameters.

None of the material in this chapter applies to nonlinear networks.

A problem with nonzero initial conditions can be reduced to a zero-state problem by using capacitance and inductance equivalent networks with step sources.

PROBLEMS

10.1 Derive the differential equation for $i(t)$ in the network of Fig. P10.1 and calculate the parameters ω_0 and Q.

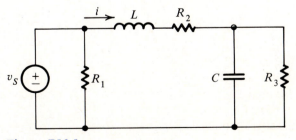

Figure P10.1

10.2 Repeat Prob. 10.1 for the network of Fig. P10.2.

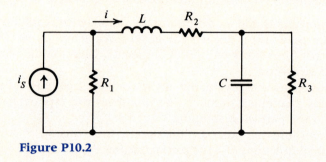

Figure P10.2

10.3 Find the parameters ω_0 and Q corresponding to the following second-order responses (t is in ms).
(a) $y(t) = 3e^{-t}\cos(t - 30°)$
(b) $y(t) = 2e^{-0.1t}\cos(2t + 45°)$
(c) $y(t) = e^{-2t}\cos(0.1t + 10°)$

10.4 For the network of Fig. P10.1, calculate ω_0 and Q for the following element values. Calculate and sketch $i(t)$, given that $i(0+) = 1$ mA, $di/dt(0+) = 0$, and $v_s(t) = 0$.
(a) $L = 0.1$ H, $C = 1$ μF, $R_2 = 0$, $R_3 = \infty$
(b) $L = 0.1$ H, $C = 1$ μF, $R_2 = 0.5$ kΩ, $R_3 = 2$ kΩ
(c) $L = 0.1$ H, $C = 1$ μF, $R_2 = 0.1$ kΩ, $R_3 = 5$ kΩ

10.5 Repeat Prob. 10.4 for the network of Fig. P10.2, with $i_s(t) = 0$. Use $R_1 = 0.1$ kΩ.

10.6 The zero-input response to initial conditions of a second-order network is observed on an oscilloscope to be as shown in Fig. P10.6. Estimate the circuit parameters Q and ω_0.

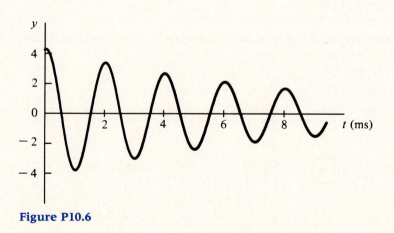

Figure P10.6

10.7 Repeat Prob. 10.6 for the response of Fig. P10.7.

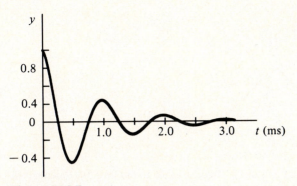

Figure P10.7

10.8 For each of the following system functions, calculate the poles and zeros and write the differential equation relating $y(t)$ to $x(t)$.
(a) $Y/X = 2(s - 1)/(s^2 + 3s + 2)$
(b) $Y/X = (s^2 + s)/(s^2 + s + 1)$
(c) $Y/X = (s^2 + 3s + 2)/(s^3 - s^2 + s)$

10.9 The system functions of Prob. 10.8(a) and (b) are to be considered as the system functions of networks. If, in each case, $x(t) = u_{-1}(t)$ and $y(0+) = 1$, $dy/dt(0+) = -1$, calculate $y(t)$.

10.10 The system function of a network is known to have two poles and one zero. Responses are given to two different inputs:

If $x(t) = e^t u_{-1}(t)$, $y(t) = -e^{-t} + 2e^{-2t} + e^t$.
If $x(t) = u_{-1}(t)$, $y(t) = 0.5e^{-t} - e^{-2t} - 2$.

Find the network poles and zeros and write the system function. (*Note:* The responses are not zero-state responses.)

10.11 Repeat Prob. 10.10, but with a system function with three poles and two zeros, and the following data:

If $x(t) = u_{-1}(t)$, $y(t) = 2e^{-t}\cos(t + 45°) - 0.5e^{-t}$.
If $x(t) = \cos t\, u_{-1}(t)$, $y(t) = 0.5e^{-t}\cos(t - 30°) + e^{-t} + 0.5\cos(t - 30°)$.

(*Hint*: Use the polynomial form rather than the factored form for the numerator.)

10.12 Calculate the Laplace transform of the following time functions:
(a) $f(t) = e^{-t} + 2e^{2t}$
(b) $f(t) = \cos t + \sin 2t$

10.13 Using the Real Differentiation Theorem, show that $\dfrac{d}{dt}\sin \omega t. = \omega \cos \omega t$.

10.14 Calculate the Laplace transform of $f(t) = te^{-at}\cos(\omega t + \theta)$.

10.15 Calculate the Laplace transform of the pulse waveform $f(t)$ of Fig. P10.15. [*Hint*: Express $f(t)$ as a sum of steps, ramps, and delayed steps and ramps.].

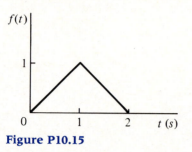

Figure P10.15

10.16 Differentiate $f(t)$ in Fig. P10.15 twice to obtain $f''(t)$. Calculate the transform of $f''(t)$ and use the integration theorem to find the transform of $f(t)$.

10.17 Repeat Prob. 10.15 for the transform $f(t)$ of Fig. P10.17.

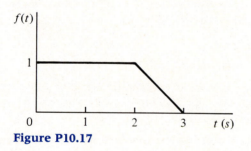

Figure P10.17

10.18 Find the inverse of the following Laplace transforms:
(a) $F(s) = s/(s + 1)(s + 2)$
(b) $F(s) = s/(s + 1)^2$

10.19 Repeat Prob. 10.18 for the following Laplace transforms:
(a) $F(s) = (2s + 1)/(s + 1)(s^2 + s + 1)$
(b) $F(s) = (1 - 2e^{-s} + e^{-2s})/s(s + 1)$

10.20 Show that the system function is the Laplace transform of the zero-state response to a unit impulse input.

10.21 Find the solution, for $t \geq 0$ of the differential equation $d^2y/dt^2 + 2dy/dt + 2y = dx/dt + x$, given $x(t) = u_{-1}(t)$, $y(0-) = 0$, $dy/dt(0-) = 1$.

10.22 Repeat Prob. 10.21 for the differential equation $d^2y/dt^2 + dy/dt = 2x$, given that $x(t) = \cos 2t\, u_{-1}(t)$, $y(0-) = dy/dt(0-) = 0$.

10.23 Show that the zero-state response of a network to a unit impulse is the derivative of the zero-state response to a unit step.

10.24 Solve for the zero-state response $i(t)$ in Fig. P10.1, if $v_S(t) = u_{-1}(t)$. Use the element values of Prob. 10.4(c).

10.25 Repeat Prob. 10.24 for the network of Fig. P10.2, with $i_S(t) = u_{-1}(t)$. Use $R_1 = 0.5$ kΩ.

10.26 Solve for the zero-state response $v(t)$ in the network of Fig. P10.26.

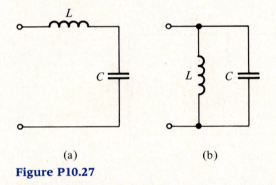

Figure P10.26

10.27 Find the driving-point impedance for the networks of Fig. P10.27.

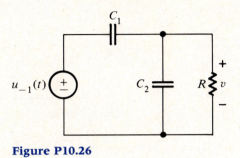

(a)   (b)

Figure P10.27

10.28 Repeat Prob. 10.27 for the network of Fig. P10.28.

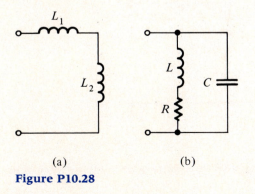

(a)   (b)

Figure P10.28

10.29 Find the driving-point impedance of the network of Fig. P10.29. (*Hint*: Use the π–T transformation.)

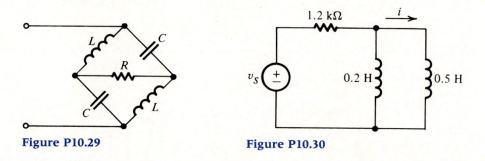

Figure P10.29 **Figure P10.30**

10.30 Calculate the system function I/V_s in the network of Fig. P10.30. Calculate the zero-state response $i(t)$ if $v_s(t) = u_{-1}(t)$.

10.31 Calculate the system function V_o/V_s in the network of Fig. P10.31. Solve for the zero-state response $v_o(t)$ if $v_s(t) = \cos t\, u_{-1}(t)$ volts (t is in milliseconds).

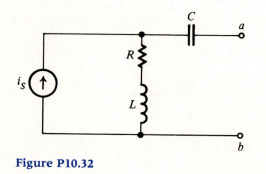

Figure P10.31

10.32 Find the Thevenin and Norton equivalents at *ab* in the network of Fig. P10.32.

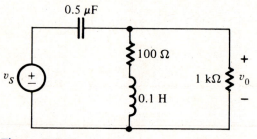

Figure P10.32

10.33 Calculate the driving-point impedance at *ab* in the network of Fig. P10.33.

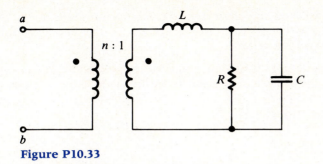

Figure P10.33

10.34 For the two-port of Fig. P10.34:
  (a) Find the Z parameters.
  (b) With port 2 terminated by a resistance R, find that value of R such that the driving-poing impedance at port 1 is R.
  (c) For R as in (b), find V_2/V_1.

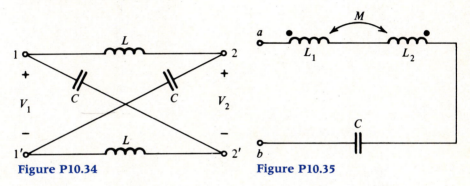

Figure P10.34   **Figure P10.35**

10.35 Calculate the driving-point impedance at ab for the network of Fig. P10.35.

10.36 The ideal transformer of Fig. P10.36, with its two windings connected in series, is

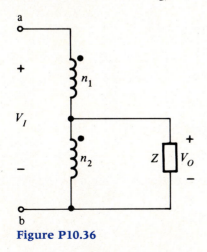

Figure P10.36

known as an autotransformer. In practice, the turns ratio $n_1 : n_2$ can be varied by moving a tap along a coil. Calculate the voltage transfer ratio V_O/V_I and the driving-point impedance at ab.

10.37 The circuit of Fig. P10.37(a) represents a BJT with a transformer-coupled load. The transformer is modeled as shown in Fig. P10.37(b). If the network is initially in the zero state, calculate $v_O(t)$ if $i_S(t) = 0.01u_{-1}(t)$ mA. (Use the small-signal model for the transistor, with $\beta = 50$.)

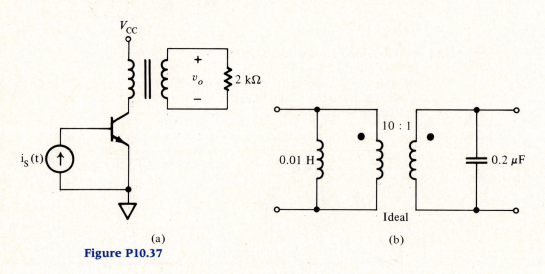

(a)   (b)

Figure P10.37

10.38 The two-port of Fig. P10.38 is the "hybrid pi" model of a BJT valid at high frequencies. Calculate the y parameters of this two-port.

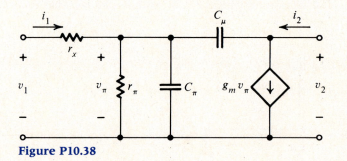

Figure P10.38

10.39 Synthesize, with one C or L and resistances, the driving-point impedances listed below. Take s to be in ms^{-1}, and Z to be in $\text{k}\Omega$.
(a) $Z(s) = 1/(s + 2)$
(b) $Z(s) = (2s + 5)/(s + 2)$
(c) $Z(s) = s/(s + 0.5)$

10.40 The network of Fig. P10.40 includes an ideal op amp, i.e., $A \to \infty$.
  (a) Calculate the system functions $V_o/V_1|_{v_2=0}$ and $V_o/V_2|_{v_1=0}$.
  (b) Find circuit elements for the impedances such that $V_o/V_1|_{v_2=0} = -Vs$.
      Take s to be in ms^{-1}.

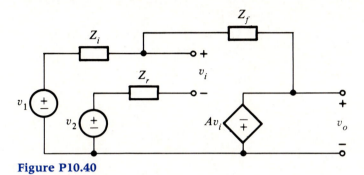

Figure P10.40

10.41 The network of Fig. P10.41 is a small-signal high-frequency model of a MOSFET source-follower amplifier.
  (a) Calculate the y parameters of the two-port within the dotted line.
  (b) Calculate the Norton equivalent at terminals 22′, given that port 1 is driven by a voltage source v_s as shown.
  (c) Calculate V_2/V_S if port 2 is terminated with a load resistance R_L.

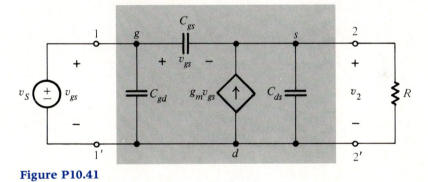

Figure P10.41

10.42 Find the characteristic polynomial for the network of Fig. P10.42. Calculate the value

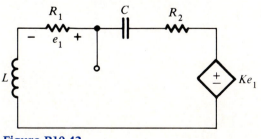

Figure P10.42

of K for which two of the natural frequencies are imaginary, and calculate the imaginary natural frequencies.

10.43 The network of Fig. P10.43 represents an astable BJT multivibrator. Using identical small-signal models with $r_\pi = 0$ for both transistors in the active region, show that this network is unstable if $\beta > 1$.

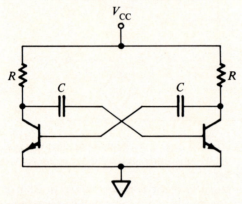

Figure P10.43

10.44 The network of Fig. P10.44 is a tuned-input oscillator. The two inductors are modeled by inductances L_1 and L_2, with series resistances R_1 and R_2, respectively, and mutual inductance M.
(a) With $L_1 = L_2 = L$, $M = kL$ and $R_1 = R_2 = R$, calculate the oscillation frequency. (Use the small-signal model of Chap. 6 for the MOSFET.) Find a condition involving L, R, k, and g_m under which oscillations can be sustained.
(b) Coils are available for which $L = 100$ μH, $Q(= \omega L/R)$ is 50 at 100 kHz and for which $k = 0.9$. Design a 100-kHz oscillator. What is the minimum value of g_m required?

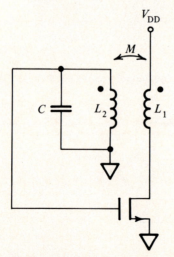

Figure P10.44

10.45 In the network of Fig. P10.45, the switch has been closed for a long time, i.e., the network is in the dc steady state. The switch is opened at $t = 0$. Calculate $i(t)$ and $v_o(t)$, $t \geq 0$.

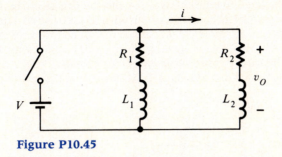

Figure P10.45

10.46 The network of Fig. P10.46 is initially at rest, with the switch in position 2. At $t = 0$, the switch is thrown to position 1, and held there until $t = 1$ ms, at which time it is thrown to position 3. Calculate $v_o(t)$.

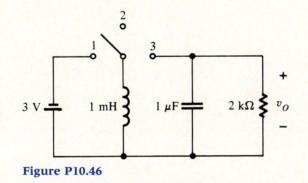

Figure P10.46

10.47 The network of Fig. P10.47(a) is subjected to the periodic input shown in Fig. P10.47(b). The periodic solution is to be calculated.
(a) Given $i_L(0-) = i_{Lo}$ and $v_C(0-) = v_{Co}$, calculate $i_L(t)$ and $v_C(t)$ at $t = 2$ ms.

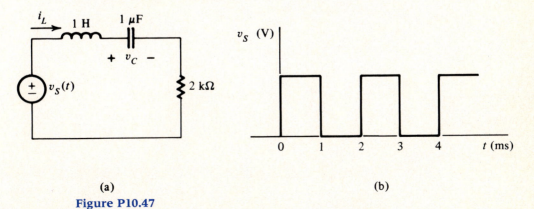

(a)   (b)

Figure P10.47

(b) Determine i_{Lo} and v_{Co} so that i_L (2 ms) = i_{Lo}, v_C (2 ms) = v_{Co}. (Since the voltages and currents are repeated, these are the conditions at the start of each cycle in the periodic steady state.)

10.48 The effect of the load inverters between A and the datum in Fig. P10.48(a) is best modeled by the parallel RC combination shown in Fig. P10.48(b). The effect of adding a capacitance C_B is to be studied. It is assumed that Q_1 has been cut off for a long time prior to $t = 0$, and that $v_S(t) = V_{CC}u_{-1}(t)$. Using a BJT model with $V_{BE} = V_{CEsat} = 0$, calculate $v_0(t)$, for (a) the case $C_B = 0$ and (b) the case $R_B C_B = RC$. Use the following values: $V_{CC} = 4$ V, $R_C = 0.3$ kΩ, $R_B = 1$ kΩ, $R = 0.1$ kΩ, $C = 500$ pF, and $\beta = 50$.

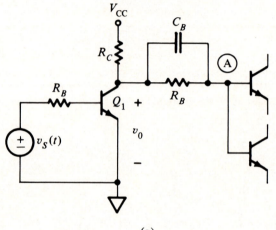

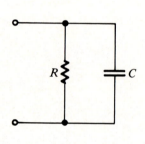

(a) (b)

Figure P10.48

CHAPTER

The Sinusoidal Steady State

11.1 INTRODUCTION

The subject of this chapter is the response of networks excited by sinusoids. The reader is no doubt aware that most electric power is transmitted and distributed in the ac form, i.e., in a sinusoidal wave shape. The frequency of power transmission is 60 Hz in North America, 50 Hz in most of the rest of the world. The reason for the use of ac in power transmission has to do with the ease of voltage transformation using transformers.

The predominance of sinusoidal waveforms in power is reason enough to study the sinusoidal steady state. However, there is another equally compelling reason. The fact is that any signal can be broken down into sinusoidal components: this applies to periodic signals (through Fourier series), nonperiodic deterministic transients (through the Fourier transform), and even random signals, under most practical conditions. The ability of a network or system to transmit a signal without distortion is essentially its ability to transmit without modification all its frequency components. By the same token, signal processing, i.e., intentional modification of a signal, is most often specified in terms of the way in which the various sinusoidal components are to be altered.

11.2 THE SINUSOIDAL STEADY STATE

11.2.1 Phasors and the Steady-State Response

The sinusoidal signal $x(t) = A \cos(\omega_p t + \theta)$ is also written as

$$x(t) = \text{Re}(Ae^{j(\omega_p t + \theta)})$$

$$= \text{Re}(Ae^{j\theta}e^{j\omega_p t})$$

(11.1)

The complex number $X = Ae^{j\theta}$, known as the *phasor* representation of the sinusoid, contains both magnitude and phase information. Given the phasor and the frequency ω_p, the signal $x(t)$ is written directly.

EXAMPLE 11.1

Give the phasor representations for the sinusoids $x_1(t) = 2 \cos(\omega_p t + 30°)$, $x_2(t) = 4 \cos(\omega_p t - 120°)$ and $x_3(t) = 3 \sin(\omega_p t + 20°)$.

Solution
For the first two signals, one writes directly

$$X_1 = 2e^{j30°} \qquad X_2 = 4e^{-j120°}$$

The signal $x_3(t)$ is first converted to a cosine, using the identity $\sin x = \cos(x - 90°)$. Thus,

$$x_3(t) = 3 \cos(\omega_p t - 70°)$$

and the phasor is

$$X_3 = 3e^{-j70°}$$

It is also possible to use a phasor representation based on sines rather than cosines, starting from the equation

$$A \sin(\omega_p t + \theta) = \text{Im}(Ae^{j\theta}e^{j\omega_p t})$$

(11.2)

If the input to a network is of the form $Ae^{j\omega t}$, the particular solution will also be of that form. It is now shown that the phasor representation of the particular solution is expressed as a rather simple function of the system function and the input phasor.

The particular solution for an input $x(t) = Xe^{j\omega_p t}$ is

$$y_p(t) = Ye^{j\omega_p t}$$

where

$$Y = H(j\omega_p)X$$

A cosine input can be written as

$$x(t) = A\cos(\omega_p t + \theta)$$

$$= \frac{A}{2}e^{j\theta}e^{j\omega_p t} + \frac{A}{2}e^{-j\theta-j\omega_p t}$$

By superposition, the particular solution is

$$y_p(t) = H(j\omega_p)\frac{A}{2}e^{j\theta}e^{j\omega_p t} + H(-j\omega_p)\frac{A}{2}e^{-j\theta}e^{-j\omega_p t} \tag{11.3}$$

If $H(s)$ is a ratio of polynomials with real coefficients, then j appears only as a member of the pair $j\omega_p$ in $H(j\omega_p)$. Therefore, replacing ω_p by $-\omega_p$ is the same as replacing j by $-j$ everywhere in $H(j\omega_p)$; this implies that $H(-j\omega_p) = H^*(j\omega_p)$. The right-hand side of Eq. (11.3) is seen to be a sum of complex conjugates, so that

$$y_p(t) = \text{Re}[AH(j\omega_p)e^{j\theta}e^{j\omega_p t}] \tag{11.4}$$

The quantity $H(j\omega_p)$ is the system function evaluated at $s = j\omega_p$. It is a complex number, and is written in polar form as

$$H(j\omega_p) = |H(j\omega_p)|e^{j\angle H(j\omega_p)}$$

Using this in Eq. (11.4) yields

$$y_p(t) = \text{Re}[A|H(j\omega_p)|e^{j[\theta+\angle H(j\omega_p)]}e^{j\omega_p t}] \tag{11.5}$$

But, this is precisely of the same form as Eq. (11.1). Therefore, the phasor representation of $y_p(t)$ is

$$Y = A|H(j\omega_p)|e^{j[\theta+\angle H(j\omega_p)]} \tag{11.6}$$

and $y_p(t)$ is written directly as

$$y_p(t) = A|H(j\omega_p)|\cos[\omega_p t + \theta + \angle H(j\omega_p)] \tag{11.7}$$

The particular solution $y_p(t)$ given by Eq. (11.7) is the *sinusoidal steady-state* solution. If $H(s)$ is stable, the exponential terms of the homogeneous solution tend to zero and the solution $y(t)$ is essentially equal to the steady-state solution, after a sufficiently long interval of time.

From Eq. (11.7), the sinusoidal steady state has the following properties:

1. It is a sinusoid, of the same frequency as the excitation.

2. The amplitude of the sinusoid is that of the excitation, multiplied by $|H(j\omega_p)|$.

3. The phase of the sinusoid is that of the excitation, shifted by $\angle H(j\omega_p)$.

EXAMPLE 11.2

Calculate the sinusoidal steady-state voltage $v_O(t)$ for the network of Fig. E11.2, given that $v_S(t) = 2\cos(\omega_p t + 45°)$ with t in seconds and (1) $\omega_p = 1$ krad/s; (2) $\omega_p = 4$ krad/s.

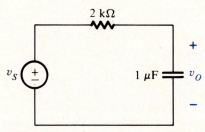

Figure E11.2

Solution

Using volts, milliamperes, and milliseconds as units,

$$\frac{V_o(s)}{V_S(s)} = H(s) = \frac{1/s}{2 + 1/s} = \frac{1}{2s + 1}$$

1. The frequency $\omega_p = 1$ krad/s is expressed in terms of ms as $\omega_p = 1$ rad/ms, so that

$$H(j\omega_p) = H(j1) = \frac{1}{2j + 1}$$

In polar form,

$$H(j1) = \frac{1}{\sqrt{5}} e^{-j63.4°}$$

The input phasor is $V_S = 2e^{j45}$, so that from Eq. (11.6), the output phasor is

$$V_O = \frac{2}{\sqrt{5}} e^{j(45°-63.4°)}$$

and the sinusoidal steady-state solution is

$$v_O(t) = (0.894)\cos(t - 18.4°)$$

where t is in milliseconds.

2. With $\omega_p = 4$ rad/ms, one writes

$$H(j\omega_p) = H(j4) = \frac{1}{8j + 1}$$

$$= \frac{1}{\sqrt{65}} e^{-j82.9°}$$

The output phasor is

$$V_0 = \frac{1}{\sqrt{65}} e^{j(45° - 82.9°)}$$

so that with t in ms,

$$v_0(t) = 0.248 \cos(4t - 37.9°)$$

This network has a pole at $s = -0.5$, i.e., a time constant of 2 ms. The homogeneous solution essentially dies out in about 5 time constants, so that, 10 ms after v_S is turned on, the output is basically the sinusoidal steady state.

11.2.2 Equivalent Networks in the Sinusoidal Steady State

When evaluated at $s = j\omega_p$, a system function is just a complex number. For an impedance, one writes

$$Z(j\omega_p) = R(\omega_p) + jX(\omega_p) \tag{11.8}$$

The real part R is called the *resistance,* and the imaginary part X is called the *reactance.* Both depend on ω_p.

For an admittance, one writes

$$Y(j\omega_p) = G(\omega_p) + jB(\omega_p) \tag{11.9}$$

The real part G is the conductance, and the imaginary part B is the *susceptance.*

For a driving-point function, since $Y = 1/Z$, it follows from Eq. (11.8) that

$$Y = \frac{1}{R + jX} = \frac{R}{R^2 + X^2} - \frac{jX}{R^2 + X^2}$$

Comparing this with Eq. (11.9) yields

$$G = \frac{R}{R^2 + X^2} \qquad B = \frac{-X}{R^2 + X^2} \tag{11.10}$$

The expressions for R and X in terms of G and B are

$$R = \frac{G}{G^2 + B^2} \qquad X = \frac{-B}{G^2 + B^2} \tag{11.11}$$

An inductance has an impedance $Z = j\omega_p L$. It is a purely reactive element, with reactance and susceptance

$$X_L = \omega_p L \tag{11.12a}$$

and

$$B_L = \frac{-1}{\omega_p L} \tag{11.12b}$$

respectively.

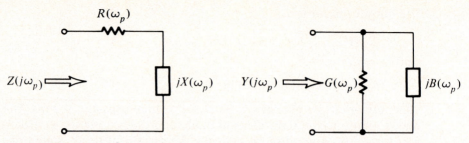

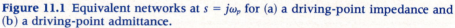

Figure 11.1 Equivalent networks at $s = j\omega_p$ for (a) a driving-point impedance and (b) a driving-point admittance.

A capacitance is also purely reactive, and its reactance and susceptance are given by

$$X_C = \frac{-1}{\omega_p C} \tag{11.13a}$$

and

$$B_C = \omega_p C \tag{11.13b}$$

The equivalent networks of Fig. 11.1 follow immediately from Eqs. (11.8) and (11.9). If $X(\omega_p) > 0(B(\omega_p) < 0)$ an inductance is used to model the reactance; the value of L is calculated from Eq. (11.12). Similarly, a capacitance is used if $X(\omega_p) < 0(B(\omega_p) > 0)$, and its value is computed from Eq. (11.13). It should be emphasized that those equivalent networks are valid *only* at $\omega = \omega_p$.

EXAMPLE 11.3

Industrial loads consist mostly of electric machines. Figure E11.3(a) shows an equivalent circuit for a 15-hp, 220-V induction motor. Calculate the admittance seen at *ab* in the sinusoidal steady state, for a frequency of 60 Hz, and derive an equivalent circuit with two parallel elements.

Solution
Rather than drawing the transform diagram as a function of s, the impedances are immediately evaluated at $s = j2\pi(60) = j377 \text{ s}^{-1}$, as shown in Fig. E11.3(b).

In Fig. E11.3(c), Eq. (11.10) is used to convert the series impedance $0.22 + j0.49 \ \Omega$ to an admittance. The figure is drawn with admittance labeling.

The admittance seen at *ab* is

$$Y = (0.767 - j1.83) \text{ S}$$

With $B_L = -1.83$ and $\omega_p = 377 \text{ s}^{-1}$, Eq. (11.12b) is used to solve for L. Figure E11.3(d) shows a network that is equivalent to the induction machine at 60 Hz.

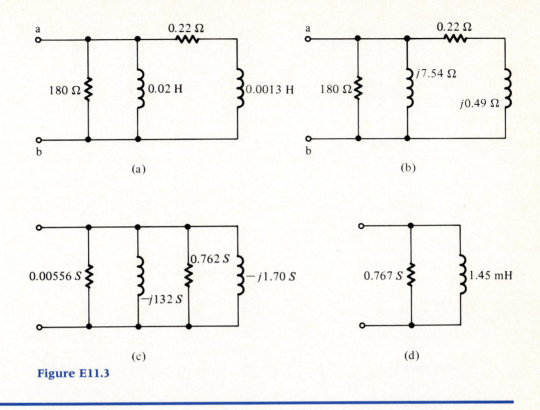

Figure E11.3

It is often useful in network analysis to study the response at very low and very high frequencies. At low frequencies, $\omega_p L$ is small while $1/\omega_p C$ is large; the converse is true at high frequencies. In words, an inductance has a small impedance at low frequencies and a high impedance at high frequencies. The converse is true for a capacitance.

These considerations often lead to simplified networks if one applies the following two basic approximation rules:

1. An element of impedance Z_1 is replaced by a short circuit if it is in series with an element of impedance Z_2 and $|Z_1| \ll |Z_2|$.

2. An element of impedance Z_1 is replaced by an open circuit if it is in parallel with an element of impedance Z_2 and $|Z_1| \gg |Z_2|$ (i.e., $|Y_1| \ll |Y_2|$).

EXAMPLE 11.4

Using approximate networks, calculate the driving-point impedance at aa' of the network of Fig. E11.4(a) at low and high frequencies.

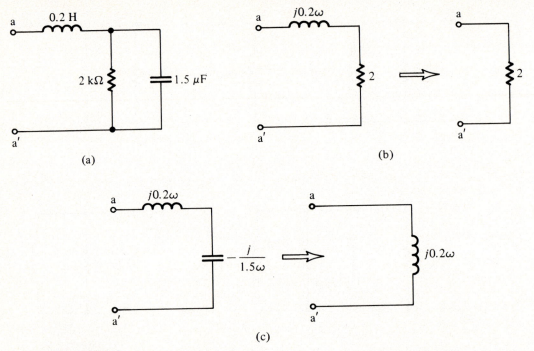

Figure E11.4

Solution

At low frequency, the capacitance impedance is large compared to that of the resistance: by rule 2, the capacitance is replaced by an open circuit. Since the inductance impedance is small compared to that of the resistance, rule 1 allows its replacement by a short circuit. The two steps are illustrated in Fig. E11.4(b). In order to justify those steps, it is necessary that

$$\frac{1}{1.5\omega} \gg 2 \quad \text{or} \quad \omega \ll \tfrac{1}{3} \text{ (ms)}^{-1}$$

and

$$0.2\omega \ll 2 \quad \text{or} \quad \omega \ll 10 \text{ (ms)}^{-1}$$

Therefore, at frequencies much less than $\tfrac{1}{3}$ krad/s the driving-point impedance is essentially 2 kΩ.

At high frequencies, the resistance is large compared to the impedance of the capacitance: this permits removing the resistance. The inductance has a high impedance compared to that of the capacitance, so the capacitance is replaced by a short circuit. The two stages are shown in Fig. E11.4(c). In order to justify those steps, it is necessary that

$$\frac{1}{1.5\omega} \ll 2 \quad \text{or} \quad \omega \gg \tfrac{1}{3} \text{ (ms)}^{-1}$$

and

$$0.2\omega \gg \frac{1}{1.5\omega} \quad \text{or} \quad \omega \gg 1.82 \ (\text{ms})^{-1}$$

Therefore, at frequencies much greater than 1.82 krad/s, the driving-point impedance is essentially $j0.2\omega$.

11.2.3 Phasor Diagrams

Because they are complex numbers, phasors may be represented as vectors in the complex plane with Kirchhoff's laws expressed in terms of vector additions. Before proceeding with KVL and KCL, it is instructive to relate the voltage and current phasors for the basic branch elements.

For a resistance,

$$V_R = RI_R$$

so that

$$|V_R| = R|I_R| \qquad (11.14a)$$

and

$$\sphericalangle V_R = \sphericalangle I_R \qquad (11.14b)$$

Thus, V_R and I_R are collinear vectors, as shown in Fig. 11.2(a), and the phasors are said to be *in phase*; this implies that the voltage and current sinusoids have identical phases.

For a capacitance,

$$I_C = j\omega_p C V_C$$

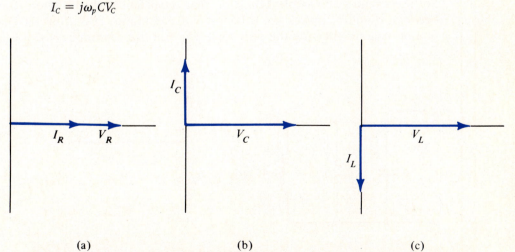

(a)   (b)   (c)

Figure 11.2 Voltage and current phasors for (a) a resistance, (b) a capacitance, and (c) an inductance.

so that

$$|I_C| = \omega_p C |V_C| \tag{11.15a}$$

and

$$\sphericalangle I_C = \sphericalangle V_C + \sphericalangle j = \sphericalangle V_C + 90° \tag{11.15b}$$

Equation (11.15b) shows that the angle of the current phasor is 90° greater than that of the voltage phasor: the current is said to *lead* the voltage by 90°. The two phasors are shown in Fig. 11.2(b).

For an inductance,

$$I_L = \frac{V_L}{j\omega_p L}$$

so that

$$|I_L| = \frac{|V_L|}{\omega_p L} \tag{11.16a}$$

and

$$\sphericalangle I_L = \sphericalangle V_L - 90° \tag{11.16b}$$

The two phasors are shown in Fig. 11.2(c); the current *lags* the voltage by 90°.

EXAMPLE 11.5

For the circuit of Example 11.3, find that capacitance which, when connected between *a* and *b*, makes the driving-point admittance real at 60 Hz. Sketch a phasor diagram showing the phasors in Fig. E11.5(a), assuming $\sphericalangle V = 0°$.

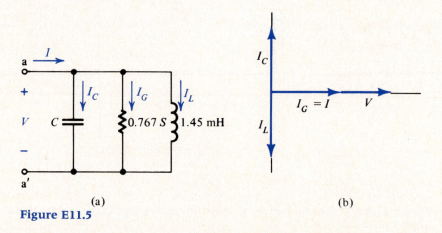

(a)   (b)

Figure E11.5

Solution

Without the capacitance, the admittance at 60 Hz is $(0.767 - j1.83)$ S. With the capacitance, it is

$$Y = 0.767 - j1.83 + j377C$$

For Y to be real,

$$C = \frac{1.83}{377} = 4.85 \text{ mF}$$

Using Eqs. (11.14) to (11.16),

$$|I_G| = 0.767|V|$$

$$|I_L| = 1.829|V|$$

and

$$|I_C| = 1.829|V|$$

The phasors are shown in Fig. E11.5(b). The inductance and capacitance currents are seen to cancel each other out, i.e., they are exactly 180° out of phase; this makes I equal to I_G, in phase with V, so the admittance is resistive.

DRILL EXERCISES

11.1 For $Y/X = H(s) = 2(s - 1)/(s + 1)(s + 2)$, find the steady-state response if $x(t) = 3 \cos(t + 45°)$.
Ans. $2.68 \cos(t + 108.44°)$

11.2 For the network of Example 11.3, find an equivalent parallel GL network at 50 Hz.
Ans. $G = 1.03$ S, $L = 1.57$ mH

11.3 Repeat Example 11.5 for a frequency of 50 Hz.
Ans. $C = 6.47$ mF

11.3 FREQUENCY RESPONSE

11.3.1 Introduction

The magnitude and phase of $H(j\omega)$, viewed as functions of frequency, together specify the *frequency response* of a network. It can be shown that the frequency response completely characterizes a network, if it is stable; this means that knowledge of the frequency response is sufficient to determine the response to any input. In other words, network specifications can be expressed in terms of the frequency response as well as the transient response to given inputs. In most cases, it is the frequency response that is specified.

As explained in Sec. 11.2.1, $H(-j\omega) = H^*(j\omega)$. This implies that

$$|H(-j\omega)| = |H(j\omega)| \tag{11.17}$$

and

$$\sphericalangle H(-j\omega) = -\sphericalangle H(j\omega) \tag{11.18}$$

In words, the magnitude is an even function of ω, while the phase is an odd function of ω. It is sufficient to know $H(j\omega)$ for $\omega > 0$, since Eqs. (11.17) and (11.18) can then be used to construct the magnitude and phase for $\omega < 0$. Nevertheless, for the time being, the magnitude and phase are plotted over $-\infty < \omega < \infty$; this will prove useful in the study of resonance (Sec. 11.3.4).

The function $H(j\omega)$ is the output complex amplitude for an input $e^{j\omega t}$; this applies for ω positive or negative. Sinusoids can be written in terms of either positive or negative ω. For example, $\cos(-2t) = \cos 2t$ and $\sin(-2t) = -\sin 2t$, so that no generality is lost if sinusoids are written in terms of positive ω.

11.3.2 Poles, Zeros, and Frequency Response

It proves to be useful to relate the frequency response to the pole-zero pattern of $H(s)$. In pole-zero form, $H(s)$ is written as

$$H(s) = \frac{K(s - z_1)(s - z_2) \cdots (s - z_m)}{(s - p_1)(s - p_2) \cdots (s - p_n)} \tag{11.19}$$

By the rules of multiplication and division for complex numbers,

$$|H(j\omega)| = \frac{|K| \, |j\omega - z_1| \, |j\omega - z_2| \cdots |j\omega - z_m|}{|j\omega - p_1| \, |j\omega - p_2| \cdots |j\omega - p_n|} \tag{11.20}$$

and

$$\begin{aligned} \sphericalangle H(j\omega) = {}& \sphericalangle K + \sphericalangle(j\omega - z_1) + \cdots + \sphericalangle(j\omega - z_m) \\ & - \sphericalangle(j\omega - p_1) - \cdots - \sphericalangle(j\omega - p_n) \end{aligned} \tag{11.21}$$

The complex numbers $j\omega$ and z_1 both represent vectors in the complex plane, as shown in Fig. 11.3. The difference $j\omega - z_1$ is obtained by the usual rules of vector subtraction; the result is seen to be a vector drawn from the tip of z_1 to the tip of $j\omega$.

This geometric construction suggests a means of visualizing the calculation of $H(j\omega)$. First, the poles (crosses) and zeros (circles) of $H(s)$ are plotted in the complex plane: there results a *pole-zero plot*. Next, the vectors $j\omega - z_1$, $j\omega - z_2$, . . . are generated, by joining $z_1, z_2, \ldots$ to the point $s = j\omega$; this is illustrated in Fig. 11.4. From Eq. (11.20),

$$|H(j\omega)| = \frac{|K| \text{ product of lengths of vectors from zeros to } j\omega}{\text{product of lengths of vectors from poles to } j\omega} \tag{11.22}$$

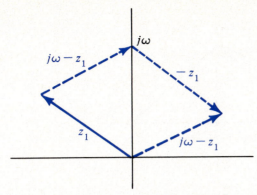

Figure 11.3 Illustrating a difference of two complex numbers.

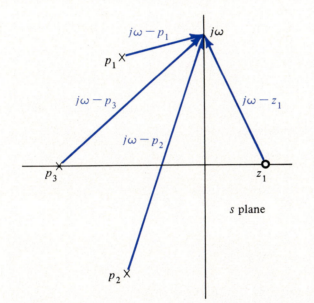

Figure 11.4 Construction of the terms of $H(j\omega)$ from the pole-zero plot.

From Eq. (11.21),

$$\angle H(j\omega) = \angle K + \text{(sum of angles of vectors from zeros to } j\omega)$$
$$- \text{(sum of angles of vectors from poles to } j\omega) \quad (11.23)$$

EXAMPLE 11.6

For $H(s) = 2(s - 1)/(s + 1)(s + 2)$, use the geometric construction to obtain $H(j\omega)$ for $\omega = 0, 1, \infty$.

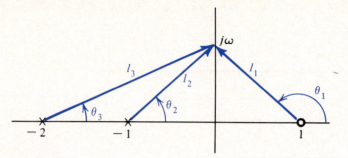

Figure E11.6

Solution

The lengths and angles are defined in Fig. E11.6. For $\omega = 0$,

$$l_1 = 1 \qquad l_2 = 1 \qquad l_3 = 2$$

$$\theta_1 = 180° \qquad \theta_2 = 0° \qquad \theta_3 = 0°$$

Using Eqs. (11.22) and (11.23),

$$|H(j0)| = \frac{2l_1}{l_2 l_3} = 1$$

$$\sphericalangle H(j0) = 0° + \theta_1 - \theta_2 - \theta_3 = 180°$$

For $\omega = 1$,

$$l_1 = \sqrt{2} \qquad l_2 = \sqrt{2} \qquad l_3 = \sqrt{5}$$

$$\theta_1 = 135° \qquad \theta_2 = 45° \qquad \theta_3 = \tan^{-1} 0.5 = 26.56°$$

Therefore,

$$|H(j1)| = \frac{2\sqrt{2}}{\sqrt{2}\sqrt{5}} = 2/\sqrt{5}$$

$$\sphericalangle H(j1) = 0° + 135° - 45° - 26.56° = 63.44°$$

Finally, as $\omega \to \infty$,

$$l_1, \, l_2, \, l_3 \to \infty$$

$$\theta_1, \, \theta_2, \, \theta_3 \to 90°$$

so that

$$\lim_{\omega \to \infty} |H(j\omega)| = 0$$

$$\lim_{\omega \to \infty} \sphericalangle H(j\omega) = 90° - 90° - 90° = -90°$$

In general, the graphical construction is not used as a calculation tool, but more as an aid in visualizing the salient features of the frequency response. This will become clear in the next few sections.

DRILL EXERCISE

11.4 Reasoning from Fig. E11.6, show that, for the system function of Example 11.6,

$$|H(j\omega)| = \frac{2}{(\omega^2 + 4)^{1/2}}$$

11.3.3 Frequency Response of Some Common Networks

FIRST-ORDER LOW-PASS RESPONSE

The system function defined by

$$H(s) = \frac{1}{sT + 1}$$

is called a *first-order low-pass system function.* Figure 11.5(a) shows a network for which $V_0/V_S = H(s)$, with $T = RC$. The frequency response is

$$H(j\omega) = \frac{1}{j\omega T + 1}$$

so that

$$|H(j\omega)| = \frac{1}{(\omega^2 T^2 + 1)^{\frac{1}{2}}} \tag{11.24}$$

and

$$\sphericalangle H(j\omega) = -\tan^{-1} \omega T \tag{11.25}$$

The magnitude and phase plots are shown in Fig. 11.6.

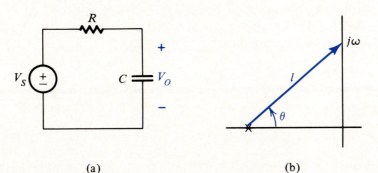

(a)   (b)

Figure 11.5 (a) A network with a first-order low-pass system function. (b) The pole-zero plot.

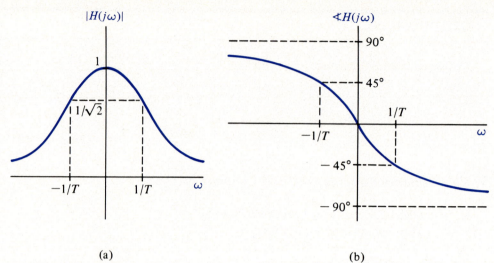

(a)   (b)

Figure 11.6 (a) Magnitude and (b) phase frequency response for the first-order low-pass system function.

This type of frequency response is termed low-pass because high frequencies are attenuated more than low frequencies. The *bandwidth* is defined as that frequency at which the magnitude is $1/\sqrt{2}$ times $|H(j0)|$. From Eq. (11.24), this occurs for $\omega = \omega_b = 1/T$.

This can also be deduced from Fig. 11.5(b). Since $H(s) = (1/T)/(s + 1/T)$ it follows that

$$|H(j\omega)| = \frac{1/T}{l} \qquad \text{and} \qquad \angle H(j\omega) = -\theta$$

Since $l = 1/T$ at $\omega = 0$, the magnitude will be down from $|H(j0)|$ by a factor of $\sqrt{2}$ for $l = \sqrt{2}/T$. This occurs for $\theta = 45°$, or $\omega = 1/T$. This also shows that $\angle H(j\omega) = -45°$ at $\omega = 1/T$, as can be verified from Eq. (11.25).

It is worth noting that the bandwidth is inversely proportional to the time constant; high bandwidth implies T small (fast step response), small bandwidth implies T large (slow step response).

EXAMPLE 11.7

A network such as that of Fig. 11.5(a) is to be designed to filter out a 60-Hz component (called "hum") from a low-frequency signal. Calculate the bandwidth such that $|H(j\omega)| = 0.01$ at a frequency of 60 Hz. For that bandwidth, what are the magnitude and phase of $H(j\omega)$ at $\omega = 2\pi \text{ s}^{-1}$?

Solution

From Eq. (11.24), with $\omega = 2\pi(60) = 377$,

$$|H(j377)| = \frac{1}{[(377T)^2 + 1]^{1/2}} = 0.01$$

or

$$(377T)^2 + 1 = 10^4$$

$$T \simeq \tfrac{100}{377} = 0.265 \text{ s}$$

and

$$\omega_b = \frac{1}{T} = 3.77 \text{ s}^{-1}$$

In hertz,

$$f_b = \frac{\omega_b}{2\pi} = 0.60 \text{ Hz}$$

At a frequency of 1 Hz, $\omega = 2\pi \text{ s}^{-1}$ and

$$|H(j2\pi)| = \frac{1}{[0.265^2(2\pi)^2 + 1]^{1/2}} = 0.515$$

and, from Eq. (11.25),

$$\angle H(j2\pi) = -\tan^{-1} 2\pi(0.265) = -59.01°$$

FIRST-ORDER HIGH-PASS RESPONSE

The system function defined by

$$H(s) = \frac{sT}{sT + 1}$$

is called a *first-order high-pass system function*. Figure 11.7(a) shows a network for which $V_O/V_S = H(s)$, with $T = RC$.

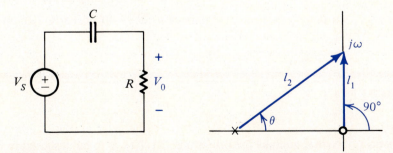

Figure 11.7 (a) A network with a first-order high-pass system function. (b) The pole-zero plot.

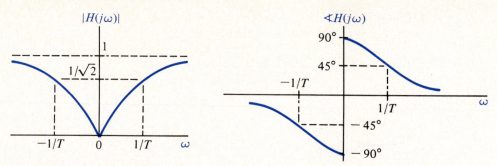

Figure 11.8 (a) Magnitude and (b) phase frequency response for the first-order high-pass system function.

The frequency response is

$$H(j\omega) = \frac{j\omega T}{j\omega T + 1}$$

so that

$$|H(j\omega)| = \frac{|\omega| T}{[\omega^2 T^2 + 1]^{\frac{1}{2}}} \tag{11.26}$$

and

$$\angle H(j\omega) = \begin{cases} 90° - \tan^{-1}\omega T & \omega > 0 \\ -90° - \tan^{-1}\omega T & \omega < 0 \end{cases} \tag{11.27}$$

The magnitude and phase are plotted in Fig. 11.8.

This type of frequency response is called high-pass because low frequencies are attenuated more than high frequencies. The *stop bandwidth* is defined as that frequency at which the magnitude is $1/\sqrt{2}$ times $|H(j\infty)|$. From Eq. (11.26), this occurs for $\omega = \omega_{sb} = 1/T$.

This can also be deducted from Fig. 11.1(b). Since $H(s) = s/(s + 1/T)$, it follows that

$$|H(j\omega)| = \frac{l_1}{l_2} \qquad \text{and} \qquad \angle H(j\omega) = 90° - \theta$$

Since $l_1/l_2 = \sin \theta$, $l_1/l_2 \to 1$ as $\omega \to \infty$, because $\theta \to 90°$. Clearly, $l_1/l_2 = 1/\sqrt{2}$ if $\omega = 45°$, which occurs at $\omega = 1/T$.

EXAMPLE 11.8

A high-pass network, such as the one in Fig. 11.7(a), is to be designed to block dc, but to pass a 20-Hz sinusoid with a phase shift of no more than 15°. Find the stop bandwidth of the filter.

Solution

From Eq. (11.27), with $\omega = 40\pi \text{ s}^{-1}$,

$$\angle H(j40\pi) = 90° - \tan^{-1} 40\pi T = 15°$$

or

$$\tan^{-1} 40\pi T = 75°$$

from which T can be computed as

$$T = 0.0297 \text{ s}$$

and, therefore,

$$\omega_{sb} = \frac{1}{T} = 33.67 \text{ s}^{-1}$$

In terms of hertz, $f_{sb} = 5.359$ Hz.

ALL-POLE SECOND-ORDER RESPONSE

The system function defined by

$$H(s) = \frac{\omega_0^2}{s^2 + s(\omega_0/Q) + \omega_0^2}$$

is a *second-order, all-pole* (i.e., no zeros) *system function*. Figure 11.9(a) shows a network for which $V_0/V_s = H(s)$, with $\omega_0 = 1/\sqrt{LC}$ and $Q = \omega_0 L/R$.

The pole-zero plot is shown in Fig. 11.9(b). The poles are $-\sigma \pm j\omega$ where, from Sec. 10.2, $\sigma = \omega_0/2Q$ and $\omega_m = \omega_0[1 - 1/4Q^2]^{1/2}$. They are complex num-

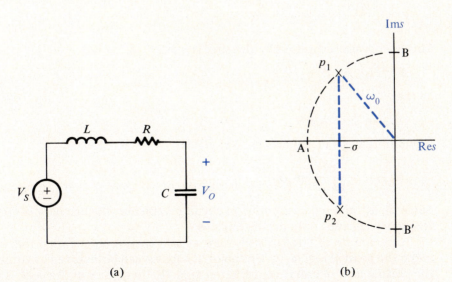

Figure 11.9 (a) A network with an all-pole transfer function. (b) The pole-zero plot.

bers if $Q > \frac{1}{2}$; it is easily verified that their magnitude is ω_0. As Q goes from $\frac{1}{2}$ to ∞, the poles move along the semicircle of radius ω_0 and go from A to B and B'.

The frequency response is given by

$$H(j\omega) = \frac{\omega_0^2}{-\omega^2 + j\omega\omega_0/Q + \omega_0^2}$$

$$= \frac{1}{1 - (\omega/\omega_0)^2 + j\omega/(\omega_0 Q)}$$

so that

$$|H(j\omega)| = \{[1 - (\omega/\omega_0)^2]^2 + (\omega/\omega_0)^2/Q^2\}^{-1/2} \tag{11.28}$$

and

$$\angle H(j\omega) = -\tan^{-1}\left[\frac{(\omega/\omega_0)/Q}{1 - (\omega/\omega_0)^2}\right] \tag{11.29}$$

Clearly,

$$|H(j0)| = 1 \qquad |H(j\infty)| = 0 \qquad |H(j\omega_0)| = Q$$

The magnitude is written as

$$|H(j\omega)| = \frac{\omega_0^2}{l_1 l_2} \tag{11.30}$$

where l_1 and l_2 are the lengths $\overline{MP_1}$ and $\overline{MP_2}$ in Fig. 11.10. It is instructive to study the behavior of $|H(j\omega)|$ by using geometric concepts. The triangle $P_1 P_2 M$ has an area A given by $\overline{P_1 P_2} \cdot \overline{OR}/2$, or

$$A = 2\frac{\omega_m \sigma}{2} = \omega_m \sigma$$

This area is also expressed by the sine law as

$$A = \frac{1}{2} l_1 l_2 \sin \phi$$

so that

$$\frac{1}{2} l_1 l_2 \sin \phi = \omega_m \sigma \tag{11.31}$$

Combining Eqs. (11.30) and (11.31) yields

$$|H(j\omega)| = \frac{\omega_0^2}{2\omega_m \sigma} \sin \phi \tag{11.32}$$

Since ω_0, ω_m, and σ depend only on circuit parameters, the behavior of $|H(j\omega)|$ vs frequency depends only on the angle ϕ.

A maximum of $|H(j\omega)|$ will occur at $\omega = \omega_{max}$, if ω_{max} is such that $\phi = 90°$ when the point M is at $j\omega_{max}$. This point, if it exists, can be localized by constructing the *peaking circle* of radius ω_m, centered at R. The frequency of the peak is just the

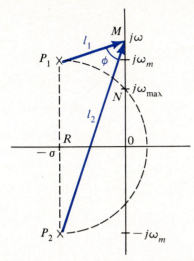

Figure 11.10 The pole-zero diagram with the peaking circle.

intersection N of the peaking circle with the imaginary axis. To see this, simply recall that an inscribed angle subtending a diameter is equal to 90°.

Using the triangle RON,

$$\omega_{max}^2 = \omega_m^2 - \sigma^2 \tag{11.33}$$

Therefore, from the expressions for σ and ω_m,

$$\omega_{max}^2 = \omega_0^2\left(1 - \frac{1}{4Q^2}\right) - \frac{\omega_0^2}{4Q^2}$$

or

$$\omega_{max} = \omega_0\sqrt{1 - \frac{1}{2Q^2}} \tag{11.34}$$

Substituting $\phi = 90°$ in Eq. (11.32) yields

$$|H(j\omega_{max})| = \frac{Q}{\sqrt{1 - 1/4Q^2}} \tag{11.35}$$

The peaking circle intersects the imaginary axis as long as $\omega_m > \sigma$. The borderline case $\omega_m = \sigma$ is called the *maximally flat* case; it corresponds to $Q = 1/\sqrt{2}$. If $Q < 1/\sqrt{2}$ then $\sigma < \omega_m$ and ϕ attains its maximum value ($< 90°$) at $\omega = 0$: $|H(j\omega)|$ decreases monotonically with ω.

If $2Q^2 \gg 1$, then $\omega_{max} \simeq \omega_0$ and $|H(j\omega_{max})| \simeq Q$.

From Eq. (11.29), it is seen that $\angle H(j0) = 0°$, $\angle H(j\infty) = -180°$ and $\angle H(j\omega_0) = -90°$. The parameter Q determines the sharpness of the change in the phase. This is illustrated in Fig. 11.11.

For large values of Q, the magnitude has *bandpass* characteristics, in that signals near $\omega = \omega_0$ are emphasized.

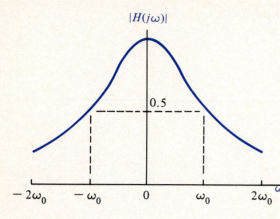
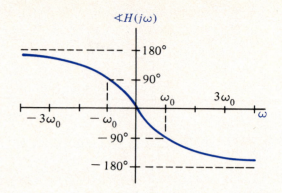

$Q = 1/2$

$Q = 1/\sqrt{2}$

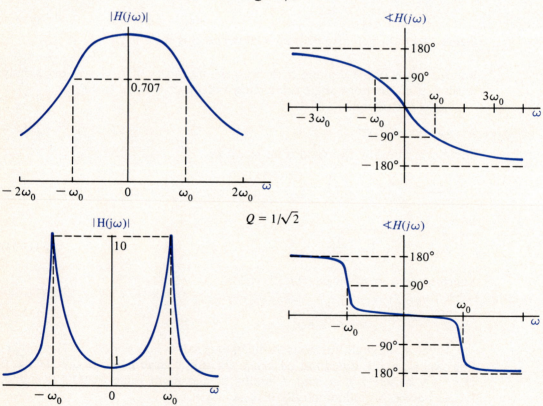

$Q = 10$

Figure 11.11 Magnitude and phase response for three values of Q.

The transient response and the frequency response are related, through Q and ω_0. A sharp peak in the frequency response magnitude indicates a high value of Q, hence an underdamped transient response. Since ω_m is proportional to ω_0, a high value of ω_m points to a fast response. (See Fig. E10.1.)

EXAMPLE 11.9

Solve the problem of Example 11.7, but with a maximally flat all-pole, second-order network, as in Fig. 11.9(a).

Solution

From Eq. (11.28), with $Q = 1/\sqrt{2}$,

$$|H(j377)| = \left\{\left[1 - \left(\frac{377}{\omega_0}\right)^2\right]^2 + 2\left(\frac{377}{\omega_0}\right)^2\right\}^{-1/2} = 0.01$$

so that

$$\left[1 - \left(\frac{377}{\omega_0}\right)^2\right]^2 + 2\left(\frac{377}{\omega_0}\right)^2 = 10^4$$

To simplify calculations, let $\Omega = (377/\omega_0)^2$. Then,

$$(1 - \Omega)^2 + 2\Omega = 10^4$$

$$\Omega^2 + 1 = 10^4$$

$$\Omega \simeq \pm 10^2$$

Therefore, taking the $+$ since $\Omega \geq 0$,

$$\frac{377}{\omega_0} = 10$$

and

$$\omega_0 = 37.7 \text{ s}^{-1}$$

Since $|H(j\omega_0)| = Q$ and since $Q = 1/\sqrt{2}$, this is also the bandwidth. It is 10 times the bandwidth obtained in Example 11.7. The reason is that the magnitude decreases more sharply in this case than in the case of the first-order low-pass; if the magnitude is the same in both cases at $\omega = 377 \text{ s}^{-1}$, it is higher at lower frequencies for the second-order system function.

At a frequency of 1 Hz, i.e., $\omega = 2\pi \text{ s}^{-1}$, Eq. (11.28) yields

$$|H(j2\pi)| = \left\{\left[1 - \left(\frac{2\pi}{37.7}\right)^2\right]^2 + 2\left(\frac{2\pi}{37.7}\right)^2\right\}^{-1/2} = 0.9996$$

and, from Eq. (11.29), the phase is

$$\angle H(j2\pi) = -\tan^{-1}\frac{\sqrt{2}\,(2\pi/37.7)}{1 - (2\pi/37.7)^2} = -13.63°$$

Because of the larger bandwidth, a 1-Hz sinusoid is transmitted with smaller changes in its magnitude and phase than in Example 10.7. This is an improvement, because the objective is normally to transmit low-frequency signals without change.

DRILL EXERCISES

11.5 Do Example 11.7, but with $|H(j\omega)| = 0.1$ at a frequency of 60 Hz.
Ans. $f_b = 6.028$ Hz, $|H(j2\pi)| = 0.986$, $\angle H(j2\pi) = -9.418°$

11.6 Design a high-pass network as in Example 11.8, such that the phase shift at 100 Hz be 10°.
Ans. $f_{sb} = 17.63$ Hz

11.7 For what values of Q are the approximations $\omega_0 = \omega_m$ and $|H(j\omega_m)| = Q$ in error by less than 1%?
Ans. $Q \geq 5.025$

11.3.4 Resonance

For large values of Q, the second-order transfer function is an example of a circuit that exhibits *resonance*. Resonance is said to occur at some frequency ω_m if the magnitude of the frequency response shows a sharp peak at ω_m. The word *sharp* is not precisely defined: the definition is a qualitative one.

In terms of poles and zeros, resonance will occur if complex poles are close to the imaginary axis. Figure 11.12 shows the pole-zero plot for $H(s) = K(s - z_1)/(s - p_1)(s - p_2)(s - p_3)$. For s near $j\omega_m$, the vectors drawn from $p_2, p_3,$ and z_1 to s are approximately equal to the vectors drawn from $p_2, p_3,$ and z_1 to $j\omega_m$; thus, ω_m can be substituted for ω in the terms corresponding to $p_2, p_3,$ and z_1 leading to

$$H(j\omega) \approx \frac{K(j\omega_m - z_1)}{(j\omega - p_1)(j\omega_m - p_2)(j\omega_m - p_3)} \tag{11.36}$$

It follows that

$$\frac{H(j\omega)}{H(j\omega_m)} = \frac{j\omega_m - p_1}{j\omega - p_1}$$

Substituting $p_1 = -\sigma + j\omega_m$ results in

$$\frac{H(j\omega)}{H(j\omega_m)} = \frac{\sigma}{j\omega - j\omega_m + \sigma}$$

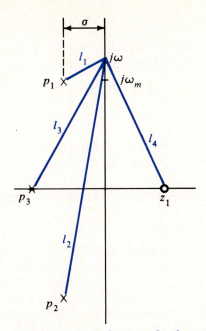

Figure 11.12 Pole-zero plot for a system function exhibiting resonance.

or

$$H(j\omega) = \frac{H(j\omega_m)}{j(\omega - \omega_m)/\sigma + 1} \qquad (11.37)$$

This should be compared to the first-order low-pass function $1/(j\omega T + 1)$; the system function $H(j\omega)$, plotted against $\omega - \omega_m$, is a first-order low-pass function, with $T = 1/\sigma$ and a (complex) gain $H(j\omega_m)$. This is illustrated in Fig. 11.13.

Clearly, this frequency response has bandpass characteristics. The magnitude peak is approximately at ω_m, called the resonant frequency. The bandwidth is defined to be the difference between the two frequencies where the magnitude is down from the peak by a factor of $\sqrt{2}$. In this case the bandwidth is

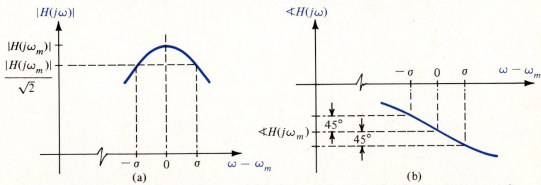

Figure 11.13 (a) Magnitude and (b) phase near $\omega = \omega_m$ for the resonant network.

$$\text{BW} = 2\sigma \tag{11.38}$$

From Eq. (10.11), given the Q factor associated with the complex poles p_1 and p_2, $\sigma = \omega_m/2Q$. From Eq. (11.38), the bandwidth, resonant frequency and Q are related by the simple expression

$$\text{BW} = \frac{\omega_m}{Q} \tag{11.39}$$

This expression is also valid if both the bandwidth and the resonant frequency are in hertz, since both sides of Eq. (11.39) can be divided by 2π.

EXAMPLE 11.10

The circuit of Fig. E11.10(a) is the simplified circuit of a JFET "front end" of a radio receiver. (Biasing considerations are postponed to Chap. 14.) The small-signal network is shown in Fig. E11.10(b). For AM broadcasting, a radio station is allocated a 8-kHz band, with center frequency between 540 kHz and 1600 kHz. In order to receive the station, it is necessary to amplify signals in the 8-kHz range covered by the station and reject the others.

For the network of Fig. E11.10(b):

1. Calculate V_o/V_i.

2. With L = 50 μH, calculate the range of the tunable capacitance C such that the resonant frequency may vary over the AM broadcast band.

3. Calculate R for a bandwidth of 8 kHz, for all resonant frequencies.

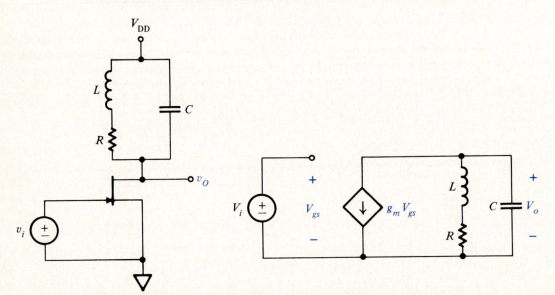

Figure E11.10 (a) A JFET tuned amplifier. (b) Its small-signal model.

Solution

$$V_o = \frac{-g_m V_{gs}}{sC + 1/(sL + R)}$$

and

$$\frac{V_o}{V_i} = \frac{-g_m(sL + R)}{s^2LC + sRC + 1}$$

The denominator is written as $s^2 + (R/L)s + 1/LC$, so that

$$\omega_0 = \frac{1}{\sqrt{LC}} \quad \text{and} \quad Q = \omega_0 \frac{L}{R}$$

For a reasonably high Q, $\omega_0 \simeq \omega_m$. (See Exercise 11.9.) Then, by Eq. (11.39),

$$BW = \frac{R}{L}$$

If L is in henries, C in microfarads, and R in kiloohms, time is in ms, hence frequency is in $(ms)^{-1}$, i.e., in kHz. The extremes of the range of C are

$$C = \frac{1}{50 \times 10^{-6}(2\pi \times 540)^2} = 1.737 \times 10^{-3} \ \mu F = 1.737 \ nF$$

and

$$C = \frac{1}{50 \times 10^{-6}(2\pi \times 1600)^2} = 1.979 \times 10^{-4} \ \mu F = 0.1979 \ nF$$

The value of R is found as

$$R = 50 \times 10^{-6} \times 2\pi \times 8 = 2.51 \times 10^{-3} \ k\Omega = 2.51 \ \Omega$$

DRILL EXERCISE

11.8 For the network of Fig. 11.9(a) with $L = 1$ mH, choose C and R such that the system function V_o/V_i has a resonant frequency at 10 kHz and a bandwidth of 100 Hz.
Ans. $C = 0.253 \ \mu F$, $R = 0.628 \ \Omega$

11.4 BODE PLOTS

The use of logarithmic scales is widespread in the display of frequency responses. A linear frequency scale makes it difficult to show, on the same graph, what happens between 1 and 10 Hz and also the behavior between 10 and 100 kHz; the interval 1 to 10 Hz would be all but invisible on a graph with a 100-kHz scale. The logarithmic scale, because it allocates the same interval to pairs of frequencies corresponding to the same ratio, makes this possible; both ranges represent a 10-to-1 increase in frequency and are allocated equal portions of the frequency axis.

A logarithmic scale is also used for the magnitude, again because of the need to display widely different values. Graphs of log magnitude and phase vs log frequency are called *Bode plots*, after Henry Bode who first proposed their use.

11.4.1 The Decibel

The magnitude of $H(j\omega)$ in *decibels* (dB) is defined as

$$|H(j\omega)|_{dB} = 20 \log|H(j\omega)| \tag{11.40}$$

where the logarithm is to the base 10.

The following table gives a few values of $|H(j\omega)|$ with the corresponding values in decibels:

$|H(j\omega)|$	$|H(j\omega)|$ dB
1	0
10	20
100	40
0.1	−20
0.01	−40
2	6
$\sqrt{2}$	3
0.5	−6
$\dfrac{1}{\sqrt{2}}$	−3

Note that the decibel scale goes up by equal *increments* if $|H(j\omega)|$ goes up by equal *factors*; for instance, a change by a factor of 10 in $|H(j\omega)|$ corresponds to an increment of 20 in the dB scale.

11.4.2 Bode Plots for Simple Terms

In this section the Bode plots are given for a few simple system functions. The next section will show how these simple plots are used to construct the Bode plots for any rational system function.

THE FACTORS s, $1/s$
Given $H(s) = s$, then

$$|H(j\omega)| = |\omega| \tag{11.41}$$

and

$$\sphericalangle H(j\omega) = 90° \tag{11.42}$$

Using Eq. (11.41), for $\omega > 0$,

$$20 \log |H(j\omega)| = 20 \log \omega \qquad (11.43)$$

For $\omega > 0$, a plot of $|H(j\omega)|$ in dB vs log ω is just a straight line of slope 20. This means that the straight line goes up 20 dB if log ω increases by 1. Since log ω increases by 1 if ω increases by a *factor* of 10, the slope is said to be 20 dB per decade.

From Eq. (11.43), $20 \log |H(j\omega)| = 0$ if $\omega = 1$; in other words, the straight line goes through 0 dB at $\omega = 1 \text{ s}^{-1}$.

For $H(s) = 1/s$, for $\omega > 0$,

$$|H(j\omega)| = \frac{1}{\omega}$$

$$20 \log |H(j\omega)| = -20 \log \omega \qquad (11.44)$$

and

$$\sphericalangle H(j\omega) = -90° \qquad (11.45)$$

Equations (11.44) and (11.45) show that the log magnitude and phase curves for $H(s) = 1/s$ are simply the negative of those for $H(s) = s$.

The Bode plots for the two system functions are shown in Fig. 11.14. The log ω axis is explicitly labeled in this figure; this is usually not necessary if one uses semilogarithmic paper.

The differential equation corresponding to $Y/X = s$ is $y(t) = dx/dt$; thus $H(s) = s$ is the system function of a differentiator.

If $Y/X = 1/s$, the differential equation is $dy/dt = x$, i.e., $y(t) = \int_{-\infty}^{t} x(\tau)\,d\tau$; thus, $H(s) = 1/s$ is the system function of an integrator.

THE FACTORS $sT + 1$, $1/(sT + 1)$

Given $H(s) = sT + 1$, with $T > 0$,

$$|H(j\omega)| = (\omega^2 T^2 + 1)^{1/2} \qquad (11.46)$$

and

$$\sphericalangle H(j\omega) = \tan^{-1} \omega T \qquad (11.47)$$

Taking logarithms on each side of Eq. (11.46),

$$20 \log |H(j\omega)| = 10 \log(\omega^2 T^2 + 1) \qquad (11.48)$$

It is not easy to plot this as it stands, so approximations are normally used, for low and high frequencies. With $\omega T \ll 1$, from Eq. (11.48),

$$20 \log |H(j\omega)| \simeq 10 \log 1 = 0 \qquad (11.49)$$

For $\omega T \gg 1$,

$$20 \log |H(j\omega)| \simeq 10 \log \omega^2 T^2 = 20 \log \omega T$$

$$= 20 \log \omega + 20 \log T \qquad (11.50)$$

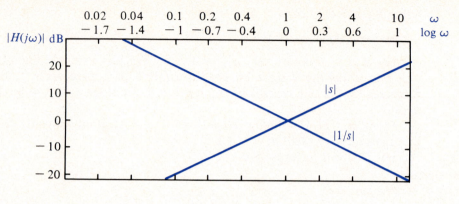

(a)

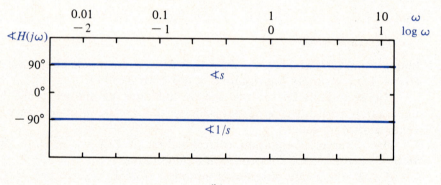

(b)

Figure 11.14 (a) Magnitude and (b) phase Bode plots for $H(s) = s$ and $H(s) = 1/s$.

Equation (11.49) represents the low-frequency asymptote, a simple horizontal line at 0 dB. The high-frequency asymptote, Eq. (11.50), is a straight line of slope 20 when plotted against log ω. This straight line crosses 0 dB at $\omega = 1/T$, called the *break frequency*.

The piecewise-linear approximation to $|H(j\omega)|$ consists of (1) the low-frequency asymptote for $\omega < 1/T$ and (2) the high-frequency asymptote for $\omega > 1/T$. The two asymptotes meet at the break frequency, where both have the value of 0 dB. The true value at $\omega = 1/T$ is, from Eq. (11.48),

$$20 \log \left| H\left(\frac{j}{T}\right) \right| = 10 \log(1 + 1) = 3 \text{ dB}$$

so that there is a 3-dB error in the piecewise-linear approximation. This turns out to be the worst-case error over the whole frequency range.

Since

$$\left| \frac{1}{j\omega T + 1} \right| = \frac{1}{|j\omega T + 1|}$$

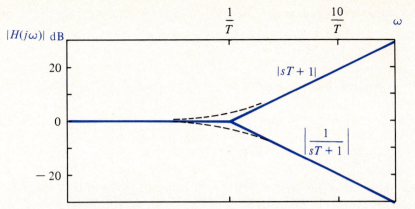

Figure 11.15 Piecewise-linear magnitude plots for $H(s) = sT + 1$ and $H(s) = 1/(sT + 1)$. The dotted curves show the exact values.

it follows that

$$20 \log \left| \frac{1}{j\omega T + 1} \right| = -20 \log |j\omega T + 1|$$

so that the magnitude plot for $H(s) = 1/(sT + 1)$ is just the negative of the plot for $H(s) = sT + 1$. Both magnitude plots are shown in Fig. 11.15.

Note that, for the first-order low-pass, $\omega = 1/T$ is that frequency at which the magnitude is $1/\sqrt{2}$, i.e., the bandwidth. Since $1/\sqrt{2}$ is equivalent to -3 dB, the term *3-dB bandwidth* is often used.

The phase plot, $\tan^{-1} \omega T$, is also difficult to plot against log ω. The low- and high-frequency values are 0 and 90°, respectively. A straight-line approximation joining those two values is shown in Fig. 11.16. It is constructed by joining the point $(0.1/T, 0°)$ to the point $(10/T, 90°)$. The approximation is exact at $\omega = 1/T$.

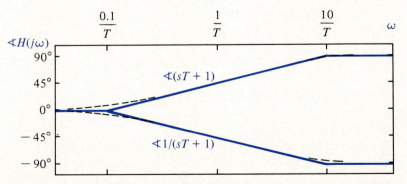

Figure 11.16 Piecewise-linear phase plots for $H(s) = sT + 1$ and $H(s) = 1/(sT + 1)$. The dotted curves show the true values.

Two other exact points may be used to improve the accuracy of the plot:

$$\tan^{-1}(\omega T) = 5.8° \quad \text{for } \omega = \frac{0.1}{T}$$

$$\tan^{-1}(\omega T) = 90° - 5.8° = 84.2° \quad \text{for } \omega = \frac{10}{T}$$

Since

$$\angle \frac{1}{j\omega T + 1} = -\angle(j\omega T + 1)$$

the phase curve for $H(s) = 1/(sT + 1)$ is the negative of that for $H(s) = sT + 1$. Both are shown in Fig. 11.16.

To be complete, one should also study the second-order term with complex roots. However, Bode plots are most useful where the poles and zeros are real. Therefore, it will not be discussed.

DRILL EXERCISE

11.9 For the factor $s^2/\omega_0^2 + s/(\omega_0 Q) + 1$, show that for the magnitude:

(a) The low-frequency asymptote is the 0-dB line.

(b) The high-frequency asymptote is a straight line of slope 40 that meets the low-frequency asymptote at $\omega = \omega_0$.

11.4.3 Bode Plots for System Functions with Real Poles and Zeros

Bode plots for system functions with real poles and zeros are easily constructed graphically. To demonstrate how this is done, consider

$$H(s) = \frac{K(sT_1 + 1)}{s(sT_2 + 1)(sT_3 + 1)}$$

By the rules of complex arithmetic,

$$|H(s)| = |K| \, |sT_1 + 1| \left| \frac{1}{s} \right| \left| \frac{1}{sT_2 + 1} \right| \left| \frac{1}{sT_3 + 1} \right| \tag{11.51}$$

and

$$\angle H(s) = \angle K + \angle(sT_1 + 1) + \angle \frac{1}{s}$$

$$+ \angle \frac{1}{sT_2 + 1} + \angle \frac{1}{sT_3 + 1} \tag{11.52}$$

Taking logarithms on each side of Eq. (11.51) yields

$$20 \log |H(s)| = 20 \log |K| + 20 \log |sT_1 + 1|$$

$$+ 20 \log \left| \frac{1}{s} \right| + 20 \log \left| \frac{1}{sT_2 + 1} \right| \qquad (11.53)$$

$$+ 20 \log \left| \frac{1}{sT_3 + 1} \right|$$

Equations (11.52) and (11.53) show that both the phase and the log magnitude are sums of first-order terms. If each term is plotted against log ω for $s = j\omega$, then the total phase and log magnitude curves are obtained by simple graphical addition. Note that it is the use of the dB measure that makes this possible, because the logarithm turns the product of magnitudes into a sum of logarithms.

EXAMPLE 11.11

Sketch the Bode plots for the high-pass function $H(s) = 10s/(10s + 1)$.

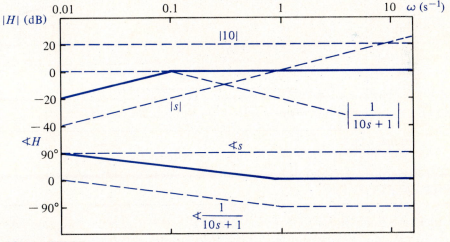

Figure E11.11

Solution

The solution is shown in Fig. E11.11. The dotted lines are the piecewise-linear approximations for the magnitude and phase of 10, s and $1/(10s + 1)$, for $s = j\omega$. The solid curves are the sums. Note in passing that, at low frequencies, the magnitude and phase are identical to those of $H(s) = s$. Thus, at low frequencies, i.e., below the break frequency $\omega = 0.1$, the first-order high-pass function behaves as a differentiator.

EXAMPLE 11.12

Sketch the piecewise-linear Bode plot for $H(s) = 2(s + 1)/s(s + 4)$.

Solution
The first step is to write $H(s)$ in such a way that the first-order terms appear in the form $sT + 1$. This yields

$$H(s) = 0.5 \frac{s + 1}{s(0.25s + 1)}$$

The Bode plots are shown in Fig. E11.12.

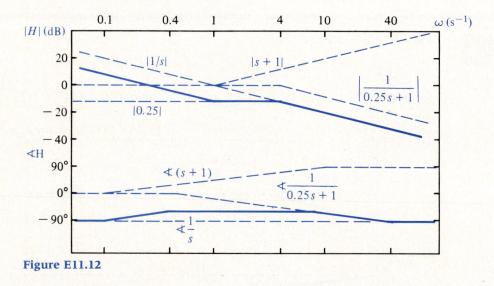

Figure E11.12

DRILL EXERCISE

11.10 Using Bode plots, verify the assertion that the first-order low-pass system function behaves as an integrator for high frequencies.

11.5 FREQUENCY RESPONSE OF SOME ELECTRONIC CIRCUITS

11.5.1 The Common-Emitter BJT Amplifier with Emitter Bypass

Figure 11.17 represents a common-emitter BJT amplifier. The resistances R_{B1}, R_{B2}, and R_E establish a stable quiescent point, as will be explained in Chap. 14. As will

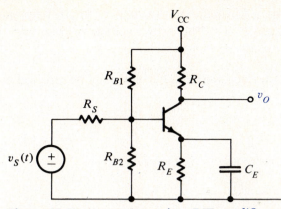

Figure 11.17 A common-emitter BJT amplifier.

be shown presently, the beneficial effect of R_E on the quiescent point is offset by an adverse effect on the small-signal gain. A *bypass capacitor C_E* is used in parallel with R_E, thus retaining the effect of R_E at dc but effectively shorting it out at sufficiently high frequencies.

The small-signal model for the amplifier is shown in Fig. 11.18(a).

Since

$$V_1 = \frac{(\beta + 1)I_b}{1/R_E + sC_E}$$

it follows that the equivalent admittance to the right of aa' in Fig. 11.18 is

$$\frac{I_b}{V_1} = \frac{1}{(\beta + 1)R_E} + \frac{sC_E}{\beta + 1}$$

As shown in Fig. 11.18(b), this is a parallel combination of a resistance $(\beta + 1)R_E$ and a capacitance $C_E/(\beta + 1)$.

The network to the left of aa' is represented in Fig. 11.18(b) by the Thevenin

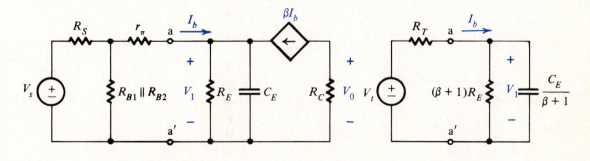

(a)   (b)

Figure 11.18 (a) Small-signal model for the common-emitter amplifier.
(b) Simplification using equivalent networks.

source V_t and resistance R_T. It is usually the case that $R_{B1} \| R_{B2}$ is large compared to R_S, so that $R_T \simeq R_S + r_\pi$ and $V_t \simeq V_s$. The output voltage is given by

$$V_o = -\beta R_C I_b \tag{11.54}$$

At low frequencies, the capacitance $C_E/(\beta + 1)$ may be replaced by an open circuit and

$$I_b = \frac{V_t}{R_T + (\beta + 1)R_E}$$

Using Eq. (11.54), the low-frequency voltage gain is

$$A_{v0} = \frac{V_o}{V_s} \simeq \frac{V_o}{V_t} = -\frac{\beta R_C}{R_T + (\beta + 1)R_E} \tag{11.55}$$

At high frequencies, the capacitance becomes a short circuit and

$$I_b = \frac{V_t}{R_T}$$

so that the high-frequency voltage gain is

$$A_{v\infty} = \frac{V_o}{V_s} \simeq \frac{V_o}{V_t} = \frac{-\beta R_C}{R_T} \tag{11.56}$$

In practice, $(\beta + 1)R_E \gg R_T$; therefore, the high-frequency gain is appreciably greater than the low-frequency gain.

The system function is now calculated from Fig. 11.18(b),

$$\frac{V_t}{I_b} = R_T + \frac{1}{1/(\beta + 1)R_E + sC_E/(\beta + 1)}$$

$$= R_T + \frac{(\beta + 1)R_E}{1 + sR_E C_E}$$

A few steps of algebra lead to

$$\frac{I_b}{V_t} = \frac{1}{R_T + (\beta + 1)R_E} \frac{sT_2 + 1}{sT_1 + 1} \tag{11.57}$$

where

$$T_1 = C_E R_E \frac{R_T}{R_T + (\beta + 1)R_E} \tag{11.58}$$

and

$$T_2 = C_E R_E \tag{11.59}$$

Using Eqs. (11.54) and (11.55),

$$\frac{V_o}{V_s} \simeq \frac{V_o}{V_t} = \frac{A_{v0}(sT_2 + 1)}{sT_1 + 1} \tag{11.60}$$

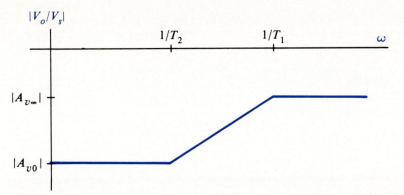

Figure 11.19 Bode magnitude plot for the BJT common-emitter amplifier.

It is apparent from Eqs. (11.58) and (11.59) that $T_1 < T_2$. Therefore, the Bode magnitude plot is as shown in Fig. 11.19. If $(\beta + 1)R_E \gg R_T$, then, from Eq. (11.58),

$$T_1 \simeq C_E \frac{R_T}{\beta + 1} \tag{11.61}$$

For good low-frequency response, the 3-dB stop bandwidth $1/T_1$ should be made as small as possible. This requires that T_1 be large, which in turn forces C_E to be a large capacitance, since $R_T/(\beta + 1)$ is usually small.

A *coupling capacitor* C_C is often added, as in Fig. 11.20(a). This has the effect of isolating the amplifier from the signal source at dc, so that the quiescent point is undisturbed by the source.

Figure 11.20(b) shows the small-signal model for the *RC*-coupled amplifier. Since C_C has a large impedance at low frequencies, the low-frequency voltage gain

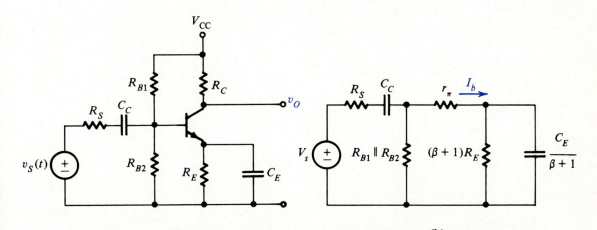

(a)   (b)

Figure 11.20 (a) An *RC*-coupled common-emitter amplifier. (b) Its small-signal model.

is small. This is of little consequence, as long as the frequency response is essentially undisturbed for ω near or above $1/T_1$. Assuming $R_{B1} \| R_{B2} \gg r_\pi$, C_C will behave as a short circuit if

$$\frac{1}{\omega C_C} \ll R_S + r_\pi$$

at $\omega = 1/T_1$, i.e., if

$$C_C(R_S + r_\pi) \gg T_1 \qquad\qquad (11.62)$$

EXAMPLE 11.13

(a) In the circuit of Fig. 11.17, $r_\pi = 100\ \Omega$, $R_C = R_S = R_E = 1\ \mathrm{k\Omega}$, $\beta = 50$, $R_{B1} \| R_{B2} = 20\ \mathrm{k\Omega}$. Calculate C_E so that the stop bandwidth is 100 Hz. Calculate the low- and high-frequency gains.

(b) Add a coupling capacitor as in Fig. 11.20(a), choosing C_C so as to leave the response essentially as in (a) for $f \geq 100$ Hz.

Solution

(a) Since $R_{B1} \| R_{B2}$ is large compared to r_π and R_S, it follows that

$$R_T \simeq R_S + r_\pi = 1.1\ \mathrm{k\Omega}$$

Since $(\beta + 1)R_E = 51\ \mathrm{k\Omega}$ is large compared to R_T, Eq. (11.61) may be used and

$$T_1 = \frac{1.1C_E}{51}$$

For a break frequency of 100 Hz,

$$T_1 = \frac{1}{2\pi 100} = 1.59\ \mathrm{ms}$$

Therefore,

$$C_E = 51\frac{1.59}{1.1} = 73.7\ \mu\mathrm{F}$$

From Eq. (11.55), the low-frequency gain is

$$A_{v0} \simeq -\frac{50}{51 + 1.1} = -0.96$$

while the high-frequency gain is given by Eq. (11.56) as

$$A_{v\infty} = -\frac{50}{1.1} = -45$$

(b) By Eq. (11.62), one must have

$$C_C(1.1) \gg 1.59$$

or

$$C_C \gg 1.45 \ \mu\text{F}.$$

In practice, $C_C \simeq 10 \ \mu\text{F}$ would be adequate.

DRILL EXERCISE

11.11 Rework Example 11.13 with $\beta = 100$, all other data remaining unchanged.
Ans. $C_E = 146 \ \mu\text{F}$, $C_C \gg 1.45 \ \mu\text{F}$, $A_{v0} = -0.98$, $A_{v\infty} = -90.9$

11.5.2 The JFET Common-Source Amplifier at High Frequencies

The transistor models of Chap. 6 are essentially low-frequency models. At frequencies beyond approximately 1 MHz the JFET model must be modified to include energy storage effects.

Figure 11.21 shows a suitable small-signal model for the JFET at high frequencies. The capacitances are of the order of 1 pF and r_d is of the order of 0.1 MΩ. Figure 11.22(a) shows an RC-coupled JFET amplifier. The small-signal model is shown in Fig. 11.22(b). Since the objective is to analyze the high-frequency behavior, the coupling capacitance C_C, a rather large capacitance, is replaced by a short circuit.

As shown in Fig. 11.23, the capacitance C_{gd} can be thought of as a two-port, with

$$I_1 = C_{gd}sV_{gs} - C_{gd}sV_o$$
$$I_2 = -C_{gd}sV_{gs} + C_{gd}sV_o$$

leading to the y-parameter description shown.

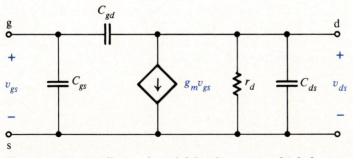

Figure 11.21 Small-signal model for the JFET at high frequencies.

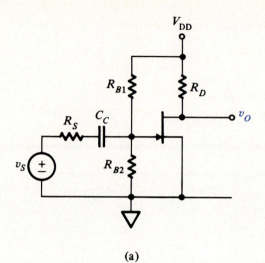

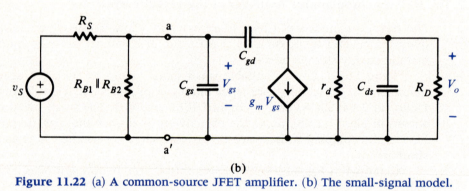

Figure 11.22 (a) A common-source JFET amplifier. (b) The small-signal model.

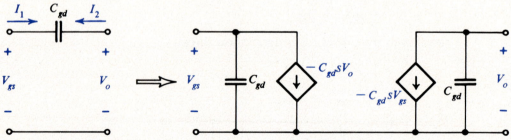

Figure 11.23 Two-port description of the gate-to-drain capacitance.

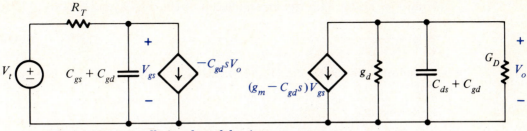

Figure 11.24 Small-signal model using y parameters.

This two-port description is incorporated in Fig. 11.24, together with a Thevenin equivalent of the network to the left of aa' in Fig. 11.22.

In order to simplify the rest of the analysis, it is assumed that ω is such that

$$g_m \gg C_{gd}\omega \qquad (11.63)$$

and

$$g_d + G_D \gg (C_{ds} + C_{gd})\omega \qquad (11.63)$$

In terms of the circuit, the output controlled source is then just $g_m V_{gs}$, and the output capacitance $C_{ds} + C_{gd}$ can be replaced by an open circuit. Under these assumptions, the voltage gain is

$$A_v = \frac{V_o}{V_{gs}} = \frac{-g_m}{g_d + G_D} \simeq -g_m R_D \qquad (11.64)$$

where the last step assumes $g_d \ll G_D$, i.e., $r_d \gg R_D$. The input side of Fig. 11.24 is shown in Fig. 11.25 with V_o replaced by $A_v V_{gs}$. The controlled-source current depends on the voltage across it, and in fact is the same as would flow in a capacitance $-A_v C_{gd}$. Accordingly, it is replaced by a capacitance, and the three parallel capacitances are added to yield

$$C_{eq} = C_{gs} + (1 - A_v)C_{gd} \qquad (11.65)$$

The capacitance C_{gd} is seen to be multiplied by the voltage gain A_v: this is known as the *Miller effect*.

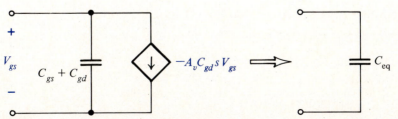

Figure 11.25 Manipulation of the input side to produce the Miller capacitance.

From Fig. 11.24, with the manipulation of Fig. 11.25, one obtains

$$\frac{V_{gs}}{V_t} = \frac{1}{sR_TC_{eq} + 1} \tag{11.66}$$

Combining this with Eq. (11.64) leads to the overall voltage gain system function

$$\frac{V_o}{V_t} = \frac{-g_mR_D}{sR_TC_{eq} + 1} \tag{11.67}$$

This is a first-order low-pass system function, with a 3-dB bandwidth of

$$f_b = \frac{1}{2\pi R_T C_{eq}} \tag{11.68}$$

If one assumes $|A_v| \gg 1$, $C_{gs} \ll |A_v|C_{gd}$, then $C_{eq} \simeq |A_v|C_{gd}$ and

$$f_b \simeq \frac{1}{2\pi R_T C_{gd}|A_v|}$$

or

$$|A_v|f_b \simeq \frac{1}{2\pi R_T C_{gd}} \tag{11.69}$$

This product is called the *gain-bandwidth product*. It depends only on the source resistance and on the gate-to-drain capacitance.

EXAMPLE 11.14

In the small-signal model of Fig. 11.22(b), $r_d = 100$ kΩ, $g_m = 10$ mS, $C_{gs} = C_{ds} = 2$ pF, $C_{gd} = 1$ pF, $R_S = R_D = 2$ kΩ, $R_{B1} \| R_{B2} = 20$ kΩ. Calculate the 3-dB bandwidth and the low-frequency voltage gain V_o/V_s.

Solution
Since $r_d \gg R_D$, Eq. (11.64) is used and yields

$$A_v = -g_m R_D = -20$$

The Miller capacitance is, from Eq. (11.65),

$$C_{eq} = 2 + (21)1 = 23 \text{ pF}$$

so that, using Eq. (11.68), with kΩ and μF,

$$f_b = \frac{1}{2\pi(2\|20)(23 \times 10^{-6})}$$

$$= 3806 \text{ ms}^{-1} = 3806 \text{ kHz} = 3.806 \text{ MHz}$$

Since $R_{B1} \| R_{B2} \gg R_S$, the Thevenin voltage $V_t \simeq V_s$ and therefore the low-frequency gain V_o/V_s is A_v.

One should verify that the assumptions of Eq. (11.63), on which the whole analysis is based, are indeed verified at $\omega = 2\pi f_b$. Since

$$C_{gd}\omega = (1 \times 10^{-6})(3806 \times 2\pi) = 0.024 \ll 10 = g_m$$

and

$$(C_{ds} + C_{gd})\omega = (3 \times 10^{-6})(3806 \times 2\pi)$$

$$= 0.072 \ll 0.502 = g_d + G_D$$

they are indeed justified.

For this amplifier, the gain-bandwidth product is, from Eq. (11.69),

$$A_v f_b = \frac{1}{2\pi(2\|20)(1 \times 10^{-6})}$$

$$= 87535 \text{ kHz} = 87.535 \text{ MHz}$$

This specifies the tradeoff between gain and bandwidth.

DRILL EXERCISE

11.12 Repeat Example 11.14, but with all capacitance values divided by 100, as might be the case if an IC MOSFET was used.
Ans. $A_v = -20$, $f_b = 381$ MHz

11.5.3 The BJT Common-Emitter Amplifier at High Frequencies

At high frequencies, the BJT model of Chap. 7 is modified to yield the so-called *hybrid-π* model of Fig. 11.26. Typical values are $r_x = 10\ \Omega$, $r_\pi = 1\ \text{k}\Omega$, $C_\pi = 100$ pF, $C_\mu = 1$ pF, $g_m = 50$ mS. Figure 11.27(a) shows the circuit of a common-emitter amplifier. The small-signal model is given in Fig. 11.27(b).

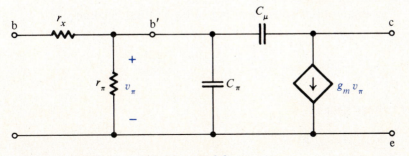

Figure 11.26 The BJT hybrid-π model.

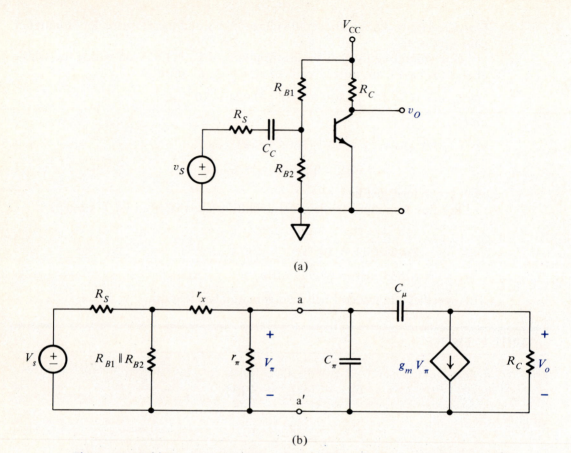

(a)

(b)

Figure 11.27 (a) A common-emitter BJT amplifier. (b) The small-signal model.

The networks to the right of aa' in both Figs. 11.22(b) and 11.27(b) are identical in structure. It suffices to replace C_{gs} by C_π, C_{gd} by C_μ, $r_d \| R_D$ by R_C and to make $C_{ds} = 0$ to generate the BJT circuit from the JFET circuit. If, in both cases, the network to the left of aa' is replaced by a Thevenin source V_t and resistance R_T, the results of the JFET analysis can be used directly for the BJT, with appropriate changes in the parameters.

The assumptions corresponding to Eq. (11.63) become

$$g_m \gg C_\mu \omega \qquad (11.70a)$$

and

$$G_C \gg C_\mu \omega \qquad (11.70b)$$

The voltage gain is

$$\frac{V_o}{V_\pi} = A_v = -g_m R_C \qquad (11.71)$$

The system function is

$$\frac{V_o}{V_t} = \frac{-g_m R_C}{s R_T C_{eq} + 1}$$

(11.72)

where

$$C_{eq} = C_\pi + (1 - A_v)C_\mu$$

(11.73)

The 3-dB bandwidth is

$$f_b = \frac{1}{2\pi R_T C_{eq}}$$

(11.74)

and the gain-bandwidth product is

$$|A_v| f_b \simeq \frac{1}{2\pi R_T C_\mu}$$

(11.75)

EXAMPLE 11.15

In the small-signal model of Fig. 11.27(b), $r_x = 10 \ \Omega$, $r_\pi = 900 \ \Omega$, $R_{B1} \| R_{B2} = 20 \ k\Omega$, $R_S = R_C = 2 \ k\Omega$, $g_m = 50 \ mS$, $C_\pi = 100 \ pF$, $C_\mu = 2 \ pF$. Find the 3-dB bandwidth and the low-frequency gain V_o/V_s.

Solution
From Eq. (11.71),

$$A_v = -50(2) = -100$$

From Eq. (11.73), the Miller capacitance is

$$C_{eq} = 100 + 101(2) = 302 \ pF$$

or

$$C_{eq} = 302 \times 10^{-6} \ \mu F$$

Since $R_{B1} \| R_{B2}$ is large compared to R_S, the Thevenin resistance is approximated by

$$R_T = r_\pi \| (r_x + R_S) = 0.9 \| 2.01 = 0.622 \ k\Omega$$

The 3-dB bandwidth is calculated from Eq. (11.74) as

$$f_b = \frac{1}{2\pi(0.622)(302 \times 10^{-6})}$$

$$= 847 \ ms^{-1} = 847 \ kHz$$

From Fig. 11.27(b), the Thevenin voltage is

$$V_t = \frac{0.9}{2.91} V_s = 0.309 V_s$$

Therefore, at low frequencies,

$$\frac{V_o}{V_s} = 0.309 \frac{V_o}{V_t} = -30.9$$

The reader will verify that $C_\mu \omega_b = 0.0105$, so that the conditions of Eq. (11.70) are verified.

DRILL EXERCISE

11.13 For the common-emitter amplifier, show that the gain-bandwidth product is always less than $(r_x + r_\pi)/2\pi C_\mu r_x r_\pi$.

11.6 SUMMARY AND STUDY GUIDE

This chapter was devoted to the study of the sinusoidal steady state in linear networks. It proved convenient to use phasors to describe sinusoidal signals: the geometric interpretation was studied through phasor diagrams. The particular solution of linear networks for sinusoidal excitation led directly to equivalent networks at a single frequency and to the concept of frequency response. Special attention was given to a few basic cases, including resonance. Bode plots were introduced and used in the study of some important electronic circuits.

There are many particular points worthy of attention in this chapter. Some of them are as follows.

A sinusoidal input leads to a steady-state sinusoidal output of the same frequency.

The magnitude of the output phasor is that of the input phasor, times the magnitude of the system function at $j\omega$; the phase of the output phasor is that of the input phasor, plus the phase of the system function.

Simple, two-element equivalent networks can be derived for an impedance or admittance, valid at one frequency only.

Phasors, as complex numbers, are added like vectors.

The main features of the frequency response can be deduced from the behavior of vectors drawn from the system function poles and zeros to points on the $j\omega$ axis.

The first-order low-pass and high-pass functions and the second-order, all-pole function are simple but useful cases that serve as building blocks for more complex cases.

In the case of resonance, it is only necessary to study the frequency response near the resonant frequency.

The plotting of Bode diagrams is basically an exercise in the addition of straight-line segments.

In an ac-coupled common-emitter BJT amplifier, the stop bandwidth is determined by the emitter bypass capacitance.

The high-frequency behavior of the JFET common-source and BJT common-emitter amplifier are similar. In both cases, there is a tradeoff between gain and bandwidth.

PROBLEMS

11.1 Find the phasor representation, in terms of cosine $[x(t) = \text{Re}(Xe^{j\omega_p t})]$ and in terms of sine $[x(t) = \text{Im}(Xe^{j\omega_p t})]$, for the following sinusoids:
(a) $x(t) = 3 \cos(2t - 30°)$
(b) $x(t) = 2 \sin(3t + 45°)$

11.2 Repeat Prob. 11.1 for the following sinusoids:
(a) $x(t) = -2 \sin t$
(b) $x(t) = -4 \cos(4t + 30°)$

'11.3 For each of the system functions $Y/X = H(s)$ given, find the steady-state output $y(t)$ for $x(t) = 3 \cos(2t - 45°)$.
(a) $H(s) = 2\dfrac{s - 1}{s(s^2 + s + 1)}$
(b) $H(s) = 2\dfrac{s(s + 1)}{(s + 2)(s + 3)}$

11.4 Repeat Prob. 11.3 for $x(t) = 4 \sin(3t + 30°)$.

11.5 A network is known to have a system function of the form $Y/X = H(s) = \dfrac{as + b}{s^2 + cs + d}$. Find a, b, c, and d, assumed real, given that

The steady-state response to $x(t) = \cos t$ is $y(t) = 1.581 \cos(t + 71.6°)$
The steady-state response to $x(t) = \cos(3t + 45°)$ is $y(t) = .798 \cos(3t - 12.34°)$

11.6 For the network of Fig. P11.6, find two-element series and parallel networks equivalent to the network at ab for $\omega = 0.5 \text{ ms}^{-1}$ and 5 ms^{-1}. Give the resistance, conductance, reactance, and susceptance in each case.

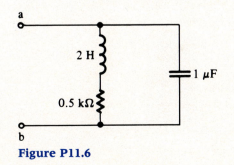

Figure P11.6

11.7 Repeat Prob. 11.6 for the network of Fig. P11.7.

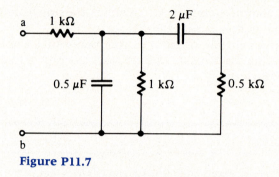

Figure P11.7

11.8 (a) Show that a mutual inductance element is represented at a frequency ω by the two-port network of Fig. P11.8(a).
   (b) Calculate the resistance and reactance seen at *ab* in Fig. P11.8(b) for $\omega = 0.5$ ms^{-1} and 5 ms^{-1}.

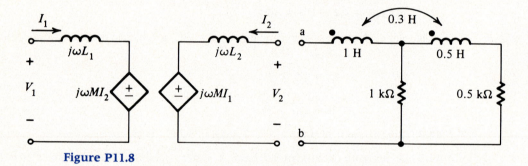

Figure P11.8

11.9 An inductor is modeled as a series combination of an inductance L and a resistance R_L. The values of L and R_L are to be determined by voltmeter measurements performed on the circuit of Fig. P11.9. Since a voltmeter only measures magnitude, phase information is not available. If $v_S(t)$ is a 60-Hz sinusoid of peak values 155 V, the peak values of $v_1(t)$ and $v_L(t)$ are 60 V and 120 V, respectively, in the sinusoidal steady state. Given that $R_1 = 50$ Ω, calculate L and R_L.

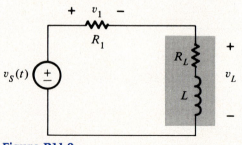

Figure P11.9

11.10 Figure P11.10 is a simple model of a power transmission system, where Z_l represents the transmission line and Z_L is the load. The source $v_S(t)$ is a 60-Hz sinusoidal voltage.
(a) For $Z_l = 40 + j200\ \Omega$ and $Z_L = 800 + j600\ \Omega$, calculate $|V_L|$.
(b) For Z_l and Z_L as in (a), calculate the value of the capacitance to be connected in parallel with Z_L so that $|V_L| = |V_S|$. This problem shows how capacitors can be used to control voltage in power systems.

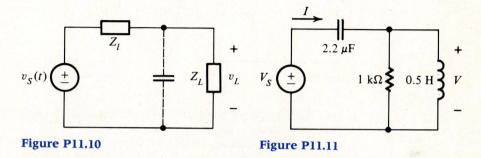

Figure P11.10   **Figure P11.11**

11.11 For the network of Fig. P11.11, calculate the system function I/V_S and V/V_S at low and high frequencies, using approximate networks. State conditions on ω for which your approximations are valid.

11.12 Repeat Prob. 11.11 for the network of Fig. P11.12.

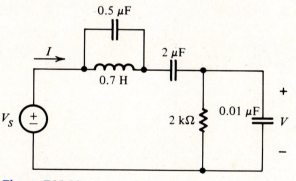

Figure P11.12

11.13 Repeat Prob. 11.11 for the network of Fig. P11.13, but calculate instead V/I_S and I/I_S.

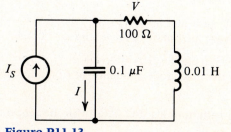

Figure P11.13

11.14 Repeat Prob. 11.13 for the network of Fig. P11.14.

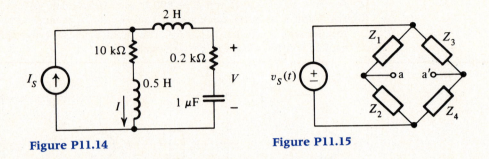

Figure P11.14   **Figure P11.15**

11.15 (a) For the bridge network of Fig. P11.15, show that the bridge is balanced, i.e., the Thevenin voltage at aa' is zero, if $Z_1/Z_2 = Z_3/Z_4$.
(b) Suppose Z_1 is an inductor, modeled as a series R and L. With $Z_2 = R_2$ and $Z_4 = R_4$, what elements should be connected as Z_3 for the bridge to be balanced at all frequencies?
(c) Show that the bridge can be balanced at any one frequency by using a parallel RL combination as Z_3, with $Z_2 = R_2$ and $Z_4 = R_4$.

11.16 In the network of Fig. P11.16, assume that $I = 1 \angle 0°$. Sketch the phasor diagram for I, V_R, V_L, V_C, and V_S at $\omega = 0.5$, $\sqrt{2}$, and 2 s^{-1}. Modify the diagram if V_S, rather than I, is given as $V_S = 5 \angle 0°$.

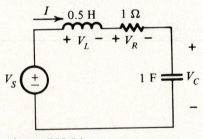

Figure P11.16

11.17 Using a phasor diagram, solve graphically Prob. 11.9. (*Hint:* Assume initially that the current phasor has zero angle.)

11.18 Draw to scale the pole-zero plot for $H(s) = (s + 1)/s(s^2 + s + 1)$ and determine graphically the magnitude and phase of $H(j\omega)$, for $\omega = 0.2$, 0.5, 1, and 10.

11.19 Draw to scale the pole-zero plot for $H(s) = 2(s^2 + 9)/(s + 1)(s^2 + 0.4s + 4.04)$ and determine graphically the magnitude and phase of $H(j\omega)$, for $\omega = 0$, 1, 1.5, 2, 2.5, 3, and 5.

11.20 Match each pole-zero pattern of Fig. P11.20(a) with the proper magnitude plot in Fig. P11.20(b).

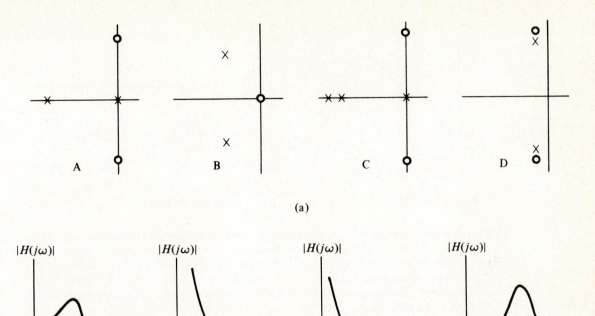

(a)

Figure P11.20

11.21 Match each pole-zero pattern of Fig. P11.20(a) with the proper phase plot in Fig. P11.21.

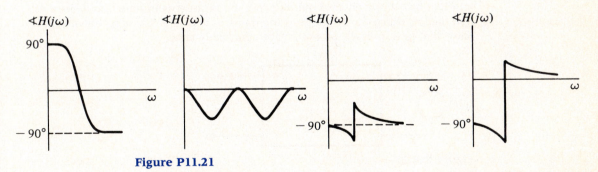

Figure P11.21

11.22 Figure P11.22 shows the response $y(t)$ of a first-order, low-pass network to a pulse $x(t)$. The rise time t_r of the response is defined as the time taken to go from 10 to 90% of the final value, as illustrated. Assume that, in a particular application, t_r can be at most $T_P/5$ for the pulse response to be called satisfactory. What is the minimum bandwidth required for satisfactory response, as a function of T_P? Calculate this bandwidth for $T_P = 1$ ms, 1 μs, and 1 ns.

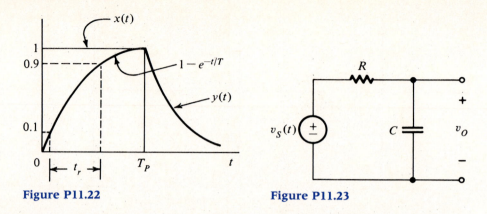

Figure P11.22

Figure P11.23

11.23 The network of Fig. P11.23 is intended to measure the constant component (i.e., the average value) of a signal $v_S(t)$. The signal consists of a constant, plus a variable part containing sinusoidal components of frequency greater than 1 kHz. Calculate the network time constant so that the frequency response V_o/V_s at 1 kHz is 0.05. For that value of the time constant, calculate the time required for the step response to reach 99% of its final value.

11.24 Repeat Prob. 11.23, but use a second-order, all-pole network with $Q = 1/\sqrt{2}$. Calculate ω_0 instead of the time constant, and the time required for the envelope of the step response to come within 1% of its final value.

11.25 The network in the box in Fig. P11.25 has an all-pole, second-order system function. Such a network may be used as a simple spectrum analyzer; if Q is high, the output will be large if $v_S(t)$ contains a sinusoid of frequency close to ω_0, small if not. The *resolution* is a measure of the ability of the filter to discriminate between two input frequencies that are close together. The bandwidth is a measure of the resolution.

In order to measure the steady-state output, it is necessary to wait until the transient has effectively died out. Suppose the measurement time t_M is defined as that time at which the envelope of the transient decays to 2% of its initial value. Show that t_M is inversely proportional to the bandwidth.

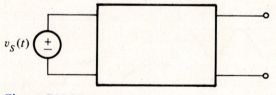

Figure P11.25

11.26 Sketch the frequency response for the so-called *all-pass* system function $H(s) = (-sT + 1)/(sT + 1)$.

11.27 A high-quality loudspeaker system normally has two different coils, each operating in its own frequency range. The audio amplifier output is fed to the speaker coils through two different filters: low-pass and high-pass. This is shown in Fig. P11.27.

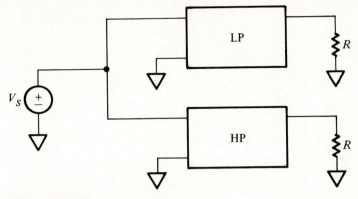

Figure P11.27

The desired system functions are

Low-pass: $\dfrac{1}{sT_1 + 1}$

High-pass: $\dfrac{sT_2}{sT_2 + 1}$

(a) Select the T's such that the low-pass bandwidth and the stop bandwidth of the high-pass filter are 1 kHz.

(b) Given $R = 8\ \Omega$, design filters with the required system functions.

11.28 Using low- and high-frequency approximations, show that the system function of the network of Fig. P11.28 behaves as a high-pass function at low frequencies and a low-pass function at high frequencies. Sketch the frequency response V/V_s.

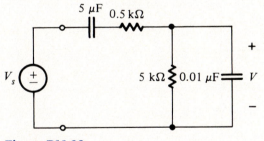

Figure P11.28

11.29 Figure P11.29(a) shows the pole-zero plot of a *double-tuned* resonant network. The two complex poles near $\pm j\omega_m$ are the important ones; they are close to each other and the j axis, compared to their distance to all other poles and zeros. The area of the pole-zero plot near $j\omega_m$ is shown in greater detail in Fig. P11.29(b).

(a) For frequencies ω near ω_m it is assumed that the magnitude and phase contributions of all poles and zeros are constant, except for the poles near ω_m. Show that the frequency response is that of an all-pole, second-order function shifted in frequency by ω_m.

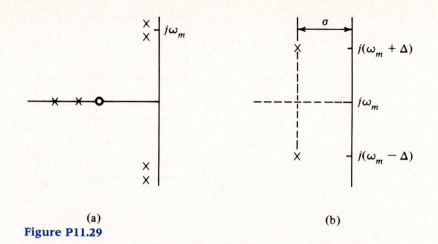

(a)   (b)

Figure P11.29

(b) Sketch the frequency response magnitude near ω_m for $\sigma = 0.2\Delta$, Δ, and 2Δ. Explain the advantage of using $\sigma = \Delta$ for a bandpass filter.

11.30 Figure P11.30 shows the pole-zero plot of a *notch filter*, designed to block the transmission of sinusoids of frequency ω_0.

(a) For frequencies ω near ω_0, it is assumed that the magnitude and phase contributions of all poles and zeros are constant, except for the pole-zero pair near ω_0. Show that the frequency response is that of a high-pass function shifted in frequency to ω_0.

(b) Sketch the frequency response for ω near ω_0.

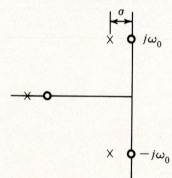

Figure P11.30

11.31 On semilogarithmic paper, sketch the Bode diagram for $H(s) = 2(s + 2)(s + 3)/s(s + 10)(s + 50)$.

11.32 Repeat Prob. 11.31 for $H(s) = 2s(s + 10)/(s + 1)^2(s + 50)$.

11.33 Repeat Prob. 11.31 for $H(s) = 3s^2/(s + 1)(s + 10)(s + 50)$.

11.34 (a) Find the magnitude and phase of $sT + 1$, for $s = j\omega$ and T *negative*, in terms of the same quantities for T positive.

(b) Sketch the Bode diagram for $H(s) = 2(s - 1)/s(s + 10)$.

11.35 Bode plots are sometimes used for second-order factors with complex roots, provided Q is relatively low. For the case of $Q = 1/\sqrt{2}$ (maximally flat), sketch the Bode plot by (a) drawing the asymptotes (Drill Exercise 11.9) and (b) calculating a few values for ω/ω_0 near 1 for the second-order all-pole function. On the same graph, plot the magnitude of the first-order low-pass function with bandwidth ω_0. Discuss the possible advantages in using the second-order network as a low-pass.

11.36 The network of Fig. P11.36 contains a model of a transformer with perfect coupling (see Chap. 4). Calculate the frequency response V_o/V_s and sketch the Bode diagram.

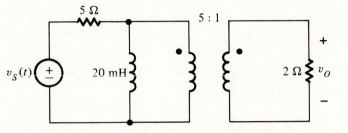

Figure P11.36

11.37 In the circuit of Fig. 11.17, $r_\pi = 75\ \Omega$, $R_C = R_S = 2\ \text{k}\Omega$, $R_E = 1\ \text{k}\Omega$, $\beta = 75$, $R_{B1} \| R_{B2} = 25\ \text{k}\Omega$.
  (a) Calculate C_E for a stop bandwidth of 700 Hz. Calculate the low- and high-frequency gains.
  (b) Add a coupling capacitor as in Fig. 11.20(a), so as to leave the frequency response essentially as in part (a) for $f \geq 700$ Hz.

11.38 Figure P11.38 represents an RC-coupled, common-source JFET amplifier. The resistances R_{B1}, R_{B2}, and R_S establish a stable quiescent point.
  (a) Calculate the system function V_o/V_i for this circuit using the small-signal model of Chap. 6.

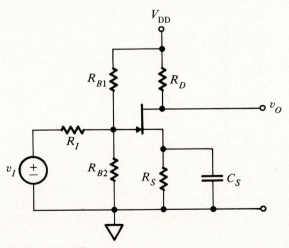

Figure P11.38

(b) For $R_{B1} \parallel R_{B2} = 20 \text{ k}\Omega$, $R_D = R_l = R_S = 1 \text{ k}\Omega$, and $g_m = 10 \text{ mS}$, find C_S for a 3-dB stop bandwidth of 100 Hz.

(c) Add a coupling capacitor C_C between R_l and the gate, and choose C_C so as to leave the response essentially as in (b) for $f \geq 100$ Hz.

11.39 Figure P11.39 represents an emitter-follower BJT amplifier with a coupling capacitor. Resistances R_{B1}, R_{B2}, and R_E establish a stable quiescent point.

(a) Using the incremental model of Sec. 11.5 for the BJT, calculate V_o/V_s.

(b) With $R_E = R_S = 1.5 \text{ k}\Omega$, $r_\pi = 75 \text{ }\Omega$, $R_{B1} \parallel R_{B2} = 20 \text{ k}\Omega$, and $\beta = 50$, choose C_C such that the stop bandwidth is 200 Hz.

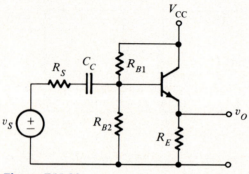

Figure P11.39

11.40 Figure P11.40 represents a source-follower JFET amplifier with a coupling capacitor. Resistances R_{B1}, R_{B2}, and R_S establish a stable quiescent point.

(a) Using the small-signal model of Chap. 6, calculate the system function V_o/V_i.

(b) With $R_S = R_i = 1 \text{ k}\Omega$, $R_{B1} \parallel R_{B2} = 25 \text{ k}\Omega$, and $g_m = 2 \text{ mS}$, choose C_C such that the stop bandwidth is 150 Hz.

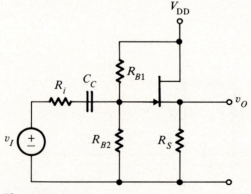

Figure P11.40

11.41 Figure P11.41(a) represents a common-base BJT amplifier with a coupling capacitor. Resistances R_{B1}, R_{B2}, and R_E establish a stable quiescent point.

(a) Show that the small-signal network to the right of ab is equivalent to that shown in Fig. P11.41(b).

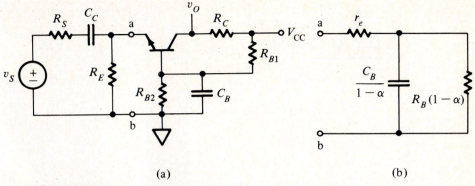

Figure P11.41

(b) Since $C_B/(1 - \alpha)$ is large, calculate V_o/V_s for $C_B \to \infty$.

(c) Choose C_C for a stop bandwidth of 100 Hz and sketch the Bode diagram for V_o/V_s, given $r_e = 2 \ \Omega$, $\beta = 50$, $R_C = R_S = R_E = 1 \ \text{k}\Omega$, and $R_{B1} \parallel R_{B2} = 20 \ \text{k}\Omega$.

(d) Choose C_B so that, at 100 Hz, the driving-point impedance seen at ab has magnitude 10% of that of R_S.

11.42 For the JFET amplifier of Fig. 11.22(a), $r_d = 100 \ \text{k}\Omega$, $g_m = 5 \ \text{mS}$, $C_{gs} = C_{ds} = 3 \ \text{pF}$, $C_{gd} = 0.75 \ \text{pF}$, $R_S = R_D = 3 \ \text{k}\Omega$, and $R_{B1} \parallel R_{B2} = 25 \ \text{k}\Omega$. Calculate the 3-dB bandwidth, the gain-bandwidth product, and the low-frequency gain V_o/V_s.

11.43 The BJT common-emitter amplifier of Fig. 11.27 has the following parameters: $r_x = 12 \ \Omega$, $r_\pi = 1 \ \text{k}\Omega$, $R_{B1} \parallel R_{B2} = 25 \ \text{k}\Omega$, $R_S = 1 \ \text{k}\Omega$, $R_C = 2.5 \ \text{k}\Omega$, $g_m = 60 \ \text{mS}$, $C_\pi = 90 \ \text{pF}$, and $C_\mu = 2.6 \ \text{pF}$. Calculate the 3-dB bandwidth, the gain-bandwidth product, and the low-frequency gain V_o/V_s.

CHAPTER

Power in the Sinusoidal Steady State

12.1 INTRODUCTION

The object of this chapter is the study of power in the sinusoidal steady state. In power engineering, one studies the transmission and distribution of electric energy in the form of sinusoidal signals. The principal reason for the use of sinusoids (i.e., ac power) is the ease of voltage transformation through transformers; it turns out that efficient transmission requires voltages which are much greater than those at either the generation or consumer ends.

Although the material in this chapter is essential to the study of power engineering, it is also a vital element in the analysis of electronic circuits. Since power is dissipated as heat, individual devices and ICs are designed with certain limits on power dissipation; it is essential that the user ensure that these limits are not exceeded.

In most cases, the fluctuations of the instantaneous power are too rapid to have a significant effect on the load. For example, the temperature of the filament in a light bulb cannot follow the power fluctuations of a 60-Hz input because the heat capacity of the filament is sufficient to absorb such fluctuations. In such cases, the average power is the important quantity.

As stated in Chap. 11, any periodic signal can be broken down into sinusoidal components. It will be shown that in some cases the average power due to a signal is the sum of the average power due to each sinusoidal component. This additive property is relatively rare, and confers upon sinusoids a special importance in the analysis of linear networks.

12.2   ROOT-MEAN-SQUARE VALUES

When dealing with periodic signals, the average power dissipated or delivered is usually more important than the time variations of the instantaneous power. The average power is simply the energy delivered during one full cycle, divided by the period, i.e.,

$$P = \frac{1}{T} \int_0^T p(t) \, dt \tag{12.1}$$

where $p(t)$ is the instantaneous power and T is the period.

The integral in Eq. (12.1) can be taken over any interval of duration T because of the periodicity of the variables. For the simple case of power delivered to a resistance R,

$$P = R\frac{1}{T} \int_0^T i^2(t) \, dt \tag{12.2}$$

The *root-mean-square (rms)* value of $i(t)$ is defined as

$$I_{\text{rms}} = \left[\frac{1}{T} \int_0^T i^2(t) \, dt \right]^{1/2} \tag{12.3}$$

so that Eq. (12.2) becomes

$$P = RI_{\text{rms}}^2 \tag{12.4}$$

The rms value is sometimes called the effective value. From Eq. (12.4), I_{rms} is that constant current which, when passed through a resistance, dissipates, on average, the same power as the given periodic current. Equations similar to Eqs. (12.3) and (12.4) apply to voltages as well, and to all periodic waveforms.

EXAMPLE 12.1

Calculate the rms value of a sinusoidal current represented by the phasor I.

Solution

With $I = |I| e^{j\theta}$, one writes

$$i(t) = |I| \cos(\omega t + \theta)$$

Since $T = 2\pi/\omega$, Eq. (12.3) yields

$$I_{rms}^2 = \frac{\omega}{2\pi} \int_0^{2\pi/\omega} |I|^2 \cos^2(\omega t + \theta) \, dt$$

Using the identity $\cos^2 x = \frac{1}{2} + (\cos 2x)/2$,

$$I_{rms}^2 = \frac{\omega}{2\pi} |I|^2 \int_0^{2\pi/\omega} \left[\frac{1}{2} + \frac{\cos(2\omega t + 2\theta)}{2} \right] dt$$

The cosine term integrated is zero: $2\pi/\omega$ is exactly one period and any sinusoid integrated over an integral number of periods is zero. Therefore,

$$I_{rms}^2 = \frac{|I|^2}{2}$$

or

$$I_{rms} = \frac{|I|}{\sqrt{2}} \qquad\qquad (12.5)$$

In words, for a sinusoid, the rms value is the peak value divided by $\sqrt{2}$.

In power engineering, one normally works with rms values. For example, the 110 V household voltage is an rms quantity: the peak value of the waveform is $110\sqrt{2} = 155.6$ V.

EXAMPLE 12.2

Calculate the rms value for the saw-tooth voltage waveform of Fig. E12.2.

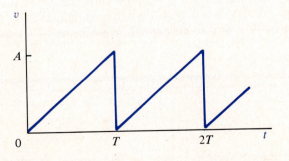

Solution

For $0 < t \leq T$,

$$v(t) = \frac{A}{T}t$$

From Eq. (12.3),

$$V_{rms}^2 = \frac{1}{T}\int_0^T \left(\frac{A}{T}\right)^2 t^2 \, dt = \frac{A^2}{3}$$

so that

$$V_{rms} = \frac{A}{\sqrt{3}}$$

For this waveform, the average value is

$$V_{av} = \frac{1}{T}\int_0^T v(t) \, dt = \frac{A}{2}$$

It is worth noting that V_{rms}^2, the average of $v^2(t)$, is *not* the square of the average of $v(t)$.

DRILL EXERCISE

12.1 Calculate the rms value for the waveforms of Figs. D12.1(a) and (b).
Ans. (a) A, (b) $A(\Delta/T)^{1/2}$

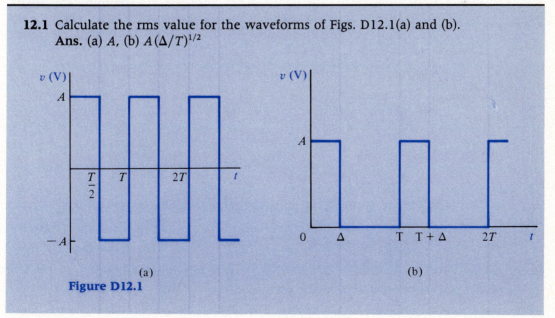

(a)   (b)

Figure D12.1

12.3 AVERAGE POWER AND STORED ENERGY

The importance of average power has already been explained in Sec. 12.1. In this section, expressions are developed for the average power in the sinusoidal steady state. Equations are derived also for the average stored energy, which turns out to bear some relation to resonance.

12.3.1 Average Power for Sinusoids

Given the current $i(t)$ through a resistance R, the voltage is $v(t) = Ri(t)$, and

$$V_{rms}^2 = \frac{1}{T} \int_0^T R^2 i^2(t) \, dt = R^2 I_{rms}^2$$

or

$$V_{rms} = R I_{rms}$$

The power delivered to a resistance R, given by Eq. (12.4) is also written

$$P = V_{rms} I_{rms} \tag{12.6}$$

Equation (12.6) holds for all periodic signals, but *only* for resistances. The key fact leading to Eq. (12.6) is that v and i are proportional to each other, i.e., related by a constant multiplier. This does not hold for elements other than resistances.

As defined thus far, the magnitude of a phasor is the peak value of the corresponding sinusoid. It is convenient in power calculations to use phasors whose magnitudes are the rms values of the corresponding sinusoids. This entails no change in the formulation of network equations, since it is only a simple scaling. A circumflex will be used to denote phasors defined in terms of rms values. Thus,

$$\hat{V} = \frac{1}{\sqrt{2}} V$$

$$V_{rms} = |\hat{V}| = \frac{1}{\sqrt{2}} |V|$$

and the time function $v(t)$ is written as

$$v(t) = |V| \cos(\omega t + \theta) = \sqrt{2} |\hat{V}| \cos(\omega t + \theta)$$

In the sinusoidal steady state, let $|\hat{V}| e^{j\phi}$ and $|\hat{I}| e^{j\theta}$ be the voltage and current phasors at a given port. Then,

$$v(t) = \sqrt{2} |\hat{V}| \cos(\omega t + \phi)$$
$$i(t) = \sqrt{2} |\hat{I}| \cos(\omega t + \theta)$$

and the power delivered to the port is

$$p(t) = 2 |\hat{V}| |\hat{I}| \cos(\omega t + \phi) \cos(\omega t + \theta)$$

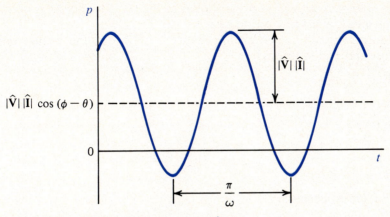

Figure 12.1 Instantaneous power vs time.

By the identity $\cos x \cos y = \frac{1}{2}[\cos(x + y) + \cos(x - y)]$,

$$p(t) = |\hat{V}||\hat{I}|\,[\cos(2\omega t + \phi + \theta) + \cos(\phi - \theta)] \qquad (12.7)$$

As shown in Fig. 12.1, the instantaneous power is a constant plus a sinusoid of frequency twice that of the excitation.

The average power is just the constant component,

$$P = |\hat{V}||\hat{I}|\cos(\phi - \theta) \qquad (12.8)$$

In the special case of a branch represented by an impedance, the term $\cos(\phi - \theta)$ is called the *power factor*. For a resistance, the current and the voltage phasors are in phase, so that $\phi = \theta$ and $\cos(\phi - \theta) = 1$; as shown in Fig. 12.2(a), the power delivered is always non-negative. In the case of a capacitance or inductance, the current and voltage phasors are 90° out of phase, and

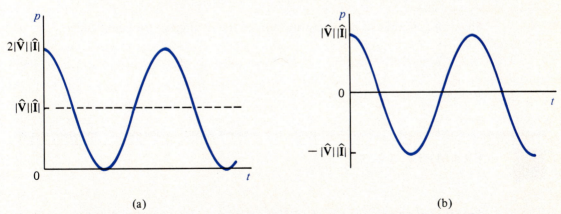

(a)   (b)

Figure 12.2 Instantaneous power delivered to (a) a resistance, (b) an inductance, or a capacitance.

$\cos(\phi - \theta) = \cos(\pm 90°) = 0$. As shown in Fig. 12.2(b), power is both delivered and received, averaging out to zero.

More generally, given an impedance $Z = |Z| \angle Z$, the voltage and current phasors are related through

$$\hat{V} = Z\hat{I}$$

so that

$$\angle \hat{V} = \angle Z + \angle \hat{I}$$

or

$$\phi - \theta = \angle Z$$

and Eq. (12.8) becomes

$$P = |\hat{V}| |\hat{I}| \cos(\angle Z) \tag{12.9}$$

The power factor is $\cos(\angle Z)$.

The relationship $|\hat{V}| = |Z| |\hat{I}|$ is combined with Eq. (12.9) to yield

$$P = |Z| |\hat{I}|^2 \cos(\angle Z) \tag{12.10}$$

and

$$P = \frac{|\hat{V}|^2}{|Z|} \cos(\angle Z) \tag{12.11}$$

It was shown in Chap. 1 that the instantaneous power is conserved. Although this was illustrated for resistive networks, the proof depended on Tellegen's theorem, itself a consequence only of Kirchhoff's laws. Therefore, conservation of power holds for all types of branches, and

$$\sum_{\text{all branches}} p_k(t) = 0$$

Since the average of a sum is the sum of the averages, it follows that

$$\sum_{\text{all branches}} P_k = 0 \tag{12.12}$$

so that average power is also conserved.

EXAMPLE 12.3

For the network of Fig. E12.3(a), $v_S(t) = 10\sqrt{2} \cos(\omega t - 45°)$, where $\omega = 2\pi(60) \text{ s}^{-1}$. Calculate the average power delivered by the source and verify that it equals the average power delivered to the resistance.

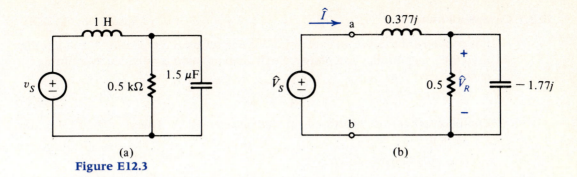

(a)   (b)

Figure E12.3

Solution

Using V, mA, kΩ, μF, H, and ms as units, the frequency is $\omega = 0.377 \text{ ms}^{-1}$ and the transform diagram, in terms of impedances in kΩ, is shown in Fig. E12.3(b).

The impedance seen at *ab* is

$$Z = 0.377j + \frac{1}{2 + 0.565j}$$

$$= 0.462 + 0.246j$$

$$= 0.523 \; \angle \; 28.03°$$

The rms value of the input voltage is 10 V. From Eq. (12.11), the average power delivered to the network by the source is

$$P = \frac{10^2}{0.523} \cos 28.03° = 168 \text{ mW}$$

Note that the power is in milliwatts, because voltage is in volts and current is in milliamperes.

By the voltage divider rule,

$$\frac{\hat{V}_R}{\hat{V}_S} = \frac{1/(2 + 0.565j)}{0.377j + 1/(2 + 0.565j)}$$

$$= 0.917 \; \angle{-43.77°}$$

Since $|\hat{V}_R| = 0.917|\hat{V}_S|$,

$$|\hat{V}_R| = 0.917|\hat{V}_S| = 9.17 \text{ V}$$

and the average power dissipated in the resistance is

$$P_R = \frac{917^2}{0.5} = 168 \text{ mW}$$

Note that the only phase information used was the angle of Z; the phase of V_S is irrelevant in a power calculation.

12.3.2 Complex Power

It is convenient to have an expression for P in terms of the phasors $\hat{V}$ and $\hat{I}$, rather than their magnitudes and phases. Such an expression is provided by defining the *complex power S* as

$$S = \hat{V}\hat{I}* \qquad (12.13)$$

Complex power is expressed in *volt-amperes* (VA). Given $\hat{V} = |\hat{V}| \angle \phi$ and $\hat{I} = |\hat{I}| \angle \theta$,

$$S = |\hat{V}|e^{j\phi}|\hat{I}|e^{-j\theta}$$

$$= |\hat{V}||\hat{I}|e^{j(\phi - \theta)}$$

or

$$S = |\hat{V}||\hat{I}| \cos(\phi - \theta) + j|\hat{V}||\hat{I}| \sin(\phi - \theta) \qquad (12.14)$$

In the case of power supplied to an impedance, $\phi - \theta = \angle Z = \phi_z$ and

$$S = |\hat{V}||\hat{I}| \cos \phi_z + j|\hat{V}||\hat{I}| \sin \phi_z \qquad (12.15)$$

Comparing Eqs. (12.14) and (12.8), it is clear that the average power $P = \text{Re}(S)$. The imaginary part of S, called the *reactive power*, is denoted by Q and given the units of *vars* (volt-ampere reactive). Thus, S is a complex number given by

$$S = P + jQ \qquad (12.16a)$$

in Cartesian form and by

$$S = |\hat{V}||\hat{I}| \angle \phi_z \qquad (12.16b)$$

in polar form. The complex power delivered to an impedance is

$$S = (Z\hat{I})\hat{I}* = Z|\hat{I}|^2 \qquad (12.17a)$$

or

$$S = \hat{V}(Y\hat{V})* = Y*|\hat{V}|^2 \qquad (12.17b)$$

Given $Z = R + jX$ and $Y = G + jB$, Eqs. (12.17) become

$$S = R|\hat{I}|^2 + jX|\hat{I}|^2 \qquad (12.18a)$$

and

$$S = G|\hat{V}|^2 - jB|\hat{V}|^2 \qquad (12.18b)$$

Thus, a resistance dissipates real power. An inductance ($X > 0$) *absorbs* reactive power, while a capacitance ($X < 0$) *generates* reactive power.

The complex amplitudes $\hat{V}_k$ and $\hat{I}_k$ of the kth branch voltages and currents in a network satisfy Kirchhoff's laws. If the $\hat{I}_k$ satisfy KCL, so do the $\hat{I}_k^*$. By Tellegen's theorem

$$\sum_{\text{all branches}} \hat{V}_k \hat{I}_k^* = 0$$

or

$$\sum_{\substack{\text{all} \\ \text{branches}}} S_k = \sum_{\substack{\text{all} \\ \text{branches}}} P_k + j \sum_{\substack{\text{all} \\ \text{branches}}} Q_k = 0 \qquad (12.19)$$

so that complex power is conserved. Since both the real and the imaginary parts must be zero, it also follows from Eq. (12.19) that both the real power and the reactive power are conserved.

EXAMPLE 12.4

For the network of Example 12.3, calculate the complex power supplied by the source and the reactive power supplied to the inductance and capacitance.

Solution

From Example 12.3, the impedance seen at ab is $0.462 + j0.246 \, k\Omega$, or $0.523 \angle 28.03° \, k\Omega$.

The magnitude of $\hat{I}$ is

$$|\hat{I}| = \frac{|\hat{V}_S|}{|Z|} = \frac{10}{0.523} = 19.12 \text{ mA}$$

Using Eq. (12.17),

$$S = (0.462 + j0.246)(19.12)^2 = 168 + j90.0 \text{ mVA}$$

The real part coincides with the result of Example 12.3, and is real power supplied by the source.

From Eq. (12.18a), the reactive power absorbed by the inductance is

$$Q_L = 0.377 \, (19.12)^2 = 138 \text{ mvars}$$

Since, from Example 12.3, $|\hat{V}_R| = 9.17 \text{ V}$, the reactive power supplied to the capacitance is found from Eq. (12.18b) to be

$$Q_C = -\frac{1}{1.77}(9.17)^2 = -48 \text{ mvars}$$

The sum of Q_L and Q_C is seen to equal the reactive power supplied by the source.

In power engineering, a load is specified not by its impedance but by its rated rms voltage, the magnitude of the complex power it absorbs at rated voltage and its power factor. Since the power factor is positive for all loads that absorb real power, it is necessary to state whether the power factor is "lagging" (inductive load) or "leading" (capacitive load). The terminology comes from the fact that the current either lags (inductance) or leads (capacitance) the voltage.

EXAMPLE 12.5

Calculate the impedance at the operating frequency of 40-kVA load with rated voltage of 2 kV and power factor of 0.8 lagging.

Solution
The current in the load, with an applied voltage of 2 kV, is

$$|\hat{I}| = \frac{40,000}{2000} = 20 \text{ A}$$

The magnitude of the load impedance is

$$|Z| = \frac{|\hat{V}|}{|\hat{I}|} = \frac{2000}{20} = 100 \ \Omega$$

and its angle is

$$\angle Z = \cos^{-1} 0.8 = 36.87°$$

The angle is positive because the current is said to lag the voltage, i.e., the load is inductive.

In cartesian coordinates

$$Z = (80 + j60) \ \Omega$$

DRILL EXERCISES

12.2 Repeat Example 12.3 with $\omega = 2\pi(100) \text{ s}^{-1}$.
Ans. $P = 114$ mW

12.3 Repeat Example 12.4 with $\omega = 2\pi(100) \text{ s}^{-1}$.
Ans. $S = 114 + j122$ mVA, $Q_L = 176$ mvars, $Q_C = -54$ mvars

12.4 Calculate the impedance of a 20-kVA load, with rated voltage of 1 kV and 0.7 power factor, leading.
Ans. $Z = 50 \ \angle{-45.6°} \ \Omega$

12.3.3 Average Stored Energy in the Sinusoidal Steady State

Although the average power supplied to a capacitance or inductance in the sinusoidal steady state is zero, the average value of the stored energy in these elements is not zero. The expression given in Chap. 1 for the energy stored in the capacitance is

$$E_C = \tfrac{1}{2} C v_C^2(t)$$

The average, denoted by E_{Cave}, is

$$E_{Cave} = \frac{1}{2} C \frac{1}{T} \int_0^T v_C^2(t) \, dt$$

$$= \tfrac{1}{2} C |\hat{V}|^2 \tag{12.20}$$

Similarly, the average energy stored in an inductance is

$$E_{Lave} = \tfrac{1}{2} L |\hat{I}|^2 \tag{12.21}$$

EXAMPLE 12.6

For the series resonant network of Fig. E12.6, show that, at resonance, the average energy stored in the inductance is equal to that stored in the capacitance. Show that

$$Q = 2\pi \frac{\text{average energy stored at resonance}}{\text{energy dissipated per cycle at resonance}}$$

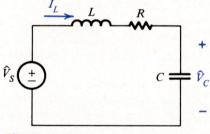

Figure E12.6

Solution
By inspection,

$$\hat{I}_L = \frac{\hat{V}_S}{j\omega L + R + 1/j\omega C}$$

At resonance, $\omega_0 L = 1/(\omega_0 C)$ and

$$\hat{I}_L = \frac{\hat{V}_S}{R}$$

Also, at resonance,

$$\hat{V}_C = \frac{\hat{I}_L}{j\omega_0 C} = \frac{\hat{V}_S}{j\omega_0 RC}$$

The average stored energies in L and C are

$$E_{L\text{ave}} = \frac{1}{2}L\frac{|\hat{V}_S|^2}{R^2}$$

and

$$E_{C\text{ave}} = \frac{1}{2}C\frac{|\hat{V}_S|^2}{(\omega_0^2 C^2 R^2)}$$

For $\omega_0^2 = 1/LC$,

$$E_{C\text{ave}} = \frac{1}{2}L\frac{|\hat{V}_S|^2}{R^2} = E_{L\text{ave}}$$

The two stored energies are equal and the total stored energy is

$$E_T = L\frac{|\hat{V}_S|^2}{R^2}$$

The energy dissipated per cycle is

$$E_{\text{diss}} = \frac{2\pi}{\omega_0}R|\hat{I}|^2$$

$$= \frac{2\pi}{\omega_0}\frac{|\hat{V}_S|^2}{R}$$

Thus,

$$2\pi\frac{E_T}{E_{\text{diss}}} = 2\pi\frac{\omega_0}{2\pi}\frac{L}{R} = \frac{\omega_0 L}{R} = Q$$

The expression for Q in terms of energy is actually valid in the context of microwave devices and circuits, where discrete-element models are no longer applicable. Resonant frequency can also be defined in this broader context as that frequency at which the stored magnetic energy ($E_{L\text{ave}}$) and electric energy ($E_{C\text{ave}}$) are equal.

DRILL EXERCISE

12.5 Repeat Example 12.6 for the parallel LRC network driven by a current source.

12.4 APPLICATIONS TO POWER TRANSMISSION

12.4.1 Efficiency of the Power Transformer

The transformer is an important component of all power distribution systems. One important parameter for power engineers is efficiency, defined as the ratio of the

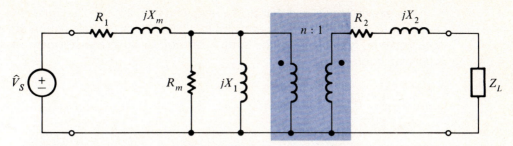

Figure 12.3 Power transformer model, with source and load.

average power delivered at the transformer *secondary* (output) to the average power supplied to the *primary* (input). Power transformers typically have efficiencies near 98%. While a 2% power loss would be negligible in many applications, this is not the case here because of the enormous amounts of power—and money—involved.

The model of Fig. 12.3 is the mutual inductance model of Chap. 4, with resistances added. The series resistances R_1 and R_2 account for the nonzero resistance of the two coils; power engineers call the power dissipated in them "copper losses." The resistance R_m accounts for the power dissipated through eddy currents and hysteresis, in the iron core around which the coils are wound; the power dissipated in R_m models the "iron losses."

In an ideal transformer, $\hat{V}_2 = \hat{V}_1/n$ and $\hat{I}_2 = -n\hat{I}_1$. Therefore,

$$\hat{V}_1 \hat{I}_1^* + \hat{V}_2 \hat{I}_2^* = \hat{V}_1 \hat{I}_1^* + (\hat{V}_1/n)(-n\hat{I}_1^*) = 0 \qquad (12.22)$$

Equation (12.22) expresses the fact that the complex power supplied to an ideal transformer is zero. Thus, the real and reactive power supplied are both zero.

The analysis is simplified if one uses the impedance transformation properties of the ideal transformer. Recall that, for the ideal transformer of Fig. 12.4(a),

$$\frac{V_1}{I_1} = n^2 Z_L \qquad (12.23)$$

In words, the impedance seen at ab is $n^2 Z_L$. Applying this impedance transformation to Fig. 12.3 yields Fig. 12.4(b).

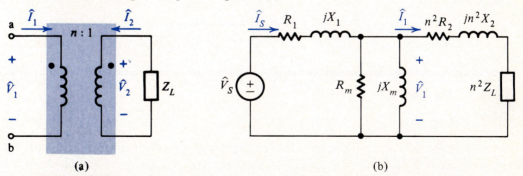

(a)   (b)

Figure 12.4 (a) Ideal transformer and load. (b) Figure 12.3 after the impedance transformation.

It is generally the case that R_1, X_1, $n^2 R_2$, and $n^2 X_2$ are small compared to $n^2 |Z_L|$ and that R_m and X_m are large compared to $n^2 |Z_L|$. Thus, the calculation of voltages and currents is reduced to the simple case of an ideal transformer. To a good approximation,

$$\hat{I}_S = \hat{I}_1 = \frac{\hat{V}_1}{n^2 Z_L}$$

$$\hat{V}_S = \hat{V}_1$$

and the power supplied to the load is given by

$$P = \frac{|\hat{V}_S|^2}{|n^2 Z_L|} \cos \phi_L \tag{12.24}$$

where $\phi_L = \sphericalangle Z_L$.

The transformer resistances and reactances are ignored in the calculation of currents and voltages because their inclusion would cause only a very small percentage change in the network variables and, hence, in the power supplied to Z_L. These resistances must be included in a calculation of transformer losses since, clearly, they account for the totality of these losses. The reactances would enter a calculation of transformer reactive power, but this will not be carried out here.

The copper losses are

$$P_c = R_1 |\hat{I}_S|^2 + n^2 R_2 |\hat{I}_1|^2$$

$$\simeq (R_1 + n^2 R_2) \frac{|\hat{V}_S|^2}{|n^2 Z_L|^2}$$

and the iron losses are given by

$$P_i = \frac{|\hat{V}_1|^2}{R_m} \simeq \frac{|\hat{V}_S|^2}{R_m}$$

The efficiency of the transformer is

$$\frac{P}{P_{\text{Total}}} = \frac{P}{P + P_i + P_c}$$

$$= \frac{n^2 |Z_L| \cos \phi_L}{n^2 |Z_L| \cos \phi_L + R_1 + n^2 R_2 + (n^2 |Z_L|)^2 / R_m} \tag{12.25}$$

EXAMPLE 12.7

Calculate the efficiency of a power transformer given the following values: $R_1 = 0.02 \ \Omega$, $R_2 = 0.5 \ \Omega$, $X_1 = 0.05 \ \Omega$, $X_2 = 1.5 \ \Omega$, $R_m = 500 \ \Omega$, $X_m = 200 \ \Omega$. The transformer is a 400/2000 V transformer, i.e., $n = \frac{1}{5}$. The load is that of Example 12.5, and the rms source voltage is 400 V.

Solution

Since $|Z_L| = 100$ Ω, $n^2|Z_L| = 4$ Ω; since $\cos \phi_L = 0.8$, Eq. (12.25) yields directly

$$\frac{3.2}{3.2 + 0.02 + 0.02 + \frac{16}{500}} = 97.8\%$$

12.4.2 A Power Transmission System

In power transmission, the power dissipated in the transmission line is proportional to $|\hat{I}|^2$. The current required to transmit a given amount of power is reduced if the voltage is raised, since power depends on the product of the two. That is why it is economical to transmit power at high voltage. Transformers are used at the generating end to "step up" from generation to transmission voltage, and at the load to "step down" to user level.

Figure 12.5(a) is a simple model of a power transmission system, where the line is modeled by a resistance R_l and a reactance X_l in series. The transformers are modeled as ideal. The turns ratios are chosen according to the desired ratios of the transmission voltage to the generation and load voltages.

The power dissipated by the line is, from Fig. 12.5(b),

$$P_{\text{diss}} = |\hat{I}_l|^2 R_l \qquad (12.26)$$

Since the transformers are ideal, all the complex power supplied by the line to the load transformer is transmitted to the load. If S is the complex power absorbed by the load, then

$$S = \hat{V}_l \hat{I}_l^*$$

and

$$|\hat{I}_l| = \frac{|S|}{|\hat{V}_l|} \qquad (12.27)$$

Therefore, from Eq. (12.26),

$$P_{\text{diss}} = R_l \frac{|S|^2}{|\hat{V}_l|^2} \qquad (12.28)$$

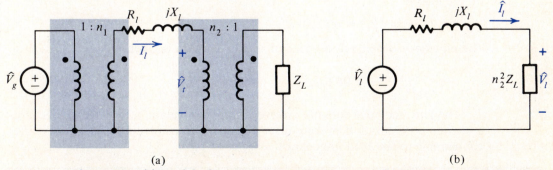

(a)   (b)

Figure 12.5 (a) Model of a power transmission system. (b) Result after transforming.

With $S = P + jQ$, Eq. (12.28) becomes

$$P_{\text{diss}} = \frac{R_l(P^2 + Q^2)}{|\hat{V}_l|^2} \tag{12.29}$$

Since S is fixed by the load, it is clear that increasing the line voltage $|\hat{V}_l|$ decreases the power lost in the line. Note also that the reactive power taken (or supplied) by the load increases the line dissipation; this is because the swapping of stored energy between the load and the line requires current, which increases dissipation.

EXAMPLE 12.8

A 200-km, 735-kV transmission line has a resistance of 0.06 Ω/km and an inductive reactance of 0.30 Ω/km. The load is rated at 500 MVA, with a power factor of 0.8 lagging. Calculate the power dissipated in the line and the efficiency of the transmission.

Solution

In general, the line resistance and reactance are both small compared to $n_2^2|Z_L|$, so that the line voltage drops are small. Therefore, it is assumed that both $|\hat{V}_r|$ and $|\hat{V}_l|$ in Fig. 12.5(b) are 735 kV.

The line resistance is given by

$$R_l = 0.06(200) = 12\ \Omega$$

By Eq. (12.28),

$$P_{\text{diss}} = \frac{12(500 \times 10^6)^2}{(735 \times 10^3)^2}$$

$$= 5.55 \times 10^6\ \text{W} = 5.55\ \text{MW}$$

From Eq. (12.16), with the power factor $\cos(\angle Z_L) = 0.8$, the real power absorbed by the load is

$$P = |S| \cos(\angle Z_L) = 400\ \text{MW}$$

The efficiency is defined in terms of real power and is given by

$$\eta = \frac{P}{P + P_{\text{diss}}} = 98.6\%$$

The line reactance does not intervene in this calculation. However, it becomes a serious consideration in long-distance power transmission, when it becomes comparable to the load impedance.

The efficiency can be increased by *power factor correction*. The real power P taken by the load is fixed by the energy requirements at the load end, e.g., lighting, heating, and mechanical work. The reactive power Q can be varied. From Eq. (12.29), the line dissipation will be least if $Q = 0$, i.e., if the load is resistive (power factor = 1). The idea behind power factor correction is to make Q as small as practically possible. This is demonstrated in the following example.

EXAMPLE 12.9

For the problem data of Example 12.8, the line dissipation is to be reduced by correcting the power factor to 1. Find the capacitance which achieves this correction at 60 Hz when connected across the load transformer on the line side, as in Fig. E12.9. Calculate the line losses.

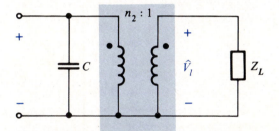

Figure E12.9

Solution

The reactive power absorbed by the load is

$$Q = \text{Im}(S) = |S| \sin(\sphericalangle Z_L)$$

Since $\cos(\sphericalangle Z_L) = 0.8$, $\sphericalangle Z_L = \pm 36.9°$. The "+" is chosen, because the power factor is lagging, i.e. the load is inductive. Thus,

$$Q = 500(0.6) = 300 \text{ Mvars}$$

The capacitance reactive power is negative, i.e., the capacitance generates vars. In order to generate 300 Mvars, from Eq. (12.18b), the capacitance must satisfy

$$\omega C |\hat{V}_l|^2 = 300 \times 10^6$$

or

$$C = \frac{300 \times 10^6}{377(735 \times 10^3)^2} = 1.47 \ \mu\text{F}$$

With $Q = 0$, $S = P = 400$ MVA and, from Eq. (12.28),

$$P_{\text{diss}} = \frac{12(400 \times 10^6)^2}{(735 \times 10^3)^2} = 3.55 \text{ MW}$$

Note that the capacitance could equally well have been connected directly across Z_L. In order to generate the same reactive power at a much lower voltage, however, it is necessary to use a much bigger capacitance, since reactive power is proportional to the square of the voltage.

DRILL EXERCISE

12.6 Calculate the impedance $n^2 Z_L$ in Example 12.9 and show that the 1.47-μF capacitance exactly cancels its susceptance.

12.4.3 Three-Phase Power

For reasons that will be explored presently, large amounts of power are usually transmitted in three phases.

A three-phase generator is represented in Fig. 12.6 as three voltage sources. The three identical load impedances Z_L form a *balanced* three-phase load. The three voltage phasors are

$$\hat{V}_a = |\hat{V}| \angle 0° \qquad \hat{V}_b = |\hat{V}| \angle 120° \qquad \hat{V}_c = |\hat{V}| \angle -120°$$

All three are produced by the same machine. The point O is ground, and $|\hat{V}|$ is called the *line-to-neutral* voltage.

The voltages $\hat{V}_{ab}$, $\hat{V}_{bc}$, $\hat{V}_{ca}$ are shown in Fig. 12.7. By symmetry, they have the same length $|\hat{V}_l|$, called the *line-to-line* voltage. The vector-addition triangle for $\hat{V}_{ab}$ is given explicitly. Clearly

$$|\hat{V}_l| = 2|\hat{V}| \cos 30° = \sqrt{3}|\hat{V}| \qquad (12.30)$$

The line-to-line voltage is the one used to specify the rating of a transmission line. For example, in a 735-kV line, the rms value of the line-to-neutral voltage is $735/\sqrt{3} = 424$ kV.

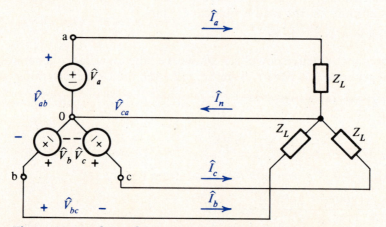

Figure 12.6 A three-phase generator with balanced load.

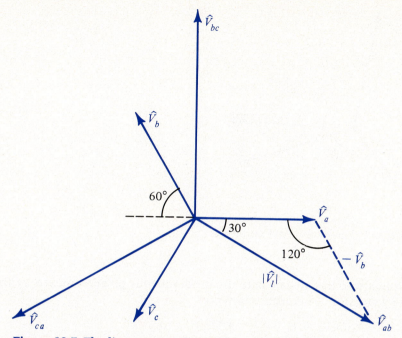

Figure 12.7 The line-to-neutral and line-to-line voltages.

From Fig. 12.7, the total average power supplied is

$$P = 3\frac{|\hat{V}|^2}{|Z_L|}\cos\phi_L = \frac{|\hat{V}_l|^2}{|Z_L|}\cos\phi_L \tag{12.31}$$

This is also the power delivered by a one-phase system with transmission voltage $|\hat{V}_l|^2$.

The three currents $\hat{I}_a$, $\hat{I}_b$, and $\hat{I}_c$ are written as

$$\hat{I}_a = \frac{\hat{V}_a}{Z_L} = \left(\frac{|\hat{V}|}{|Z_L|}\right) \measuredangle -\phi_L$$

$$\hat{I}_b = \frac{\hat{V}_b}{Z_L} = \frac{|\hat{V}|}{|Z_L|} \measuredangle (120° - \phi_L)$$

$$\hat{I}_c = \frac{\hat{V}_c}{Z_L} = \frac{|\hat{V}|}{|Z_L|} \measuredangle (-120° - \phi_L)$$

where $\phi_L = \measuredangle Z_L$. The three corresponding phasors are shown in Fig. 12.8; it should be clear from the laws of vector addition that

$$\hat{I}_n = \hat{I}_a + \hat{I}_b + \hat{I}_c = 0 \tag{12.32}$$

Equation (12.32) is a statement of the fact that the return current $I_n = 0$ in a balanced three-phase system. In practice, the earth is a good enough conductor to be used as a return line.

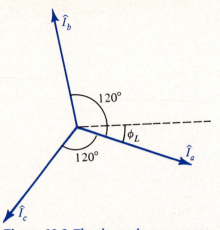

Figure 12.8 The three phase currents.

It is instructive to calculate the instantaneous power supplied by the three-phase generator. To do this one writes

$$p(t) = v_a(t)i_a(t) + v_b(t)i_b(t) + v_c(t)i_c(t)$$

Writing the time functions from the phasors in the usual manner yields

$$p(t) = |V| \cos \omega t \frac{|V|}{|Z_L|} \cos(\omega t - \phi_L)$$

$$+ |V| \cos(\omega t + 120°) \frac{|V|}{|Z_L|} \cos(\omega t + 120° - \phi_L)$$

$$+ |V| \cos(\omega t - 120°) \frac{|V|}{|Z_L|} \cos(\omega t - 120° - \phi_L)$$

$$= \frac{|V|^2}{|Z_L|} [\cos \omega t \cos(\omega t - \phi_L)$$

$$+ \cos(\omega t + 120°) \cos(\omega t + 120° - \phi_L)$$

$$+ \cos(\omega t - 120°) \cos(\omega t - 120° - \phi_L)]$$

Using the identity $\cos x \cos y = \frac{1}{2} \cos(x + y) + \frac{1}{2} \cos(x - y)$ yields

$$p(t) = \frac{1}{2} \frac{|V|^2}{|Z_L|} [\cos(2\omega t - \phi_L) + \cos \phi_L$$

$$+ \cos(2\omega t + 240° - \phi_L) + \cos \phi_L$$

$$+ \cos(2\omega t - 240° - \phi_L) + \cos \phi_L]$$

Next, the identity $\cos(x + y) = \cos x \cos y - \sin x \sin y$, plus the fact that $|V|^2/2 = |\hat{V}|^2$, results in

$$p(t) = \frac{|\hat{V}|^2}{|Z_L|}[3 \cos \phi_L + \cos(2\omega t - \phi_L)$$

$$- 0.5 \cos(2\omega t - \phi_L) - \frac{\sqrt{3}}{2} \sin(2\omega t - \phi_L)$$

$$- 0.5 \cos(2\omega t - \phi_L) + \frac{\sqrt{3}}{2} \sin(2\omega t - \phi_L)$$

$$= 3\frac{|\hat{V}|^2}{|Z_L|} \cos \phi_L = \frac{|\hat{Ve}|^2}{|Z_L|} \cos \phi_L \qquad (12.33)$$

which is the same as Eq. (12.31). Thus the instantaneous power supplied is seen to be constant. This is important for the following reason. The mechanical power supplied by the prime mover to the generator is the product of the torque and the velocity. If, as in the one-phase case, the power fluctuates within a cycle, the torque must fluctuate because there is not enough time for the velocity to change; these fluctuations in torque induce fatigue and vibrations in the generator. These effects are not present in the three-phase case, because $p(t)$ is constant. Another reason for using three-phase power is that it is a convenient way of generating several sinusoidal voltages with one machine.

EXAMPLE 12.10

A three-phase 735-kV transmission system has a balanced 1500 MVA load, with a power factor of 0.8 lagging. The resistance and reactance of the line, per phase, are as in Example 12.8. Calculate the power delivered to the load.

Solution
The load per phase is 500 MVA, with the same power factor of 0.8. As suggested by Eq. (12.31), an equivalent one-phase problem is solved, with line-to-line voltage replacing line-to-neutral. In this case, the equivalent problem is exactly that of Example 12.8.

12.5 SUPERPOSITION AND POWER

In general, power does not obey the principle of superposition. For example, consider the simple network of Fig. 12.9.

Superposition is used to calculate current components $i_1(t)$ asnd $i_2(t)$, obtained by driving the network with the sources turned on one at a time. The instantaneous power supplied to the network is

$$p(t) = [v_1(t) + v_2(t)][i_1(t) + i_2(t)]$$

$$= v_1(t)i_1(t) + v_2(t)i_2(t) + v_1(t)i_2(t) + v_2(t)i_1(t) \qquad (12.34)$$

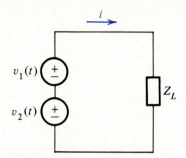

Figure 12.9 Network with two sources.

The first two terms on the right-hand side are, respectively, the power supplied by $v_1(t)$ acting alone and the power supplied by $v_2(t)$ acting alone. Unfortunately, the last two terms are not zero in general, so that superposition does not apply to instantaneous power.

The situation is more favorable if one considers average power. From Eq. (12.34), the average power supplied over a time interval 0 to T is

$$P = \frac{1}{T} \int_0^T v_1(t)i_1(t) \, dt + \frac{1}{T} \int_0^T v_2(t)i_2(t) \, dt$$

$$+ \frac{1}{T} \int_0^T v_1(t)i_2(t) \, dt + \frac{1}{T} \int_0^T v_2(t)i_1(t)$$

$$= P_1 + P_2 + \frac{1}{T} \int_0^T [v_1(t)i_2(t) + v_2(t)i_1(t)] \, dt \tag{12.35}$$

Here, P_1 and P_2 are the average power supplied by v_1 and v_2, respectively. Superposition of average power holds if the integral in Eq. (12.35) is zero. This will be true in particular if v_1 and i_1 are dc and v_2 and i_2 are sums of sinusoids, if T is a common multiple of the period of each sinusoid. This case is of great practical importance in electronic analog circuits, where signals have both a quiescent dc component and an ac component.

More generally, the integral in Eq. (12.35) can be shown to vanish if $v_1(t)$, $i_1(t)$ are sinusoids of frequency n, $v_2(t)$, $i_2(t)$ are sinusoids of frequency m, where $m \neq n$ and T is chosen to be a common multiple of the two periods (this could be a large number). In this sense, the average power in the steady state satisfies superposition for sinusoids of different frequencies.

EXAMPLE 12.11

The transistor of Fig. E12.11(a) is to have a quiescent operating point at $I_C = 10$ mA and $V_{CE} = 10$ V; this, it turns out, is achieved by connecting a 100-kΩ

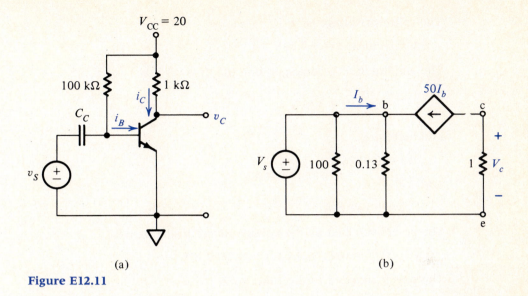

Figure E12.11

resistor from the supply to the base, as shown. The transistor has $\beta = 50$, so $r_\pi = 130\ \Omega$ at the given quiescent point. The input $v_S(t)$ is a sinusoid of frequency high enough for the coupling capacitance to be considered a short circuit. Find the dc and ac power dissipated in the transistor and in the resistances when $v_S(t)$ is such that $v_c(t)$ is a sinusoid of amplitude 10 V.

Solution

Quiescent conditions. The power supplied to the 1-kΩ load resistance is

$$P_L = \frac{(20 - 10)^2}{1} = 100\ \text{mW}$$

The power supplied to the collector of the transistor is

$$P_{CE} = V_{CE}I_C = 10(10) = 100\ \text{mW}$$

Because 100 k$\Omega \gg$ 1 kΩ and $I_B \ll I_C$, the power dissipated in the bias resistance is negligible. Similarly, because V_{BE} and I_B are small, the power supplied to the base of the transistor is also negligible.

The power supplied by V_{CC} is

$$P_{CC} = 20(I_B + I_C) \simeq 20(10) = 200\ \text{mW}$$

which is the sum of P_L and P_{CE}.

Incremental conditions. The small-signal model is shown in Fig. E12.11(b). The base current in mA is

$$I_b = \frac{V_s}{(0.13)(51)} = 0.15V_s$$

The collector voltage is

$$V_c = -50I_b(1) = -7.54V_s$$

The amplitude of V_s corresponding to $|V_c| = 10$ V is

$$|V_s| = \frac{10}{7.54} = 1.326 \text{ V}$$

and therefore

$$|\hat{V}_s| = 0.9376 \text{ V}$$

and

$$|\hat{I}_b| = 0.141 \text{ mA}$$

The power supplied by the voltage source to the base is

$$P_b = \hat{V}_s\hat{I}_b^* = 0.9376(0.141) = 0.1322 \text{ mW}$$

The power supplied to the collector is

$$P_c = \hat{V}_c(50\hat{I}_b)^* = -7.07(50)(0.141) = -50 \text{ mW}$$

The negative sign implies that the transistor supplies ac power. The ac power supplied to the 1-kΩ load is

$$P_l = \frac{|\hat{V}_c|^2}{R} = 50 \text{ mW}$$

The power dissipated in the bias resistance is negligible.
The ac power gain is

$$\frac{P_l}{P_b} = \frac{50}{0.1322} = 378.2$$

It is instructive to study the power balance. The power supplied by the signal source is only a small fraction of the total and is of interest only in the calculation of the power gain. Summing dc and ac power, the major contributions are as follows.

Supplied by V_{CC}: $200 + 0 = 200$ mW
Supplied to the collector of the transistor: $100 - 50 = 50$ mW
Supplied to the load: $100 + 50 = 150$ mW

The power supplied is at dc, but there is a 50 mW output of ac power to the 1-kΩ load. The transistor can thus be viewed as a device that receives power from a dc source and converts it to ac power, which it supplies to a load.

While it is instructive to trace out the flow of dc and ac power in the circuit, it is usually only necessary to calculate the total power supplied and received. In this example,

$$P_{CC} = V_{CC}I_C = 200 \text{ mW}$$

$$P_{\text{Load}} = P_L + P_l = 150 \text{ mW}$$

and, by conservation of power, the collector dissipation is

$$P_{\text{Ctot}} = 200 - 150 = 50 \text{ mW}$$

DRILL EXERCISE

12.7 Repeat Example 12.11 with a 2-kΩ load resistance.
   Ans. $P_L = P_{\text{CE}} = 50$ mW, $P_{\text{CC}} = 100$ mW, $P_b = 32.7$ μW, $P_l = -P_c = 24.6$ mW

12.6   POWER TRANSFER

There are many instances when it is desirable to transfer the greatest possible amount of power from a source to a load. In Sec. 12.6.1, the transfer of power between a source Thevenin equivalent and an impedance load is studied. In Sec. 12.6.2, a two-port is interposed between the source and the load: this gives rise to a two-port characterization in terms of power, known as the scattering parameters.

12.6.1 Maximum Power Transfer

In Fig. 12.10, the source is represented by its Thevenin equivalent and the load by an impedance. The object of the analysis is to determine which factors affect the power received by the load.

Given $Z_L = R_L + jX_L$, $R_L \geq 0$, the average power absorbed by the load is

$$P_L = R_L |\hat{I}|^2 = R_L \frac{|\hat{V}_S|^2}{|Z_S + Z_L|^2}$$

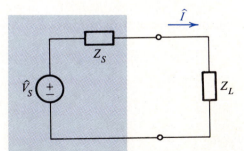

Figure 12.10 Thevenin equivalent of a source transferring power to a load.

If $Z_S = R_S + jX_S$, $R_S \geq 0$, then

$$P_L = \frac{R_L |\hat{V}_S|^2}{(R_S + R_L)^2 + (X_S + X_L)^2} \tag{12.36}$$

PROBLEM 1: Z_L GIVEN

There are two different problems of interest. The first one is as follows: Given Z_L, how should Z_S be chosen to maximize P_L? Since the numerator in Eq. (12.36) is independent of Z_S, P_L is maximized by making the denominator as small as possible. Given the constraint that $R_S \geq 0$, this occurs if $R_S = 0$ and $X_S = -X_L$. For example, given a resistive load, maximum power is transferred by making $R_S = 0$. The maximum power transferred is, by Eq. (12.36),

$$P_L = \frac{|\hat{V}_S|^2}{R_L} \tag{12.37}$$

In practice, R_S cannot be made zero but should be made as small as possible.

PROBLEM 2: Z_S GIVEN

In the second problem, Z_S is fixed and Z_L is the variable. One must maximize P_L with respect to both R_L and X_L. Since the numerator in Eq. (12.36) is independent of X_L, the best choice of X_L is that which minimizes the denominator, i.e., $X_L = -X_S$. With this choice, Eq. (12.36) becomes

$$P_L = R_L \frac{|\hat{V}_S|^2}{(R_S + R_L)^2}$$

Taking the derivative with respect to R_L,

$$\frac{dP_L}{dR_L} = |\hat{V}_S|^2 \frac{(R_S + R_L)^2 - 2R_L(R_S + R_L)}{(R_S + R_L)^4}$$

The derivative is zero if

$$R_S + R_L - 2R_L = 0$$

or

$$R_L = R_S$$

Therefore, the choice of Z_L that maximizes power transfer is

$$Z_L = R_S - jX_S = Z_S^* \tag{12.38}$$

With $X_L = -X_S$ and $R_L = R_S$, Eq. (12.34) yields

$$P_L = \frac{|\hat{V}_S|^2}{4R_S} \tag{12.39}$$

This is the *maximum power available* from the source.

If both the load and source impedances are fixed, a network is often placed between the two to achieve maximum power transfer.

EXAMPLE 12.12

A power amplifier with output impedance of 560 Ω (resistive) is available to drive a loudspeaker with impedance of 8 Ω (also resistive). A transformer is to be used to match the source and load. Assuming an ideal transformer, find the required turns ratio n.

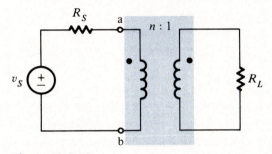

Figure E12.12

Solution

The impedance seen at ab must be equal to R_S, or

$$n^2 R_L = R_S$$

Therefore,

$$n = \left(\frac{R_S}{R_L}\right)^{1/2} = \left(\frac{560}{8}\right)^{1/2} = 8.37$$

Note that, since the ideal tranformer dissipates no power, all the power drawn from the source is transmitted to the load. Furthermore, because n is chosen to satisfy maximum power transfer, the load receives all the power available from this source.

DRILL EXERCISE

12.8 Repeat the derivations of Sec. 12.6.1 if the source is represented by its Norton equivalent and the load by an admittance.

12.6.2 Scattering Parameters of a Two-Port

The scattering description of a two-port is primarily used when power considerations are important. It is also used at high (microwave) frequencies, when it becomes difficult to measure open-circuit and short-circuit two-port parameters. Small capacitances between ostensibly open-circuited terminals allow appreciable current to flow, and the so-called skin effect makes it difficult to have an effective short circuit. At high frequencies, it is usual for a two-port to operate between two sources with given Thevenin resistances, as in Fig. 12.11; more often than not, each port is connected to a transmission line. Since scattering parameter measurements are performed with the ports terminated, they are easier to do than open- or short-circuit tests.

The variables used in the scattering description are linear combinations of the voltages and currents, defined as follows:

$$\hat{A}_1 = \frac{\hat{V}_1/\sqrt{R_1} + \hat{I}_1\sqrt{R_1}}{2} \tag{12.40a}$$

$$\hat{B}_1 = \frac{\hat{V}_1/\sqrt{R_1} - \hat{I}_1\sqrt{R_1}}{2} \tag{12.40b}$$

with similar expressions for $\hat{A}_2$ and $\hat{B}_2$. These variables are always defined with respect to given resistances R_1 and R_2. From Fig. 12.11, $\hat{V}_1 = \hat{V}_{S1} - R_1\hat{I}_1$, and Eq. (12.40a) yields

$$\hat{A}_1 = \frac{\hat{V}_{S1}}{2\sqrt{R_1}} \tag{12.41}$$

and

$$|\hat{A}_1|^2 = \frac{|\hat{V}_{S1}|^2}{4R_1} \tag{12.42}$$

From Eq. (12.39a), $|A_1|^2$ is the maximum power that can be delivered to port 1 and is called the *incident* power. The phasor $\hat{A}_1$ is called the *incident variable* at port 1.

Equation (12.40) is easily solved for $\hat{V}_1$ and $\hat{I}_1$, yielding

$$\hat{V}_1 = \sqrt{R_1}\,(\hat{A}_1 + \hat{B}_1)$$

and

$$\hat{I}_1 = \frac{\hat{A}_1 - \hat{B}_1}{\sqrt{R_1}}$$

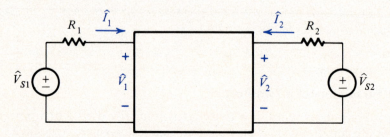

Figure 12.11 Terminated two-port, for the study of scattering parameters.

These are used to calculate the power delivered to port 1, i.e.,

$$P_1 = \text{Re}(\hat{V}_1 \hat{I}_1^*)$$

$$= \text{Re}[(\hat{A}_1 + \hat{B}_1)(\hat{A}_1 - \hat{B}_1)^*]$$

$$= \text{Re}[\hat{A}_1 \hat{A}_1^* - \hat{B}_1 \hat{B}_1^* - (\hat{A}_1 \hat{B}_1^* - \hat{B}_1 \hat{A}_1^*)]$$

Since $\hat{A}_1 \hat{B}_1^* - \hat{B}_1 \hat{A}_1^* = 2j \text{Im}(\hat{A}_1 \hat{B}_1^*)$, it follows that

$$P_1 = |\hat{A}_1|^2 - |\hat{B}_1|^2 \tag{12.43}$$

The power delivered to port 1 is the power available from the source, less some power which may be considered as "reflected back"; this is why $|B_1|^2$ is called the *reflected power* and B_1, the *reflected variable* at port 1. If the two-port is to draw the maximum available power into port 1, it is necessary that the reflected power be zero.

The *scattering parameters* of the two-port are defined by the following equations:

$$\hat{B}_1 = S_{11} \hat{A}_1 + S_{12} \hat{A}_2 \tag{12.44a}$$

$$\hat{B}_2 = S_{21} \hat{A}_1 + S_{22} \hat{A}_2 \tag{12.44b}$$

By Eq. (12.41), $\hat{V}_{S1} = 0$ implies $\hat{A}_1 = 0$ and, similarly, $\hat{V}_{S2} = 0$ implies $\hat{A}_2 = 0$. Therefore, from Eq. (12.44a),

$$S_{11} = \left. \frac{\hat{B}_1}{\hat{A}_1} \right|_{\hat{A}_2=0} = \left. \frac{\hat{B}_1}{\hat{A}_1} \right|_{V_{S2}=0} \tag{12.45}$$

Thus, $|S_{11}|^2$ is the ratio of reflected to incident power at port 1 with port 2 shorted, and S_{11} is called the input *reflection* coefficient. Similarly, S_{22} is the output reflection coefficient.

With $\hat{V}_{S1} = 0$, $\hat{A}_1 = 0$, and according to Eq. (12.43), the power delivered to port 1 is $-|\hat{B}_1|^2$; this is tantamount to saying that the power supplied *by* port 1, i.e., to R_1, is $|\hat{B}_1|^2$. From Eq. (12.44a),

$$S_{12} = \left. \frac{B_1}{A_2} \right|_{V_{S1}=0} \tag{12.46}$$

and $|S_{12}|^2$ is seen to be the ratio of power delivered to R_1 to the power incident at port 2 when $\hat{V}_{S1} = 0$; $|S_{12}|^2$ is called the *reverse transducer gain*.

Similarly, $|S_{21}|^2$ is the ratio of power delivered to R_2 to the power incident at port 1 when $\hat{V}_{S2} = 0$; it is called the *forward transducer gain*. S_{21} and S_{12} are often expressed in decibels.

EXAMPLE 12.13

Given $|S_1| = 0.8$, $|S_{21}| = 0.5$, with $\hat{V}_{S2} = 0$, $R_1 = 50\ \Omega$ and $|\hat{V}_{S1}| = 10$ V, calculate the following:

(a) The power delivered to port 1
(b) The percentage of the available power delivered to port 1
(c) The power delivered to R_2
(d) The power dissipated in the two-port

Solution
By Eq. (12.42),

$$|\hat{A}_1|^2 = \frac{|\hat{V}_{S1}|^2}{4R_1} = 0.5 \text{ W}$$

This is the incident, or available, power. Since $\hat{V}_{S2} = 0$, $\hat{A}_2 = 0$ and

$$\hat{B}_1 = S_{11}\hat{A}_1$$

$$|\hat{B}_1|^2 = |S_{11}|^2|\hat{A}_1|^2 = 0.64(0.5) = 0.32 \text{ W}$$

Using Eq. (12.43), the power delivered to port 1 is

$$P_1 = 0.5 - 0.32 = 0.18 \text{ W}$$

This represents $0.18/0.5 = 36\%$ of the available power. Since $\hat{A}_2 = 0$

$$\hat{B}_2 = S_{21}\hat{A}_1$$

$$|\hat{B}_2|^2 = |S_{21}|^2|\hat{A}_1|^2 = 0.25(0.5) = 0.125 \text{ W}$$

This is the power delivered to R_2.

The power dissipated in the two-port is just the difference between the power into port 1 and the power out of port 2, i.e.,

$$\text{Power lost} = 0.18 - 0.125 = 0.055 \text{ W}$$

DETERMINATION OF SCATTERING PARAMETERS
The scattering parameters are determined by voltage and current measurements. From Eqs. (12.45) and (12.40),

$$S_{11} = \left.\frac{\hat{V}_1/\sqrt{R_1} - \hat{I}_1\sqrt{R_1}}{\hat{V}_1/\sqrt{R_1} + \hat{I}_1\sqrt{R_1}}\right|_{\hat{V}_{S2}=0}$$

or

$$S_{11} = \left.\frac{\hat{V}_1/\hat{I}_1 - R_1}{\hat{V}_1/\hat{I}_1 + R_1}\right|_{\hat{V}_{S2}=0} \tag{12.47}$$

Defining Z_{i1} as the impedance seen at port 1 when $\hat{V}_{S2} = 0$ yields

$$S_{11} = \frac{Z_{i1} - R_1}{Z_{i1} + R_1} \tag{12.48}$$

Note that $S_{11} = 0$ if $Z_{i1} = R_1$, i.e., if port 1 presents a match to the source. In that case, $B_1 = 0$ and all the incident power is absorbed at port 1. The parameter S_{22} is expressed by a similar expression.

With $\hat{V}_{s2} = 0$, it is clear from Fig. 12.11 that $\hat{I}_2 = -\hat{V}_2/R_2$. From Eq. (12.40b), $\hat{B}_2$ is written as

$$\hat{B}_2 = \frac{\hat{V}_2/\sqrt{R_2} + \hat{V}_2/\sqrt{R_2}}{2} = \frac{\hat{V}_2}{\sqrt{R_2}}$$

From Eq. (12.41), $\hat{A}_1 = \hat{V}_{s1}/2\sqrt{R_1}$, and it follows that

$$S_{21} = \frac{\hat{B}_2}{\hat{A}_1}\bigg|_{\hat{V}_{s2}=0} = 2\sqrt{\frac{R_1}{R_2}}\frac{\hat{V}_2}{\hat{V}_{s1}}\bigg|_{\hat{V}_{s2}=0} \tag{12.49}$$

The parameter S_{12} has a similar expression, with the subscripts 1 and 2 interchanged.

EXAMPLE 12.14

The network of Fig. E12.14(a) contains a transformer, modeled by an ideal transformer with a magnetizing inductance. Calculate the transformer's scattering parameters S_{11} and S_{21} as functions of frequency, with respect to $R_1 = 50\ \Omega$ and $R_2 = 2\ \Omega$.

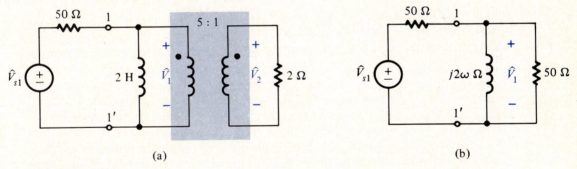

(a)   (b)

Figure E12.14

Solution
The 2-Ω resistance is reflected to the primary as $2(5)^2 = 50\ \Omega$, as in Fig. E12.14(b). Since $\hat{V}_{s2} = 0$, the impedance seen at $11'$ is Z_{i1} and is given by

$$Z_{i1}(j\omega) = \frac{50(j2\omega)}{50 + j2\omega} = \frac{j100\omega}{50 + j2\omega}$$

The input reflection coefficient is, by Eq. (12.48),

$$S_{11}(j\omega) = \frac{j100\omega/(50 + j2\omega) - 50}{j100\omega/(50 + j2\omega) + 50} = \frac{-1}{j0.08\omega + 1}$$

From Fig. E12.14(a),

$$\hat{V}_2 = \tfrac{1}{5}\hat{V}_1 = \frac{\tfrac{1}{5}\hat{V}_{S1}\,Z_{i1}(j\omega)}{Z_{i1}(j\omega) + 50} = \frac{j0.008\omega\hat{V}_{S1}}{j0.08\omega + 1}$$

From Eq. (12.49),

$$S_{21} = 2\sqrt{\frac{50}{2}}\left(\frac{\hat{V}_2}{\hat{V}_{S1}}\right) = \frac{j0.08\omega}{j0.08\omega + 1}$$

At low frequencies, i.e., $|0.08\omega| \ll 1$, $S_{11} = -1$ and $S_{21} = 0$. Since $|S_{11}| = 1$, all the power is reflected, i.e., none enters the two-port. This is clear from the network, if one realizes that the inductance is a short circuit at low frequencies.

At high frequencies, $S_{11} = 0$ and $S_{21} = 1$: all the available power is transmitted to the load. The reason, of course, is that the inductance can be replaced by an open circuit at high frequencies and the turns ratio of the ideal transformer is such as to provide a match between source and load.

DRILL EXERCISE

12.9 Repeat Example 12.14, but with $R_2 = 1\ \Omega$.
   Ans. $S_{11} = -(j0.04\omega + 1)/(j0.12\omega + 1)$, $S_{21} = j0.113\omega/(j0.12\omega + 1)$

12.7 SUMMARY AND STUDY GUIDE

The subject of this chapter was power in the sinusoidal steady state. The rms value of a periodic signal was defined, and specialized to sinusoids. Average power was discussed, followed by the notion of complex power. The average stored energy was then derived for inductances and capacitances. This was followed by a section concerned with applications to power transmission, including discussions of transformer efficiency, transmission systems and three-phase power. It was then shown that average power obeys superposition for sinusoids: this was used to study power flow in an amplifier. Finally, the problem of power transfer was studied; this involved the concept of impedance matching and also the notion of scattering parameters.

The student should pay particular attention to the following points.

An rms value can be calculated for any periodic waveform; in general, it is *not* related to the peak value by a factor of $\sqrt{2}$, as in the case of a sinusoid.

Resistances dissipate real average power; reactive elements only absorb or generate reactive power.

Reactive elements have average stored energy.

Practical transformers can often be modeled as ideal transformers in the calculation of network variables; dissipated power is calculated in a second step.

Power calculations for a balanced three-phase system can be reduced to calculations with a single-phase system with a voltage equal to the three-phase
line-to-line voltage.

In general, superposition does not apply to power. One notable exception is
the case of average power for sinusoidal sources of different frequencies.

In order to transfer maximum power, make the source resistance as small as
possible and match the load to the source impedance.

The study of scattering parameters takes the view that the power delivered to
a port by a source is the difference between the power available from the
source (incident power) and the power not taken from the source
(reflected power). The scattering coefficients are essentially parameters
describing the relationships between these power quantities.

PROBLEMS

12.1 Calculate the average and rms values of each signal in Fig. P12.1. The signal of Fig.
P12.1(b) is a rectified sine wave of frequency ω_0 rad/s.

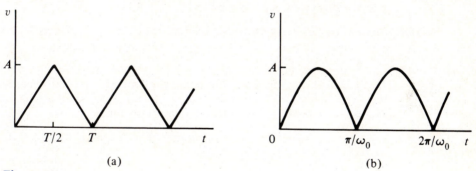

(a)   (b)

Figure P12.1

12.2 Repeat Prob. 12.1 for the waveform in Fig. P12.2.

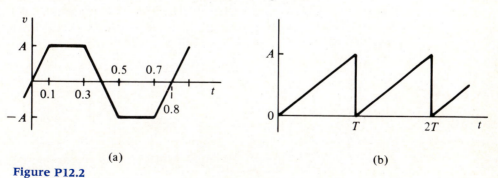

(a)   (b)

Figure P12.2

12.3   In an inexpensive voltmeter, the signal $x(t)$ is first rectified, and the average, $\bar{x}$, of the rectified signal drives the meter. Suppose the meter reads full scale for $\bar{x} = 10$ V.
   (a) If $x(t)$ is a sinusoid, what is the rms value of $x(t)$ that produces a full-scale reading?
   (b) Repeat (a) if $x(t)$ is a rectangular wave as in Fig. D12.1(a).

12.4   For the network of Fig. P12.4, calculate the complex power supplied by the source, if $v_S(t)$ is a sinusoid of frequency 2 kHz and rms amplitude 10 V. Calculate the complex power delivered to each branch and verify that power is conserved.

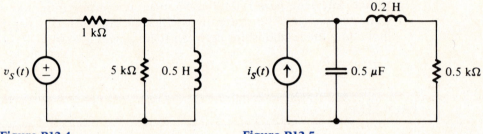

Figure P12.4   **Figure P12.5**

12.5   Repeat Prob. 12.4 for the network of Fig. P12.5, where $i_S(t)$ is a sinusoid of frequency 1 kHz and rms amplitude 50 mA.

12.6   The source $v_S(t)$ in Fig. P12.4 is a sinusoid of frequency ω and rms amplitude of 1 V. Calculate as a function of ω, the real average power supplied by the source. Sketch the curve of power vs ω.

12.7   Let $Z(j\omega)$ be the driving-point impedance of a network composed of passive elements (i.e., R's, L's, and C's). Show that $-90° \leq \angle Z(j\omega) \leq 90°$ for all ω. (*Hint:* The average real power supplied at the network terminals is nonnegative.)

12.8   Calculate the admittance at the operating frequency of a 100-kVA load with rated voltage 2 kV and power factor 0.6 lagging.

12.9   Repeat Problem 12.8 for a 150-kVA load with rated voltage 2 kV and power factor 0.8 leading.

12.10   For the network of Fig. P12.10, calculate the average energy stored in the inductance and the capacitance, for a sinusoidal voltage $v_S(t)$ of frequency ω. Calculate the resonant frequency by equating the electric and magnetic stored energies and the circuit Q by the formula of Example 12.6.

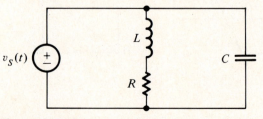

Figure P12.10

12.11 Repeat Prob. 12.10 for the network of Fig. P12.11.

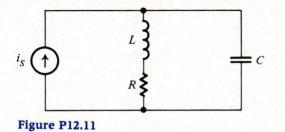

Figure P12.11

12.12 Show that, if the load impedance of a transformer is allowed to vary, the value of Z_L that maximizes efficiency, as given by Eq. (12.25), leads to the condition that the copper losses equal the iron losses.

12.13 Calculate the efficiency of a power transformer, given the following values: $R_1 = 0.03\ \Omega$, $R_2 = 0.7\ \Omega$, $X_1 = 0.03\ \Omega$, $X_2 = 2.1\ \Omega$, $R_m = 750\ \Omega$, $X_m = 750\ \Omega$. The turns ratio, as defined in Fig. 12.3, is $n = \frac{1}{4}$, and Z_L is a 100-Ω resistance.

12.14 Transformer parameters are often evaluated by means of open-circuit and short-circuit tests. The following data are given, for the transformer in Fig. P12.14, modeled by the network of Fig. 12.3. Open-circuit test $(Z_L = \infty)$: $\hat{V}_1 = 550\ \angle 0°\,\mathrm{V}$, $|\hat{V}_2| = 110$ V, $\hat{I}_1 = 2.236\ \angle{-63.4°}$ A. Short-circuit test $(Z_L = 0)$: $\hat{V}_1 = 10\ \angle 0°\,\mathrm{V}$, $\hat{I}_1 = 3.12\ \angle{-67.0°}$ A.
  (a) Assuming R_m, $X_m \gg R_1$, $R_2\ X_1$, $n^2 R_2$, and $n^2 X_2$, calculate n, R_m, X_m, $R_1 + n^2 R_2$, $X_1 + n^2 X_2$.
  (b) Calculate the efficiency of the transformer for a resistive load $Z_L = 2\ \Omega$.

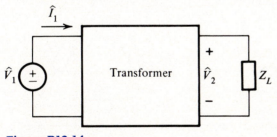

Figure P12.14

12.15 A 100-MVA load, with a power factor of 0.6, lagging, is supplied by a 550-kV line with a line resistance of 20 Ω. The line frequency is 60 Hz.
  (a) Calculate the capacitance needed in parallel with the load to correct the power factor to unity.
  (b) Calculate the line losses with and without the capacitance. (Assume, as in Example 12.8, that the line voltage drop is negligible.)
  (c) If C is the capacitance obtained in (a), calculate the line losses and the corrected load power factor for capacitances of $C/4$ and $C/2$.

12.16 Show that, given constant real average power delivered to the load, the efficiency of a transformer is greatest for a load power factor of 1.

12.17 For power transmission over long distances, the line impedance is an important factor. It turns out that, at 60 Hz, the line impedance is essentially reactive. In Fig. P12.17, $\hat{V}_1 = |\hat{V}_1| e^{j\phi}$ and $\hat{V}_2 = |\hat{V}_2| e^{-j\phi}$.

(a) Show that the average real power transmitted from node 1 to node 2 is $P = |\hat{V}_1| |\hat{V}_2| \sin \phi / X$.

(b) Power engineers use the phase difference ϕ to control the flow of power in a network. Under what conditions will power flow (i) from 1 to 2; (ii) from 2 to 1? What is the maximum amount of power that can be transmitted between 1 and 2, given fixed values of $|\hat{V}_1|$, $|\hat{V}_2|$ and X?

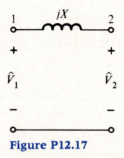

Figure P12.17

12.18 Calculate the average real power delivered to the so-called delta-connected load in Fig. P12.18 by a three-phase source with an rms line-to-line voltage $|\hat{V}_l|$ rms. Define an equivalent single-phase problem.

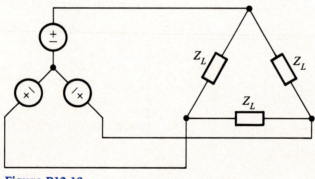

Figure P12.18

12.19 Derive analytically Eq. (12.32).

12.20 A 5-kV, three-phase line is supplying power to a balanced 1500-kVA load with a power factor of 0.85, lagging. If the line resistance is 0.20 Ω per phase, calculate the line losses and the efficiency of the transmission. Assume negligible line voltage drop.

12.21 Calculate the total average power delivered to the resistance in the network of Fig. P12.21, given that $v_S(t) = 1 + \cos t$ and $i_S(t) = \sin 3t$.

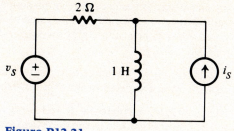

Figure P12.21

12.22 Repeat Prob. 12.21 for the same $v_S(t)$, but with $i_S(t) = 2 \sin(t - 45°)$.

12.23 Repeat Prob. 12.21 for the network of Fig. P12.23, given v_S is a sinusoid of frequency 1 kHz and rms amplitude 2 V.

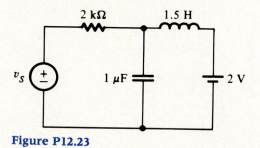

Figure P12.23

12.24 Although superposition does not generally apply to instantaneous power, there are certain special cases where it does. The network of Fig. P12.24(a) is driven, as shown,

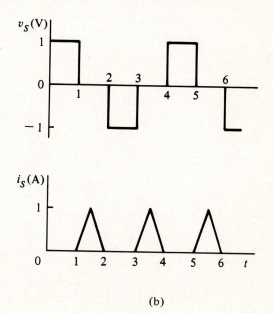

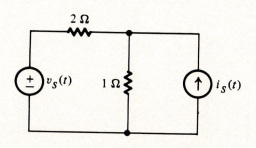

(a)

(b)

Figure P12.24

by a voltage source $v_S(t)$ and a current source $i_S(t)$, both periodic and shown in Fig. P12.24(b). Show that the instantaneous power dissipated in the network is the sum of the power dissipated when v_S acts alone, plus the power dissipated when i_S acts alone.

12.25 In the network of Fig. P12.24(a), $v_S(t) = 2$ V and $i_S(t)$ is a sinusoid of rms amplitude 0.5 A. Calculate the average power dissipated in the two resistances.

12.26 For the JFET amplifier of Fig. P12.26, calculate the dc and ac power supplied by the battery V_{DD} and the ac and dc power dissipated in the 2-kΩ load and in the transistor. The transistor is biased such that $I_D = 3$ mA, and $g_m = 2$ mS. The source $v_S(t)$ is a sinusoid of amplitude 0.5 V rms. The coupling capacitor C_c is essentially a short circuit at the signal frequency, and the power dissipated in R_{B1} and R_{B2} is negligible.

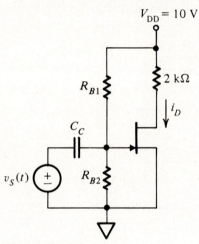

Figure P12.26

12.27 Figure P12.27 shows a BJT common-emitter amplifier with an RC-coupled load and

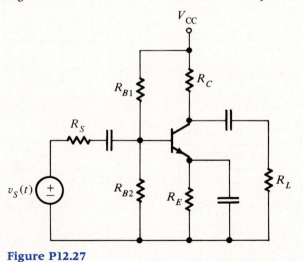

Figure P12.27

an emitter bypass capacitor. Resistors R_{B1}, R_{B2}, and R_E are chosen so as to yield a quiescent $I_C = 10$ mA. if $v_S(t)$ is a sinusoid of amplitude 0.2 V rms, calculate the power delivered to R_L, the power supplied by the battery and the power dissipated in the collector, for the following data: $\beta = 50$, $r_\pi = 125\ \Omega$, $R_C = 580\ \Omega$, $V_{CC} = 12$ V, $R_E = 120\ \Omega$, $R_L = 580\ \Omega$, $R_S = 50\ \Omega$. The resistances R_{B1} and R_{B2} dissipate negligible power and the capacitors can be replaced by short circuits at the signal frequency.

12.28 The 40-kΩ resistors in the emitter follower of Fig. P12.28 are such as to yield a quiescent $V_O = 5$ V. The BJT has a β of 49, $r_\pi = 100\ \Omega$ and $V_{CC} = 10$ V. The signal input $v_S(t)$ is a sinusoid with an rms value of 2 V and a frequency sufficiently high to make C_C essentially a short circuit.

  (a) Calculate the average power supplied by V_{CC}, the ac and dc power supplied to the 2-kΩ load and the power dissipated in the BJT.

  (b) Calculate the power supplied by the signal source v_S.

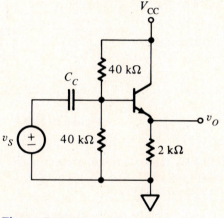

Figure P12.28

12.29 The 50-kΩ and 100-kΩ resistors in the source follower of Fig. P12.29 are such as to

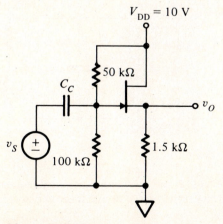

Figure P12.29

yield a quiescent $V_O = 5$ V. The JFET has a g_m of 2 mS at this quiescent point. The signal input $v_S(t)$ is as in Prob. 12.28.
(a) Calculate the average power supplied by V_{DD}, the ac and dc power supplied to the 1.5-kΩ load and the power dissipated in the JFET.
(b) Calculate the power supplied by the signal source v_S.

12.30 Calculate the maximum power available from a source represented by a Thevenin equivalent circuit with a sinusoidal Thevenin voltage of 10 V rms and an impedance of (i) 1 Ω, (ii) 1 kΩ, and (iii) $(1 + j2)$ Ω.

12.31 In the network of Fig. P12.31, v_S is a sinusoidal voltage with rms amplitude $\hat{V}_S$. Calculate, as a function of ω, the maximum power available at ab.

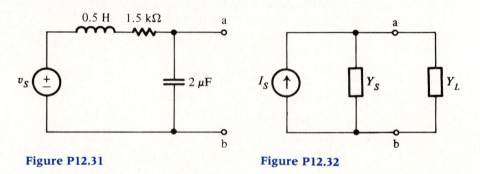

Figure P12.31   **Figure P12.32**

12.32 For the network of Fig. P12.32, calculate the maximum power available at ab, in terms of the real and imaginary parts of Y_S. What must Y_L be in order that this maximum available power be delivered to Y_L?

12.33 Choose the capacitance C and the turns ratio n in Fig. P12.33 so that the average power transferred from the source to the 2-Ω load is maximized, given that $v_S(t)$ is a 100-Hz sinusoid.

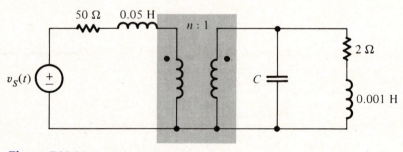

Figure P12.33

12.34 In the network of Fig. P12.34, R_S can be chosen as any value greater than or equal to 100 Ω. Choose R_S and the turns ratio n to maximize the power delivered to the 8-Ω load.

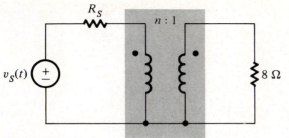

Figure P12.34

12.35 A two-port has the following S parameters, defined with respect to $R_1 = R_2 = 50 \ \Omega$: $S_{11} = 0.8 \ \angle 180°$, $S_{22} = 0.8 \ \angle 135°$, $S_{12} = 0.4 \ \angle 45°$, $S_{21} = 0.4 \ \angle -45°$. With $\hat{V}_{S1} = 1 \ \angle 0°$ volts and $\hat{V}_{S2} = 0$, calculate the power delivered to ports 1 and 2 and the power dissipated in the two-port.

12.36 Repeat Prob. 12.35 with $\hat{V}_{S1} = \hat{V}_{S2} = 1 \ \angle 0°$.

12.37 (a) For the two-port within the dotted box in Fig. P12.37, calculate the S parameters with respect to the two resistances shown:
   (b) Choose X_1 and X_2 so that $|S_{11}| = 0$. Show that, under these conditions, all the power available from the source is delivered to the 200-Ω load.
   (c) Select reactive elements that have the reactances of part (b) at 1 kHz. For $\hat{V}_{S1} = 1 \ V$, calculate, as a function of frequency, the power delivered to the load.

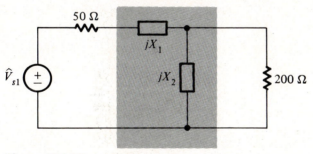

Figure P12.37

12.38 Calculate the S parameters of the high-frequency MOSFET model of Fig. P12.38, with respect to $R_1 = R_2 = R \ll r$. [The capacitance C_{eq} is the Miller capacitance, $C_{eq} = C_{gs} + (1 - A_v)C_{gd}$, where $A_v = \hat{V}_{ds}/\hat{V}_{gs}$.]

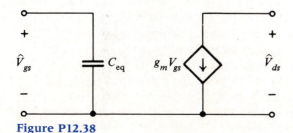

Figure P12.38

CHAPTER

Analog Circuits

13.1 INTRODUCTION

At this point in the development of the subject of circuits and electronics the reader should be in a position to undertake a reasonably thorough study of a very wide range of practical electronic circuits. The subject of this chapter is analog circuits, i.e., those circuits in which the signal may have any value within some practical range. Although the logic circuits discussed in Chap. 8 are also analog in the sense that their instantaneous currents and voltages can have any value within some practical range, the information or signal which they process is contained only in the two states of the ideal logic elements which they approximate. Included under the general title of analog circuits are signal processing and generation circuits as well as those used for measurement and dc power generation.

Just as the most useful logic families were seen in Chap. 8 to be those which are based on active elements, so most useful analog circuits also contain elements which exhibit gain. The operational amplifier (op amp) is by far the most important active element in analog circuit design. This is so because of its versatility and its near ideal behavior, which greatly simplifies the design process. Consequently this chapter is devoted to simple but practical circuits based on the ideal op amp. Where appropriate and where the level of the text permits, situations are included in which the op amp's departure from ideal behavior has to be considered.

13.2 THE IDEAL OP AMP

The ideal op amp was introduced in Sec. 4.4.6. Figure 4.18, which is reproduced in Fig. 13.1, shows the model of an ideal op amp driven by two voltage sources. Most of the ideal op amp's properties, which are listed below, can be inferred from the figure.

1. It behaves as an ideal transvoltage element with an infinite input impedance, zero output impedance, and a voltage gain A. The infinite input impedance means that the input currents can always be assumed to be zero. For ideal behavior, A, which is often called the *open-loop gain*, has to be infinite.

2. It is sensitive to differential-mode excitation only (see Sec. 3.6). Thus the output does not respond to the common-mode component of the input signal, $(v_1 + v_2)/2$. This property is referred to in the technical literature as *common-mode rejection* which, for the ideal op amp, is infinite.

3. Since the model is purely resistive, the bandwidth of the op amp is infinite.

Some other points which apply to op amps in general should also be noted. The output of the op amp, in contrast to its input, is *single-ended*, which is the term used to denote the fact that the output appears at a single terminal referred to ground. Note, however, that the input can be made single-ended as well by grounding one of the input terminals. If terminal 1 is grounded in Fig. 13.1 then

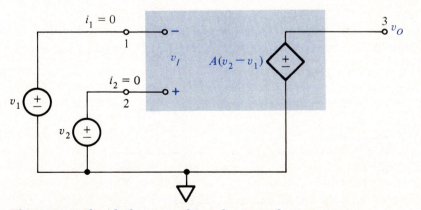

Figure 13.1 The ideal op amp driven by two voltage sources.

$v_O = Av_2$, while grounding terminal 2 results in $v_O = -Av_1$. Because of the input-output polarity relationship in the two cases, terminal 1 is called the *inverting* input terminal while terminal 2 is referred to as being *noninverting*.

Although this is not apparent from Fig. 13.1, the op amp is a *direct-coupled* device which means that its input and output can be connected directly to the sources and load, respectively. This is in contrast, for example, to discrete transistor amplifiers which, in general, have to be connected via coupling capacitors. The op amp thus belongs to the category of what are called dc amplifiers. No ambiguity arises from using this abbreviation both for "direct-coupled" and "direct current" because a direct-coupled amplifier can amplify signals with frequencies down to zero.

It is customary to represent op amps in circuit diagrams by the symbol shown in Fig. 13.2(a). So, for example, the circuit in Fig. 13.1 will appear as shown in Fig. 13.2(b). Notice that the connection to ground of the VCVS in Fig. 13.1 is not shown. This is because a practical op amp does not have a ground terminal. Being an active element, the op amp has to be connected to power supplies, and it is the ground connection of the power supply which establishes the datum for the op amp. In Fig. 13.3 the op amp is represented by a more complete symbol which shows the power supply terminals. Usually the op amp is operated with two power supplies as shown in the figure.

Henceforth the simple symbol in Fig. 13.2(a) will be used to denote the op amp, whether it is assumed ideal or not. Unless stated otherwise, the op amp will be assumed to be ideal.

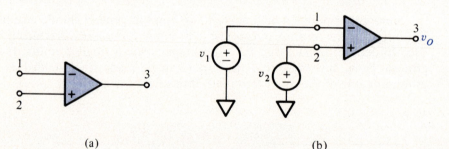

(a)   (b)

Figure 13.2 (a) The op amp symbol. (b) The circuit in Fig. 13.1 redrawn using the symbol in (a).

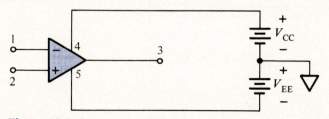

Figure 13.3 Op amp symbol showing connection to power supplies.

13.3 LINEAR CIRCUITS BASED ON THE OP AMP

In Sec. 4.4.6 the reader was shown how an amplifier with a precisely known gain A_f can be designed by connecting a simple resistive feedback network between the output of an op amp and its inverting input terminal. The two amplifier configurations which are based on this approach, namely the inverting amplifier shown in Fig. 4.19 and the noninverting amplifier in Fig. 4.20, are, in fact, special cases of a broad class of circuits which have topologies similar to those of the two amplifiers, but where the resistors R and R_f are, in general, impedances $Z(s)$ and $Z_f(s)$.

The circuits which are analyzed in the following sections are divided on the basis of the inverting and the noninverting configuration.

13.3.1 Circuits Using the Inverting Configuration

The inverting amplifier circuit in Fig. 4.19 is redrawn in Fig. 13.4(a) using the op amp symbol in Fig. 13.2(a). The corresponding general form of the inverting configuration appears in Fig. 13.4(b).

The voltage gain of the amplifier is, according to Eq. (4.35),

$$A_f = \frac{v_O}{v_I} = \frac{-R_f}{R} \tag{13.1}$$

For the circuit in Fig. 13.4(b), the system function is, by analogy to Eq. (13.1),

$$H(s) = \frac{V_o}{V_i} = \frac{-Z_f(s)}{Z(s)} \tag{13.2}$$

THE INVERTING AMPLIFIER

Since the voltage gain of the inverting amplifier in Fig. 13.4(a) depends only on the ratio of two resistors, the accuracy with which it can be set is limited only by the tolerances on the values of R_f and R.

The input impedance of the inverting amplifier is, from Eq. (4.34)

$$Z_{\text{in}} = R \tag{13.3}$$

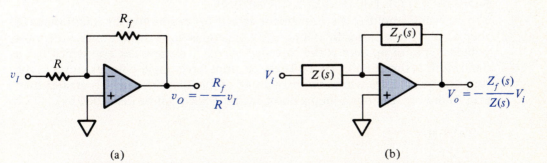

Figure 13.4 (a) The inverting amplifier. (b) The general form of the inverting configuration.

From Eqs. (13.1) and (13.3) it is apparent that R can be chosen to yield a convenient value of Z_{in}, whereupon R_f determines A_f. The choice of Z_{in} is dictated by the fact that the source of the input voltage v_i may not be able to deliver more than a specified value of current, which sets a lower limit on R. The fact that a very large value of R magnifies the op amp's deviation from ideal behavior puts an upper limit on R.

Since the ideal op amp has zero output impedance, it follows that the inverting amplifier also has an output impedance of zero.

EXAMPLE 13.1

A certain signal source, which has zero output impedance, delivers an output voltage of 1 V and a maximum current of 10 mA. Design an amplifier which will develop an output voltage of 10 V when driven by that source.

Solution
Since the source delivers 1 V the gain must be $\frac{10}{1} = 10$. Therefore, from Eq. (13.1)

$$\frac{R_f}{R} = 10$$

The maximum source output current of 10 mA limits R to

$$R \geq \frac{1 \text{ V}}{10 \text{ mA}} = 0.1 \text{ k}\Omega$$

If $R = 0.1$ kΩ is chosen then R_f has to be

$$R_f = 10R = 1 \text{ k}\Omega$$

BANDWIDTH OF THE INVERTING AMPLIFIER
The inverting amplifier is a convenient vehicle for examining the implications of the fact that the bandwidth of a practical op amp is finite. In fact one of the most popular types of op amp is the so-called *internally compensated* one which has a bandwidth of the order of 10 Hz. These op amps have single-pole responses in the range of frequencies where the magnitude of the open-loop gain is greater than one. Therefore, the open-loop gain of such a device can be described by

$$A(s) = \frac{A}{1 + Ts}$$

If the analysis in Sec. 4.4.6 is repeated without allowing the open-loop gain to go to infinity, the result is, using $A(s)$ in the place of A,

$$\frac{V_o}{V_i} = -\frac{R_f}{R}\frac{A(s)}{1 + R_f/R + A(s)}$$

Substituting for $A(s)$ yields

$$\frac{V_o}{V_i} = \frac{A_f}{D(s)}$$

where

$$A_f = -\frac{R_f}{R}$$

and

$$D(s) = 1 + \frac{1}{A}\left(1 + \frac{R_f}{R}\right)(1 + Ts)$$

The associated pole p is found to be

$$p = -\frac{1}{T}\left(\frac{A}{1 - A_f} + 1\right)$$

which, for $|A_f| \ll A$, can be approximated by

$$p = -\frac{1}{T}\frac{A}{1 - A_f} \qquad (13.4)$$

Equation (13.4) shows that, as the low-frequency closed-loop gain A_f is reduced, the associated closed-loop bandwidth $f_p = |p|/2\pi$ increases. In fact, for $|A_f| \gg 1$, Eq. (13.4) simplifies to

$$p = -\frac{1}{T}\frac{A}{|A_f|}$$

revealing that the gain-bandwidth product $|A_f f_p|$, given by

$$|A_f f_p| = \frac{A}{2\pi T} \qquad (13.5)$$

is a constant. Therefore, there is a direct trade-off between closed-loop gain and bandwidth.

EXAMPLE 13.2

An op amp with an open-loop voltage gain $A = 10^4$ and a bandwidth of 20 Hz is used in an inverting amplifier with $R_f = 10$ kΩ, $R = 1$ kΩ. Determine the closed-loop gain and bandwidth.

Solution

The closed loop gain A_f is

$$A_f = -\frac{R_f}{R} = -10$$

Since $R_f/R = 10 \gg 1$, the error in using Eq. (13.5) is small. Therefore

$$|A_f f_p| = 10^4 \times 20 = 200 \text{ kHz}$$

which from the bandwidth is found to be

$$f_p = \tfrac{200}{10} = 20 \text{ kHz}$$

which would certainly be more than acceptable for any audio frequency application.

Equation (13.5) is very general and applies to all negative feedback circuits which have simple negative-real-axis-pole frequency responses. It thus applies to all the circuits considered in this section except for the negative impedance converter.

THE SUMMING AMPLIFIER

The virtual short between the op amp input terminals in the inverting amplifier makes it very simple to obtain the scaled sum of a number of inputs. Figure 13.5 shows how this is done.

The output v_O is easily computed by superposition since, for any v_i that is zero, the corresponding resistor R_i appears across the op amp's input and, since $v = 0$, has no effect. Therefore the output v_O is found by inspection to be

$$v_O = -\left(\frac{v_1}{R_1} + \frac{v_2}{R_2} + \cdots + \frac{v_N}{R_N}\right)R_f$$

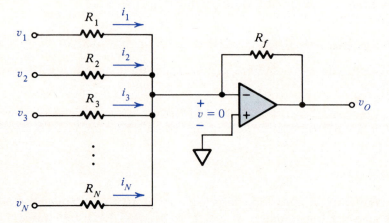

Figure 13.5 The summing amplifier.

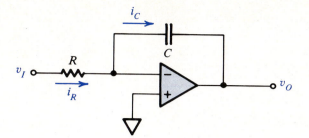

Figure 13.6 The inverting integrator.

This circuit is used in analog computation and, as will be seen further on, it also affords a convenient means of producing offsets in transfer functions.

THE INTEGRATOR
By specifying $Z_f(s)$ in Fig. 13.4(b) to be

$$Z_f(s) = \frac{1}{sC}$$

and letting

$$Z(s) = R$$

leads to

$$H(s) = -(sCR)^{-1} \tag{13.6}$$

which is the system function of an integrator with gain $-(CR)^{-1}$. The corresponding circuit appears in Fig. 13.6. This is, in fact, the same circuit as the one which is shown in Fig. 4.22 and which was analyzed in the time domain in Sec. 4.4.6.

As in the case of the inverting amplifier, R is selected to limit the current drawn from the source, after which C is chosen to yield the appropriate integration time constant $T = RC$.

DRILL EXERCISE

13.1 Design an integrator which will produce a ramp with a slope of -1 V/msec in response to a step input of 1 V. The integrator must not draw more than 10 mA from the input source.
Ans. $R = 0.1$ kΩ, $C = 10$ μF

The preceding approach to the design of an integrator is not very attractive in integrated circuits because of the relatively large resistors and capacitors that are needed. These drawbacks can be circumvented by the use of a rather ingenious

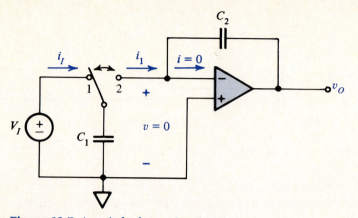

Figure 13.7 A switched capacitor integrator.

technique which is used in *switched capacitor* filters. The simplified circuit of a switched capacitor integrator is shown in Fig. 13.7.

Assume, for simplicity, that the input voltage is a constant V_I as shown in the figure, and that initially the switch is at position 2 and both capacitors are uncharged. When the switch is thrown to 1, an impulse of current i_I occurs and the charge in capacitor C_1 undergoes a step increase Δq given by

$$\Delta q = C_1 V_I$$

If the switch is now returned to 2, the virtual short at the op amp's input terminals results in an impulsive current i_1 which discharges C_1. Since the op amp's input current $i = 0$, $i_2 = i_1$ and the charge in C_2 undergoes a step change Δq. Because the impulse in i_2 is positive, the output voltage v_O, which is also the voltage across C_2, falls by an amount

$$\Delta v_O = -\frac{q}{C_2} = -\frac{C_1}{C_2} V_I$$

When the switch returns to 1, C_1 receives another charge Δq, but the charge on C_2 does not change since $i_2 = -i_1 = 0$. If the switch is thrown back and forth with a frequency $f_c = 1/T_c$, but output voltage v_O will vary with time as shown in Fig. 13.8.

Evidently, as T_c becomes very short the graph in Fig. 13.8 tends to a ramp with a slope given by

$$\frac{\Delta v_O}{\Delta t} = -\frac{C_1}{C_2 T_c} V_I$$

This corresponds to a constant current i_2

$$i_2 = \frac{C_1}{T_c} V_I$$

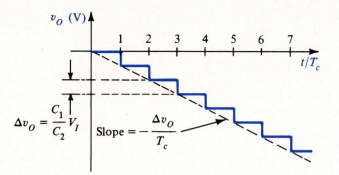

Figure 13.8 Time dependence of v_O in Fig. 13.7 for a switching frequency of $1/T_c$.

In the integrator in Fig. 13.6 if $v_I = V_I$ then the current i_C is given by

$$i_C = \frac{V_I}{R}$$

Comparing this equation with that for i_2 it is clear that the combination of the switch and capacitor C_1 behaves as an equivalent resistance R_{eq} given by

$$R_{eq} = \frac{T_c}{C_1}$$

So, for a given value of C_1, R_{eq} can be controlled by means of the switching frequency $1/T_c$.

In conclusion, for a sufficiently high switching frequency f_c, the circuit in Fig. 13.7 behaves as an integrator with time constant

$$T = R_{eq} C_2 = T_c \frac{C_2}{C_1}$$

This is a very important result because not only is the circuit resistorless but T depends on the *ratio* of two capacitors. The IC technology can control ratios much more accurately than absolute values.

DRILL EXERCISE

13.2 Repeat Exercise 13.1 using the circuit in Fig. 13.7. Choose $T_c = T/10^2$.
    Ans. $f_c = 0.1$ MHz, $C_1 \leq 0.1 \ \mu$F, $C_2/C_1 = 100$

The reader is invited to consider how the switch in Fig. 13.7 could be implemented in an NMOS IC using pass transistors (see Prob. 13.8).

LOW-PASS FILTER

The discussion will be limited to the simple first-order low-pass filter which is characterized by the system function

$$H(s) = (sT + 1)^{-1} \tag{13.7}$$

Equation (13.7) can be realized with an op amp circuit by specifying $Z_f(s)$ in Fig. 13.4(b) to be

$$Z_f(s) = (G_f + sC)^{-1}$$

and $Z(s)$ to be

$$Z(s) = R$$

In that case the system function $H(s)$ is

$$H(s) = \frac{-Z_f(s)}{Z(s)} = -[R(G_f + sC)]^{-1}$$

$$= \frac{-(R_f/R)}{1 + sT} \tag{13.8}$$

where $R_f = 1/G_f$ and $T = R_f C$. The resulting circuit is shown in Fig. 13.9.

Not only does Eq. (13.8) reduce to Eq. (13.7), except for a sign change, when $R_f = R$, but the circuit in Fig. 13.9 possesses the advantage of providing gain, if required, by virtue of having the ratio R_f/R selectable.

Thus the circuit in Fig. 13.9 behaves as an inverting amplifier with a constant gain of $-R_f/R$ for frequencies below $1/T$ while for $\omega > 1/T$ the gain rolls off with a slope of -20 dB per decade.

By using more complicated networks for $Z_f(s)$ and $Z(s)$ in Fig. 13.4(b) one can generate other filter functions. In general, the order of system functions synthesized in this manner is practically limited to two. Higher order functions are obtained by cascading first-order and second-order stages.

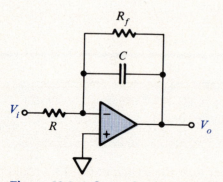

Figure 13.9 A first-order, low-pass filter.

DRILL EXERCISE

13.3 Specify the component values in Fig. 13.9 to yield a low frequency gain of
-10 and a break frequency in the Bode plot at 4 kHz. The input voltage is
0.1 V and the input must not draw more than 1 mA.
Ans. $R = 100\ \Omega$, $R_f = 1\ k\Omega$, $C = 0.04\ \mu F$

13.3.2 Circuits Using the Noninverting Configuration

The simplest circuit using the noninverting configuration of the op amp is the
noninverting amplifier which appears in Fig. 4.20 and which was analyzed in Sec.
4.4.6. Figure 13.10 shows this circuit redrawn in a more compact form.

Since, according to Eq. (4.37), the voltage gain of this amplifier is

$$A_f = \frac{v_O}{v_I} = \frac{R_f + R}{R} \tag{13.9}$$

the approach of generalizing R_f and R to $Z_f(s)$ and $Z(s)$ that was used with such
success with the inverting configuration is not as useful here. The reason is
suggested by Eq. (13.9), where R appears both in the numerator and the denominator. Consequently the synthesis of system functions

$$H(s) = \frac{N(s)}{D(s)}$$

with specified numerators $N(s)$ and denominators $D(s)$ is not so straightforward.
As a simple example, it is not possible to achieve the noninverting form of the
system function of an ideal integrator given in Eq. (13.6).

THE NONINVERTING AMPLIFIER

Because the noninverting input of the op amp draws no current from the source,
the input impedance of the amplifier $Z_{in} \to \infty$. As in the case of the inverting

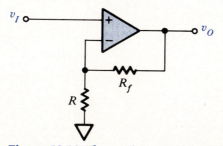

Figure 13.10 The noninverting amplifier.

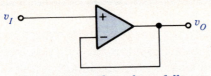

Figure 13.11 The voltage follower.

amplifier the output impedance is zero. So in the noninverting amplifier there is an explicit restriction only on the ratio of the two resistor values.

The high Z_{in} of this amplifier results in its frequent use as a *buffer* where the primary requirement is not voltage gain but negligible loading of the source. In such a case the voltage gain is set to unity which, as can be seen from Eq. (13.9), is achieved by letting $R \rightarrow \infty$, i.e., by simply removing R in Fig. 13.10. Since, in that case, $A_f = 1$ for all values of R_f, this resistor is usually replaced by a short circuit. Figure 13.11 shows the resulting simple circuit which is usually called a *voltage follower* by analogy to the emitter follower.

EXAMPLE 13.3

Because of its high input impedance the noninverting amplifier is a logical candidate for the design of a voltmeter. Figure E13.3 shows a simple analog voltmeter in which a meter having an internal resistance R_M is connected in the place of R_f. If the meter requires a current of 0.1 mA for a full-scale deflection and if $R_M = 10$ kΩ, select R for the voltmeter to function in the 0- to 1-V range.

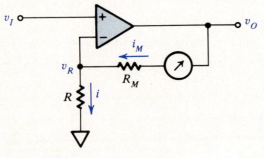

Figure E13.3

Solution
Because of the virtual short across the op amp's input terminals

$$v_I = V_R$$

Also, since the op amp inputs draw no current, $i_M = i$, and since

$$i = \frac{v_R}{R} = \frac{v_I}{R}$$

it follows that

$$i_M = \frac{v_I}{R}$$

irrespective of the value of R_M. Therefore

$$R = \frac{v_I}{i_M} = \frac{1}{0.1} = 10 \text{ k}\Omega$$

13.3.3 Other Linear Op Amp Circuits

THE DIFFERENTIAL AMPLIFIER

Practical situations often arise in which a signal which appears between two terminals, neither of which is grounded, must be amplified by a controlled amount. One way of doing this is by means of the circuit shown in Fig. 4.21 which is redrawn in a more compact form in Fig. 13.12.

As the analysis in Sec. 4.4.6 shows, the circuit is a superposition of the inverting and the noninverting amplifiers, with an attenuator, consisting of R_4 and R_3, added to the noninverting input terminal for greater freedom in design. In particular, as that analysis shows, if $R_2 = R_4 = R$ and $R_1 = R_3 = R_f$ in Fig. 13.12, then the output voltage of the amplifier is

$$v_O = \frac{R_f}{R}(v_B - v_A) \tag{13.10}$$

Since the amplifier has two input terminals there are three possible ways of defining an input impedance. The differential input impedance is that which the

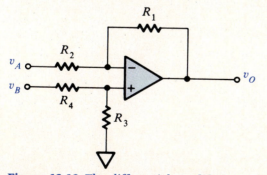

Figure 13.12 The differential amplifier.

amplifier presents between its two input terminals. As explained in Sec. 4.4.6, because of the virtual short between the input terminals of the op amp, the impedance between the differential amplifier's input terminals is simply $R_2 + R_4 = 2R$ if $R_2 = R_4 = R$.

It is left as an exercise for the reader to show that the driving point impedance at each of the two input ports is R_2 for the inverting port and $R_3 + R_4$ for the noninverting.

DRILL EXERCISE

13.4 Verify that the driving point impedance of the inverting and noninverting ports of the amplifier in Fig. 13.12 is, respectively, R_2 and $R_3 + R_4$.

THE NEGATIVE IMPEDANCE CONVERTER (NIC)

This example differs markedly from all preceding ones in that connections are made between the output of the op amp and the noninverting as well as the inverting input terminal. This leads to some rather interesting and useful terminal characteristics.

Figure 13.13 shows the circuit of a negative impedance converter (NIC). Interest here is focused on its driving point impedance R_{in} which is determined by connecting the voltage source shown and computing the resulting current i.

Because of the virtual short across the op amp's input terminals, $v_1 = v$ and

$$i_R = \frac{v_1}{R} = \frac{v}{R}$$

Since the op amp input currents are zero, $i_f = i_R$ so that the voltage v_2 is

$$v_2 = i_f R_f + i_R R = i_R(R_f + R) = \frac{R_f + R}{R} v$$

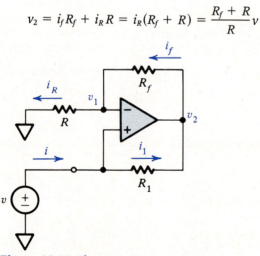

Figure 13.13 The negative impedance converter (NIC).

The currents i_1 and i are equal, so

$$i = \frac{v - v_2}{R_1} = \left(1 - \frac{R_f + R}{R}\right)\frac{v}{R_1}$$

$$= -\frac{R_f}{RR_1}v$$

Therefore the driving point impedance is

$$R_{in} = \frac{v}{i} = -R_1\frac{R}{R_f}$$

a negative quantity.

A possible application of this one-port is as a voltage-to-current converter. This is shown in Fig. 13.14(a) where the details of the NIC circuit are suppressed since it can be viewed as a branch element with a known i-v characteristic. From Fig. 13.14(b), where the circuit is redrawn in a more convenient form, the current i_L to the load is given by

$$i_L = \frac{G_S G_L v_I}{G_S + G_N + G_L} \tag{13.11}$$

where $G_N = -R_f/RR_1$. If $G_N = -G_S$, then

$$i_L = G_S v_I$$

which is independent of G_L. Notice that G_L does not even have to be linear.

Another possible application of the NIC is as a *negative resistance* amplifier. Thus if G_N is adjusted so that

$$G_S + G_N + G_L < G_L$$

then $i_L/G_S v_I > 1$.

Notice that these applications require close control over the resistor values to

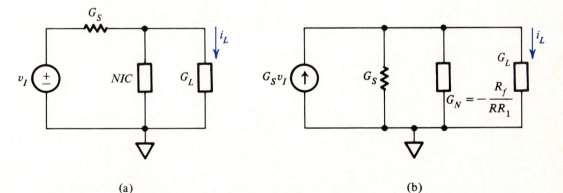

(a)   (b)

Figure 13.14 The NIC as a voltage-to-current converter: (a) circuit and (b) model for analysis.

achieve the required cancellation. If this is not done the circuit may, in fact, become unstable. Thus if a capacitance exists in parallel with G_L in Fig. 13.14, then if $G_S + G_N + G_L < 0$, the impedance of the circuit will have a pole in the right half-plane.

13.4 NONLINEAR CIRCUITS BASED ON THE OP AMP

The nonlinear circuits which are considered in this section owe their nonlinearity to the presence of one or more diodes. Passive nonlinear circuits which perform such important functions are rectification and limiting were encountered by the reader in Chap. 5. The objective here is to examine a larger number of useful nonlinear functions and to show how the use of an op amp results in character-istics which are closer to the ideal than what can be achieved with a passive circuit.

13.4.1 The Half-wave Rectifier

One of the most basic and important nonlinear circuit functions is *rectification* which involves the conversion of a periodic signal which has both positive and negative excursions to another periodic signal which has a single polarity at all times. The half-wave rectifier which was analyzed in Sec. 5.5.4 is the simplest type of rectifier. Its circuit appears in Fig. 5.15 and is redrawn in Fig. 13.15(a).

Figure 13.15(b) shows the shape of the output voltage v_O when the input v_I is a sinusoid of amplitude V_i and frequency $f = 1/T$. The solid curve corresponds to the situation where the diode forward voltage drop V_D is negligibly small compared to V_i and, therefore, a voltage transfer characteristic like the one in Fig. 5.16(b), but with $V_B = 0$, applies. The broken curve, on the other hand, is obtained when the effect of V_D is significant.

In an application where, say, the objective is to rectify the voltage distributed by the local power utility, V_D is negligible since V_i is of the order of 100 V. However, if a precise half-wave rectifier is required and V_i is of the order of a few volts or less, then the simple circuit in Fig. 13.15(a) is unacceptable.

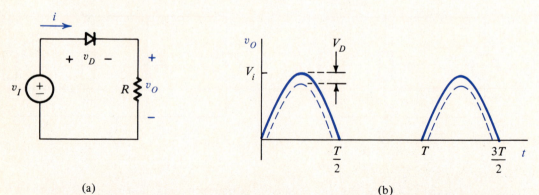

(a) (b)

Figure 13.15 The half-wave rectifier (a) circuit and (b) output voltage neglecting the diode voltage drop V_D (solid curve), and including V_D (broken curve).

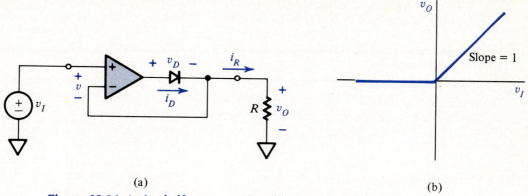

(a)   (b)

Figure 13.16 Active half-wave rectifier: (a) circuit and (b) voltage transfer characteristic.

Figure 13.16(a) shows the circuit of an active half-wave rectifier which exhibits a voltage transfer characteristic close to that of an ideal half-wave rectifier. The transfer characteristic, shown in Fig. 13.16(b), is verified by solving the circuit using the assumed state analysis method introduced in Chap. 4.

DIODE ON
In that case the feedback path between the op amp's output and the negative input terminal is closed and, therefore, $v = 0$. Hence

$$v_O = v_I$$

Since the op amp's inputs draw no current

$$i_D = i_R$$

The diode will remain ON providing $i_D > 0$, which translates into

$$i_R = i_D > 0$$

or, since $v_O = i_R R$ and $v_I = v_O$,

$$v_D > 0$$

This verifies the right-half plane of the characteristic in Fig. 13.16(b).

DIODE OFF
From the preceding analysis this occurs when $v_I < 0$. Since $i_R = i_D = 0$

$$v_O = i_R R = 0$$

which verifies the characteristic in Fig. 13.16(b) for $v_I < 0$.

The part of the circuit in Fig. 13.16(a) between the source and the load R thus behaves as a voltage follower for $v_I > 0$ and as an open circuit for $v_I < 0$. Not only does it behave as an ideal diode, but it also does not load the source the way the passive rectifier does. Thus the current drawn from the source in Fig. 13.16(a) is zero, and the load current i_R is supplied by the op amp. Typical op amps have maximum output currents of the order of 100 mA.

DRILL EXERCISE

13.5 Show that if the diode in Fig. 13.16(a) is reversed, the resulting voltage transfer characteristic is $v_O = v_I$ for $v_I < 0$ and $v_O = 0$ for $v_I > 0$.

13.4.2 The Peak Detector

In many applications, the half-wave rectifier discussed above represents an intermediate step in the process of obtaining a dc voltage. Therefore, the rectifier is usually followed by a filter which extracts the constant component out of the rectifier's output signal. The simplest filter is a capacitor connected across the rectifier's load. In Chap. 15 this problem is discussed in the context of power supplies where the effective load resistance is quite low. In the case of the *peak detector*, which is shown in Fig. 13.17(a), the load resistance can be assumed initially to be negligibly high.

Let v_I be an arbitrary function of time, as shown by means of the broken curve in Fig. 13.17(b), and let $v_O(0) = 0$. Then, assuming an ideal diode for simplicity, as v_I becomes positive the diode turns on and v_O also increases as the capacitor charges up. The capacitor voltage v_O follows v_I until the latter reaches the first peak, $v = V_1$, at $t = t_1$. Now, as v_I begins to fall below V_1 the diode cuts off. This occurs because v_O remains equal to V_1 since, to discharge C, the current i_C would have to be negative. When v_I next rises above V_1 the diode again turns on and the capacitor charges up until the next peak at $t = t_2$ is reached, and so on.

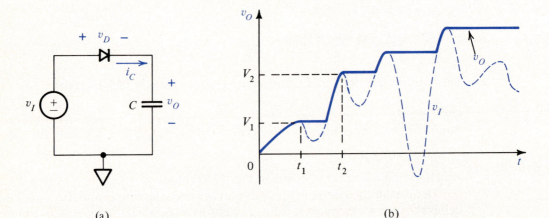

(a)   (b)

Figure 13.17 The peak detector: (a) circuit and (b) response to an input signal of arbitrary shape with $v_O(0) = 0$. An ideal diode is assumed.

Thus the capacitor in the circuit in Fig. 13.17(a) memorizes or *detects* the most positive value of the input voltage v_I.

DRILL EXERCISE

13.6 Show that if the diode in Fig. 13.17(a) is reversed, the circuit detects the most negative input voltage.

THE ENVELOPE DETECTOR

An interesting application of the peak detector is as an envelope detector in a radio receiver. One of the oldest and still widely used methods of transmitting information is a technique called amplitude modulation (AM) which consists of varying or *modulating* the amplitude of a high frequency *carrier* signal in proportion to the amplitude of the information which is to be transmitted. The waveform v_I in Fig. 13.18(b) shows the result of this AM process for the case where the information is a low frequency sinusoid of frequency f_m. Evidently, the information is contained in the envelope of the high frequency carrier, indicated by means of the broken line in Fig. 13.18(b).

If the signal v_I in Fig. 13.18(b) is applied to the input of the peak detector in Fig. 13.17(a), then the steady-state output is simply a constant voltage equal to the maximum value of v_I. On the other hand, if a resistor R is connected across the capacitor, as shown in Fig. 13.18(a), then, for a range of values of the time constant RC, and, in particular, if $RC \gg 1/f_c$, where f_c is the frequency of $v_I(t)$, the output voltage approximates the envelope of $v_I(t)$. The circuit is then said to be *detecting* the modulating signal.

The reason for this behavior is that every time v_I falls after reaching a peak, the diode still cuts off, but now the capacitor can discharge through R with a time constant RC. If $RC \gg T_c = 1/f_c$, then v_o changes very little during the time interval T_c following the instant when the diode cuts off. Consequently the diode remains off for most of this time interval.

Defining $t = t_o$ to be some time when v_I reaches its peak as shown in Fig. 13.18(c), the capacitor voltage begins to decay according to the equation

$$v_O(t) = v_O(t_o)e^{-t/T}$$

where $T = RC$. If $RC \gg T_c$ and $(t - t_o) \simeq T_c$, then this equation can be approximated by

$$v_O(t) \simeq v_O(t_o)\left(1 - \frac{t}{T}\right) \tag{13.12}$$

This explains why the portions of the output voltage in Fig. 13.18(b), corresponding to the time during which C discharges, are shown as straight lines.

At the point where the decay curve of $v_O(t)$ intersects $v_I(t)$ the diode turns on and C is charged up to the next peak of $v_I(t)$.

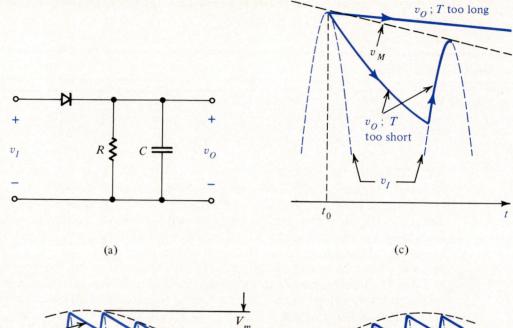

(a)

(c)

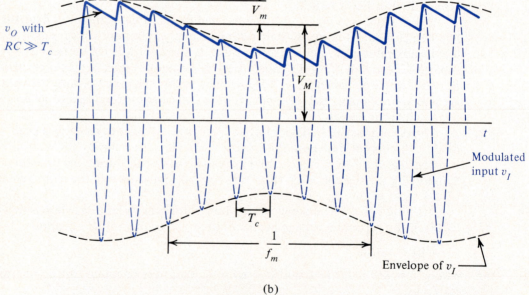

(b)

Figure 13.18 (a) Peak detector with finite load resistor R. (b) Output of peak detector (heavy solid line) for an amplitude-modulated signal input (light solid line) in the case where $RC \gg T_c = 1/f_c$. (c) Magnified view of a small portion of (b) showing effect of $T = RC$ on v_O in (a) when v_I has the form shown in (b).

It should be apparent from the preceding analysis and from Fig. 13.18(c) that there is an optimum decay rate for v_O if it is desired to recover the information in the envelope of v_I. The slope of the $v_O(t)$ decay, according to Eq. (13.12), is equal to

$$\frac{dv_O}{dt} = \frac{-v_O(t_o)}{T}$$

and the magnitude of this slope must clearly be not less than the magnitude of the greatest slope in the envelope of v_I.

For the particular case shown in FIg. 13.18(b), where the envelope is a sinusoid described by

$$v_M = V_M + V_m \sin \omega_m t$$

the slope of v_M at $t = t_o$ is

$$\left. \frac{dv_M}{dt} \right|_{t_o} = \omega_m V_m \cos \omega_m t_o$$

while that of v_O is

$$\left. \frac{dv_O}{dt} \right|_{t_o} = \frac{-v_M(t_o)}{T}$$

From Fig. 13.18(c) the condition which must be fulfilled is

$$\left. \frac{dv_O}{dt} \right|_{t_o} \leq \left. \frac{dv_M}{dt} \right|_{t_o}$$

Substituting the expressions for these derivatives and for v_M at $t = t_o$ yields

$$T \leq \frac{-(V_M + V_m \sin \omega_m t_o)}{\omega_m V_m \cos \omega_m t_o}$$

Straightforward differentiation of the right-hand side of this equation with respect to t_o shows that it is a minimum when

$$\sin \omega_m t_o = -m$$

or

$$\cos \omega_m t_o = -(1 - m^2)^{1/2}$$

where $m = V_m/V_M$ and is called the *modulation index*. Therefore, the worst-case limit on T is given by

$$T \leq \frac{V_M - mV_m}{\omega_m V_m (1 - m^2)^{\frac{1}{2}}}$$

or

$$T \leq \frac{(1 - m^2)^{1/2}}{m\omega_m}$$

In order to avoid distortion of the modulating signal, $m = 1$ is the upper limit on the modulation index. From the above equation, as $m \to 1$ the right hand side tends to zero, meaning that T has to be made very small; so small, in fact, that the requirement that $T \gg 1/f_c$ is violated. However, as the following drill exercise illustrates, practical designs can still be achieved with the simple circuit in Fig. 13.18(b) even with m quite close to unity. This is important because, to maximize the distance over which AM transmisison can be achieved, m has to be as close to unity as possible.

DRILL EXERCISE

13.7 Refer to Fig. 13.18. If $C = 1$ nF, compute the value of R required for the circuit in Fig. 13.18(a) to be useful as an envelope detector in a standard broadcast band AM receiver, where f_c is in the range of 540 to 1600 kHz. Assume the highest modulation frequency to be 4 kHz and $m \le 0.95$.
Ans. 13 kΩ maximum

A PRECISION PEAK DETECTOR

If the half-wave rectifier in Fig. 13.16(a) is used in place of the diode in the peak detector circuit in Fig. 13.17(a), the resulting circuit, shown in Fig. 13.19(a), will evidently behave in the same manner as depicted in Fig. 13.17(b). Note that, for a practical diode, the capacitor in Fig. 13.17(a) will charge up to only $V_i - V_D$.

Since in an actual application of the peak detector there will always be a finite resistance across C, the arrangement in Fig. 13.19(b) can be used to minimize its effect. The added buffer presents a very high input resistance to the capacitor, thus minimizing its discharge current, and at the same time supplies R with a current equal to v_I/R when $v_I > 0$.

In the case of an envelope detector, a separate resistor can be connected across C in Fig. 13.19(b), and therefore, a practically small value of C can be chosen to yield an appropriate time constant, independent of the load R.

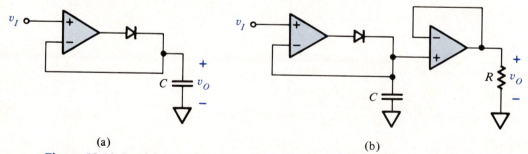

(a)   (b)

Figure 13.19 Precision peak rectifier: (a) basic circuit and (b) circuit modified to minimize discharging of C.

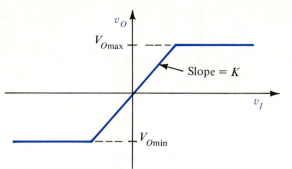

Figure 13.20 Characteristic of an ideal limiter.

13.4.3 The Limiter

Figure 13.20 shows the characteristic of an ideal *limiter*. The name is derived from the fact that, as can be seen from this characteristic, the output voltage v_O is proportional to the input voltage v_I for some range of v_I, while for v_I outside this range v_O is limited to some V_{Omax} at one end of the range and V_{Omin} at the other.

The simplest type of limiter or clipper was analyzed in Sec. 5.6.2, and its circuit diagram and transfer characteristic are shown in Figs. 5.18 and 5.19(a) respectively. Figure 5.19(b) illustrates the effect of the limiter on a sinusoidal input voltage having an amplitude greater than either of the two constant voltage sources. The requirement for two constant voltage sources is impractical and, in general, unneccessary because of the existence of *Zener diodes*. The piecewise-linear characteristic of a Zener diode is shown in Fig. 13.21 together with its symbol. A model based on ideal diodes appears in Fig. 13.21(c). From the characteristic and the model it should be clear that for $v < 0$ the Zener diode behaves like an ordinary diode with $V_D \simeq 0.7$ V while for $v > 0$ it also behaves like a diode but

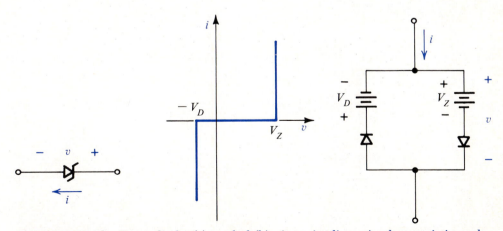

Figure 13.21 The Zener diode: (a) symbol (b) piecewise-linear *i-v* characteristic, and (c) model.

with a forward voltage drop of V_z. Zener diodes are available with V_z ranging from 1 or 2 V to 100 V or so.

The circuit of a simple Zener diode limiter appears in Fig. 13.22(a). Because the Zeners are connected in series opposition, limiting occurs when both are conducting, with a voltage V_D across one and V_z across the other. Therefore, if the Zeners are identical and if v_I is in the range

$$-(V_z + V_D) < v_I < V_z + V_D$$

both diodes are off and $v_O = v_I$. Outside this range limiting occurs and $v_O = \pm(V_z + V_D)$ depending on the polarity of v_I. The resulting transfer characteristic is shown in Fig. 13.22(b).

DRILL EXERCISE

13.8 Show that if the output of the limiter in Fig. 13.22(a) is terminated by a resistor R_L, the slope of the central region in the voltage transfer characteristic is reduced to $R_L/(R + R_L)$.

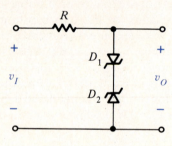

(a)

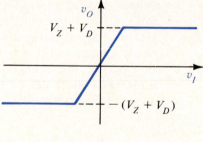

(b)

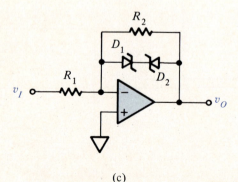

(c)

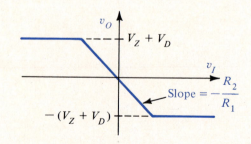

(d)

Figure 13.22 Limiter using Zener diodes: (a) passive circuit and (b) its transfer characteristic; (c) active design and (d) its voltage transfer characteristic.

By placing the two Zener diodes in Fig. 13.22(a) in the feedback path of an inverting amplifier, one obtains the added degree of freedom that permits the gain v_O/v_I to be adjusted. The resulting active limiter is shown in Fig. 13.22(c).

Because of the virtual short at the op amp's input, the output voltage v_O is given by $-(R_2/R_1)v_I$ if

$$-(V_Z + V_D) < v_O < V_Z + V_D$$

since, in this range, the voltage across the two diodes is insufficient for either to conduct. Consequently the corresponding range for v_I is

$$\frac{R_1}{R_2}(V_Z + V_D) > v_I > -\frac{R_1}{R_2}(V_Z + V_D)$$

Outside this range v_O is clamped at either $V_Z + V_D$ or at $-(V_Z + V_D)$. Figure 13.22(d) shows the transfer characteristic of the active limiter.

DRILL EXERCISES

13.9 Show that if v_I in Fig. 13.22(c) is given by $v_I = -10 \sin \omega t$ volts and if $R_1 = R_2 = 5$ kΩ and $V_Z = 4.3$ V the output waveform will appear as shown in Fig. D13.9.

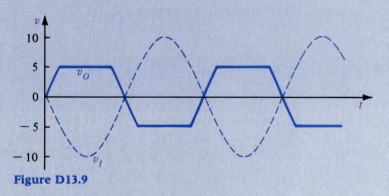

Figure D13.9

13.10 Show that if a constant negative voltage $-V_R$ is connected to the negative input terminal of the op amp in Fig. 13.22(c) through a resistor equal in value to R_1, the transfer characteristic in Fig. 13.22(d) will be shifted to the right by $+V_R$. (*Hint*: Recall the summing amplifier.)

Limiters are used, for example, in radio circuits designed for the reception of frequency-modulated (FM) signals. Any amplitude fluctuations in such signals are undesirable and, consequently, limiters are used to convert these signals to constant amplitude ones.

13.4.4 The Comparator

The characteristic of an ideal comparator is shown in Fig. 13.23. Depending on whether the input voltage is less than or greater than some reference V_R the output voltage takes on one of the two values.

Figure 13.24(a) shows the simplest comparator circuit, which is just an op amp operated open-loop. If an op amp is operated with two power supplies, as shown explicitly in the figure, then its output voltage range is limited to approximately V_{CC} and $-V_{EE}$. Since there is no feedback in the comparator circuit, the very high open-loop gain of the op amp causes v_O to saturate at $V_{Omax} \simeq V_{CC}$ when v_I exceeds V_R and at $V_{Omin} \simeq -V_{EE}$ when v_I falls below V_R. The resulting transfer characteristic is as shown in Fig. 13.24(b). Such comparators are relatively slow because of the time it takes to bring the op amp out of saturation. Switching also tends to be rather underdamped.

For improved performance and flexibility an inverting configuration can be used with two Zener diodes in the feedback path. Figure 13.24(c) shows such a comparator. Its transfer characteristic can be derived by letting $R_2 \to \infty$ in the limiter circuit shown in Fig. 13.22(c). This causes the slope of the amplifying part of the characteristic in Fig. 13.22(d) to tend to infinity. The resultant transfer characteristic is as shown in Fig. 13.24(d) for identical Zener diodes.

From the analysis of the summing amplifier in Sec. 13.3.1 it follows that the modification shown in Fig. 13.24(e) moves the characteristic in Fig. 13.24(d) bodily to the right by an amount V_R, as shown in Fig. 13.24(f).

THE COMPARATOR WITH HYSTERESIS

One of the disadvantages of the preceding comparators is that their behavior is erratic in the presence of noise. If the input voltage to a comparator has noise superimposed on it then, in the vicinity of V_R, the output voltage may change state several times. This is illustrated in Fig. 13.25(a). On the other hand, if the comparator has *hysteresis* in its characteristic, as illustrated in Fig. 13.25(b), then this erratic behavior can be suppressed. The arrows on the characteristic mean that, if $v_O = V_{Omax}$, then v_I must be reduced to below V_{RD} for v_O to fall to V_{Omin} and, conversely, if $v_O = V_{Omin}$, it will not rise to V_{Omax} until v_I exceeds V_{RU}. If $V_{RU} - V_{RD}$ is made

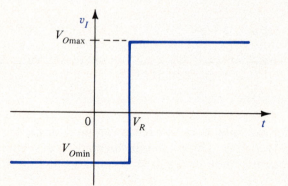

Figure 13.23 Characteristic of an ideal comparator.

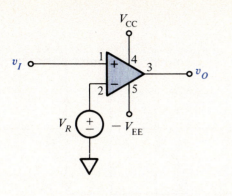

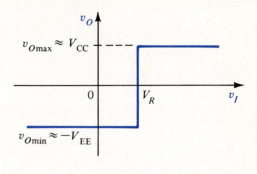

(a)

(b)

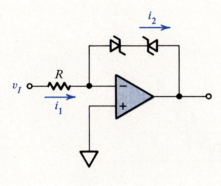

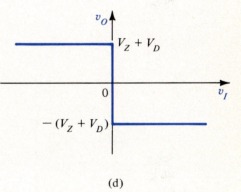

(c)

(d)

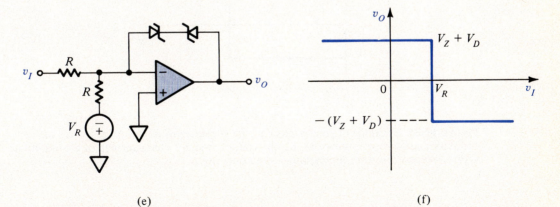

(e)

(f)

Figure 13.24 Comparator circuits: (a) simple op amp comparator and (b) its transfer characteristic; (c) Inverting comparator with Zener diodes and (d) its characteristic; (e) same as (c) but with voltage reference; (f) voltage characteristic of (e).

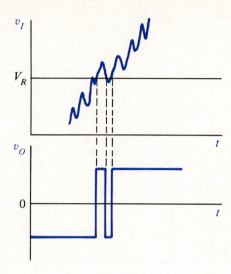
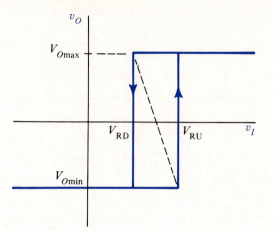

(a)

(b)

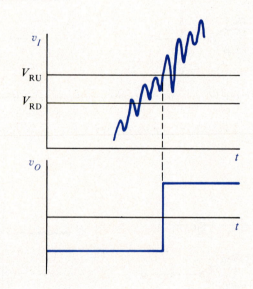

(c)

Figure 13.25 Behavior of comparators in the presence of noise: (a) response of ordinary comparator to noisy input, (b) voltage transfer characteristic of a comparator with hysteresis, and (c) its response to a noisy input.

larger than the noise amplitude then the erratic behavior is eliminated as shown in Fig. 13.25(c).

The price that is paid for the noise immunity is the uncertainty in the threshold for switching which the hysteresis introduces. Thus if the input voltage in Fig. 13.25(c) were falling rather than rising, v_O would change state at $v_I = V_{RD}$ instead of switching when V_{RU} was crossed, as shown in Fig. 13.25(c). So the hysteresis replaces a sharp reference level V_R by a band of width $V_{RU} - V_{RD}$.

Figure 13.26(a) shows a comparator with hysteresis, based on the op amp. Notice that the circuit is identical to the inverting amplifier except for one crucial difference, namely, that the roles of the inverting and noninverting terminals are interchanged. The reader is invited to subject this circuit to the same analysis that was used on the inverting amplifier in Sec. 4.4.6. The result will be that the voltage gain is also $-R_2/R_1$. This result is wrong because it is derived on the basis of a purely resistive model, which implicitly presupposes that the circuit has a stable solution. In fact, if the op amp is assumed to have a finite open-loop gain A and a single negative real axis pole as was done in Sec. 13.3.1 for the inverting amplifier, straightforward analysis shows that the system function for the circuit in Fig. 13.26(a) has a right-half-plane pole which means that the circuit is unstable.

Just like the latch, the circuit in Fig. 13.26(a) has two stable states. As in the case of the op amp comparator's characteristic in Fig. 13.24(b) the output saturates at V_{Omax} or V_{Omin}, and the op amp's output becomes independent of its input voltage.

If the op amp is saturated, the small signal gain is zero and, therefore, the virtual short approximation at its input is not valid. Hence the circuit in Fig. 13.26(a) can be replaced by the model in Fig. 13.26(b). In order to analyze this circuit it is necessary to determine the value of v_I which will reduce v_1 to zero, thereby putting the op amp back in the active, high gain region of its operating characteristics; i.e., it is necessary to determine the threshold for switching between states.

Applying superposition to the circuit in Fig. 13.26(b)

$$v_1 = \frac{R_2}{R_1 + R_2} v_I + \frac{R_1}{R_1 + R_2} v_O$$

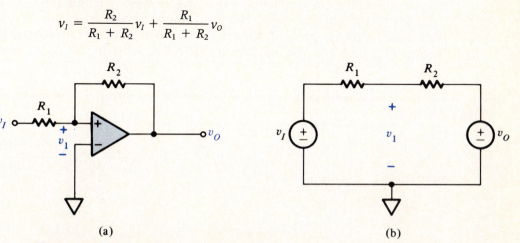

(a)   (b)

Figure 13.26 The comparator with hysteresis: (a) circuit and (b) model for analysis.

The logical procedure is to examine the circuit for

$$v_I \gg V_{RU}, V_{RD} \text{ and } v_I \ll V_{RU}, V_{RD}$$

1. $v_I \gg V_{RU}, V_{RD}$

As in the op amp comparator in Fig. 13.24(a), as soon as v_1 becomes slightly positive v_O will saturate at V_{Omax} because of the op amp's high open-loop gain. To bring the op amp out of this condition it is necessary to reduce v_1 to zero. If $v_1 = 0$, then from the above equation,

$$v_I = V_{RD} = -\frac{R_1}{R_2} V_{Omax}$$

If v_I is reduced below V_{RD}, the circuit switches to the opposite state with $v_O = V_{Omin}$.

2. $v_I \ll V_{RU}, V_{RD}$

Using an argument similar to the one in (1), $v_1 = 0$ if

$$v_I = V_{RU} = -\frac{R_1}{R_2} V_{Omin}$$

Normally V_{Omax} is positive while V_{Omin} is negative so the transfer characteristic of the comparator in Fig. 13.26(a) will be as shown in Fig. 13.25(b) except for the fact that here V_{RD} is negative.

Thus, for the op amp to leave the $v_O = V_{Omax}$ state v_I must be lowered below V_{RD} while to leave the other state v_I has to be raised above $V_{RU} > V_{RD}$.

The circuit possesses memory because, when $V_{RD} < v_I < V_{RU}$, the value of v_O is determined by the direction from which v_I entered this range.

Note that the slope of the broken line in Fig. 13.25(b) is equal to $-R_2/R_1$ which is what the op amp's gain would be if it had a stable operating point in the amplifying mode. Thus when the thresholds for switching, V_{RU} and V_{RD}, are crossed, the circuit enters this amplifying mode but, because it is unstable, the circuit switches to the opposite state. The switching is very fast because, just as the bandwidth of the inverting amplifier is greater than that of the op amp, as shown in Sec. 13.3.1, so here the transient response is faster than that of the op amp.

DRILL EXERCISE

13.11 The comparator in Fig. 13.26(a) is to be used in the presence of input noise having a peak amplitude of 50 mV. If $V_{Omax} = 12$ V, $V_{Omin} = -12$ V, select appropriate values of R_1 and R_2.

Ans. $\dfrac{R_2}{R_1} = 240$

Other applications of the op amp in nonlinear circuits will be encountered in the problems for this chapter.

13.5   SUMMARY AND STUDY GUIDE

The reader should now be in a position to design, build, and test some simple but useful analog circuits using the op amp. Not only is the design of such circuits relatively straightforward but, as the student's measurements should show, their characteristics are very near to those that are predicted over a wide range of operating conditions.

Several important concepts and design techniques were covered. Listed below are some of the more important points which the reader should note.

Most useful signal processing functions require the use of one or more active elements for their implementation in circuit form. Even those functions that can be achieved with passive circuits can often be performed more accurately with active circuits.

The operational amplifier is an extremely powerful analog circuit building block because, for a wide range of applications, its characteristics approach those of an ideal differential input, voltage-controlled voltage source.

The virtual short approximation can always be made in the analysis of op amp circuits based on either the inverting or the noninverting configuration. If the circuit contains diodes, the approximation is only valid if the connection between the inverting input terminal and the op amp output is not broken, as by a cutoff diode.

The summing property of the inverting amplifier can be conveniently exploited for introducing offsets in transfer functions.

Circuits with feedback from the op amp's output to its noninverting terminal are potentially unstable and their analysis in general is not straightforward.

PROBLEMS

13.1   Design an op amp circuit with a voltage gain of -20 and an input resistance of $10\ k\Omega$.

13.2   The circuit in Fig. P13.2 shows how an op amp can be used to design a current-to-voltage converter. Specify values of R_f and R_s for the circuit to have a transresistance v_o/i_s of -5 V/mA.

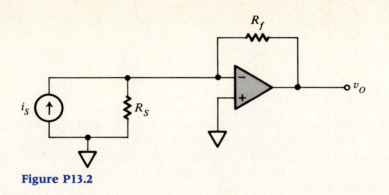

Figure P13.2

13.3 Figure P13.3 shows a technique that is used when it is desired to avoid using very large resistor values.

(a) Show that the voltage gain is given by

$$\frac{v_O}{v_S} = \frac{-(R_2 + R_3 + R_2 R_3 / R_4)}{R_1}$$

(b) If the resistor values have an upper limit of 1 MΩ, design an amplifier with a voltage gain of -10 and $R_1 = 1$ MΩ.

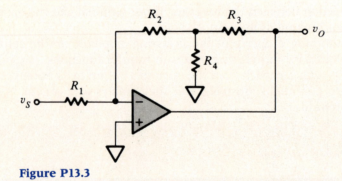

Figure P13.3

13.4 Show that the circuit in Fig. P13.4 behaves as a current amplifier with a gain of

$$\frac{i_L}{i_S} = \frac{-(R_1 + R_2)}{R_1}$$

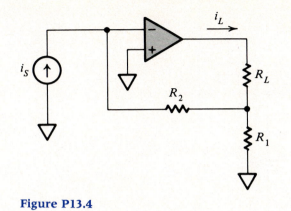

Figure P13.4

13.5 The integrator in Fig. 13.6 has $R = 10\,k\Omega$ and $C = 0.5\,\mu F$. Plot the output voltage v_O for the input shown in Fig. P13.5.

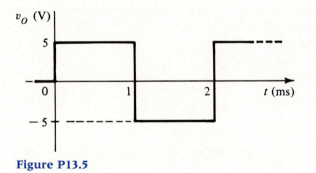

Figure P13.5

13.6 Show that if R_f in Fig. 13.5 is replaced by a capacitor C_f, the resulting circuit functions as a summing integrator.

13.7 Design a first-order active low-pass filter circuit having a bandwidth of 5 kHz, an input resistance of 1 $k\Omega$ and a dc voltage gain of −5.

13.8 The smallest capacitance that can be achieved reliably in a MOS IC is 0.1 pF.
   (a) Design an IC switched capacitor integrator with a time constant of 0.1 ms using a switching frequency of 100 kHz and the smallest possible capacitance values.
   (b) Figure P13.8 shows how the switch in Fig. 13.7 can be realized in a MOS IC. Sketch the waveforms for v_C and $\bar{v}_C$ paying careful attention to their timing relationship. Indicate on your diagram the switching frequency f_c.

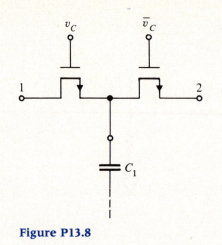

Figure P13.8

13.9 Specify $Z_f(s)$ and $Z(s)$ in Eq. (13.2) for a differentiator with a time constant T. Draw the circuit and specify the values of the components for $T = 1$ ms.

13.10 An inverting amplifier has a low-frequency voltage gain of -20. The op amp has an open-loop voltage gain of 10^4 and a first-order, low-pass response with a 3-dB frequency of 10 Hz. If the input is a step of amplitude 1 V, compute the rise-time t_r of the output, defined as the time for the output voltage to go from 10 to 90% of its final value.

13.11 Specify the value of the resistors in Fig. P13.11 for the circuit to behave as a summing amplifier with

$$v_O = v_1 + 2v_2$$

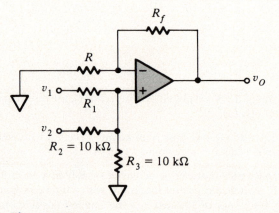

Figure P13.11

13.12 Specify the values of the resistors in Fig. 13.12 for the differential amplifier to have a voltage gain of 20 and a differential input resistance of 20 kΩ. What are the respective driving-point resistances of the inverting and the non inverting input of the amplifier?

13.13 Consider the circuit in Fig. P13.11 with the resistor R connected to an input voltage v_3 instead of ground. If $R_f = 20 \text{ k}\Omega$, $R = 10 \text{ k}\Omega$, $R_1 = R_2 = R_3 = (20/3) \text{ k}\Omega$, determine the transfer function of the circuit.

13.14 Show that if the load G_L in Fig. 13.14 is replaced by a capacitor C the circuit can be made to behave as an integrator with gain $1/RC$. What will the transfer function v_2/v_I be in such a case if $R_f = R$ in Fig. 13.13?

13.15 For the circuit in Fig. 13.14(a), with G_L replaced by a capacitor C, show that
(a) The circuit will be unstable if $G_S + G_N < 0$.
(b) The circuit may be adjusted to have a first-order, low-pass filter characteristic. Derive an expression for the break frequency and the low-frequency gain if the output is v_2 in Fig. 13.13.

13.16 Show that the circuit in Fig. P13.16 behaves as a half-wave rectifier with $v_O = -(R_2/R_1)/v_I$ for $v_I < 0$ and $v_O = 0$ for $v_I > 0$.

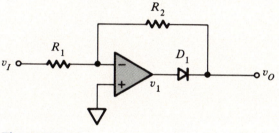

Figure P13.16

13.17 When the input voltage to the rectifier in Fig. P13.16 is positive, the op amp's output v_1 saturates at a negative voltage V_{Omin} determined by the supply voltage V_{EE} in Fig. 13.3. Show that by connecting a second diode D_2 between the inverting input terminal and the output of the op amp in Fig. P13.16, with the anode at the input terminal, the voltage v_1 is clamped at $-V_D$ when $v_I > 0$, and that D_2 is off when $v_I < 0$.

13.18 The resistor R_f in Fig. 13.4(a) is replaced by a diode described by

$$i \simeq I_s \exp\left(\frac{v}{V_T}\right)$$

Show that the resulting circuit behaves as a logarithmic amplifier with transfer function

$$v_O = -V_T\left[\ln\left(\frac{v_I}{R}\right) - \ln I_s\right]$$

13.19 If v_I in Fig. 13.17(a) is a square wave which switches between $+3$ and -7 V, sketch and label the corresponding voltage variation of v_D. When this circuit is used in this manner, i.e., the output is taken across the diode, it is called a *clamping* circuit.

13.20 Draw the circuit diagram of the clamping circuit described in Prob. 13.19 but with the diode replaced by the op amp-diode combination in Fig. 13.16.

13.21 If $v_I = 2 \sin 2\pi f t$ volts in Fig. 13.18 with $f = 100$ kHz and $C = 1$ nF, determine what value R must have if v_O is to vary by not more than approximately 5%.

13.22 Commercial radio stations which transmit using AM, operate with carrier frequencies between 540 and 1600 kHz. The transmitted modulation frequencies are in the range of 100 to 4000 Hz. If the modulation index is limited to 0.98 design a suitable envelope detector using a 500 pF capacitance.

13.23 Using the approach on which the summing amplifier is based, modify the limiter in Fig. 13.22(c) such that $v_O = 0$ corresponds to $v_I = V_R$, where V_R is a positive quantity.

13.24 The comparator in Fig. 13.26 is modified by connecting a resistor $R = 10$ kΩ between the noninverting input of the op amp and a $+10$ V supply. If the op amp output limits at ± 12 V, determine the thresholds V_{RU} and V_{RD} if $R_2 = 10$ kΩ and $R_1 = 1$ kΩ.

CHAPTER

Transistor Amplifiers

14.1 INTRODUCTION

In this chapter the focus is on analog circuits at the transistor level. This is necessary firstly because the op amp, which formed the basis of the study in Chap. 13, is not useful under extreme signal conditions, i.e., very high frequency, very small signal, as well as high voltage and current. Such applications require the use of discrete devices. Moreover, unusual applications of the op amp make it desirable for the designer to have some understanding of the insides of the op amp package, which means understanding how the transistor circuits which form the op amp operate. Finally, there is still ample room for the design of new and improved op amps.

Since transistors are used in both discrete and integrated circuits, such issues as the biasing of transistor amplifiers and the interconnection of transistors are examined here from both the discrete and the IC point of view. Circuits in which JFETs or MOSFETs are employed are usually integrated, consequently these de-

613

vices will not be included in the discussion of discrete circuit techniques. Only small signal amplifiers are considered since power amplifiers are treated separately in Chap. 15.

14.2 BIASING DISCRETE AMPLIFIERS

Figure 14.1 shows the complete circuit diagram of a discrete, small-signal, common-emitter BJT amplifier. The input and output coupling capacitors, C_1 and C_2, are required to prevent any dc interaction between the amplifier and its source R_S and load R_L. Such interaction is undesirable for several reasons which include:

1. The BJT's quiescent point may be disturbed.

2. The source and/or load may be another amplifier stage with its own separately set quiescent point.

3. The load, such as a loudspeaker, may not function properly if the amplifier's output has a dc component.

4. Some sources, such as electrostatic microphones, will not support any dc current, being effectively infinite resistance devices.

Resistors R_E, R_{B1}, R_{B2}, and R_C establish the transistor's Q point. Because R_E would reduce the voltage gain of the amplifier it is bypassed with a capacitor C_E, chosen to be large enough so that, at signal frequencies, the impedance between the emitter of the BJT and ground is negligibly small.

Figure 14.2(d) shows the dc circuit corresponding to the amplifier in Fig. 14.1. This arrangement is widely used not only for common-emitter but for common-collector and common-base amplifiers as well. It can be justified on the basis of the following argument. If the BJT's Q point, defined by I_C and V_{CE} on its output

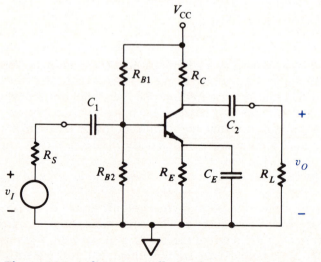

Figure 14.1 A discrete, small-signal, common-emitter BJT amplifier.

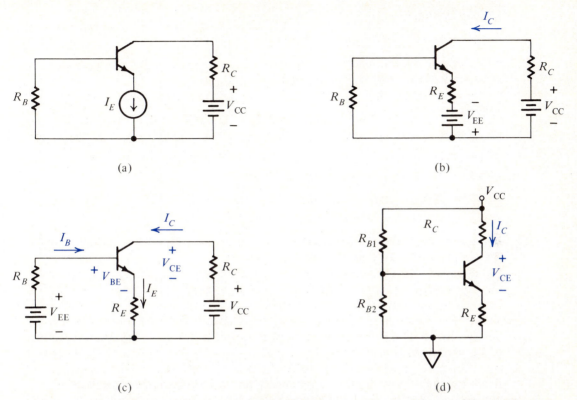

(a)

(b)

(c)

(d)

Figure 14.2 Steps in the evolution of the biasing circuit of the amplifier in Fig. 14.1.

characteristics, is to be insensitive to parameter variations then I_C should be determined by a current source and V_{CE} by a voltage source. This, and, in particular, the solution for V_{CE} is very impractical.

Figure 14.2(a) shows a step in the right direction, where the current source determines I_E which is related to I_C through the relatively stable parameter $\alpha \simeq 1$. The next step is shown in Fig. 14.2(b) where the current source is approximated by the combination of V_{EE} and R_E. It is a simple matter to show that moving V_{EE} from the emitter to the base, as shown in Fig. 14.2(c), has no effect on the dc solution for the currents. Finally, recognizing that the combination of V_{EE} and R_B is the Thevenin equivalent of the network consisting of V_{CC}, R_{B1}, and R_{B2}, provided that

$$V_{EE} = \frac{V_{CC}\, R_{B2}}{R_{B1} + R_{B2}} \tag{14.1a}$$

and

$$R_B = \frac{R_{B1}\, R_{B2}}{R_{B1} + R_{B2}} \tag{14.1b}$$

leads to Fig. 14.2(d).

Notice that

$$V_{CE} = V_{CC} - I_C(R_C + R_E)$$ (14.2)

where $\alpha = 1$ is assumed. Therefore, providing the power supply voltage and the resistors are constant, V_{CE} will be as stable as I_C.

The equations defining the operating point are most easily obtained with reference to Fig. 14.2(c). From KVL around the input loop

$$I_E R_E + I_B R_B = V_{EE} - V_{BE}$$

which is solved for I_C by substituting $I_C = \beta I_B$ and $I_E = [(\beta + 1)/\beta]I_C$, yielding the result

$$I_C = \frac{\beta(V_{EE} - V_{BE})}{(\beta + 1)R_E + R_B}$$ (14.3)

Equations (14.2) and (14.3), together with Eq. (14.1) define the operating point of the circuit in Fig. 14.2(d).

Practical considerations, some of which are considered in the following example, restrict the value of V_{EE} and especially R_E to ranges where this combination behaves as a relatively poor approximation to a current source. Nevertheless, it should be noted that if $(\beta + 1)R_E \gg R_B$, then since $\beta \gg 1$, the collector current I_C defined by Eq. (14.3) is independent of β. Moreover, if $V_{EE} \gg V_{BE}$, then the effect of V_{BE} variations is also reduced.

The most important attribute of the biasing network in Fig. 14.2(d) is the flexibility that it provides the designer by virtue of the existence of three adjustable parameters, R_{B1}, R_{B2}, and R_E. This is in contrast to the simple biasing networks which were considered in Chap. 7, such as the one in Fig. E7.7(a), where R_B is the sole parameter. The following example illustrates the advantage of the biasing arrangement in Fig. 14.2(d).

EXAMPLE 14.1

An amplifier with a small-signal voltage gain of not less than 10 is to operate with source and load resistors of 50 Ω. The design is being picked up at the point where it has been decided that the quiescent operating point will be $I_C = 10$ mA, $V_{CE} = 5$ V. The selected BJT has $\beta = 50$ and $V_{BE} = 0.75$ V as nominal parameters, and $V_{CC} = 12$ V.

1. Design the amplifier in Fig. 14.1 to meet the above specifications. Then determine the change in I_C that would occur if β was tripled.

2. Repeat the design using the circuit in Fig. E14.1(a) which is a practical version of the BJT amplifier analyzed in Chap. 7.

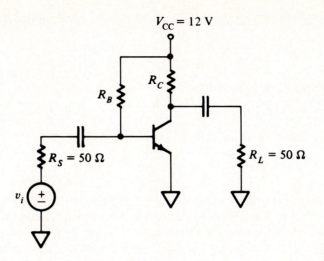

(a)

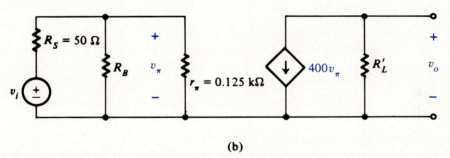

(b)

Figure E14.1

Solution

1. All capacitors are assumed to be effectively short circuits at signal frequencies. The small-signal parameters of the BJT are

$$g_m = \frac{I_C}{V_T} = \frac{10}{0.025} = 400 \text{ mS}$$

$$r_\pi = \frac{\beta}{g_m} = \frac{50}{400} = 0.125 \text{ k}\Omega$$

Figure E14.1(b) shows the small-signal model of the amplifier in Fig. 14.1 with R_B defined by Eq. (14.1b) and $R'_L = R_C R_L / (R_C + R_L)$.

The output voltage is given by

$$v_o = -400 R'_L v_\pi \text{ volts}$$

with R'_L in kΩ units. Since v_π is related to v_i through the voltage divider relationship

$$v_\pi = \frac{v_i R_I}{R_S + R_I}$$

where R_I is defined by

$$R_I = R_B \parallel r_\pi$$

the small-signal gain is

$$\frac{v_o}{v_i} = -400 \frac{R'_L R_I}{R_S + R_I}$$

This expression shows that the maximum voltage gain is obtained by maximizing R'_L and R_I. Since R_L is fixed at 50 Ω, this means choosing $R_C \gg R_L$ so that $R'_L \rightarrow R_L$. Also, R_I approaches its upper limit r_π when $R_B \gg r_\pi$. Since, as can be seen from Eq. (14.2), R_C also determines V_{CE}, a sensible approach is to begin by attempting to set $R_B \gg r_\pi$; say, $R_B = 10r_\pi = 1.25$ kΩ. In that case the voltage gain equation yields

$$R'_L = \frac{10}{400} \frac{R_S + R_I}{R_I} \approx \frac{10}{400} \frac{R_S + r_\pi}{r_\pi} = \frac{10}{400} \frac{0.05 + 0.125}{0.125}$$

$$= 0.035 \text{ k}\Omega$$

as a lower limit on R'_L since 10 is the lower limit on the voltage gain.
From the expression relating R'_L to R_C

$$R_C = \frac{R'_L R_L}{R_L - R'_L} = \frac{35(50)}{50 - 35}$$

$$= 116 \ \Omega$$

which is the corresponding lower limit on R_C.
Equation (14.3) yields the following relationship between V_{EE} and R_E:

$$V_{EE} - \frac{I_C(\beta + 1)}{\beta} R_E = V_{BE} + \frac{I_C}{\beta} R_B$$

or, substituting values,

$$V_{EE} - \frac{10(51)}{50} R_E = 0.75 + \frac{10}{50} 1.25$$

$$V_{EE} - 10.2 R_E = 1.0$$

where, from Eq. (14.1a), V_{EE} must be less than $V_{CC} = 12$ V.
Finally, from Eq. (14.2)

$$R_C + R_E = \frac{V_{CC} - V_{CE}}{I_C} = \frac{12 - 5}{10} = 0.7 \text{ k}\Omega$$

It was observed that, for good stability of the Q point, R_E should be as large as possible in order better to approximate a current source. A reasonable approach is, therefore, to select R_C to be as small as possible. Let $R_C = 120 \ \Omega$ which is within the limit set by the gain equation. Hence,

$$R_E = 0.7 - R_C = 0.7 - 0.12 = 0.58 \text{ k}\Omega$$

whence

$$V_{EE} = 1.0 + 10.2R_E = 6.9 \text{ V}$$

There now remains only to solve for R_{B1} and R_{B2} which, from Eq. (14.1), are found to be $R_{B1} = 2.17 \text{ k}\Omega$ and $R_{B2} = 2.93 \text{ k}\Omega$.

If β is tripled to 150, then from Eq. (14.3),

$$I_C = \frac{150(6.9 - 0.75)}{151(0.58) + 1.25} = 10.4 \text{ mA}$$

Thus, despite the various design constraints, the biasing network holds the collector current constant to within 4% for a threefold variation in β.

2. The small-signal model in Fig. E14.1(b) applies to the amplifier in Fig. 14.1(a) as well, so again R_B should be large compared to r_π. However, the biasing equation in this case is

$$R_B = \frac{V_{CC} - V_{BE}}{I_B}$$

Since $I_C = 10$ mA and $\beta = 50$, it follows that $I_B = 0.2$ mA which fixes R_B at

$$R_B = \frac{12 - 0.75}{\frac{10}{50}} \text{ k}\Omega = 56 \text{ k}\Omega$$

which is certainly much larger than r_π. Because $V_{CE} = 5$ V,

$$R_C = \frac{V_{CC} - V_{CE}}{I_C} = \frac{12 - 5}{10}$$

$$= 0.7 \text{ k}\Omega$$

which is much larger than the $116 \ \Omega$ lower limit arrived at in (1).

Now if $\beta = 150$,

$$I_C = \beta I_B = 150(0.2) = 30 \text{ mA}$$

and

$$V_{CE} = V_{CC} - I_C R_C = 12 - 30(0.7) = -9 \text{ V}$$

which means that the BJT saturates since V_{CE} will fall only to $V_{CEsat} \simeq 0.2$ V. Thus the simple biasing arrangement in Fig. E14.1(a) does not stabilize the Q point against variations in β.

The preceding example not only illustrates the stabilization of the Q point which occurs when the three-resistor biasing network in Fig. 14.2(d) is used, but also the way the three adjustable parameters R_{B1}, R_{B2}, and R_E permit the designer to satisfy simultaneously other constraints, such as the required gain and output signal swing, the latter depending on V_{CE} as well as I_C.

This biasing arrangement is not restricted to the common-emitter amplifier. In later sections it will be seen that the BJT operated common-base or common-collector can also be biased using the same circuit.

14.3 LOW-FREQUENCY PROPERTIES OF THE THREE BJT CONFIGURATIONS

The amplifying property of the BJT operated common-emitter and the impedance transforming property of the common-collector configuration were illustrated in Chap. 7 using a first-order, low-frequency, small-signal model of the BJT and conceptually simple but impractical circuits. In this section these two configurations, as well as the common-base, are examined in more practical circuits. The discussion is limited to their low-frequency properties.

14.3.1 The Common-Emitter Configuration

A practical circuit for a one transistor, discrete circuit amplifier was introduced in Fig. 14.1. Figure 14.3 shows the *mid-band*, small-signal model of the amplifier, which is valid at frequencies which are high enough for the capacitors to behave as effective short circuits, but which, at the same time, are low enough for energy storage effects in the BJT to be negligible.

The voltage gain of this amplifier v_o/v_i was calculated in Example 14.1 and was found to be

$$\frac{v_o}{v_i} = \frac{-g_m R_L' R_I}{R_I + R_S} \qquad (14.4)$$

where

$$R_I = \frac{R_B r_\pi}{R_B + r_\pi}$$

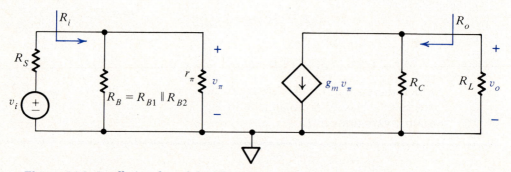

Figure 14.3 Small-signal model of the common-emitter amplifier.

and

$$R'_L = \frac{R_C R_L}{R_C + R_L}$$

Notice that, as $R_s \to 0$, Eq. (14.4) reduces to the same form as Eq. (7.20), that is, the voltage gain is the product of g_m and the effective load resistance R'_L.

With the BJT operating, for example, at $I_C = 1$ mA and, assuming $R'_L = 1$ kΩ, the voltage gain has an upper limit of -40 since $g_m = 40$ mS. In general, R_s is not negligibly small and, to prevent further attenuation of the signal at the input to the BJT, R_B is chosen large in comparison to r_π so that $R_I \to r_\pi$.

The amplifier's input resistance R_i, as defined in Fig. 14.3, is clearly given by

$$R_i = R_I \simeq r_\pi \tag{14.5}$$

Since $r_\pi = \beta/g_m = \beta V_T/I_C$, R_i can be increased by reducing I_C, although this involves a tradeoff with the voltage gain. The latter can be maintained high by means of R_L, which, in turn, enters operating point considerations via the influence of R_C on V_{CE} and, therefore, on the output voltage swing.

From Fig. 14.3 the output resistance R_o is simply

$$R_o = R_C \tag{14.6}$$

Following the approach discussed in Sec. 11.5.1, the capacitors in Fig. 14.1 can be chosen to yield an acceptable low-frequency 3-dB point, although, as this point is reduced toward zero, the required capacitors become impractically large and, alternate designs have to be considered, such as the differential amplifier which is discussed later in this chapter.

DRILL EXERCISES

14.1 The circuit in Fig. 14.1 has $R_S = 10$ kΩ, $R_{B1} = 50$ kΩ, $R_{B2} = 10$ kΩ, $R_C = 3.8$ kΩ, $R_E = 2.2$ kΩ, $R_L = 1$ kΩ, and $V_{CC} = 24$ V. If $\beta = 50$ and $V_{BE} = 0.7$ V, compute I_C and V_{CE}.
Ans. $I_C = 1.37$ mA, $V_{CE} = 15.8$ V

14.2 For the circuit in Drill Exercise 14.1 determine the small-signal voltage gain v_o/v_i and the input resistance R_i.
Ans. -3.3, 822 Ω

14.3.2 The Common-Collector Configuration

The common-collector or emitter-follower configuration was introduced in Chap. 7. Figure 14.4 shows the complete circuit of an emitter follower which is more practical than the arrangement in Fig. 7.14. A comparison of the circuits in

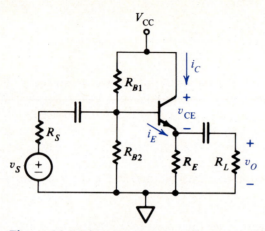

Figure 14.4 Discrete common-collector amplifier (emitter follower).

Fig. 14.4 and 14.1 shows that the corresponding dc circuits are the same except for the absence of R_C in the former. However, since R_C does not enter into the calculation of I_C, as can be seen from Eq. (14.3), the collector current will have the same value and stability for the same values of V_{CC}, R_{B1}, R_{B2}, and R_E in the two circuits.

Figure 14.5 shows the midband small-signal model corresponding to the complete emitter follower in Fig. 14.4.

Comparing Figs. 14.5 and 7.15, it is seen that the resistance R'_i defined in the former figure is given by Eq. (7.26) with R_E replaced by R'_L, where

$$R'_L = \frac{R_L R_E}{R_L + R_E}$$

that is,

$$R'_i = r_\pi + (\beta + 1)R'_L \qquad (14.7)$$

Clearly the amplifier's input resistance R_i, also defined in Fig. 14.5, will be

$$R_i = R'_i \parallel R_B \qquad (14.8)$$

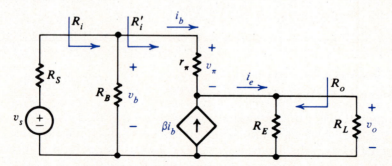

Figure 14.5 Midband, small-signal model of the emitter follower in Fig. 14.4.

The voltage gain is calculated by noting that

$$v_o = i_e R_L' = (\beta + 1)R_L' i_b$$

Since the voltage v_b is related to v_s through the voltage divider equation

$$v_b = \frac{v_s R_i}{R_i + R_S}$$

and the base current i_b is given by

$$i_b = \frac{v_b}{R_i'}$$

the voltage gain v_o/v_s is

$$\frac{v_o}{v_s} = \frac{(\beta + 1)R_L' R_i}{R_i'(R_i + R_S)} \qquad (14.9)$$

If R_B is relatively large, then $R_i' = R_i$ and Eq. (14.9) simplifies to

$$\frac{v_o}{v_s} = \frac{(\beta + 1)R_L'}{R_S + r_\pi + (\beta + 1)R_L'} \qquad (14.10)$$

which is the same as Eq. (7.25) except for the loading due to R_L.

The output resistance R_o, defined in Fig. 14.5, can be obtained quickly with the help of the results of Sec. 7.5. Thus, from Figs. 7.16 and 14.5, R_o is seen to be the parallel combination of R_E and the quantity defined by Eq. (7.27) with R_B replaced by $R_B \| R_S$, i.e.,

$$R_o = R_E \left\| \left[\frac{(R_B \| R_S) + r_\pi}{(\beta + 1)} \right] \right.$$

$$= \frac{R_E[(R_B \| R_S) + r_\pi]}{(R_B \| R_S) + r_\pi + R_E(\beta + 1)} \qquad (14.11)$$

From Eq. (14.11) one concludes that, in order to exploit the impedance transforming properties of the emitter follower to the fullest, R_B should be chosen large compared to R_i' so that $R_i \to R_i'$. Having done that, one notes next from Eq. (14.11) that R_S and R_E should be small for a low R_o. However, since R_E must be large for good bias stability, it is R_S that will limit R_o. Equation (14.10) supports this argument since, for a voltage gain close to unity, R_L' and, therefore, R_E must be large while R_S should be low.

DRILL EXERCISES

14.3 The circuit in Fig. 14.4 has $R_{B1} = R_{B2} = 50$ kΩ, $R_E = 1.2$ kΩ, $V_{CC} = 10$ V. If the BJT has $\beta = 100$ and $V_{BE} = 0.7$ V determine I_C and V_{CE}.
Ans. 2.94 mA, 6.47 V

14.4 The emitter follower described in Exercise 14.3 has $R_S = 2.5$ kΩ and $R_L = 700$ Ω. Find the voltage gain v_o/v_s, the input resistance R_i and output resistance R_o.
Ans. 0.85, 16.1 kΩ, 30 Ω

An important advantage of the emitter follower is the fact that it can be operated with much larger input signals than can either of the other two configurations. This can be deduced from Fig. 14.5 where one notes that the BJT input voltage v_π is given by

$$v_\pi = \frac{v_b r_\pi}{r_\pi + (\beta + 1)R_L'}$$

Since R_S is normally small compared to R_i so that $v_b \simeq v_s$, and because $r_\pi \ll (\beta + 1)R_L'$, it follows that

$$\frac{v_\pi}{v_s} \ll 1$$

So most of the input signal appears across R_L'. Observe also that, provided the voltage on the base of the transistor does not rise above V_{CC}, the BJT will not saturate. The reader will see in Chap. 15 how these properties are exploited in the design of power amplifiers.

14.3.3 The Common-Base Configuration

The circuit for a practical common-base amplifier using a discrete *npn* BJT is shown in Fig. 14.6. Again observe that the biasing network is the same as in Fig. 14.1 so Eqs. (14.1) and (14.3) apply here too. Since the transistor operates with its base common, the pertinent set of output characteristics are in the i_C-v_{CB} plane and the operating point in question is, therefore, defined by I_C and V_{CB}. The latter is determined by

$$V_{CB} = V_{CC} - V_{BE} - I_C(R_C + R_E) \tag{14.12}$$

The midband, small-signal model corresponding to the circuit in Fig. 14.6 is shown in Fig. 14.7. For convenience, the common-base model in Fig. 7.8(b) is used.

In the following analysis it will be assumed that $R_E \gg r_e$, which is reasonable since $r_e = 1/g_m$ is of the order of 10 Ω or less for I_C in the milliampere range, while R_E will typically be of the order of 100 Ω or more.

From Fig. 14.7 the input resistance R_i is simply

$$R_i \simeq r_e$$

while the output resistance R_o is

$$R_o = R_C$$

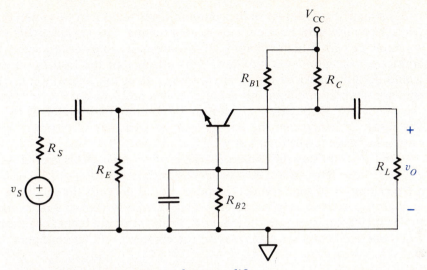

Figure 14.6 Discrete common-base amplifier.

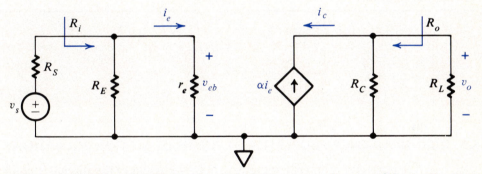

Figure 14.7 Midband small-signal model of the common-base amplifier in Fig. 14.6.

Defining

$$R'_L = R_L \parallel R_C$$

the voltage gain is found to be by inspection

$$\frac{v_o}{v_s} = \frac{R'_L}{R_S + r_e} \qquad (14.13)$$

Because r_e is so small and $\alpha \simeq 1$, the above equation can often be approximated by

$$\frac{v_o}{v_s} \simeq \frac{R'_L}{R_S}$$

which shows that, in contrast to conditions in the common-emitter amplifier, here the voltage gain is approximately independent of the BJT parameters and of β in particular.

The main disadvantage of the common-base circuit is its low input impedance which severely limits its usefulness as an amplifier. The reason is that to achieve an amplifier with a high voltage gain it is necessary to cascade several common-emitter or common-base amplifiers. This means that the load R_L on each stage would be influenced by the input resistance R_i of the following one. With the common-emitter amplifier this is not a serious problem since $R_i \simeq r_\pi$ which, for, say, $I_C = 1$ mA, is roughly 1 kΩ. But in the common-base circuit $R_i \simeq r_e = r_\pi/\beta$. The result is that in fact the voltage gain of a cascade of common-base stages cannot be made larger than unity.

On the credit side the common-base configuration yields greater bandwidth, as will be seen later, and is able to operate with larger values of V_{CC} which means a larger output voltage swing. The latter has to do with the fact that, for large enough V_{CC}, transistors break down, which manifests itself as a rapid and uncontrolled increase in i_C. It turns out that a BJT can withstand higher collector voltages when operated common-base than in the common-emitter configuration.

DRILL EXERCISE

14.5 The circuit in Fig. 14.6 has $V_{CC} = 15$ V, $R_{B1} = 10$ kΩ, $R_{B2} = 5$ kΩ, $R_E = 8.6$ kΩ, $R_C = 16$ kΩ, $R_L \to \infty$, and $R_S = 50$ Ω. Assuming $V_{BE} = 0.7$ V, $\beta \gg 1$, calculate I_C, V_{CB}, R_i, v_o/v_s.
Ans. 0.5 mA, 2 V, 50 Ω, 160

14.4 HIGH-FREQUENCY PROPERTIES OF THE THREE BJT CONFIGURATIONS

A reasonably complete comparison of the behavior of the three BJT configurations at frequencies where energy storage effects in the transistor cannot be neglected is beyond the scope of this text. The following discussion presents some relatively simple ideas which should help to place the high-frequency limitations of the three configurations in at least a qualitative perspective.

FETs are not included in this section. In contrast to the high-frequency response of the common-source FET, which was analyzed in Sec. 11.5.2, the responses of the other two configurations generally tend to be better because they do not suffer from the Miller effect. For the same reason, the common-base and common-collector configurations have dynamic responses which are superior to that of the common-emitter BJT.

It should be emphasized at the outset that the following analyses are approximate, and the resulting equations in general yield only approximate values of the high-frequency 3-dB point for the amplifier in question. Results should be treated with particular suspicion if they correspond to frequencies which are beyond the range of validity for the transistor model. Computer programs for circuit analysis are used where accurate results are needed, such as during the more advanced phases of a design.

14.4.1 The Hybrid-π Model of the BJT

Figure 14.8 shows the *hybrid-π* model of the BJT. It is probably the most widely used small-signal model because it faithfully describes the small-signal behavior of the BJT over a major part of the frequency range where the device is useful as an amplifier.

A comparison of Figs. 7.7(b) and 14.8 reveals that the hybrid-π model consists of the simple, low frequency first-order model augmented with the capacitances C_π and C_μ and the resistance r_x, which are needed to account for the behavior of the BJT at high frequencies. A complete evaluation of the model would involve a comparison of the measured and the computed values of, say, some convenient set of two-port parameters. The present exposition is more limited but, in fact, leads to the same conclusions that the more detailed study would yield.

Since the main interest is in the amplifying properties of the BJT, attention is focused on the parameter β which, it will be recalled, is the small-signal, two-port parameter h_{21} of the common-emitter BJT.

Figure 14.9 shows the hybrid-π model connected for calculating $H_{21}(s)$. By inspection, the output short-circuit current I_o is given by

$$I_o = g_m V_\pi - I_\mu = g_m V_\pi - s C_\mu V_\pi$$

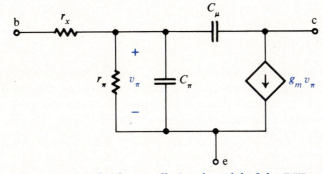

Figure 14.8 Hybrid-π small-signal model of the BJT.

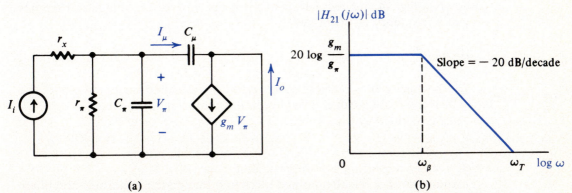

(a) (b)

Figure 14.9 (a) Circuit for computing $H_{21}(s)$ of the hybrid-π BJT model. (b) Magnitude Bode plot of $H_{21}(j\omega)$.

Because of the current source I_i, r_x has no effect, and therefore, V_π is given by

$$V_\pi = \frac{I_i}{g_\pi + s(C_\mu + C_\pi)}$$

Substituting this expression into the one above for I_o yields

$$H_{21}(s) = \frac{I_o}{I_i} = \frac{g_m - sC_\mu}{g_\pi + s(C_\mu + C_\pi)} \tag{14.14}$$

In a practical BJT $C_\pi \gg C_\mu$ and, because $\beta \gg 1$, $g_m \gg g_\pi$. Consequently the zero

$$z = \frac{g_m}{C_\mu}$$

has a much higher break frequency than the pole

$$p = \frac{-g_\pi}{C_\pi + C_\mu} \simeq \frac{-g_\pi}{C_\pi} \tag{14.15}$$

Equation (14.14) can hence be approximated, for $\omega \ll |z|$, by the first-order low-pass function

$$H_{21}(j\omega) \simeq \frac{g_m}{g_\pi + j\omega(C_\mu + C_\pi)} \tag{14.16}$$

The Bode plot of $|H_{21}(j\omega)|$ appears in Fig. 14.9(b). Observe that at low frequencies $H_{21}(j\omega) \to g_m/g_\pi = \beta$, as expected. It is instructive to examine $H_{21}(j\omega)$ at frequencies where its magnitude approaches unity, since the device is surely of little practical use when its current gain is less than one. For frequencies $\omega \gg |p|$ Eq. (14.16) is approximated by

$$|H_{21}(j\omega)| \simeq \frac{g_m}{\omega(C_\pi + C_\mu)}$$

As shown in Fig. 14.9(b), this falls with a slope of -20 dB per decade and reaches unity at a frequency ω_T given by

$$\omega_T = \frac{g_m}{C_\pi + C_\mu} \tag{14.17}$$

The frequency ω_T is called the *unity-gain-bandwidth* and is a parameter of the BJT which the manufacturer provides on the transistor data sheets.

Notice, by the way, that

$$\omega_T \ll |z|$$

which justifies neglecting z in the preceding analysis.

A comparison of calculations and measured data on the magnitude and phase of the BJT's two-port parameters reveals generally good agreement up to frequencies extending well beyond the 3-dB frequency ω_β of $H_{21}(j\omega)$ given by

$$\omega_\beta = |p| = \frac{g_\pi}{C_\mu + C_\pi} \qquad (14.18)$$

However, as $\omega \to \omega_T$ significant discrepancies arise and more detailed models have to be used.

DRILL EXERCISE

14.6 A typical general purpose BJT has $f_T = \omega_T/2\pi = 400$ MHz, $C_\mu = 2$ pF, $r_x = 20\ \Omega$. If $\beta = 50$ and the BJT quiescent collector current is 1 mA, compute $f_\beta = \omega_\beta/2\pi$.
Ans. 8 MHz

14.4.2 The Common-Emitter Amplifier

The high-frequency response of the common-emitter amplifier was studied in Sec. 11.5.3. Figure 14.10 shows the high frequency, small-signal model of the amplifier in Fig. 14.1. Comparing Figs. 11.27(a) and 14.10, the high-frequency 3-dB point is, from Eqs. (11.74), (11.73), and (11.71),

$$\omega_h \simeq [R_t\{C_\pi + C_\mu(1 + g_m R'_L)\}]^{-1} \qquad (14.19)$$

where

$$R_t = r_\pi \,\|\, [r_x + (R_S \,\|\, R_B)] \qquad (14.20)$$

A comparison of Eqs. (14.4) and (14.20) reveals that there is a tradeoff between gain and bandwidth since increasing g_m or R'_L raises the voltage gain and lowers ω_h. This observation could have been made, in fact, on the basis of Eq. (14.17).

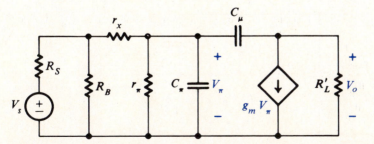

Figure 14.10 High-frequency, small-signal model of the common-emitter amplifier in Fig. 14.1.

DRILL EXERCISE

14.7 The BJT in Exercise 14.1 has $C_\mu = 3$ pF, $r_x = 25$ Ω, and $f_T = 400$ MHz. Compute $f_h = \omega_h/2\pi$, and f_β.
Ans. 1.38 MHz, 8 MHz

The preceding exercise illustrates an important point which is that ω_h in a common-emitter amplifier is, in general, less than ω_β defined by Eq. (14.18). This can be inferred from Eqs. (14.19) and (14.10) which show that, unless $(R_S \| R_B) \ll r_\pi$, $\omega_h < [r_\pi(C_\pi + C_\mu)]^{-1}$ because of the Miller effect due to C_μ.

14.4.3 The Emitter Follower

Figure 14.11 shows the high-frequency small-signal model for the circuit in Fig. 14.4. For simplicity, the loading by the resistors R_{B1} and R_{B2} is assumed to be negligibly small. In the following analysis it will also be assumed that r_x is small compared to R_S, which is almost always true. In that case the model in Fig. 14.11 simplifies to the one shown in Fig. 14.12.

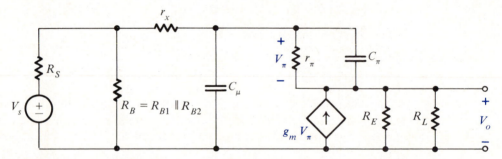

Figure 14.11 High-frequency, small-signal model of the emitter follower in Fig. 14.4.

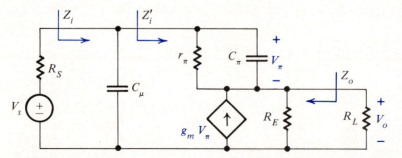

Figure 14.12 Approximate high-frequency model of the emitter follower.

The analysis of the emitter follower model shown in Fig. 14.12 is significantly more complicated than that of the common-emitter amplifier, even with the above simplifications. Since there are two capacitances in the model the system function is expected to have two poles. The following calculation yields an approximate expression for the *dominant* pole, i.e., the lowest frequency pole which, therefore, dominates in determining the high-frequency 3-dB point.

Noting that Eq. (14.7) can also be expressed in the form

$$R_i' = r_\pi + (1 + g_m r_\pi)R_L'$$

it follows from a comparison of Figs. 14.5 and 14.12 that Z_i', as defined in the latter figure, is given by the above expression with r_π replaced by Z_π, where

$$Z_\pi = (g_\pi + sC_\pi)^{-1}$$

Hence

$$Z_i' = Z_\pi + (1 + g_m Z_\pi)R_L' \tag{14.21}$$

Assume that, at frequencies of interest, $|g_m Z_\pi| \gg 1$. Since the real part of $y_\pi = 1/Z_\pi$ is $g_\pi = g_m/\beta$, the validity of the assumption rests on whether the magnitude of the susceptance ωC_π is much smaller than g_m. However, $g_m/C_\pi \simeq \omega_T$ so the assumption is that $\omega \ll \omega_T$. In that case

$$Z_i' \simeq Z_\pi(1 + g_m R_L')$$

$$= \frac{1 + g_m R_L'}{g_\pi + sC_\pi} \tag{14.22}$$

Equation (14.22) is the impedance of a parallel combination of a resistance R_{eq} and a capacitance C_{eq}, where

$$R_{eq} = \frac{1 + g_m R_L'}{g_\pi} \tag{14.23a}$$

and

$$C_{eq} = \frac{C_\pi}{1 + g_m R_L'} \tag{14.23b}$$

Hence the circuit in Fig. 14.12 can be approximated by that in Fig. 14.13. The

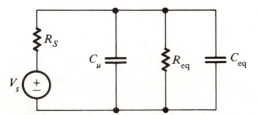

Figure 14.13 Circuit in Fig. 14.12 with Z_i' replaced by approximate model.

resulting network has a single pole ω_h given by

$$\omega_h^{-1} = (C_\mu + C_{eq})(R_S \parallel R_{eq})$$

$$= \left(C_\mu + \frac{C_\pi}{1 + g_m R_L'} \right)[R_S \parallel r_\pi(1 + g_m R_L')] \qquad (14.24)$$

As can be seen from this equation, the value of ω_h depends in particular on R_S and R_L'. Thus, for small R_L' and large R_S, which is a frequent situation,

$$\omega_h^{-1} \simeq \left(C_\mu + \frac{C_\pi}{1 + g_m R_L'} \right) r_\pi(1 + g_m R_L')$$

and if, in addition,

$$C_\mu < \frac{C_\pi}{1 + g_m R_L'}$$

then

$$\omega_h^{-1} \simeq C_\pi r_\pi \simeq \omega_\beta$$

On the other hand, if R_L' is large and R_S is small then

$$\omega_h^{-1} \simeq C_\mu R_S$$

In both instances the dominant pole ω_h is normally at a significantly higher frequency than what can be achieved with a common-emitter amplifier designed for appreciable voltage gain.

The network in Fig. 14.12 can be shown to have one finite zero located near ω_T. Although the hybrid-π is not accurate at frequencies approaching ω_T, nevertheless the break frequency of the zero can be assumed to be negligibly high.

DRILL EXERCISE

14.8 The BJT in Exercise 14.4 has $C_\mu = 3$ pF and $f_T = 400$ MHz. Compute its approximate bandwidth.
Ans. 17.5 MHz

14.4.4 The Common-Base Amplifier

The high-frequency, small-signal model of the common-base amplifier, with R_E neglected, is shown in Fig. 14.14(a). As in the case of the emitter follower, it is necessary to make some reasonable approximations to gain some insight into the circuit's frequency response. If r_x is assumed to be negligibly small, then the circuit simplifies to the form shown in Fig. 14.14(b). In that case the analysis is very

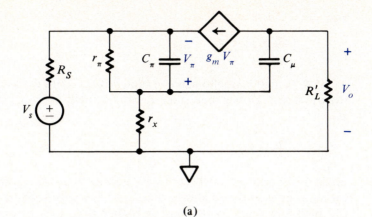

(a)

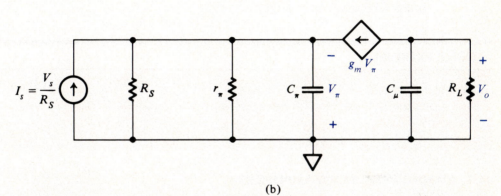

(b)

Figure 14.14 (a) The high-frequency, small-signal model of the common-base amplifier and (b) an approximate model valid if r_x can be neglected.

simple if one realizes that the controlled current source decouples the two sub-networks which it connects in the sense that the poles of each can be independently calculated. Thus,

$$\frac{V_o}{V_\pi} = \frac{-g_m}{G'_L + sC_\mu}$$

where $G'_L = 1/R'_L$, and

$$\frac{V_\pi}{V_s} = \frac{-G_S}{G_S + g_\pi + g_m + sC_\pi}$$

where $G_S = R_S^{-1}$.

Hence V_o/V_s, which the product of the preceding two equations, has two real poles with break frequencies

$$\omega_{p1} = (C_\mu R'_L)^{-1} \tag{14.25}$$

and

$$\omega_{p2} = \frac{G_S + g_\pi + g_m}{C_\pi} \tag{14.26}$$

If R'_L is made large to achieve high voltage gain, then ω_{p1} dominates, which is the usual case. Notice that, since g_m will normally be much larger than both G_S and g_π,

$$\omega_{p2} \simeq \frac{g_m}{C_\pi} \simeq \omega_T$$

Since the hybrid-π model is not valid near $\omega \simeq \omega_T$ the significance of the equation for ω_{p2} is only in that it shows that ω_{p2} is at a very high frequency. In either case the high-frequency 3-dB point is higher than in a common-emitter amplifier, unless R'_L is made very large.

DRILL EXERCISE

14.9 The transistor in Exercise 14.5 has $\beta = 50$, $C_\mu = 3$ pF, and $f_T = 400$ MHz. Compute the approximate value of the high-frequency 3-dB point.
Ans. 3.3 MHz

14.5 BIASING INTEGRATED AMPLIFIERS

The biasing techniques that are used in discrete transistor circuits are not suitable for ICs. This is primarily due to the fact that resistors occupy a disproportionately large amount of area in the IC. The monolithic IC technologies also can only yield very small capacitances, and even these consume an IC area in proportion to the capacitance. Although circuit techniques exist for using fewer capacitors than the number shown in, say, Fig. 14.1, it is not easy to eliminate all of them and still achieve reliable designs. For these reasons the biasing methods used in ICs are quite different from the ones encountered so far.

The basic approach consists of the use of constant current sources to establish the quiescent collector currents in the circuit, as in Fig. 14.2(a). This frees the designer from having to satisfy a multitude of often conflicting constraints, as illustrated in Example 14.1. By exploiting the fact that a BJT in the active mode or a MOSFET in pinch-off are reasonably good approximations to a current source, and by taking advantage of the close matching of transistors that integration permits, simple efficient current sources can be designed.

14.5.1 The Simple IC Current Source

Figure 14.15 shows a circuit of what is probably the simplest useful bipolar IC current source. In the circuit Q_2 acts as the current source. Since the topology

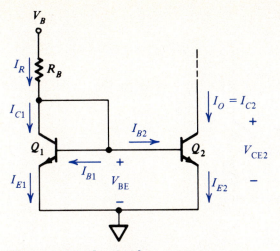

Figure 14.15 The simple current source.

forces the two transistors to have the same V_{BE}, the condition $I_{E1} = I_{E2}$ will be true if the BJTs are closely matched, which is achievable in a monolithic IC. Now the current I_R is determined by

$$I_R = \frac{V_B - V_{\mathrm{BE}}}{R_B}$$

But, by KCL,

$$I_R = I_{C1} + I_{B1} + I_{B2}$$

$$= I_{E1} + \frac{I_{E2}}{\beta + 1}$$

and, because $\beta \gg 1$ and $I_{E1} = I_{E2}$, hence

$$I_R \simeq I_{E2}$$

So the collector current of Q_2 is

$$I_{C2} \simeq \frac{V_B - V_{\mathrm{BE}}}{R_B}$$

which is independent of β provided only that $\beta \gg 1$.

The above, of course, will only be true if Q_2 is in the active mode where the device does behave as a current source insofar as the collector terminal is concerned. This means that the circuit in Fig. 14.15 will behave as a current source provided that $V_{\mathrm{CE2}} > V_{\mathrm{CEsat}}$.

14.5.2 The Widlar Current Source

The preceding current source is practical where the required current is relatively large, and therefore, the resistor R_B is not too big. However, the input transistors

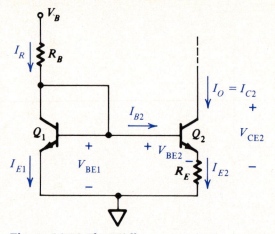

Figure 14.16 The Widlar current source.

of some op amps for example, operate with quiescent collector currents of a few microamperes. With typical values of V_B in the 10 V range, the required value of R_B is of the order of 100 kΩ which would require an unacceptably large area in the IC. In such a case the Widlar current source shown in Fig. 14.16 could be used.

The following equations can be written down from an inspection of the circuit:

$$I_R = I_{E1} + I_{B2} \tag{14.27a}$$

$$I_R = \frac{V_B - V_{BE1}}{R_B} \tag{14.27b}$$

$$I_{E2} = \frac{V_{BE1} - V_{BE2}}{R_E} \tag{14.27c}$$

Now, from Eq. (7.3a), if the BJTs are matched with respect to their parameter I_S, then

$$\frac{I_{E1}}{I_{E2}} = \exp\left(\frac{V_{BE1} - V_{BE2}}{V_T}\right) \tag{14.28}$$

In a practical design situation, $I_O = I_{C2} \simeq I_{E2}$ is known, as is the practical upper limit on R_B. Since V_B is usually fixed, the value of I_{E1} is also known from Eqs. (14.27a) and (14.27b). Equation (14.28) then yields $V_{BE1} - V_{BE2}$, which is used in Eq. (14.27c) to compute R_E.

EXAMPLE 14.2

The circuit in Fig. 14.16 is used to produce an output current $I_O = 100\ \mu$A. The BJTs are matched with $\beta \gg 1$ and $V_{BE} = 0.7$ V at $I_E = 100\ \mu$A. What value of R_E is needed to limit R_B to 10 kΩ if $V_B = 10$ V?

Solution

Since I_{E1} will be greater than I_{E2}, V_{BE1} will be more than 0.7 V which is specified at $I_E = 100$ μA $= I_{E2}$. However, in view of the exponential dependence of I_E on V_{BE}, V_{BE1} should be sufficiently close to 0.7 V to justify the use of that value in Eq. (14.27b), at least initially.

From Eq. (14.27b)

$$I_R = \frac{10 - 0.7}{10} = 0.93 \text{ mA}$$

Since $\beta \gg 1$, $I_{E1} \simeq I_R$ according to Eq. (14.27a). Hence $I_{E1}/I_{E2} = 0.93/0.1 = 9.3$. Solving Eq. (14.28) for $V_{BE1} - V_{BE2}$ results in

$$V_{BE1} - V_{BE2} = V_T \ln\left(\frac{I_{E1}}{I_{E2}}\right)$$

$$= 25 \ln 9.3 = 55.7 \text{ mV}$$

Substitution of this result in Eq. (14.27) yields the required value of R_E as

$$R_E = \frac{V_{BE1} - V_{BE2}}{I_{E2}} = \frac{55.7}{0.1} = 557 \text{ } \Omega$$

It should be noted that since

$$V_{BE1} = V_{BE2} + 55.7 \text{ mV} = 0.7557 \text{ V}$$

the error in computing I_R using $V_B = 0.7$ V was less than 10%. One could of course improve on the accuracy of the above solution by iterating the calculation with this improved estimate of V_{BE1}.

14.5.3 The Current Source as an Active Load

It was pointed out in Chap. 7 that the use of active loads was not restricted to the MOS technology, and that similar techniques of avoiding the use of space-occupying load resistors are used in the design of bipolar ICs as well. The basic approach consists of exploiting the high Norton resistance of an active current source, such as the one in Fig. 14.15, by using the current source in the place of an amplifier's collector load resistance. Figure 14.17 shows how this can be done.

A comparison of Figs. 14.15 and 14.17 reveals that the network consisting of Q_2, Q_3, and R_B in Fig. 14.17 is the same as the circuit in Fig. 14.15 except for the fact that the transistors in the former are *pnp*. Consequently Q_2 in Fig. 14.17 behaves as a current source provided that it does not saturate, which means that $V_{CC} - v_O$ must not be less than V_{ECsat} of Q_2.

The small-signal output resistance r_o of a BJT biased in the active region so far has been assumed to be infinite. In fact it can be shown to be given approximately by

$$r_o = \frac{k}{g_m}$$

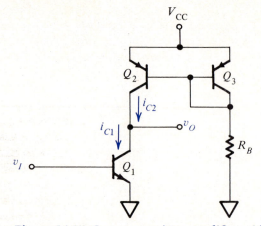

Figure 14.17 Common-emitter amplifier with an active load.

where k is typically of the order of 10^4. So if the quiescent collector current I_C is of the order of a milliampere, $r_o \sim 100$ kΩ.

Therefore, if the current source transistor Q_2 in Fig. 14.17 were biased to deliver a dc collector current I_{C2} of, say, 0.1 mA, the common-emitter amplifier Q_1 would also operate with a quiescent collector current of $I_{C1} = 0.1$ mA, but its effective collector load resistance, R_{eff}, would be $R_{\text{eff}} = 10^4/g_m = 2.5$ MΩ. Hence the voltage gain of Q_1 would be approximately equal to

$$\frac{v_o}{v_i} \simeq -g_m R_{\text{eff}} \simeq -10^4$$

Although the circuit in Fig. 14.17 contains a resistor, R_B, its value is significantly lower than R_{eff}. Thus if $V_{CC} = 10$ V, R_B will be of the order of 100 kΩ, so it will occupy an area approximately ten times smaller than what would be needed for a 2.5 MΩ resistor. Moreover, the reference circuit in Fig. 14.17 consisting of Q_3 and R_B can be shared between several active loads.

14.5.4 A MOS Current Source

Figure 14.18 shows an arrangement similar to the bipolar current mirror but based on n-channel MOSFETs. As in almost all MOS ICs, the circuit contains no resistors. Transistor Q_3 is the current source, while Q_1 and Q_2 constitute the biasing network.

Since both Q_1 and Q_2 operate with $v_{GD} = 0$ it follows that both are in the pinch-off mode. Also, because they are connected with their channels in series, their drain currents are equal. Therefore, from Eq. (6.3)

$$\beta_1(v_{GS1} - V_T)^2 = \beta_2(v_{GS2} - V_T)^2$$

where V_T is assumed to be the same for both devices, which is true in an IC. This equation shows that v_{GS2}, which is also the gate-to-source voltage of Q_3, can be chosen by adjusting the relative magnitudes of β_1 and β_2 which are device-

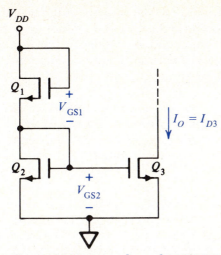

Figure 14.18 An *n*-channel MOSFET current source.

geometry-dependent. Thus, noting that

$$V_{DD} = v_{GS1} + v_{GS2}$$

it is a straightforward matter to show that

$$v_{GS2} = \frac{V_{DD} + V_T[(\beta_2/\beta_1)^{1/2} - 1]}{1 + (\beta_2/\beta_1)^{1/2}}$$

which, for $\beta_2/\beta_1 \gg 1$, simplifies to

$$v_{GS2} = (V_{DD} - V_T)\left(\frac{\beta_1}{\beta_2}\right)^{1/2} + V_T$$

Hence Q_3 will have a constant drain current

$$I_{D3} = \frac{\beta_3(v_{GS2} - V_T)^2}{2}$$

provided that its drain voltage remains high enough for Q_3 to operate in pinch-off, i.e., $v_{GD} < V_T$.

14.6 THE DIFFERENTIAL AMPLIFIER

Most analog circuits and, in particular, amplifiers, can be reduced to an interconnection of a relatively small number of standard "building blocks," many of which contain as few as two active elements, be they BJTs, JFETs, MOSFETs, or a mixture thereof. Probably the most important of these building blocks in the differential amplifier which, for example, is a key component of the op amp.

The principal advantage of the differential amplifier is the fact that it has all the characteristics of a common-emitter amplifier including, in particular, its relatively high voltage gain, and yet these are achieved without the use of the large emitter bypass capacitor which is needed in the common-emitter stage. Moreover, the differential amplifier is a dc amplifier and, consequently, the input coupling capacitors are not needed either. As an added bonus, the differential amplifier has, as its name implies, a differential input and provides both an inverting and a noninverting output.

Because bipolar analog circuits are still the most important, emphasis in this section is on the BJT differential amplifier. However, the use of the JFET in this configuration is also briefly considered because of its high input impedance and because JFETs can be included with BJTs in the same IC.

14.6.1 The BJT Differential Amplifier

A simple BJT differential amplifier is shown in Fig. 14.19. The reader will recognize the similarity of this circuit to that of the basic ECL gate in Fig. 8.29. However, in contrast to the gate, the circuit in Fig. 14.19 is used as a linear amplifier with two inputs and two outputs.

QUIESCENT CONDITIONS IN THE BJT DIFFERENTIAL AMPLIFIER

Assuming that the two input voltages have no average component, the Q point of the amplifier is computed from the circuit in Fig. 14.20(a). Because of the complete symmetry of the circuit, all corresponding currents and voltages in the left and right halves must be equal. In that case the quiescent conditions can be determined from the half-circuit in Fig. 14.20(b). The emitter resistor has to have a value of $2R_E$ because the voltage drop across it is $R_E(I_{E1} + I_{E2}) = 2R_E I_E$ since $I_{E1} = I_{E2} = I_E$.

A comparison of Figs. 14.20(b) and 14.2(b) reveals that they are the same as

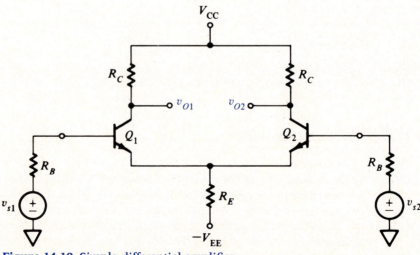

Figure 14.19 Simple differential amplifier.

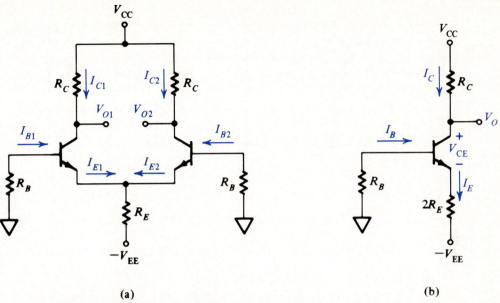

Figure 14.20 (a) Differential amplifier circuit under dc conditions and (b) corresponding half-circuit.

far as the determination of the collector current I_C is concerned. Therefore Eq. (14.3), with R_E replaced by $2R_E$, defines I_C in Fig. 14.20(b). The BJT will be in the active mode provided that $V_{CE} > V_{CEsat}$, where

$$V_{CE} = (V_{CC} + V_{EE}) - I_C(R_C + 2R_E) \tag{14.29}$$

DRILL EXERCISE

14.10 The circuit in Fig. 14.19 has $V_{CC} = V_{EE} = 12$ V, $R_C = 6$ kΩ, $R_E = 56$ kΩ, $R_B = 0$. Determine I_C and V_{CE} for both transistors with $v_{s1} = v_{s2} = 0$.
Ans. 0.1 mA, 12.2 V

SMALL-SIGNAL CONDITIONS

The low-frequency, small-signal model of the differential amplifier is shown in Fig. 14.21(a). Because of the circuit's symmetry, the techniques developed in Sec. 3.6 can be used. Figure 14.21(b) shows the resulting differential-mode half-circuit while the common-mode half-circuit appears in Fig. 14.21(c). The input excitations are as defined in Eqs. (3.40) and (3.41), i.e.,

$$v_c = \frac{v_{s1} + v_{s2}}{2} \tag{14.30a}$$

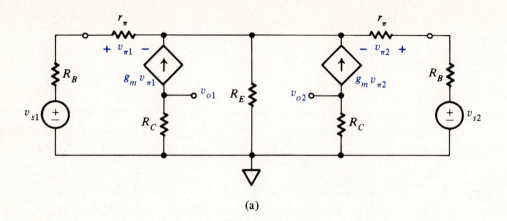

(a)

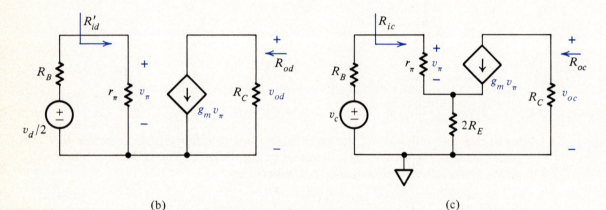

(b)   (c)

Figure 14.21 (a) Small-signal model of differential amplifier in Fig. 14.20; corresponding (b) differential-mode and (c) common-mode half-circuits.

and

$$v_d = v_{s1} - v_{s2} \tag{14.30b}$$

The differential-mode half circuit is recognized as that of a common-emitter amplifier with no emitter resistance. Therefore the output voltage v_{od} depends on v_d according to

$$\frac{v_{od}}{v_d} = \frac{-g_m R_C r_\pi}{2(r_\pi + R_B)} \tag{14.31}$$

which tends to

$$\frac{v_{od}}{v_d} \simeq \frac{-g_m R_C}{2} \tag{14.32}$$

as $R_B \to 0$.

Evidently the input resistance of this circuit is simply

$$R'_{id} = r_\pi \tag{14.33}$$

while the output resistance is

$$R_{od} = R_C \tag{14.34}$$

On the other hand, the common-mode half-circuit in Fig. 14.21(c) looks like a common-emitter amplifier with no bypass capacitor across the emitter resistor. Its input resistance R_{ic} is, except for a factor of two, the same as that of an emitter follower, namely,

$$R_{ic} = R_\pi + (\beta + 1)2R_E \tag{14.35}$$

Therefore v_π is related to v_c through the voltage divider equation

$$\frac{v_\pi}{v_c} = \frac{r_\pi}{r_\pi + R_B + 2R_E(\beta + 1)}$$

whence the voltage gain v_{oc}/v_c is

$$\frac{v_{oc}}{v_c} = \frac{-g_m R_C r_\pi}{r_\pi + R_B + 2R_E(\beta + 1)} \tag{14.36}$$

which tends to zero as $2R_E(\beta + 1) \rightarrow \infty$. Finally, the output resistance is

$$R_{oc} = R_c \tag{14.37}$$

Now, making use of the relationships expressed in Eqs. (3.42) and (3.43), the output voltages are found to be, by superposition,

$$v_{o1} = v_{oc} + v_{od} \tag{14.38a}$$

and

$$v_{o2} = v_{oc} - v_{od} \tag{14.38b}$$

These are related to the input voltage components v_c and v_d through the *common-mode voltage gain* A_c, defined as

$$A_c = \frac{v_{oc}}{v_c} \tag{14.39a}$$

where v_{oc}/v_c is described by Eq. (14.36), and the *differential-mode voltage gain* A_d, defined by

$$A_d = \frac{v_{od}}{v_d} \tag{14.39b}$$

where v_{od}/v_d is given by Eq. (14.31). Hence Eqs. (14.38) can be rewritten in the form

$$v_{o1} = A_c v_c + A_d v_d \tag{14.40a}$$

$$v_{o2} = A_c v_c - A_d v_d \tag{14.40b}$$

An ideal differential amplifier has a common-mode gain A_c of zero, and if it is to be used as the basis for the design of an op amp, its differential-mode gain A_d should be as large as possible. A figure of merit called the *common-mode rejection ratio* (CMRR) is used to quantify the extent to which these ideal conditions are approached. It is defined as follows

$$\text{CMRR} = \frac{A_d}{A_c} \tag{14.41}$$

and is usually expressed in decibels (dB).

DRILL EXERCISES

14.11 The circuit in Fig. 14.19 has $V_{CC} = V_{EE} = 10$ V, $R_B = 0$, $R_E = 900$ Ω, and $R_C = 200$ Ω. If the BJT is characterized by $\beta = 50$ and $V_{BE} = 0.7$ V, compute I_C, A_d, and CMRR.
Ans. 5.07 mA, -20.3, 186 (45.4 dB)

14.12 In Exercise 14.11 compute the small-signal output voltages, v_{o1}, v_{o2}, and $v_{o1} - v_{o2}$, when v_{s1} and v_{s2} have the following values, respectively: (a) 5 mV, -5 mV; (b) 105 mV, 95 mV; (c) 10 mV, 0 mV.
Ans. (a) -203 mV, 203 mV, -406 mV; (b) -214 mV, 192 mV, -406 mV; (c) -204 mV, 202 mV, -406 mV

The above exercises illustrate that, for a sufficiently high CMRR, the output voltages v_{o1} and v_{o2} respond only to the differential input component v_d. However, they also demonstrate that if it were practical to take the output differentially between the collectors of the two transistors, that is, to measure $v_{o1} - v_{o2}$, then there would not be any common-mode output component even if the CMRR were poor.

14.6.2 A Practical BJT Differential Amplifier

One of the weaknesses of the circuit in Fig. 14.19 is its low CMRR. Because the signals which differential amplifiers are called upon to amplify often have large unwanted common-mode components superimposed on them, a useful differential amplifier must have a CMRR which is orders of magnitudes greater than what can be practically achieved with this simple circuit.

An examination of Eqs. (14.31) and (14.36) reveals that, for a given A_d, the CMRR depends on R_E. Thus

$$\frac{A_d}{A_c} = \frac{r_\pi + R_B + 2R_E(\beta + 1)}{2(r_\pi + R_B)} \simeq \frac{R_E(\beta + 1)}{r_\pi + R_B}$$

for large R_E. The solution to the problem is the use of a current source such as the Widlar circuit in Fig. 14.16. This not only produces a well stabilized output current of $I_O = 2I_E$, but also yields a very high effective R_E due to the high resistance which is presented by the collector of a BJT in the active mode. As explained in Sec. 14.5.3, for typical IC BJTs a figure of 100 kΩ is not unreasonable.

Figure 14.22 shows the circuit of a differential amplifier with a Widlar current source. It is important to note that the use of a current source to set the quiescent currents of the amplifier transistors Q_1 and Q_2 renders the amplifier's characteristics insensitive to the values of the source resistances R_{S1} and R_{S2}. This should be clear from Eq. (14.3). However, there must be a dc path between the inputs of the differential amplifier and ground to support the quiescent base currents $I_B = I_O/2\beta$.

Because of the current-source biasing, input coupling capacitors are not needed, although if v_{S1} and v_{S2} have large enough dc components either Q_1 and Q_2 or Q_3 can saturate.

Another disadvantage of the circuit in Fig. 14.19 is its low *differential input resistance R_{id}* defined as the resistance measured between the input terminals of the amplifier. From Eq. (14.33)

$$R_{id} = 2R'_{id} = 2r_\pi$$

In some differential amplifiers this characteristic is improved upon by the use of so-called Darlington circuits which are considered in the next section. Most

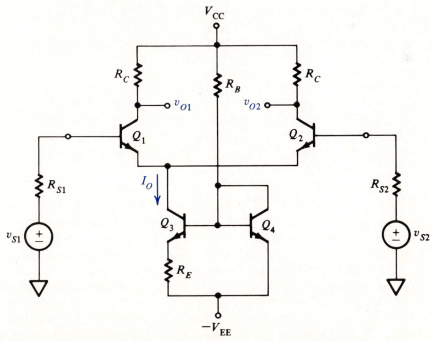

Figure 14.22 Differential amplifier with current-source biasing.

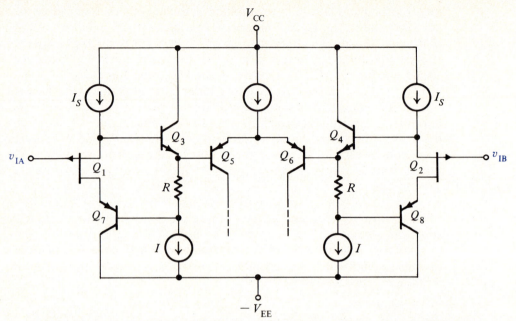

Figure 14.23 Partial circuit of a bipolar differential amplifier with a JFET input stage.

modern IC designs, however, achieve this by using JFETs. This is illustrated in Fig. 14.23.

The p-channel JFETs operate in the common-drain configuration with the current sources I_S establishing their quiescent drain currents. Notice that the load on both JFETs is the input impedance of an emitter follower (Q_3 and Q_4) and, therefore, high. Notice also that, because the common-drain amplifier, like the emitter-follower, has superior high-frequency characteristics, the bandwidth of the amplifier will, in all likelihood, be limited by the basic differential amplifier, consisting of Q_5 and Q_6, since its frequency response will be that of a common-emitter BJT. The collector loads of Q_5 and Q_6 are normally active ones, based on designs such as that shown in Fig. 14.17.

14.6.3 The MOS Differential Amplifier

The circuit for a simple MOSFET differential amplifier is derived by direct analogy to the BJT implementation in Fig. 14.20. The result in the case of n-channel MOSFETs is shown in Fig. 14.24(a). Following an analysis very similar to the one in Sec. 14.6.1, half-circuits corresponding to quiescent conditions and to differential-mode and common-mode excitation are derived. These are shown in Fig. 14.24(b), (c), and (d) respectively.

Example 6.2 illustrated the analysis of circuits such as the one in Fig. 14.24(b). From Fig. 14.24(c) and (d) the differential-mode gain A_d and the CMRR can be

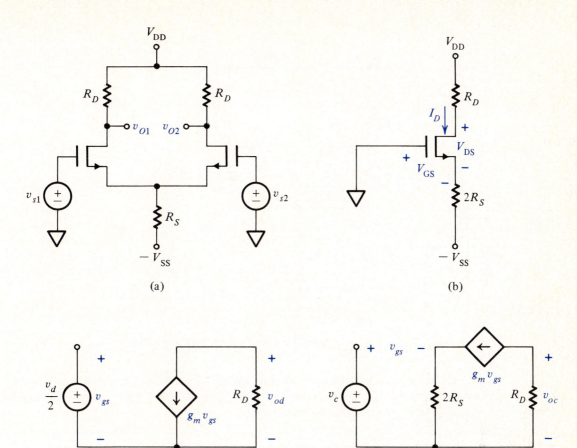

Figure 14.24 Simple *n*-channel MOSFET differential amplifier: (a) circuit, (b) dc half-circuit, (c) differential-mode half-circuit, and (d) common-mode half-circuit.

shown to be given by, respectively,

$$A_d = \frac{v_{od}}{v_d} = \frac{-g_m R_D}{2} \tag{14.42}$$

and

$$\text{CMRR} = \frac{A_d}{A_c} = \frac{1 + 2g_m R_S}{2} \tag{14.43}$$

Because of the relatively low g_m of FETs the CMRR of this amplifier would tend to be even lower than that of the corresponding BJT circuit. Therefore, there is an even greater incentive to replace R_S by an active current source such as the one in Fig. 14.18. Also the drain load resistors R_D would be replaced by active pull-ups in

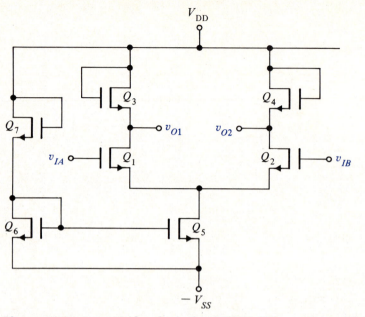

Figure 14.25 A practical *n*-channel MOSFET differential amplifier.

an IC design. Figure 14.25 shows a more practical MOSFET differential amplifier. Transistors Q_1 and Q_2 constitute the basic amplifier, Q_3 and Q_4 are the load pull-ups and Q_5, Q_6, and Q_7 form the current source.

DRILL EXERCISE

14.13 Verify Eqs. (14.42) and (14.43).

14.7 OTHER MULTIPLE-TRANSISTOR CIRCUITS

The following circuits represent important building blocks for analog circuit design, and are also excellent illustrations of how quantities such as impedance level, bandwidth, and maximum output signal can be improved upon with respect to what can be achieved with one transistor.

14.7.1 The Darlington Circuit

Figure 14.26(a) illustrates a two-transistor Darlington circuit, and the corresponding small-signal model is shown in Fig. 14.26(b). From Fig. 14.26(a)

$$i_C = i_{C1} + i_{C2} = \beta_1 i_{B1} + \beta_2 i_{B2}$$

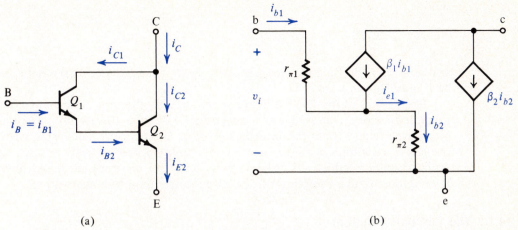

Figure 14.26 (a) Two-transistor Darlington circuit and (b) its small-signal model.

but, since $i_{B2} = i_{E1} = (\beta_1 + 1)i_{B1}$, therefore,

$$i_C = [\beta_1 + \beta_2(\beta_1 + 1)]i_{B1}$$

Noting that $i_B = i_{B1}$ and using the fact that $\beta \gg 1$, leads to the result

$$i_C \simeq \beta_1 \beta_2 i_B$$

In other words the circuit behaves like a BJT with an effective β equal to $\beta_1 \beta_2$.

The effective r_π is also very high which can be shown by noting that since, from Fig. 14.12(b),

$$v_i = [r_{\pi 1} + (\beta_1 + 1)r_{\pi 2}]i_b$$

$$= \left[\frac{\beta_1}{I_{C1}} + (\beta_1 + 1)\frac{\beta_2}{I_{C2}}\right]V_T i_b$$

and, because

$$I_{C2} = \beta_2 I_{E1} \simeq \beta_2 I_{C1}$$

therefore

$$v_i \simeq \frac{2\beta_1 \beta_2 V_T}{I_{C2}} i_b$$

Since

$$I_C = I_{C1} + I_{C2} \simeq \frac{I_{C2}}{\beta_2} + I_{C2} \simeq I_{C2}$$

the effective input driving-point resistance is given by

$$\frac{v_i}{i_b} = r_{\pi \text{eff}}$$

where

$$r_{\pi\,\text{eff}} = \frac{2\beta_1\beta_2}{g_m}$$

and $g_m = V_T/I_C$. So the input resistance $r_{\pi\,\text{eff}}$ is $2\beta_1$ times greater than what it would be if only Q_2 was used.

Since the β of the BJT influences very strongly the input impedance of both a common-emitter amplifier and an emitter follower, the Darlington circuit is often used where the input impedance which is practically achievable with one transistor is not high enough. As mentioned in the preceding section, some differential amplifiers which are used as input stages of op amps are based on these Darlington circuits, sometimes using three rather than two transistors.

14.7.2 The Cascode Amplifier

The *cascode* amplifier is a clever combination of a common-emitter and a common-base amplifier which has a voltage gain approximately equal to that of a common-emitter stage but with a considerably greater bandwidth. Figure 14.27 shows a simplified circuit of the cascode amplifier together with its low-frequency, small-signal model.

As in most cascaded amplifier circuits, the small-signal voltage gain is most easily obtained by solving the circuit starting at the output. The output voltage v_o is

$$v_o = \alpha R_L i_{e2} = -\alpha R_L g_m v_\pi$$

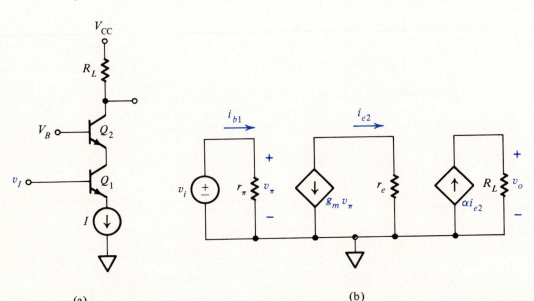

(a)   (b)

Figure 14.27 The cascode amplifier: (a) simplified circuit and (b) low-frequency, small-signal model.

and, since $v_\pi = v_i$,

$$\frac{v_o}{v_i} = -\alpha R_L g_m \simeq -R_L g_m$$

This is exactly the same gain that would be obtained if the common-emitter transistor was used by itself with a load R_L. The benefits that accrue from using the more complicated cascode circuit are twofold. One is that significantly larger output voltages can be obtained with the cascode and the other is an appreciable improvement in high-frequency response.

If the output characteristics of a BJT, or FET for that matter, are extended out to higher voltages than those for which the curves presented in Chap. 6 and 7 are valid, the characteristics turn up relatively sharply. The very large currents that flow under these conditions result in power dissipation levels which destroy the transistor. It turns out that the *avalanche breakdown* voltage at which this occurs is higher in the common-base BJT than in the common-emitter. Consequently the cascode can be operated at higher values of V_{CC} and, therefore, larger output voltage swings are possible.

The improved high frequency response is due to the fact that the effective load on the common-emitter transistor is $r_e = 1/g_m$, which is very low. Therefore the Miller effect due to the capacitance C_μ is greatly reduced. The common-base transistor also exhibits greater bandwidth than a common-emitter stage with the same R_L. In practice the response is usually limited by the time constant $C_L R_L$, where C_L includes C_μ as well as any other parasitic capacitance across the output [see Eq. (14.25)].

14.8 SUMMARY AND STUDY GUIDE

The reader should now have an appreciation for the characteristics of some of the more important "building blocks" that are used in the design of such complex circuits as those which give rise to the near ideal behavior of the op amp.

Several important concepts and design techniques were covered. Of these, the reader should note, in particular, the points listed below.

The basic approach used in achieving a stable quiescent point is the design of a current source which determines the collector current of the BJT or the drain current of the FET. In ICs the design is aimed at minimizing the number and size of resistors and exploiting the close matching of transistor parameters which the monolithic technology achieves.

At the transistor level the basic building block is the common-emitter or the common-source amplifier. However, because even at low frequencies neither approaches the near ideal behavior of the op amp, they often have to be combined with, respectively, the emitter follower or the source follower because of the latter's impedance transforming properties and good frequency response. The common-base amplifier is used when a large output voltage and good frequency response are needed.

Frequency response calculations rely on often justifiable approximations which reduce the analysis to that of a single time constant circuit. It is important to verify the validity of the approximations after the calculations are made.

The monolithic IC technology facilitates the design of current source circuits that not only yield superior quiescent point stability but also improve the CMRR of differential amplifiers and help to conserve chip area.

BJT current sources can be used as active loads.

The use of the differential amplifier with current-source biasing eliminates the need for any of the coupling or bypass capacitors that are normally required in ordinary common-emitter or common-source amplifiers.

Because the JFET can be integrated with the BJT it is a useful element in bipolar ICs because of its very high input impedance.

PROBLEMS

14.1 Determine the Q point, I_C and V_{CE}, for the amplifier in Fig. 14.1 given that $V_{CC} = 12$ V, $R_E = 3$ kΩ, $R_{B2} = 40$ kΩ, $R_{B1} = 80$ kΩ, $R_C = 4$ kΩ, $\beta = 100$, and $V_{BE} = 0.7$ V.

14.2 Assuming $\beta \gg 1$, design the biasing network for the amplifier in Fig. 14.1 to yield a quiescent point at $I_C = 0.5$ mA, $V_{CE} = 8$ V with $V_{CC} = 18$ V and $R_C = 8$ kΩ. Select R_{B1} and R_{B2} so that their parallel resistance is not less than $10r_\pi$ for a nominal β of 50. Then compute the variation in I_C which would occur if the circuit you designed was mass-produced with transistors having β in the range of 30 to 150.

14.3 If $R_S = R_L = 4$ kΩ for the amplifier in Prob. 14.2, compute the range of small signal voltage gain v_o/v_i for the specified variation in β.

14.4 The emitter-follower in Fig. 14.4 operates from $V_{CC} = 10$ V. If the transistor is characterized by $\beta = 100$ and $V_{BE} = 0.7$ V, and if $R_{B1} = R_{B2} = 100$ kΩ, $R_E = 10$ kΩ, $R_S = 5$ kΩ and $R_L = 500$ Ω compute:
(a) The quiescent condition, I_C, V_{CE}
(b) The input resistance R_i, output resistance R_o, and voltage gain v_o/v_s
(c) The maximum output voltage swing for operation of the BJT in the active mode

14.5 The BJT in Fig. 14.6 is operating with a quiescent collector current of 1.0 mA. If $R_S = 25$ Ω, $R_C = R_L = 5$ kΩ, compute the small-signal input resistance R_i and the voltage gain v_o/v_s assuming R_E to be negligibly large, and the capacitors to be effective short circuits at frequencies of interest.

14.6 The amplifier in Fig. 14.1 is modified in such a way that not all of R_E is bypassed, as shown in Fig. P14.6. If $R_{E1} = 200$ Ω and $R_{E2} = R_E - R_{E1}$, compute the input resistance R_i and voltage gain v_o/v_i for the parameters specified in Probs. 14.1 and 14.3.

14.7 Compute the small-signal voltage gain v_o/v_s of the circuit in Fig. P14.7. Assume $\beta \gg 1$ and $V_{BE} = 0.7$ V. (*Hint:* The assumption that $\beta \gg 1$ means that, for example, when computing quiescent conditions, the base current of the emitter follower can be taken to be negligibly small compared to the collector current of the common-emitter stage.)

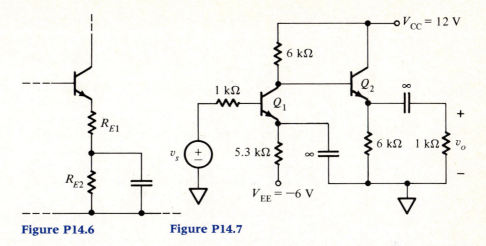

Figure P14.6   **Figure P14.7**

14.8 The two-port parameter $H_{21}(s)$ of a common-emitter BJT, biased to a quiescent collector current of 10 mA, is measured, and it is found that at 50 MHz $|H_{21}(j\omega)| = 10$. If $C_\mu = 3$ pF, $r_x = 50$ Ω, and $\beta = 100$, compute C_π and f_T.

14.9 A BJT with $f_T = 300$ MHz, $\beta = 50$, $C_\mu = 2$ pF, and $r_x = 50$ Ω is used in an emitter follower with $R_E = 1$ kΩ. If the quiescent collector current is 10 mA and $R_S = 1$ kΩ, determine the midband voltage gain and the high-frequency 3-dB point.

14.10 The transistor described in Prob. 14.9 is used in Prob. 14.5 but with $R_L = 1$ kΩ. If there is 20 pF of parasitic capacitance across R_L, determine the midband voltage gain and the high-frequency 3-dB point.

14.11 Compare the midband voltage gains and high-frequency 3-dB points of the amplifiers in Figs. 14.1, 14.4, and 14.6. Neglect the biasing networks including R_C and, in each case, assume that $R_L = R_S = 1$ kΩ. The BJT is biased at $I_C = 1$ mA and has the following parameters: $\beta = 50$, $f_T = 400$ MHz, $C_\mu = 2$ pF, $r_x = 50$ Ω.

14.12 Determine the output current I_O in Fig. P14.12. Assume $V_{EB} = 0.7$ V and $\beta \gg 1$.

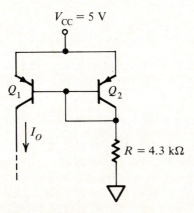

Figure P14.12

14.13 Repeat Prob. 14.12 for the improved current mirror in Fig. P14.13.

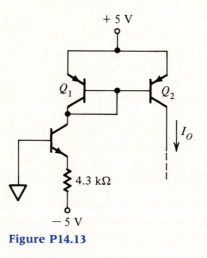

+ 5 V

Q_1 Q_2

I_O

4.3 kΩ

− 5 V

Figure P14.13

14.14 The FETs in Fig. 14.18 have $V_T = 1.5$ V, $\beta_2 = \beta_3 = 70$ μA/V², $\beta_1 = 0.7$ μA/V². If $V_{DD} = 30$ V, determine I_O.

14.15 The differential amplifier in Fig. 14.19 has $V_{CC} = 16$ V, $V_{EE} = 8$ V, $R_C = 4$ kΩ, $R_B = 1$ kΩ, and $R_E = 1.8$ kΩ. If $\beta = 100$ and $V_{BE} = 0.7$ V, compute the quiescent conditions, I_C and V_{CE}, in the BJTs as well as A_d and CMRR. Express the CMRR in decibels.

14.16 Repeat Prob. 14.15 for the case where a resistor $R_L = 2$ kΩ is connected between the collectors of the BJTs. (*Hint:* Treat R_L as two 1kΩ resistors in series.)

14.17 Repeat Prob. 14.15 for the case where there is a resistor $R_e = 50$ Ω connected between each emitter and the upper end of R_E.

14.18 The BJTs in Fig. 14.19 are replaced by n-channel JFETs with $g_m = 1$ mS. If $R_C = R_E = 10$ kΩ, and $R_B = 100$ kΩ, compute A_d and the CMRR in decibels.

14.19 The BJT which is used in the differential amplifier described in Prob. 14.15 has $f_T = 400$ MHz, $r_x = 50$ Ω, and $C_\mu = 3$ pF. Determine its bandwidth with respect to differential-mode input signals.

14.20 The circuit in Fig. P14.20 shows a technique for measuring temperature. The element R_T is a temperature-sensitive resistor. When $R_T = R$, the voltage v_d can be shown to be zero. As the resistance of R_T deviates from R, v_d changes from zero. If the characteristics of R_T are known then the value of v_d can be converted to a temperature. Since v_d is normally of the order of millivolts, it has to be amplified. If v_d was applied to the inputs of a differential amplifier, what would the CMRR have to be for the output to be proportional to v_d with an error of not more than 1% when $v_d = 1$ mV? Assume that the changes in R_T are sufficiently small that its value can be assumed to remain approximately equal to R.

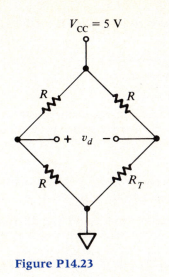

$V_{CC} = 5\text{ V}$

R R

$+$ v_d $-$

R R_T

Figure P14.23

14.21 The resistor R_E in Fig. 14.19 is replaced by a 1 mA current source. If $V_{CC} = 12$ V, $R_C = 12$ kΩ, and $R_B = 0$, what would be the largest common-mode voltage that could be applied to the inputs without driving the BJTs into saturation?

14.22 Show that if the current-source circuit in Fig. 14.22 behaves as an ideal current source, then even if R_{S1} and R_{S2} are not equal, the differential gain A_d can be computed from the corresponding half-circuit with R_{S1} or R_{S2} replaced by $(R_{S1} + R_{S2})/2$. Assume identical BJTs with $\beta \gg 1$ and $V_{BE1} = V_{BE2}$.

14.23 The disadvantage of the differential amplifier is that both output voltages normally contain large dc components due to quiescent conditions. Since the use of coupling capacitors has to be eschewed, especially in ICs, the technique that is normally used involves shifting the dc level to zero. Figure P14.23 is a simple arrangement for doing this. Assuming $\beta \gg 1$ determine the value of I_S for the output voltage v_O to be zero.

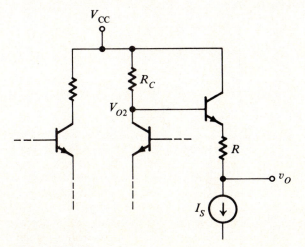

Figure P14.20

14.24 Making reasonable assumptions, determine an expression for the V_{SD} of the JFETs in Fig. 14.23.

14.25 The MOSFETs in Fig. 14.25 are characterized by $V_T = 1.5$ V, $\beta_1 = \beta_2 = 720$ μA/V^2 and $\beta_3 = \beta_4 = 35$ μA/V^2. Determine A_d assuming that the drain current of Q_5 is 0.2 mA.

14.26 Q_1 and Q_2 in Fig. 14.22 are each replaced with the two-transistor Darlington circuit shown in Fig. 14.26. If $\beta = 100$ and the current source produces an $I_O = 0.2$ mA, what is A_d and the differential input resistance R_{id} with $R_C = 100$ kΩ and $R_{S1} = R_{S2} = 0$?

14.27 The cascode circuit in Fig. 14.27 is operated with $V_{CC} = 30$ V, $I = 10$ mA, $R_L = 1$ kΩ. Determine its small-signal voltage gain v_o/v_i if the BJTs have $\beta = 50$, $V_{BE} = 0.7$ V.

14.28 The input of the cascode amplifier described in Prob. 14.27 is driven by a source that has a 50-Ω Thevenin resistance. If the BJTs have $f_T = 400$ MHz, $C_\mu = 3$ pF, $r_x = 50$ Ω, compute:
(a) The high-frequency 3-dB point of the common-emitter stage assuming that the load impedance due to Q_2 is resistive;
(b) The high-frequency 3-dB point of the common-base stage assuming that the output impedance of Q_1 is resistive.

From your results in (a) and (b) suggest what the bandwidth of the cascode might be.

14.29 Q_1 in Fig. 14.27 is replaced by a common-source n-channel JFET with $g_m = 10$ mS. Determine the low-frequency, small-signal voltage gain v_o/v_i using the data from Prob. 14.27.

14.30 The circuit in Fig. P14.30 consists of an emitter follower driving a common-base stage. It can thus be expected to have a high input impedance, good high-frequency response and a large output voltage swing. If $R_C = R_S = 1$ kΩ and $\beta = 100$, determine its small-signal voltage gain v_o/v_i and input resistance R_i with both BJTs operating at $I_C = 1$ mA. Compare your result with the voltage gain of a common-emitter amplifier having the same R_S, R_C, and I_C.

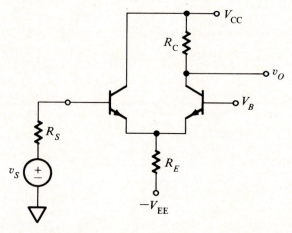

Figure P14.30

Power Applications

15.1 LINEAR POWER AMPLIFIERS

In many systems applications significant amounts of power are needed to operate various devices such as loudspeakers for audio reproduction, servomotors for control systems, and acoustic transducers for sonar systems. The large current swings and voltage swings required in power applications are provided by large and expensive transistors. It is, therefore, economically important to design a power amplifier with the smallest transistor capable of delivering the required power and to minimize the dc power dissipated in the amplifier. The efficiency of power amplifiers is a critical measure of their performance and, in this section, the operation and efficiency of some basic amplifier configurations are examined.

15.1.1 The Class A Transistor Amplifier

A *class A* amplifier is one in which the output (collector or drain) current is never zero. This means that the transistor, whether it is a BJT or FET, never cuts off.

Usually class A amplifiers operate over the most linear portion of the transistor's characteristics, i.e., the active region for a BJT or pinch-off in a FET.

In Fig. 15.1(a) the circuit diagram of a single stage transistor amplifier is shown, and in Fig. 15.1(b) a set of idealized output characteristics are given. Although the points to be discussed are applicable to any of the transistor types, the BJT is used for illustration. For simplicity, biasing details are omitted. The Q point, for the idealized characteristics assumed, is chosen to be at the midpoint of the load line in order to get maximum symmetrical output signal swing. Thus the quiescent conditions are

$$V_{CE} = \frac{V_{CC}}{2} \tag{15.1}$$

and

$$I_C = \frac{V_{CC}}{2R_L} \tag{15.2}$$

Under sinusoidal operation, the voltage waveform across R_L will be a sinusoid of *maximum* amplitude $V_{CC}/2$, so that the ac power, P_l, delivered to the load R_L has a *maximum* value $P_{l,max}$ given by

$$P_{l,max} = \frac{(V_{CC}/2\sqrt{2})^2}{R_L} = \frac{V_{CC}^2}{8R_L} \tag{15.3}$$

The power supplied by the source V_{CC} is

$$P_{CC} = V_{CC} \, \text{ave}\{i_C\} = \frac{V_{CC}^2}{2R_L} \tag{15.4}$$

since $\text{ave}\{i_C\} = I_C$.

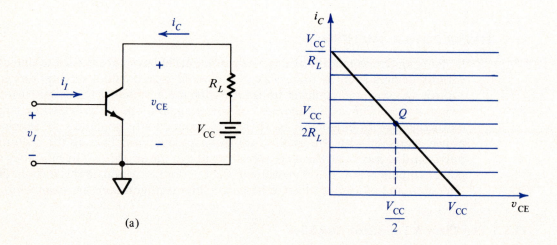

(a)   (b)

Figure 15.1 Class A amplifier: (a) circuit and (b) load line.

The *efficiency* of the amplifier, as measured by its ability to convert dc power from the supply to ac power to the load, is defined as

$$\eta = \frac{\text{signal power to load}}{\text{dc power supplied}} = \frac{P_l}{P_{CC}} \tag{15.5}$$

For the maximum output given by Eq. (15.3)

$$\eta_{max} = \tfrac{1}{4} \tag{15.6}$$

or 25%. This is the idealized maximum: it ignores power losses in biasing circuits, and assumes the amplifier is delivering the maximum power. In practice the efficiency is much lower than 25%.

To calculate the power balance for this amplifier one calculates the dc power dissipated in R_L as

$$P_L = I_C^2 R_L = \frac{V_{CC}^2}{4R_L} \tag{15.7}$$

The power dissipated in the transistor's collector P_C is, by conservation,

$$P_C = P_{CC} - P_L - P_l \tag{15.8}$$

namely the power supplied by V_{CC} less the total power dissipated in R_L. The transistor dissipation is least when maximum ac power is delivered to the load and is given by

$$P_{C,min} = \frac{V_{CC}^2}{8R_L}$$

It has a maximum value of

$$P_{C,max} = \frac{V_{CC}^2}{4R_L}$$

for zero signal amplitude.

EXAMPLE 15.1

A power transistor is driving a 10-Ω load and the quiescent point is at $I_C = 0.5$ A, $V_C = 5$ V, and $V_{CC} = 10$ V. Calculate the maximum ac power delivered to the load when the amplifier circuit of Fig. 15.1 is used. What is the collector dissipation? What is the efficiency?

Solution
The maximum ac power in R_L is given by Eq. (15.3)

$$P_{l,max} = \frac{(5/\sqrt{2})^2}{10} = 1.25 \text{ W}$$

the dc power $P_L = 0.5^2(10) = 2.5$ W, and the power supplied by the V_{CC} source is $P_{CC} = 5$ W. The collector dissipation is, by conservation,

$$P_c = P_{CC} - P_L - P_l = 5 - 3.75 = 1.25 \text{ W}$$

and the maximum efficiency

$$\eta_{max} = \frac{P_l}{P_{CC}} = \frac{1.25}{5} = 0.25$$

Transformer-coupled load

Equation (15.7) shows that the dc power dissipated in R_L accounts for half the power supplied by the source V_{CC}. Since this is not signal power, it does not serve any useful purpose and for bandpass or high-pass systems applications it can be eliminated by means of a transformer which couples the transistor to the load as shown in Fig. 15.2(a). The transformer is adequately modeled (see Chap. 4) by the network shown within the *screened* contour. The dc quiescent current I_C flows through the inductance L and there is thus no dc power dissipated in R_L. Under zero signal conditions $v_{CE} = V_{CC}$, so that the dc operating point is constrained to lie on a vertical line in the i_C-v_{CE} plane; this line is called the *dc load line* and is shown dashed in Fig. 15.2(b). Under ac conditions and at frequencies such that $\omega L \gg n^2 R_L$, the small-signal components of i_C and v_{CE} are related by

$$i_c = \frac{-v_{ce}}{n^2 R_L}$$

so that the instantaneous signal values lie on the straight line with slope $-1/n^2 R_L$ passing through Q. This *ac load line* is shown in Fig. 15.2(b). From the construction it is seen that the maximum peak-to-peak voltage swing for v_{CE} is $2V_{CC}$. In Fig.

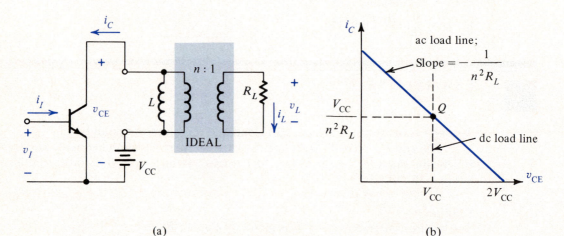

(a)   (b)

Figure 15.2 Transformer coupled class A amplifier: (a) circuit and (b) ac and dc load lines.

15.2(b), the quiescent point is shown at $V_{CE} = V_{CC}$ which corresponds to $I_C = V_{CC}/n^2 R_L$ and results in a maximum symmetrical swing.

Under sinusoidal operation, the maximum ac power to the load, expressed in terms of the maximum excursions in i_C and v_{CE} about the quiescent point, is given by

$$P_{l,\max} = \frac{v_{CE,\max} i_{C,\max}}{2} = \frac{V_{CC}^2}{2n^2 R_L} \tag{15.9}$$

The power supplied by the source V_{CC} is

$$P_{CC} = V_{CC} I_C = \frac{V_{CC}^2}{n^2 R_L} \tag{15.10}$$

and the power dissipated in the transistor, according to Eq. (15.8) with $P_L = 0$, has a minimum value of

$$P_{C,\min} = P_{CC} - P_{l,\max}$$

or

$$P_{C,\min} = \frac{V_{CC}^2}{2n^2 R_L} \tag{15.11}$$

The efficiency, as given by Eq. (15.5), is at most 50%. The maximum efficiency is therefore doubled by using a transformer to couple the load, thereby eliminating dc power wasted in the load. The maximum dissipation is P_{CC} when $P_l = 0$.

15.1.2 The Maximum-Dissipation Hyperbola

Once the load power requirements are specified in a large-signal application, the next step is the selection of an appropriate transistor. Clearly the transistor must not only be capable of handling the required voltage and current swings but must also handle the expected collector dissipation. Transistor ratings as specified by manufacturers, will, in general, include the maximum allowed collector current $i_{C,\max}$, the maximum allowed collector voltage $v_{CE,\max}$, and the maximum average collector dissipation $P_{C,\max}$.

The maximum ratings restrict the allowed operating region of the transistor to that shown in Fig. 15.3. For safe operation Q must lie on or below the maximum *collector-dissipation hyperbola* which is the locus of all points satisfying the equation

$$v_{CE} i_C = P_{C,\max} \tag{15.12}$$

Exceeding $P_{C,\max}$ overheats the transistor. This, it turns out, causes the collector current to rise, which causes the collector dissipation to increase further. This chain of events, which results in the rapid destruction of the transistor, is called *thermal runaway*.

In Fig. 15.3 the allowed operating region, as specified by the maximum ratings, is shown. It is clear that the load line must intersect the v_{CE} axis at a value less than $v_{CE,\max}$ and must intersect the i_C axis at a current less than $i_{C,\max}$. Any safe

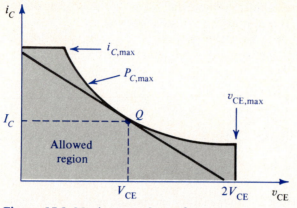

Figure 15.3 Maximum ratings of transistor.

design must constrain the load line to lie within this region and, at most, be tangential to the maximum dissipation hyperbola. Figure 15.3 shows the situation such that a tangential load line exists which satisfies the maximum allowable i_C, v_{CE}. This situation exists for $v_{CE,max}\, i_{C,max} \geq 4P_{C,max}$.

For a Q point chosen to yield the maximum symmetrical swing in current and voltage as shown in Fig. 15.3, it follows that, for a class A amplifier, such as the one shown in Fig. 15.1(a),

$$V_{CE} = I_C R_L \tag{15.13}$$

where R_L must be chosen such that $V_{CE} < v_{CE,max}/2$ and $I_C < i_{C,max}/2$ are satisfied. From Eqs. (15.12) and (15.13) the Q point is found to be

$$I_C = \sqrt{\frac{P_{C,max}}{R_L}} \tag{15.14}$$

and

$$V_{CE} = \sqrt{P_{C,max} R_L} \tag{15.15}$$

EXAMPLE 15.2

A transistor with the following ratings is used in the circuit of Fig. 15.1(a):

$$P_{C,max} = 40\ \text{W} \qquad V_{CE,max} = 80\ \text{V} \qquad i_{C,max} = 5\ \text{A}$$

The load is a 10-Ω resistance. Design a symmetrical-swing power amplifier to deliver maximum power to the load. What is the required value of V_{CC}? What is the load power? What is the efficiency?

Solution

From Eqs. (15.14) and (15.15) the quiescent point is at

$$I_C = \sqrt{\frac{40}{10}} = 2 \text{ A}$$

$$V_{CE} = \sqrt{40(10)} = 20 \text{ V}$$

The required supply is, from Eq. (15.1),

$$V_{CC} = 2V_{CE} = 40 \text{ V}$$

and thus $P_{CC} = V_{CC}I_C = 80$ W. The maximum ac load power is given by Eq. (15.3)

$$P_{l,max} = \frac{V_{CC}^2}{8R_L} = \frac{1600}{80} = 20 \text{ W}$$

and the efficiency

$$\eta = \frac{P_{l,max}}{P_{CC}} = 0.25 \qquad \text{or} \qquad 25\%$$

EXAMPLE 15.3

Repeat Example 15.2 using the circuit of Fig. 15.2 given a 2:1 transformer. Calculate the collector dissipation with full output and with zero output.

Solution

With the given transformer the ac load is $n^2R_L = 40$ Ω and the quiescent point is therefore at

$$I_C = \sqrt{\frac{40}{40}} = 1 \text{ A}$$

$$V_{CE} = \sqrt{40(40)} = 40 \text{ V}$$

This is also V_{CC} for the transformer coupled design and thus $P_{CC} = 40(1) = 40$ W, half that required for the design in the preceding example. The maximum ac load power is given by Eq. (15.9)

$$P_{l,max} = \frac{V_{CC}^2}{2n^2R_L} = \frac{1600}{80} = 20 \text{ W}$$

as before; however, the efficiency is now 50% since P_{CC} is 40 W. The collector dissipation P_C, given by $P_{CC} - P_l$, is 20 W under full output and 40 W with zero output.

EXAMPLE 15.4

Using the transistor specifications given in Example 15.2, design a transformer coupled amplifier to deliver maximum power to a 10-Ω load. Specify the required supply voltage V_{CC}, the turns ratio n, the power delivered to the load, and the efficiency.

Solution

The quiescent point is given by Eqs. (15.14) and (15.15) with $R_L = n^2 10\ \Omega$ and $P_{l,max} = 20$ W (n being the turns ratio to be determined), i.e.,

$$I_C = \sqrt{\frac{P_{C,max}}{n^2 R_L}} = \frac{2}{n}\ \text{A}$$

and

$$V_{CE} = \sqrt{(P_{C,max} n^2 R_L)} = 20n\ \text{V}$$

As long as the maximum allowed current and voltage are respected there is some design flexibility in the choice of n. For a symmetric swing

$$I_C = \frac{2}{n} \le \frac{i_{C,max}}{2} = 2.5\ \text{A}$$

and

$$V_{CE} = 20n \le \frac{v_{CE,max}}{2} = 40\ \text{V}$$

These inequalities are satisfied for

$$0.8 < n < 2$$

This flexibility permits the choice of n to be made with respect to other aspects that were not part of the original specifications. In this example, the availability, "off the shelf," of $n = 1$ or $n = 2$ transformers is one consideration; another is the fact that power supplies tend to be cheaper for lower current capacities or may be cheaper for certain combinations of voltage and current. For $n = 1$, $I_C = 2$ A, $V_{CE} = 20$ V. The maximum power and efficiency will be the same as in the previous example.

15.1.3 Class B Complementary-Symmetry Ampliflier

The low maximum efficiency of 50% for the class A transformer-coupled configuration is due to the presence of a large quiescent current which must be supplied even under zero-signal conditions. In a *class B* amplifier a transistor is biased at cutoff and therefore no power needs to be supplied under quiescent conditions. Since a single transistor biased at cutoff will produce a rectified half-wave sinusoidal collector current when driven by a sinusoidal signal, circuits with two transistors are needed to produce sinusoidal currents in a load. Fortunately because both *pnp* and *npn* transistors are available, relatively simple circuits which use only two transistors are easily designed to supply the sinusoidal signals to a load.

In Fig. 15.4(a) a typical amplifier which utilizes this *complementary symmetry* is shown. The amplifier consists of two class B emitter followers. For $v_i > 0$, Q_1, the *npn* transistor, is active and Q_2 is cut off; thus during this half-cycle the emitter follower circuit shown in Fig. 15.4(b) applies. The roles of Q_1 and Q_2 interchange for the other half-cycle when v_i is negative. The transistors are a matched pair and this configuration requires both a positive and a negative supply voltage.

In Fig. 15.4(c) the various idealized circuit current waveforms are shown when the transistors are driven by a sinusoidal input.

The current through R_L is a sinusoid of amplitude I_l; with Q_1 supplying the positive half-cycle of current and Q_2 the negative half-cycle of current. The ac power delivered to R_L is $\hat{I}_l^2 R_L$ or, in terms of amplitude,

$$P_l = \frac{I_l^2 R_L}{2} \tag{15.16}$$

and the power supplied by the batteries is, from Fig. 15.4, twice that supplied by each V_{CC} source, i.e.,

$$P_{CC} = 2V_{CC}\,\text{ave}\{i_{C1}\} \tag{15.17}$$

Since i_{C1} is a half-wave sinusoid of peak value I_l, the average value is, as in the

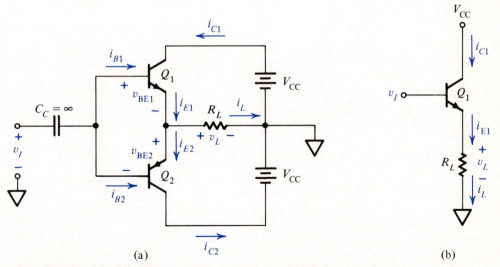

(a)   (b)

Figure 15.4 Class B complementary-symmetry amplifier: (a) circuit, (b) circuit for $v_I > 0$ half-cycle, and (c) ideal waveforms for class B complementary-symmetry class B amplifier.

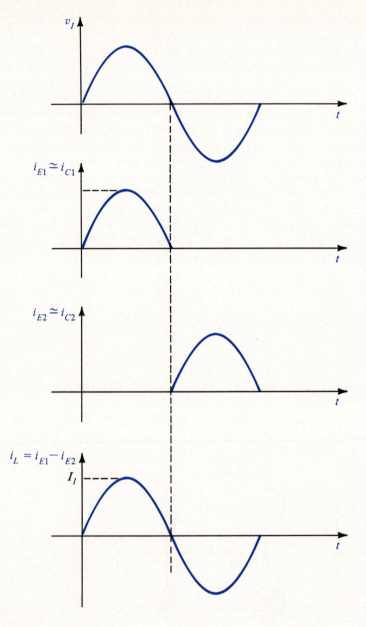

(c)

Figure 15.4 *(continued)*

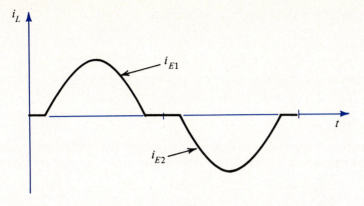

(d)

Figure 15.4 *(continued)*. (d) Waveform for class B amplifier showing crossover distortion.

half-wave rectifier of Drill Exercise 5.5,

$$\text{ave}\{i_{c1}\} = \frac{I_l}{\pi}$$

Thus

$$P_{cc} = \frac{2V_{cc}I_l}{\pi} \tag{15.18}$$

and hence the collector dissipation *per transistor* is

$$P_c = \frac{P_{cc} - P_l}{2}$$

$$= \frac{V_{cc}I_l}{\pi} - \frac{I_l^2 R_L}{4} \tag{15.19}$$

The amplifier efficiency η is given by

$$\eta = \frac{P_l}{P_{cc}} = \frac{\pi}{4}\frac{I_l R_L}{V_{cc}} \tag{15.20}$$

The following questions need to be answered with respect to this amplifier's performance.

1. What is the maximum available output power?

2. What is the maximum collector dissipation per transistor?

3. What is the maximum efficiency?

From Eq. (15.18), for a fixed V_{cc}, the output power is seen to increase linearly with I_l, hence the maximum power occurs when I_l is at its largest value. From Fig.

15.4(b) and (c) this is the current at saturation,

$$I_{l,\text{max}} \simeq \frac{V_{CC}}{R_L} \tag{15.21}$$

The maximum ac load power, from Eq. (15.16), is therefore

$$P_{l,\text{max}} = \frac{V_{CC}^2}{2R_L} \tag{15.22}$$

Maximum collector dissipation occurs when the right-hand side of Eq. (15.19), for fixed V_{CC}, is a maximum. This occurs at a peak current given by

$$I_l\,(P_{C,\text{max}}) = \frac{2V_{CC}}{\pi R_L} \tag{15.23}$$

yielding a maximum collector dissipation per transistor of

$$P_{C,\text{max}} = \frac{V_{CC}^2}{\pi^2 R_L} \tag{15.24}$$

$$\simeq 0.1\,\frac{V_{CC}^2}{R_L}$$

Notice that this maximum in collector dissipation does *not* occur at maximum output power but at a value given by Eq. (15.16) evaluated with I_l as given by Eq. (15.23). Comparing Eqs. (15.22) and (15.24) shows that

$$P_{C,\text{max}} = \frac{2P_{l,\text{max}}}{\pi^2} \simeq \frac{P_{l,\text{max}}}{5} \tag{15.25}$$

and thus, for a given output power requirement, transistors for class B operation need have about $\frac{1}{10}$ the required power rating of a transistor for a class A design.

The maximum efficiency, from Eq. (15.20), occurs at maximum I_l, and is therefore given by

$$\eta_{\text{max}} = \frac{\pi}{4} = 78.5\% \tag{15.26}$$

The efficiency is dramatically improved over class A operation and, more significantly, the collector dissipation is, from Eq. (15.19), zero for zero signal and, for maximum power, is given by

$$P_C\!\left(\eta = \frac{\pi}{4}\right) = P_{l,\text{max}}\frac{(1/\eta - 1)}{2} \tag{15.27}$$

In practical class B amplifiers the waveforms for the currents and, in particular the load current i_L, will deviate slightly from those shown in Fig. 15.4(c). This is mainly due to the fact that, for $|v_l| < |V_{BE}|$, the base currents are negligible and therefore the load current will not be truly sinusoidal but will be distorted as shown in exaggerated fashion in Fig. 15.4(d). This phenomenon, called *crossover*

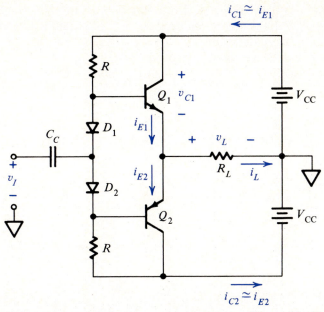

$i_{C1} \simeq i_{E1}$

Figure 15.5 Complementary symmetry amplifier with diode biasing.

distortion, can be eliminated by biasing the EBJs of Q_1 and Q_2 at 0.7 V. A simple method for doing this is to use diodes with $v_D \simeq V_{BE}$ connected to the power supplies via current limiting resistors as shown in Fig. 15.5. With this biasing arrangement i_{E1} and i_{E2} are exact half-sinusoidal waveforms and the load voltage is sinusoidal.

DRILL EXERCISES

15.1 Show that when I_l in Fig. 15.4 is adjusted for maximum collector dissipation, the efficiency is 50%.

15.2 Show, from Fig. 15.5, that when Q_1 is cut off, $v_{C1} = V_{CC} - v_L$ ranges from V_{CC} (when $v_L = 0$) to $2V_{CC}$ when $v_L = -V_{CC}$. Hence the largest allowable V_{CC} for a class B transistor amplifier is $v_{CE,max}/2$.

EXAMPLE 15.5

Using a transistor with ratings: $P_{C,max} = 8$ W, $v_{CE,max} = 50$ V, $i_{C,max} = 4$ A, and the amplifier configuration of Fig. 15.5, calculate the V_{CC} needed, the maximum power possible into a 10-Ω load, the collector dissipation at maximum output power, the efficiency, the peak voltage input required for the maximum output, and the maximum collector dissipation for this design. Neglect bias circuit losses and assume ideal transistor characteristics with $V_{BE} = 0$.

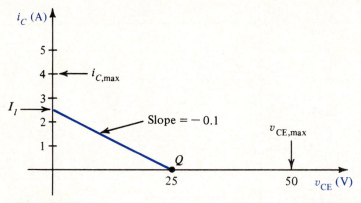

Figure E15.5

Solution

Since the operation is class B, each transistor can operate along the locus shown in Fig. E15.5. The load line slope is fixed at -0.1 and, for the transistors specified, according to the result of Exercise 15.2, one can thus use $V_{CC} = 25$ V with $I_l = 2.5$ A. The maximum power to the load is thus given by Eq. (15.22) as

$$P_{l,max} = \frac{V_{CC}^2}{2R_L} = 31.25 \text{ W}$$

The power supplied by the two batteries, given by Eq. (15.18) with I_l as in Eq. (15.21) is

$$P_{CC} = \frac{2}{\pi} \frac{V_{CC}^2}{R_L} = 39.8 \text{ W}$$

Hence the efficiency at maximum power output is $\pi/4$ or 78.5% as expected for a class B amplifier, and the average collector dissipation per transistor at full output is, therefore, given by Eq. (15.19) as

$$P_C = \frac{39.8 - 31.25}{2} = 4.27 \text{ W}$$

Since this is an emitter-follower configuration the peak input voltage will be 25 V, the same as the peak voltage across R_L. The maximum collector dissipation is given by Eq. (15.24)

$$P_{C,max} = \frac{V_{CC}^2}{\pi^2 R_L} = 6.33 \text{ W}$$

indicating that the 8-W power-handling capability of the transistors is not being fully utilized.

In this example it was noted that the transistors were being underutilized and that the load R_L was not getting as much power as is implied by Eq. (15.25). For the 8-W rating this equation implies that a load power of almost 40 W should be possible. The reason why this is not obtained in this example is that the 10-Ω load resistor specified is too large for $P_{l,max}$, as given by Eq. (15.22), to equal 40 W even when the V_{CC} chosen is the maximum allowable value of $v_{CE,max}/2$. This limitation can be overcome by coupling the load using a transformer. In the next section the operation of class B, transformer-coupled amplifiers, is discussed.

DRILL EXERCISE

15.3 What is the smallest value of R_L which can be used in Example 15.5 given $V_{CC} = 25$ V.
Ans. 7.9 Ω

15.1.4 Class B Transformer-Coupled Amplifier

In many applications it is often necessary to deliver power to a resistive load by a transformer. One of the main reasons for doing this is that it provides the designer with an extra degree of freedom, the turns ratio, making it easier to achieve maximum efficiency. In Fig. 15.6 two transistors, biased at cutoff, are driving a load via a center-tapped transformer. The input signals v_{B1} and v_{B2} are 180° out of phase so that, during the positive half-cycles of v_{B1}, the inputs will cause Q_1 to be active while Q_2 is cut off. During the negative half-cycles of v_{B1}, Q_2 is active and Q_1 remains cut off. Under maximum drive conditions, when the transistors go from the edge of cutoff to the edge of saturation, the peak sinusoidal signal voltage v_{p1} at the primary of the output transformer will be V_{CC}. The battery supply current i_C will be a full-wave rectified sinusoid of peak value I_c which, under the maximum output conditions, is given by

$$I_{c,max} = \frac{V_{CC}}{n^2 R_L} \tag{15.28}$$

and I_C has a maximum average value, $I_{C,max}$, of twice that of the half-wave rectifier of Drill Exercise 5.5, i.e.,

$$I_{C,max} = \frac{2I_{c,max}}{\pi} \tag{15.29}$$

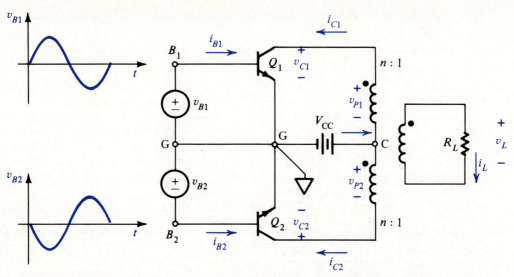

Figure 15.6 Simplified circuit of a class B transformer-coupled amplifier.

The maximum average power supplied by the battery is, therefore,

$$P_{CC,max} = V_{CC}\,I_{c,max} = \frac{2V_{CC}^2}{\pi n^2 R_L} \tag{15.30}$$

The maximum ac load power, neglecting any losses in the transformer, is

$$P_{l,max} = \frac{V_{CC}}{\sqrt{2}}\,\frac{I_{c,max}}{\sqrt{2}} \tag{15.31}$$

$$= \frac{V_{CC}^2}{2n^2 R_L}$$

which is identical to Eq. (15.22) with R_L replaced by $n^2 R_L$. The maximum efficiency is the ratio $P_{l,max}/P_{CC,max}$

$$\eta_{max} = \frac{\pi}{4} = 78.5\% \tag{15.32}$$

which is the same as for the dc coupled class B amplifier of the previous section. Again the collector dissipation per transistor is zero for zero signal. Thus, whether transformer-coupled or dc-coupled, for a given power requirement, transistors for a class B amplifier need only have about $\frac{1}{10}$ the power rating of the transistor needed for a class A design. The maximum collector dissipation occurs at a slightly lower output and is given by Eq. (15.24) with R_L replaced by $n^2 R_L$, i.e.,

$$P_{C,\max} = \frac{V_{CC}^2}{\pi^2 n^2 R_L} \tag{15.33}$$

DRILL EXERCISES

15.4 Show that the per-collector dissipation in a transformer-coupled class B amplifier, for arbitrary output, is given by

$$P_C = \frac{V_{CC} I_c}{\pi} - \frac{I_c^2 n^2 R_L}{4}$$

15.5 Show from Fig. 15.6 that, when Q_1 is cut off, $v_{C1} = V_{CC} + n v_L$ ranges from V_{CC} (when $v_L = 0$) to $2V_{CC}$ (when $n v_L = V_{CC}$).

15.6 Find the value of I_c at which the collector dissipation is greatest, and verify Eq. (15.33). What is the efficiency under those circumstances?
Ans. $2V_{CC}/\pi n^2 R_L$, $V_{CC}^2/\pi^2 n^2 R_L$, 50%

The disadvantages of the design of Fig. 15.6 are the need for the two out-of-phase signals v_{B1}, v_{B2}, the two *matched* transistors, and a center-tapped transformer. The out-of-phase signals can be obtained by using either a second center-tapped transformer at the input, as shown in Fig. 15.7, or by using the phase splitter circuit of Fig. P6.15. In the circuit of Fig. 15.7 the crossover distortion is eliminated by biasing the EBJs of Q_1 and Q_2 at 0.7 V by the diode and current limiting resistor R.

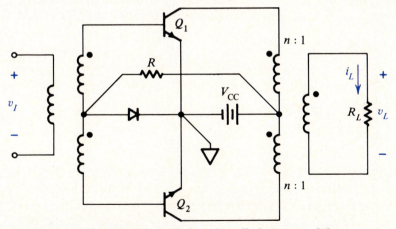

Figure 15.7 Transformer-coupled push-pull class B amplifier.

EXAMPLE 15.6

Design a class B push-pull amplifier as in Fig. 15.6 to deliver maximum power to a 10-Ω load using two transistors having the ratings of Example 15.5, i.e.,

$$P_{C,\max} = 8 \text{ W} \qquad v_{CE,\max} = 50 \text{ V} \qquad i_{C,\max} = 4 \text{ A}$$

Specify V_{CC}, the output transformer turns ratio n, the maximum output power, the efficiency, and the maximum collector dissipation.

Solution

From Eq. (15.31) the maximum load power is

$$P_{l,\max} = \frac{V_{CC}^2}{2n^2 R_L}$$

From Exercise 15.6 the maximum collector dissipation is

$$P_{C,\max} = \frac{V_{CC}^2}{\pi^2 n^2 R_L} \tag{15.34}$$

hence the maximum value of load power cannot exceed

$$P_{l,\max} \leq \frac{\pi^2 P_{c,\max}}{2} \simeq 40 \text{ W}$$

(approximately 5 times $P_{C,\max}$) and, according to Eq. (15.31), this means that

$$V_{CC} I_{c,\max} \leq 80 \text{ W}$$

Making $V_{CC} = v_{CE,\max}/2 = 25$ V is the low current solution for which $I_{c,\max} = 3.2$ A. Making $I_{c,\max} = i_{C,\max} = 4$ A is the low voltage solution for which $V_{CC} = 15$ V.

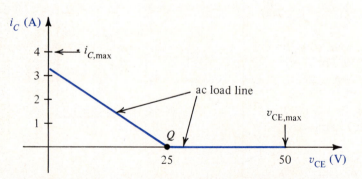

Figure E15.6 Load line and transistor ratings for Example 15.6.

Since, according to Eq. (15.28)

$$n^2 = \frac{V_{CC}}{R_L I_{c,\,max}}$$

the required n value can be between 0.612 and 0.88. Choosing the design at $V_{CC} = 25$ V and $I_{c,\,max} = 3.2$ A, the output power is 40 W, the battery power is $\simeq$ 51 W, and the efficiency is 78.5%. The maximum per-collector dissipation is given by Eq. (15.34) as $\simeq 8$ W, the maximum allowed. The ac load line and output ratings are shown in Fig. E15.6. The power constraint is that the *average* collector dissipation must be less than $P_{C,\,max}$. The instantaneous power $p_C(t)$ can, however, exceed $P_{C,\,max}$.

DRILL EXERCISES

15.7 Show that if, in Example 15.6 transistors rated at $P_{C,\,max} = 4$ W, $v_{CE,max} = 40$ V, $i_{c,\,max} = 1$ A, are used, the design results in a maximum load power of 10 W with $V_{CC} = 20$ V and $n^2 = 2$.

15.8 Show that, when $v_{CE,max} i_{c,\,max} < 20 P_{Cmax}$ for a transistor, that the maximum load line power for a class B amplifier is $P_{l,max} = v_{CE,max} i_{c,\,max}/4$.

EXAMPLE 15.7

Design a class B transformer-coupled amplifier to deliver 10 W to a 10-Ω load using transistors which have $v_{CE,max} = 50$ V. Specify $P_{C,\,max}$, $i_{c,\,max}$, V_{CC}, and the turns ratio n.

Solution
From Example 15.6 the collector dissipation must be at least $\frac{1}{5}$ the load power or $P_{C,max} = 2$ W. For the required 10 W in the load, using Eq. (15.31)

$$V_{CC} I_{c,\,max} \simeq 20$$

and for $V_{CC} = v_{CE,max}/2 = 25$ V, $I_{c,\,max} = 0.8$ A. Thus $i_{C,\,max} = 0.8$ A and the turns ratio from Example 15.6 is given by

$$n^2 = \frac{V_{CC}}{R_L I_{c,\,max}} = \frac{25}{10(0.8)} = 3.125$$

15.2 POWER SUPPLIES

It should be evident at this point that most electronic circuits require power supplies with constant voltages. These dc voltages are usually obtained from ac power outlets by *rectification, filtering,* and *regulation.* In this section basic full-wave rectifier circuits and typical filter circuits are examined, and the operation of a regulated power supply is outlined. Ideal diodes will be assumed throughout this section.

15.2.1 Rectifiers

In Chap. 5 the operation of a *half-wave* diode rectifier circuit was discussed and it was shown that the load current was a half-wave sinusoid. In Fig. 15.8(a) the circuit of a full-wave diode rectifier is shown. For positive input half-cycles D_1 conducts and D_2 is OFF, while for negative half-cycles D_2 conducts and D_1 is OFF. The load voltage is, therefore, as shown in Fig. 15.8(b), where the numerical values will depend on the ac input voltage and the transformer turns ratio. Note that, when one diode is ON, the voltage across the OFF diode, as given by KVL, is $2V_l$. The diodes must therefore withstand this *peak inverse voltage.* The average value of load voltage V_{dc} is just the average of $v_L(t)$, namely,

$$V_{dc} = \frac{2V_l}{\pi} \tag{15.35}$$

The network of Fig. 15.9, which is called a full-wave bridge rectifier, does not require a tapped transformer but uses four diodes. During positive input half-cycles D_1 and D_3 are ON, D_2 and D_4 are OFF; during negative half-cycles D_2 and D_4 are ON, D_1 and D_3 are OFF. The output v_L is thus given by a full-wave rectified sinusoid.

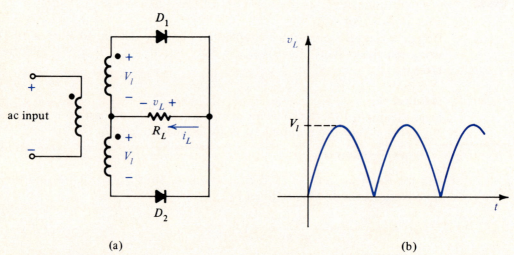

(a)   (b)

Figure 15.8 (a) Full-wave rectifier and (b) the load voltage.

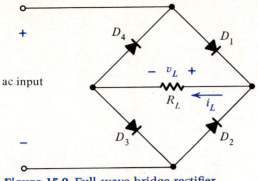

Figure 15.9 Full-wave bridge rectifier.

15.2.2 Filters

Reduction of the level of unwanted time varying components in v_L is accomplished by filters. These are low-pass in nature and one of the simplest examples of a filtered rectifier circuit is the capacitance filter shown in Fig. 15.10, consisting of a capacitor shunting the load of a half-wave rectifier. If the circuit is turned on at $t = 0$, v_L will follow v_S and the capacitance will charge up to V_m at $t_1 \simeq \pi/2\omega$, during this time D will be ON. As $v_S(t)$ starts to fall below V_m, D becomes reverse biased and C discharges exponentially through R_L as a zero-input problem with "initial" condition $v_L = V_m$ on the capacitance voltage. The waveforms are as shown in Fig. 15.11(a) and, for a large enough RC time constant, the exponential decay is approximately linear. The diode D remains OFF until t_2 when $v_L(t) = v_S(t)$, D switches ON again, and C charges up again to V_m. The drop in output voltage V_r is called the *ripple* voltage. An approximate analysis, leading to useful design criteria, is straightforward.

Assuming the exponential voltage decay can be approximated by a linear relation, i.e., for small x, $e^{-x} \simeq 1 - x$, then

$$v_L(t_2) = V_m \left(1 - \frac{t_2 - t_1}{R_L C} \right) = V_m - V_r \tag{15.36}$$

Assuming also that the diode OFF time, $t_2 - t_1$ is approximately one period $T = 1/f$ and, that the diode ON time is negligibly small, the $v_L(t)$ approximation

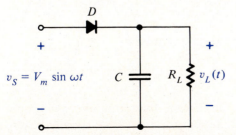

Figure 15.10 Capacitor-filtered half-wave rectifier.

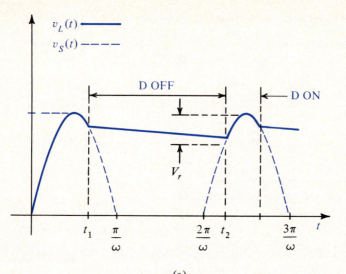

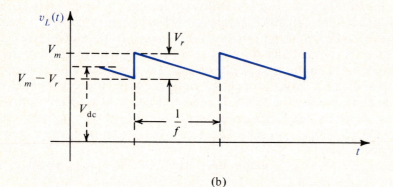

(b)

Figure 15.11 (a) Load voltage and input voltage waveforms for the circuit of Fig. 15.10. (b) Approximate load waveform for half-wave rectifier.

corresponding to these assumptions is shown in Fig. 15.11(b). V_r is given by Eq. (15.36) as

$$V_r = \frac{V_m}{R_L C f} = \frac{I_{dc}}{fC} \tag{15.37}$$

since $I_{dc} \simeq V_m/R_L$. The dc voltage is now written as

$$V_{dc} = V_m - \frac{I_{dc}}{2fC} \tag{15.38}$$

The ripple is proportional to the dc current drain and inversely proportional to C and the supply frequency. Thus, raising the frequency usually results in smaller capacitors and transformers so that, for example, aircraft use power supplies operating at 400 Hz to take advantage of the volume and weight reduction.

DRILL EXERCISES

15.9 Show that, for a full-wave rectifier with a capacitance filter, $V_r \simeq I_{dc}/2fC$, and $V_{dc} = V_m - I_{dc}/4fC$.

15.10 Show that R_L for the conditions of Exercise 15.9 sees a dc Thevenin circuit with $V_T = V_m$ and $R_{eq} = 1/(4fC)$.

15.11 Show that the fractional ripple for a capacitance filtered full-wave supply is $V_r/V_{dc} \simeq 1/(4fR_LC)$.

From the preceding analysis and, in particular, from Exercise 15.10 it is evident that a filtered power supply has a Thevenin representation and, as a consequence, V_{dc} will decrease as I_{dc} increases. This property is clearly undesirable in electronic, constant-voltage, power supplies, and consequently, feedback circuits are used to maintain the supply voltage constant, independent of the current drain. Such supplies are called regulated power supplies. The output voltage-output current characteristics of a power supply is called its *voltage regulation*.

The capacitance filtering technique illustrated in this section is only one of many possible filters; some others are shown in Fig. 15.12. They all operate

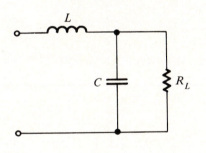

(a)

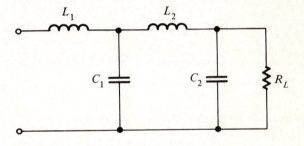

(b)

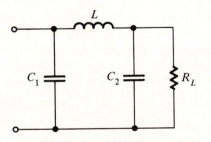

(c)

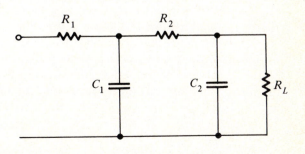

(d)

Figure 15.12 Samples of higher-order filters.

basically by establishing low-impedance parallel and high-impedance series branches for the ripple frequency components.

EXAMPLE 15.8

Design a transformer-coupled, two-diode, full-wave rectifier power supply to operate at a nominal output of 20 V and capable of supplying up to 2 A with a fractional ripple not to exceed 5%. Give the regulation characteristics assuming a 60 Hz ac supply is available.

Solution
The required filter capacitance is given by the ripple requirement, i.e.,

$$0.05 = \frac{1}{4 f R_L(\min) C}$$

where $R_L(\min)$ is taken as $20/2 = 10\ \Omega$, hence

$$C = 8.3\ \text{mF}$$

The Thevenin resistance is

$$R_{eq} = \frac{1}{4 f C} = 0.05 R_L(\min) = 0.5\ \Omega$$

The regulation will be 1-V drop from 20 V at maximum load current.

15.2.3 Regulated Power Supplies

The purpose of a regulated dc power supply is to provide a fixed voltage v_L which is not dependent on the load current i_L, nor on changes in ac supply-line voltages. A properly regulated power supply will have a low output resistance and will maintain v_L at a predetermined value independent of i_L. A simple series regulated power supply is shown in Fig. 15.13. The Thevenin circuit of the rectifier and filter combination, enclosed within the *screened boundary*, is usually called the unregulated power supply. The circuit shown uses an op amp to compare a fraction of the load voltage v_L with a voltage reference V_R, usually derived from a Zener diode. The amplified difference v_L, drives the base of the emitter follower.

Assuming an ideal op amp the virtual short property makes $v_A = 0$ hence

$$K v_L = V_R \tag{15.39}$$

or

$$v_L = \frac{V_R}{K} \tag{15.40}$$

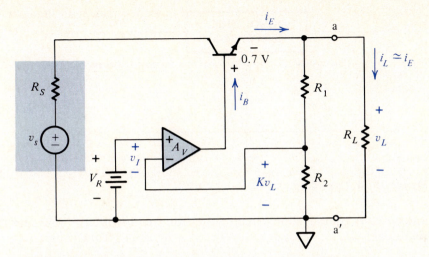

Figure 15.13 Series-regulated power supply circuit.

where K is the voltage divider ratio $R_2/(R_1 + R_2)$. The load voltage is, therefore, given by

$$v_L = \frac{V_R}{K} = \frac{V_R(R_1 + R_2)}{R_2} \qquad (15.41)$$

determined only by the reference source V_R and a resistance ratio independent of v_S or R_S. The role of Q_1 is to supply the load current, and as long as v_S is large enough to keep Q_1 in the active mode the circuit will maintain v_L constant as given by Eq. (15.41).

The measure of regulation and stabilization is expressed in terms of the sensitivity of v_L to changes in unregulated supply voltage v_S load current i_L, and temperature:

The input sensitivity is

$$Sv_S = \frac{\partial v_L}{\partial v_S}$$

the output resistance is

$$S_{i_L} = \frac{\partial v_L}{\partial i_L} = R_o$$

and the temperature sensitivity is

$$S_T = \frac{\partial v_L}{\partial T}$$

Monolithic (single chip) regulators are currently available as three-terminal packages. Figure 15.14 shows a standard circuit configuration. These regulators require no adjustments and are available in a wide range of preset output voltages

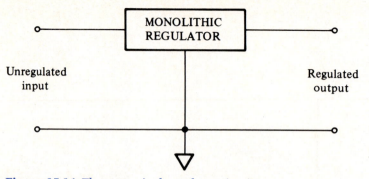

Figure 15.14 Three-terminal regulator circuit.

and typically contain short-circuit protection circuits. Typical values for the sensitivities are

$$Sv_S = 5 \times 10^{-3} \qquad R_o = 0.05 \ \Omega \qquad S_T \simeq 1 \ \text{mV/°C}$$

15.3 THE SILICON CONTROLLED RECTIFIER (SCR)

For very large power applications where linearity is not important, even efficiencies of class B amplifiers are inadequate, because of excessive power dissipation in the semiconductor devices. Fortunately, in many applications, load waveforms are unimportant and power or average current are the only important parameter. As a result, devices which behave as *controlled switches* are used in high-power applications. The most common controlled switch is the *silicon controlled rectifier* (SCR), which has the circuit symbol shown in Fig. 15.15(a). Its current voltage characteristics are shown in Fig. 15.15(b) for three values of gate current. The characteristic for a given gate current, say i_{G2}, is that of a very high

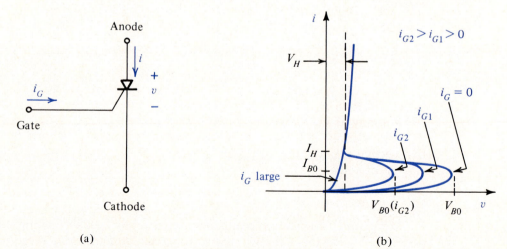

(a)   (b)

Figure 15.15 (a) Symbol and (b) characteristics of an SCR.

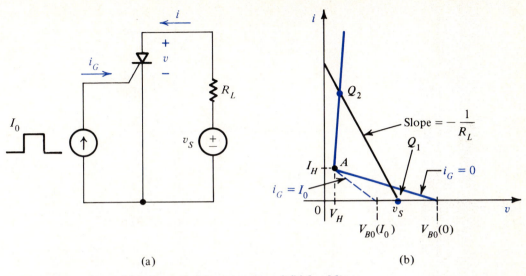

(a)   (b)

Figure 15.16 (a) SCR switching circuit and (b) load line.

resistance (OFF) for values of i less than I_{BO}, and v less than the *breakover voltage* V_{BO}; that of a negative resistance for i between I_H and I_{BO} and v between V_H and V_{BO}; that of a low resistance (ON) for $i > I_H$ and $v > V_H$. I_H and V_H are called the holding or *latching* current and voltage, respectively.

In Fig. 15.16(a) the SCR is shown connected in series with a voltage source v_S and a load R_L. In Fig. 15.16(b) the load line and the ideal PL characteristic of the SCR (idealized by taking $I_{BO} = 0$) are shown for $i_G = 0$ (solid) and for $i_G = I_O$ (dashed). For $i_G = 0$ the breakover voltage $V_{BO}(0)$ is greater than v_S and the load line intersects the characteristic at three points, but only those labeled Q_1 and Q_2 are stable solutions. For $i_G = I_O$ the breakover voltage $V_{BO}(I_O)$ is less than v_S, and the only solution is Q_2. The Q_1 solution is a low-current situation where the SCR is considered OFF, while the Q_2 solution is a low-voltage one and the SCR is considered ON.

Switching the SCR from OFF to ON

With the circuit initially at Q_1 as defined in Fig. 15.16(b), then as i_G is increased V_{BO} decreases until it is sufficiently below v_S for the operating point to jump to Q_2 and for the SCR to switch ON (low resistance) as shown in Fig. 15.16(c). The ON current may be very large but the SCR voltage drop will be essentially V_H, which is typically about 1 V. At this point the gate loses control since Q_2 is the solution irrespective of i_G and the SCR can only be turned OFF by moving the load line to the left of point A in Fig. 15.16(b).

Turning the SCR OFF

With the circuit initially at Q_2, the gate current has no effect and the SCR is turned OFF by making

$$v_S < V_H + I_H R_L \simeq V_H$$

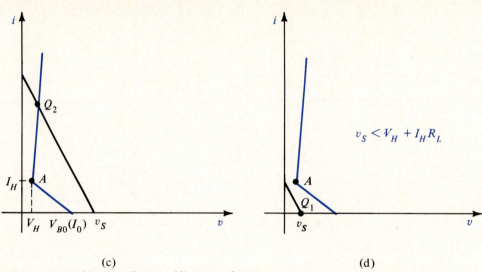

(c)   (d)

Figure 15.16 (c) ON solution. (d) OFF solution.

since usually $I_H R_L$ is negligible, which moves the load line to the left of A causing the operating point to jump from a high current position to a zero current position Q_1 as shown in Fig. 15.16(d).

15.3.1 SCR Power Control

If the voltage source in the circuit of Fig. 15.16(a) is sinusoidal, and the gate is turned on at the same frequency, then the SCR will be turned ON and OFF each alternate half-cycle. The ON time of the SCR is determined by the gate trigger-current timing and, as a result, the average current through R can be varied over a wide range. In Fig. 15.17 the operation of such a *phase-controlled "power supply"* is illustrated by showing the gate and v_s waveforms and the resulting load current waveform $i(t)$. Assuming $V_s < V_{BO}(0)$ the SCR is OFF until triggered by the gate driving-pulse at $\omega t = \phi$, *the firing angle*. The SCR turns ON and the load current becomes

$$i = \frac{V_s \sin \omega t - V_H}{R_L} \tag{15.42}$$

since $v = V_H$ once the SCR is ON. This relation holds until $v_s = V_H$ when the load line moves to the left of A and the SCR switches OFF. This occurs at $\omega t = \pi - \phi_0$, and i remains zero until the next cycle as shown by the screened waveform in Fig. 15.17. Since, typically, $V_H \ll V_s$, one can usually take $\phi_0 \approx 0$.

By controlling the firing angle ϕ, the *conduction time* during which the SCR is ON is adjusted and, therefore the average current or power delivered to R_L are varied. For the circuit shown, assuming $\phi_0 = 0$, the average power delivered to R_L as obtained by calculating the mean squared value of i, is

$$P_{av} = \hat{I}^2 R_L \tag{15.43}$$

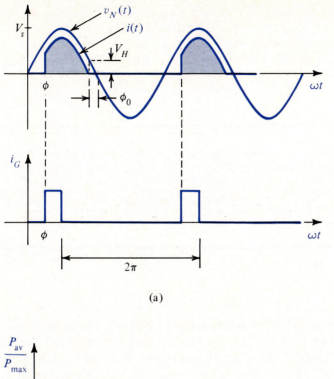

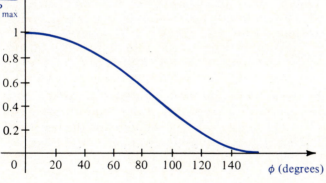

Figure 15.17 (a) $i(t)$ waveform for $v_S = V_s \sin \omega t$ and $i_G(t)$ as shown applied to the Fig. 15.16(a) circuit. (b) Power-angle characteristic for the circuit of Fig. 15.16.

or

$$P_{av} = \left[\frac{1}{2\pi} \int_\phi^\pi \frac{v_S^2(t)}{R_L^2} \, d(\omega t) \right] R_L$$

$$= V_s^2 \frac{\pi - \phi + \sin \phi \cos \phi}{4\pi R_L} \tag{15.44}$$

which can be varied from 0 to $V_s^2/4R_L$ by changing the firing angle ϕ. The variation of power in R_L as a function of ϕ is illustrated in Fig. 15.17(b).

The power dissipated in the SCR, P_D, is given by the average value of the product vi; since the off-state current is so small this is dominated by the contribution during the ON time and is therefore given by

$$P_D = V_H I_{dc} \tag{15.45}$$

DRILL EXERCISES

15.12 Verify Eqs. (15.43) and (15.44).

15.13 Show that $I_{dc} = V_s(1 + \cos \phi)/(2\pi R_L)$ for the circuit of Fig. 15.16.

EXAMPLE 15.9

A light-dimmer switch circuit consisting of an SCR connected as in Fig. 15.16 supplies power to a 100-Ω lighting load from a 110-V 60-Hz supply. The gate triggering is set to make the firing angle 90° after the start of each cycle. Calculate the dc and the rms current through the load and the power delivered to the load. What is the dissipation in the SCR for $V_H = 1$ V? Assume $\phi_o = 0$. What is the efficiency?

Solution
With $\phi = 90°$ the current flows for only one quarter cycle and the dc current is given by

$$I_{av} = \frac{1}{2\pi} \int_{\pi/2}^{\pi} \frac{110\sqrt{2}}{100} \sin \theta \, d\theta = 0.248 \text{ A}$$

The rms current is given by

$$\hat{I}^2 = \frac{1}{2\pi} \int_{\pi/2}^{\pi} (1.1\sqrt{2})^2 \sin^2 \theta \, d\theta = 0.30 \text{ A}^2$$

or

$$\hat{I} = 0.549 \text{ A}$$

The total load power is

$$\hat{I}^2 R_L = 30 \text{ W}$$

and the SCR dissipation as given by Eq. (15.45) is, for $V_H = 1$ V, barely 0.25 W. The efficiency is therefore 99.2%.

15.3.2 SCR Motor Control

Another typical application of the SCR is in the control of a dc motor's speed by controlling the *phase* at which an SCR is driven. In Fig. 15.18(a) a dc motor is shown with its armature connected to an ac source in series with an SCR. The motor field winding is driven by a separate dc current I_f generating a constant magnetic field. The force between the two current-carrying windings produces an average rotational torque which drives the motor shaft and is proportional to the average armature current. In Fig. 15.18(b) the motor, as seen at its armature terminals, is replaced by its Thevenin model consisting of a resistance R and a dc voltage source (*back emf*) which is proportional to the angular speed and is given by

$$E = K\omega \tag{15.46a}$$

where K is a motor constant.

With a load torque T_L on the motor shaft the average of the armature current I_A is given by

$$\text{ave}\{i_A\} = I_A = \frac{T_L}{K} \tag{15.46b}$$

where K is the same motor constant. If the gate trigger is applied at a phase ϕ then the current, given by

$$i_A = \frac{V_m \sin \theta - K\omega}{R} \tag{15.47}$$

is as shown in Fig. 15.19 and is only nonzero during the interval between ϕ and θ_1. The conduction ceases when θ_1 satisfies the condition

$$V_m \sin \theta_1 = K\omega \tag{15.48}$$

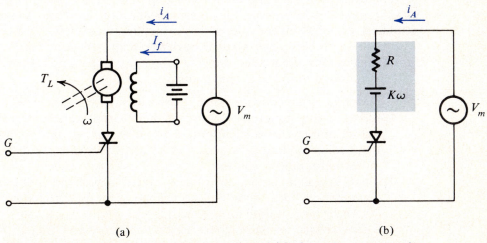

(a) (b)

Figure 15.18 (a) SCR speed control circuit and (b) Thevenin representation.

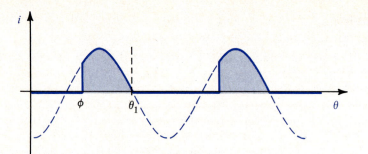

Figure 15.19 Motor armature current.

The average current is given by

$$I_A = \frac{1}{2\pi R} \int_{\phi}^{\theta_1} (V_m \sin\theta - K\omega) \, d\theta$$

$$= \frac{1}{2\pi R}[-V_m(\cos\theta_1 - \cos\phi) - K\omega(\theta_1 - \phi)] \qquad (15.49)$$

Using Eqs. (15.46) and (15.48) to eliminate I_A, Eq. (15.49) can be rewritten as

$$\frac{T_L}{K} = \left(\frac{V_m}{2\pi R}\right)[\cos\phi - \cos\theta_1 - \sin\theta_1(\theta_1 - \phi)] \qquad (15.50)$$

Thus for a given firing angle ϕ and load torque T_L Eq. (15.50) will be satisfied at some θ_1 and hence, through Eq. (15.48), for some angular speed ω. This speed can therefore be controlled by the firing phase angle ϕ.

15.4 SUMMARY AND STUDY GUIDE

The purpose of this chapter has been to present a brief introduction to some basic power applications. The common classification of electronic power amplifiers was introduced and some design guidelines established for class A and class B amplifiers. Rectification and filtered power supplies were briefly described and the operation of a simple op amp voltage regulator circuit was examined. Finally some simple applications of silicon controlled rectifiers were illustrated. The attention of students is drawn to the following points.

A power amplifier converts power from a dc supply into ac power to a load.
For sinusoidal operation the class B amplifier consumes zero power under zero drive and has an efficiency of 78.5% at maximum output power.
The maximum collector dissipation in a class B amplifier occurs at less than maximum output power.
The ripple component in a capacitance-filtered rectifier circuit is estimated by assuming a linear decay in load voltage when the diode(s) are OFF, and a negligible ON time for the diodes.
The basic voltage regulator circuit is essentially a noninverting op amp with a reference voltage input.

The SCR is a device which makes it possible to control the duration of unidirectional current flow per cycle in a circuit. It can therefore be used to
control the power delivered to resistive loads by an ac source, and in the
speed control of many types of motors.

PROBLEMS

15.1 A transformer-coupled power amplifier is required to deliver 10 W to an 8-Ω load.
The quiescent point is adjusted for symmetrical swing class A operation and $V_{CC} = 30$
V. Find the Q point, the transformer turns ratio, the peak collector current I_{Cm}, and
the efficiency. Calculate the minimum and maximum collector dissipation for the
amplifier.

15.2 Derive an expression for the collector dissipation P_C of an ideal class B amplifier
which is valid from zero signal to maximum signal. The supply V_{CC}, transformed load
impedance $n^2 R_L$ are all specified.

15.3 Show that the collector dissipation P_C for the amplifier of Prob. 15.2 has an extremum
value. Plot the collector dissipation as a function of output power.

15.4 A push-pull transformer-coupled class B amplifier has $V_{CC} = 30$ V, $n = 2$, and
$R_L = 30$ Ω. The transistors are identical. Calculate the maximum output signal
power and the per-transistor collector dissipation under maximum output conditions.

15.5 A class B transformer-coupled amplifier has $V_{CC} = 20$ V, $n = 0.5$, and $R_L = 20$ Ω.
Calculate the maximum load power achievable and the maximum collector dissi-
pation in each transistor.

15.6 Using a transistor with ratings $P_{C,max} = 40$ W, $v_{CE,max} = 50$ V, $i_{C,max} = 5$ A in the complementary configuration of Fig. 15.5. Calculate the maximum power that can be
delivered to a load R_L. For what value of R_L is this greatest?

15.7 Design a class B complementary symmetry amplifier to deliver 50 W to an 8-Ω load
with transistors having $V_{CE,max} = 80$ V. Specify V_{CC}, $P_{C,max}$, and $i_{C,max}$ for the transistors.

15.8 The complementary-symmetry amplifier of Fig. 15.5, is to be used to deliver 10 W
(maximum) of power to an 8-Ω loudspeaker.
(a) Calculate the average emitter current for each transistor at maximum power.
(b) Assuming identical values of $\beta = 70$, calculate the input resistance and then the
power gain for this amplifier.

15.9 Show that the average value of a full wave rectified waveform is given by Eq. (15.8).

15.10 A half-wave rectifier is used to design a nominal 20-V dc supply which must be
capable of supplying up to 2 A. The ripple must be kept below 2%. Design a circuit
such as the one in Fig. 15.10 to achieve these specifications using a 60-Hz ac source.
What is R_{eq}?

15.11 Repeat Prob. 15.10 using a 400-Hz source.

15.12 Repeat Prob. 15.10 using a full-wave rectifier circuit.

15.13 Design a transformer-coupled full-wave rectifier power supply with a capacitance filter to supply 10 V at 500 mA with a 3% ripple. What is R_{eq} for this supply at low frequency? The ac source is at 60 Hz.

15.14 For Prob. 15.10 with the capacitance filter, calculate the maximum voltage across the diode.

15.15 If in the regulator circuit of Fig. 15.13 the op amp gain is 10^4 and the output resistance is 100 Ω, calculate the dependence of v_L on i_L if Q_1 has $\beta = 50$ and $K = 0.01$. Neglect the current in R_1 and R_2.

15.16 The circuit of Fig. 15.16(a) is adjusted so that the conduction angle is 90°. The lamp load is equivalent to a 50-Ω resistance, and v_S is a 220-V rms sinusoid. Calculate the dc and rms load currents and the power delivered to the load.

15.17 For the circuit of Prob. 15.16 plot the load power vs the conduction angle.

15.18 Figure P15.18 illustrates a chopper circuit, using four SCRs. The gate currents, defined with respect to ground are $i_{g1}(t) = A \sin \omega t$ and $i_{g2} = -A \sin \omega t$. Sketch $v_O(t)$.

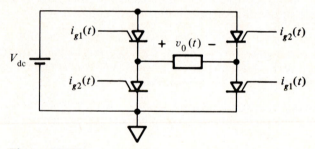

Figure P15.18

APPENDIX

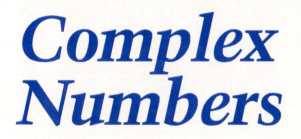

Complex Numbers

A.1 DEFINITIONS

A complex number $z = (x, y)$ is a vector in the xy plane, as shown in Fig. A.1. Addition, subtraction, multiplication, and division are all defined for complex numbers; it turns out that these operations may be performed in the usual way if one writes

$$z = x + jy \tag{A.1}$$

where $j^2 = -1$. Accordingly, x and y are known as the *real* and *imaginary* parts of z, respectively. One writes

$$\text{Re}(z) = x \tag{A.2}$$

$$\text{Im}(z) = y \tag{A.3}$$

Equation (A.1) is the representation of z in *Cartesian, or rectangular, form.*

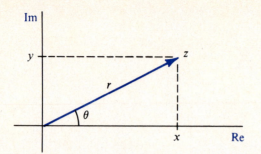

Figure A.1 Complex number in the complex plane.

A complex number may also be represented in *polar form*, i.e., $z = (r, \theta)$. One writes

$$|z| = r = (x^2 + y^2)^{\frac{1}{2}} \tag{A.4}$$

and

$$\sphericalangle z = \theta = \tan^{-1}\frac{y}{x} \tag{A.5}$$

Here $|z|$ is called the *magnitude* and $\sphericalangle z$ the *argument*, or *angle*, of z.
From Fig. A.1, it is clear that

$$\text{Re}(z) = x = r \cos \theta \tag{A.6}$$

and

$$\text{Im}(z) = y = r \sin \theta \tag{A.7}$$

Equations (A.4) and (A.5) convert rectangular to polar form, while Eqs. (A.6) and (A.7) implement the polar-to-rectangular conversion. Many electronic calculators are equipped to do both.
It can be shown that

$$e^{j\theta} = \cos \theta + j \sin \theta \tag{A.8}$$

This is known as De Moivre's theorem.
From Eqs. (A.1), (A.6), (A.7), and (A.8),

$$z = r \cos \theta + jr \sin \theta = re^{j\theta} \tag{A.9}$$

The *complex conjugate* of z is

$$z^* = x - jy \tag{A.10}$$

or

$$z^* = re^{-j\theta} \tag{A.11}$$

In either representation, the complex conjugate is obtained simply by replacing j by $-j$.

DRILL EXERCISES

A.1 Convert the following complex numbers to polar form: (a) $z = 1 + j$,
(b) $z = -1 - j$, (c) $z = -2 + 3j$.
Ans. (a) $\sqrt{2} \angle 45°$, (b) $\sqrt{2} \angle 225°$, (c) $\sqrt{13} \angle 123.7°$

A.2 Convert the following complex numbers to Cartesian form: (a) $z = 3 \angle 75°$,
(b) $z = 0.5 \angle 45°$, (c) $z = 1 \angle 155°$.
Ans. (a) $0.776 + j2.90$, (b) $0.354 + j0.354$, (c) $-0.906 + j0.423$

A.3 Find the complex conjugate, in polar form, for each of the following complex
numbers: (a) $z = 3 \angle 75°$, (b) $z = -1 + j$.
Ans. (a) $z = 3 \angle -75°$, (b) $\sqrt{2} \angle 225°$

A.2   OPERATIONS WITH COMPLEX NUMBERS

A.2.1 Addition and Subtraction

Complex numbers are added and subtracted exactly like vectors. Specifically, let
$z_1 = x_1 + jy_1$ and $z_2 = x_2 + jy_2$. Then,

$$z_1 + z_2 = (x_1 + x_2) + j(y_1 + y_2) \tag{A.12}$$

$$z_1 - z_2 = (x_1 - x_2) + j(y_1 - y_2) \tag{A.13}$$

Numbers given in polar form must first be converted to cartesian form in order
to carry out addition and subtraction.

It is of interest to consider the special case $z_2 = z_1^*$:

$$z + z^* = (x + x) + j(y - y) = 2x$$

$$z - z^* = (x - x) + j(y + y) = j2y$$

or

$$z + z^* = 2\ \text{Re}(z) \tag{A.14}$$

$$z - z^* = 2j\ \text{Im}(z) \tag{A.15}$$

A.2.3 Multiplication and Division

Multiplication may be done in cartesian form:

$$
\begin{aligned}
z_1 z_2 &= (x_1 + jy_1)(x_2 + jy_2) \\
&= x_1 x_2 + jy_1 x_2 + jx_1 y_2 + j^2 y_1 y_2 \\
&= x_1 x_2 - y_1 y_2 + j(y_1 x_2 + x_2 y_1) \tag{A.16}
\end{aligned}
$$

However, it is more easily done in polar form, as follows:

$$z_1 z_2 = r_1 e^{j\theta_1} r_2 e^{j\theta_2} = r_1 r_2 e^{j(\theta_1 + \theta_2)} \tag{A.17}$$

This is especially useful when several terms are to be multiplied. For example,

$$z_1 z_2 z_3 z_4 = r_1 e^{j\theta_1} r_2 e^{j\theta_2} r_3 e^{j\theta_3} r_4 e^{j\theta_4} = r_1 r_2 r_3 r_4 e^{j(\theta_1 + \theta_2 + \theta_3 + \theta_4)} \tag{A.18}$$

Since the right-hand side of Eq. (A.18) is in the form $|z| e^{j \angle z}$, it follows that

The magnitude of a product is the product of the magnitudes.
The angle of a product is the sum of the angles.

It is of interest to consider the special case of a complex number times its complex conjugate,

$$zz^* = r e^{j\theta} r e^{-j\theta} = r^2 \tag{A.19}$$

or

$$zz^* = |z|^2 \tag{A.20}$$

In words, the product of a complex number and its conjugate is the square of its magnitude.

Division may be handled in cartesian form, if the denominator is turned into its magnitude by multiplication with its conjugate, as follows: write

$$\frac{z_2}{z_1} = \frac{x_2 + jy_2}{x_1 + jy_1}$$

and multiply the numerator and denominator by $x_1 - jy_1$. This yields

$$\frac{z_2}{z_1} = \frac{x_2 + jy_2}{x_1 + jy_1} \frac{x_1 - jy_1}{x_1 - jy_1} = \frac{(x_2 + jy_2)(x_1 - jy_1)}{x_1^2 + y_1^2}$$

The division has been reduced to a multiplication, with the result

$$\frac{z_2}{z_1} = \frac{x_1 x_2 + y_1 y_2}{x_1^2 + y_1^2} + j\frac{x_1 y_2 - x_2 y_1}{x_1^2 + y_1^2}$$

Like multiplication, division is more easily done with numbers in polar form. For example,

$$\frac{z_1 z_2}{z_3 z_4} = \frac{r_1 e^{j\theta_1} r_2 e^{j\theta_2}}{r_3 e^{j\theta_3} r_4 e^{j\theta_4}} = \frac{r_1 r_2}{r_3 r_4} e^{j(\theta_1 + \theta_2 - \theta_3 - \theta_4)} \tag{A.21}$$

Since the right-hand side of Eq. (A.21) is in the form $|z| e^{j \angle z}$, the following rules apply to a product/quotient of complex numbers:

The magnitude of the result may be obtained by replacing each complex number by its magnitude in the operation.
The angle of the result is the sum of the angles of the numerator quantities, minus the sum of the angles of the denominator quantities.

DRILL EXERCISES

A.4 Calculate $w_1 = z_1 + z_2$, $w_2 = z_1 - z_2$, and $w_3 = 2z_1 + z_2$, for the following values of z_1 and z_2: (a) $z_1 = 1 + 2j$, $z_2 = -0.5 + j$; (b) $z_1 = 0.5 \sphericalangle 70°$, $z_2 = 1 \sphericalangle 135°$.
Ans. (a) $w_1 = 0.5 + 3j$, $w_2 = 1.5 + j$, $w_3 = 1.5 + 5j$; (b) $w_1 = -0.536 + j1.177$, $w_2 = 0.878 - j0.238$, $w_3 = -0.365 + j1.647$

A.5 Calculate $w_1 = z_1 z_2$ and $w_2 = z_1/z_2$, for $z_1 = 1 + 2j$, $z_2 = -0.5 + j$. Give your answer in rectangular form.
Ans. $w_1 = -2.5$, $w_2 = 1.2 - j1.6$

A.6 Calculate $w_1 = z_1 z_2 z_3$ and $w_2 = z_1/z_2 z_3$, for $z_1 = 0.5 \sphericalangle 50°$, $z_2 = 1.5 \sphericalangle -75°$, $z_3 = 2 \sphericalangle 150°$. Give your answer in polar form.
Ans. $w_1 = 1.5 \sphericalangle 125°$, $w_2 = 0.167 \sphericalangle -25°$

A.7 Repeat Exercise A.6 for $z_1 = 1 + 2j$, $z_2 = -0.5 + j$, $z_3 = 2 + 0.5j$.
Ans. $w_1 = 5.15 \sphericalangle 193.4°$, $w_2 = 0.971 \sphericalangle -66.6°$

Index

Page numbers in italics indicate that the Contents should be consulted for topic details.